Ancient Pathways, Ancestral Knowledge

VOLUME TWO

McGILL-QUEEN'S INDIGENOUS AND NORTHERN STUDIES

(In memory of Bruce G. Trigger)

John Borrows, Sarah Carter, and Arthur J. Ray, Editors

The McGill-Queen's Indigenous and Northern Studies series publishes books about Indigenous peoples in all parts of the northern world. It includes original scholarship on their histories, archaeology, laws, cultures, governance, and traditions. Works in the series also explore the history and geography of the North, where travel, the natural environment, and the relationship to land continue to shape life in particular and important ways. Its mandate is to advance understanding of the political, legal, and social relations between Indigenous and non-Indigenous peoples, of the contemporary issues that Indigenous peoples face as a result of environmental and economic change, and of social justice, including the work of reconciliation in Canada. To provide a global perspective, the series welcomes books on regions and communities from across the Arctic and Subarctic circumpolar zones.

1 *When the Whalers Were Up North*
Inuit Memories from the Eastern Arctic
Dorothy Harley Eber

2 *The Challenge of Arctic Shipping*
Science, Environmental Assessment, and Human Values
Edited by David L. VanderZwaag and Cynthia Lamson

3 *Lost Harvests*
Prairie Indian Reserve Farmers and Government Policy
Sarah Carter

4 *Native Liberty, Crown Sovereignty*
The Existing Aboriginal Right of Self-Government in Canada
Bruce Clark

5 *Unravelling the Franklin Mystery*
Inuit Testimony
David C. Woodman

6 *Otter Skins, Boston Ships, and China Goods*
The Maritime Fur Trade of the Northwest Coast, 1785–1841
James R. Gibson

7 *From Wooden Ploughs to Welfare*
The Story of the Western Reserves
Helen Buckley

8 *In Business for Ourselves*
Northern Entrepreneurs
Wanda A. Wuttunee

9 *For an Amerindian Autohistory*
An Essay on the Foundations of a Social Ethic
Georges E. Sioui

10 *Strangers Among Us*
David Woodman

11 *When the North Was Red*
Aboriginal Education in Soviet Siberia
Dennis A. Bartels and Alice L. Bartels

12 *From Talking Chiefs to a Native Corporate Elite*
The Birth of Class and Nationalism among Canadian Inuit
Marybelle Mitchell

13 *Cold Comfort*
My Love Affair with the Arctic
Graham W. Rowley

14 *The True Spirit and Original Intent of Treaty 7*
Treaty 7 Elders and Tribal Council with Walter Hildebrandt, Dorothy First Rider, and Sarah Carter

15 *This Distant and Unsurveyed Country*
A Woman's Winter at Baffin Island, 1857–1858
W. Gillies Ross

16 *Images of Justice*
Dorothy Harley Eber

17 *Capturing Women*
The Manipulation of Cultural Imagery in Canada's Prairie West
Sarah A. Carter

18 *Social and Environmental Impacts of the James Bay Hydroelectric Project*
Edited by James F. Hornig

19 *Saqiyuq*
Stories from the Lives of Three Inuit Women
Nancy Wachowich in collaboration with Apphia Agalakti Awa, Rhoda Kaukjak Katsak, and Sandra Pikujak Katsak

20 *Justice in Paradise*
Bruce Clark

21 *Aboriginal Rights and Self-Government*
The Canadian and Mexican Experience in North American Perspective
Edited by Curtis Cook and Juan D. Lindau

22 *Harvest of Souls*
The Jesuit Missions and Colonialism in North America, 1632–1650
Carole Blackburn

23 *Bounty and Benevolence*
A History of Saskatchewan Treaties
Arthur J. Ray, Jim Miller, and Frank Tough

24 *The People of Denendeh*
Ethnohistory of the Indians of Canada's Northwest Territories
June Helm

25 *The Marshall Decision and Native Rights*
Ken Coates

26 *The Flying Tiger*
Women Shamans and Storytellers of the Amur
Kira Van Deusen

27 *Alone in Silence*
European Women in the Canadian North before 1940
Barbara E. Kelcey

28 *The Arctic Voyages of Martin Frobisher*
An Elizabethan Adventure
Robert McGhee

29 *Northern Experience and the Myths of Canadian Culture*
Renée Hulan

30 *The White Man's Gonna Getcha*
The Colonial Challenge to the Crees in Quebec
Toby Morantz

31 *The Heavens Are Changing*
Nineteenth-Century Protestant Missions and Tsimshian Christianity
Susan Neylan

32 *Arctic Migrants/Arctic Villagers*
The Transformation of Inuit Settlement in the Central Arctic
David Damas

33 *Arctic Justice*
On Trial for Murder - Pond Inlet, 1923
Shelagh D. Grant

34 *The American Empire and the Fourth World*
Anthony J. Hall

35 *Eighteenth-Century Naturalists of Hudson Bay*
Stuart Houston, Tim Ball, and Mary Houston

36 *Uqalurait*
An Oral History of Nunavut
Compiled and edited by John Bennett and Susan Rowley

37 *Living Rhythms*
Lessons in Aboriginal Economic Resilience and Vision
Wanda Wuttunee

38 *The Making of an Explorer*
George Hubert Wilkins and the Canadian Arctic Expedition, 1913–1916
Stuart E. Jenness

39 *Chee Chee*
A Study of Aboriginal Suicide
Alvin Evans

40 *Strange Things Done*
Murder in Yukon History
Ken S. Coates and William R. Morrison

41 *Healing through Art*
Ritualized Space and Cree Identity
Nadia Ferrara

42 *Coyote and Raven Go Canoeing*
Coming Home to the Village
Peter Cole

43 *Something New in the Air*
The Story of First Peoples Television Broadcasting in Canada
Lorna Roth

44 *Listening to Old Woman Speak*
Natives and Alternatives in Canadian Literature
Laura Smyth Groening

45 *Robert and Francis Flaherty*
A Documentary Life, 1883–1922
Robert J. Christopher

46 *Talking in Context*
Language and Identity in Kwakwaka'wakw Society
Anne Marie Goodfellow

47 *Tecumseh's Bones*
Guy St-Denis

48 *Constructing Colonial Discourse*
Captain Cook at Nootka Sound
Noel Elizabeth Currie

49 *The Hollow Tree*
Fighting Addiction with Traditional Healing
Herb Nabigon

50 *The Return of Caribou to Ungava*
A.T. Bergerud, Stuart Luttich, and Lodewijk Camps

51 *Firekeepers of the Twenty-First Century*
First Nations Women Chiefs
Cora J. Voyageur

52 *Isuma*
Inuit Video Art
Michael Robert Evans

53 *Outside Looking In*
Viewing First Nations Peoples in Canadian Dramatic Television Series
Mary Jane Miller

54 *Kiviuq*
An Inuit Hero and His Siberian Cousins
Kira Van Deusen

55 *Native Peoples and Water Rights*
Irrigation, Dams, and the Law in Western Canada
Kenichi Matsui

56 *The Rediscovered Self*
Indigenous Identity and Cultural Justice
Ronald Niezen

57 *As affecting the fate of my absent husband*
Selected Letters of Lady Franklin Concerning the Search for the Lost Franklin Expedition, 1848–1860
Edited by Erika Behrisch Elce

58 *The Language of the Inuit*
Syntax, Semantics, and Society in the Arctic
Louis-Jacques Dorais

59 *Inuit Shamanism and Christianity*
Transitions and Transformations in the Twentieth Century
Frédéric B. Laugrand and Jarich G. Oosten

60 *No Place for Fairness*
Indigenous Land Rights and Policy in the Bear Island Case and Beyond
David T. McNab

61 *Aleut Identity*
Tradition and Modernity in an Indigenous Fishery
Katherine L. Reedy-Mascher

62 *Earth into Property*
Aboriginal History and the Making of Global Capitalism
Anthony J. Hall

63 *Collections and Objections*
Aboriginal Material Culture in Southern Ontario, 1791–1914
Michelle A. Hamilton

64 *These Mysterious People*
Shaping History and Archaeology in a Northwest Coast Community
Susan Roy

65 *Telling It to the Judge*
Taking Native History to Court
Arthur J. Ray

66 *Aboriginal Music in Contemporary Canada*
Echoes and Exchanges
Edited by Anna Hoefnagels and Beverley Diamond

67 *In Twilight and in Dawn*
A Biography of Diamond Jenness
Barnett Richling

68 *Women's Work, Women's Art*
Nineteenth-Century Northern Athapaskan Clothing
Judy Thompson

69 *Warriors of the Plains*
The Arts of Plains Indian Warfare
Max Carocci

70 *Reclaiming Indigenous Planning*
Edited by Ryan Walker, Ted Jojola, and David Natcher

71 *Setting All the Captives Free*
Capture, Adjustment, and Recollection in Allegheny Country
Ian K. Steele

72 *Before Ontario*
The Archaeology of a Province
Edited by Marit K. Munson and Susan M. Jamieson

73 *Becoming Inummarik*
Men's Lives in an Inuit Community
Peter Collings

74 *Ancient Pathways, Ancestral Knowledge*
Ethnobotany and Ecological Wisdom of Indigenous Peoples of Northwestern North America
Nancy J. Turner

Ancient Pathways, Ancestral Knowledge

ETHNOBOTANY AND ECOLOGICAL WISDOM OF INDIGENOUS PEOPLES OF NORTHWESTERN NORTH AMERICA

VOLUME TWO

The Place and Meaning of Plants in Indigenous Cultures and Worldviews

Nancy J. Turner

McGill-Queen's University Press | Montreal & Kingston | London | Chicago

ISBN 978-0-7735-4380-5 (cloth)
ISBN 978-0-7735-8539-3 (ePDF)
ISBN 978-0-7735-8540-9 (ePUB)

Legal deposit second quarter 2014
Bibliothèque nationale du Québec

Reprinted 2015, 2021

Printed in Canada on acid-free paper that is 100% ancient forest free (100% post-consumer recycled), processed chlorine free

This book has been published with the help of a grant from the Canadian Federation for the Humanities and Social Sciences, through the Awards to Scholarly Publications Program, using funds provided by the Social Sciences and Humanities Research Council of Canada. Funding has also been received from the McLean Foundation.

Funded by the Government of Canada | Financé par le gouvernement du Canada | Canada | Canada Council for the Arts | Conseil des arts du Canada

We acknowledge the support of the Canada Council for the Arts.

Nous remercions le Conseil des arts du Canada de son soutien.

LIBRARY AND ARCHIVES CANADA CATALOGUING IN PUBLICATION

Turner, Nancy J., 1947–, author
Ancient pathways, ancestral knowledge : ethnobotany and ecological wisdom of Indigenous peoples of northwestern North America / Nancy J. Turner.

(McGill-Queen's Indigenous and northern studies ; 74)
Includes bibliographical references and indexes.
Contents: Volume one. The history and practice of indigenous plant knowledge – Volume two. The place and meaning of plants in indigenous cultures and worldviews.
Issued in print and electronic formats.
ISBN 978-0-7735-4380-5 (v. 1 : bound). – ISBN 978-0-7735-4380-5 (v. 2 : bound).–
ISBN 978-0-7735-8539-3 (v. 1 : ePDF). – ISBN 978-0-7735-8540-9 (v. 1 : ePUB).–
ISBN 978-0-7735-8539-3 (v. 2 : ePDF). – ISBN 978-0-7735-8540-9 (v. 2 : ePUB)

1. Indians of North America – Ethnobotany – Canada, Western. 2. Indians of North America – Ethnobotany – Canada, Northern. 3. Traditional ecological knowledge – Canada, Western. 4. Traditional ecological knowledge – Canada, Northern. 5. Indians of North America – Ethnobotany – Northwestern States. 6. Traditional ecological knowledge – Northwestern States. I. Title. II. Series: McGill-Queen's Indigenous and northern studies ; 74

E98.B7T87 2014 581.6'3097 C2014-900602-0
C2014-900603-9

Set in 10/13 Gentium Plus with Sina Nova, Cyntho Pro, and Cyntho Slab Pro
Book design and typesetting by Garet Markvoort, zijn digital
Botanical illustrations by Hein Nouwens/shutterstock.com, based on Meyers Konversations-Lexik, 1897

CONTENTS

Volume One

Volume Two

FIGURES AND TABLES

FIGURES

TABLES

PREFACE

This is the second of two integrated volumes on how knowledge about plants and environments has been acquired, transmitted, and adapted, over time and space, by Indigenous peoples in northwestern North America. People have travelled through and made their homes in this amazingly diverse region for thousands of years – possibly 14,000 years or longer – since the Pleistocene times when most of the land and much of the coastline were smothered by vast expanses of ice. Since these ancient times, many different peoples have learned to survive here, using their skills and ingenuity. Inventing, modifying and diversifying, exploring, experimenting, practising, and sharing, they not only subsisted across times of immense change but also created rich and rewarding lives for themselves, lives strong with social bonds, with artistic expression, with ritual and ceremony, and with enduring relationships with other species – the plants and animals of their territories. They developed an immense knowledge of, respect for, and appreciation for these species, fostered not only by need but also by bonds of kinship, recognizing the lives of these entities as being parallel to and as important as those of humans. Throughout the millennia, people have interacted with and integrated their lives with these species, such that their oral traditions have often compellingly blended the lives of the plant and animal actors, together with river and mountain personae, into the lives and activities of human actors in ways that defy the "us" and "them" dichotomy of predominant society.

One of the most remarkable facets of Indigenous peoples' knowledge and relationships with plants and environments is that there are many congruities in knowledge systems across linguistic, cultural, geographical, and temporal boundaries. These similarities include common origins of many plant names, which evolved or were borrowed from some central ancestral form, common methods and tools for harvesting and processing plant foods and materials, and common themes in narratives told by peoples living at great distance from each other. A fundamental focus of these volumes, then, is to try to better understand how peoples' systems of knowledge, practice, and belief – their Traditional Ecological Knowledge systems – have been developed, disseminated, and adjusted and how they have evolved across the great landscapes and seascapes of northwestern North America. What did people bring with them? What were their legacies? How did old and new ideas and practices combine and shift? How did languages develop and reflect the nuances and changes of different environments and cultures? As a follow-up question, it is equally important in these times of global declines in biodiversity and cultural diversity to ask what lessons can be learned from this understanding that can be applied in maintaining and sustaining this type of knowledge, both for the benefit of those who first developed it and as a significant part of the common heritage of all humanity.

Volume 1 of this book provides the background and context for the study, examining the historical aspects of ethnobotanical and ethnoecological knowledge in the area. Plant names are presented and discussed in detail because they can give many clues about the history of relationships and sharing of goods and ideas between and among different groups, both linguistically related and linguistically disparate. Changes effected by European newcomers – traders, settlers, missionaries, government officials, and industrialists – to Indigenous peoples' lifeways, languages, and knowledge systems are also described, providing insights into the ways that new knowledge and new circumstances may be accommodated in cultural systems, as well as the ways that knowledge can be transformed or lost. The last three chapters of volume 1 focus on three major categories of plant use – food, materials, and medicine – setting the stage for the more integrated and transcending approaches in this second volume.

There are so many strands and so many themes to be recognized and addressed that the task is daunting. Most of these will never be fully deciphered, but even the clues offered in the evidence presented can give some significant insights and lessons. Hopefully, too, some of the examples revealed in this research will identify new areas and directions of inquiry. Ultimately, the lessons point to immense and sometimes overwhelming complexity, to entire shifting, transforming, and evolving systems choreographed simultaneously by the landscapes, the life forms, and the human spirit, all connected in an endless and magnificent dance.

My efforts to understand and describe these trajectories of ethnobotanical knowledge have been a work of intense fascination and enjoyment over a span of four and a half decades. The Indigenous botanical and cultural experts who have been my teachers and guides on this journey are people who have truly lived surrounded by and interacting with the plants of their environments in the context of their own languages and cultures. They have actually participated in harvesting the root vegetables, inner bark of trees, wild greens, and other wild foods that are little known at present, and because of this, they are like open windows to knowledge otherwise shrouded in dim vestiges of people's collective memories. Ida Jones of the Ditidaht, Dr Margaret Siwallace of the Nuxalk, Florence Davidson of the Haida, Clan Chief Adam Dick (Kwaxsistalla) of the Kwakwaka'wakw, Elsie Claxton of the Saanich, Annie York of the Nlaka'pamux, Sam Mitchell and Edith O'Donaghey of the Stl'atl'imx/St'at'imc, and Mary Thomas of the Secwepemc are such people. Many more are acknowledged by name in the preface to the first volume. As children and youths, and even into adulthood, all of them participated in traditional plant harvesting and processing and learned the applications of plant resources under the guidance of their parents, grandparents, aunts, uncles, and other relatives. Although they did not know each other, they all shared the common experience of a deep and abiding relationship with their home place, and they have all felt the heartbreak of environmental degradation and the accompanying cultural losses. Witnessing these losses, and being convinced of the value of their own teachings, all of these individuals have attempted to strengthen, renew, and pass on their knowledge and experiences to younger generations. I sincerely hope that, in writing these two volumes and highlighting the richness and importance of their teachings, I have done them justice and have honoured them. I would like to dedicate this work to all of the Indigenous teachers and environmental knowledge holders, to those who have benefited from their teachings, and to those of generations to come who will be able to understand a little more clearly the vitality and fundamental significance of their ancestral knowledge.

PART THREE

Integration and Management

8

Moving for the Harvest: Seasonal Rounds and Plant Knowledge

A whole bunch of us would go up ... on top of the mountain, way up high, close to the snow line, by horseback every year ... for a few days, three of four days or a week, mainly to pick huckleberries [*Vaccinium membranaceum*]. These berries are ripe in August, and that's when people would go up ... We got huckleberries, and dug roots while we were up there, wild potatoes [*Claytonia lanceolata*], wild onions [*Allium cernuum*], whatever we could find up there on the mountain. The men would hunt too ... I haven't been up there since I was young, but that used to be a tradition in the olden days ... They'd gather [glacier lily bulbs], a whole bunch of tiger lily bulbs ... that black moss [black tree lichen] ... We'd bring down six or eight baskets [large coiled cedar-root pack baskets] with pack horses ... baskets and sacks of mountain potatoes ... They also dried cow-parsnip roots for medicine. (Edith O'Donaghey, cited in Turner 1992b, 418)

INTRODUCTION

Edith O'Donaghey, a Stl'atl'imx/St'at'imc elder, was born and raised at Shalalth on Seton Lake, nestled in the Coast Mountains of British Columbia. The mountain where she went as a child with her family and other family groups, called Mission Ridge, was above their village. Although the trips Edith described took place around the late 1920s, the actual pattern of travel to different destinations within a people's territory over the course of a season would have been established probably millennia ago – in fact, possibly as early as when people first existed in the region. Furthermore, travelling from place to place in an annual cycle for resource harvesting was a widespread practice over the entire area, both on the coast and in the interior, in the north and in the south. This cyclical movement to different points around the landscape over the course of the seasons was essential to acquire the diverse resources that people needed for survival, not only food but various materials and medicines as well. The travels were also simply part of a way of life.

The "seasonal rounds," or patterns of seasonal movement and seasonal residence – camps are considered "home," too (R.E. Ignace 2008) – within and across diverse geographic and ecological areas, reflect and embrace the complex systems of knowledge and practice that integrate all of the different aspects of peoples' lifeways: harvesting food, gathering wood and using fire for cooking and landscape modification, constructing shelters, making and using canoes, honing skills in basketry, in making fishing gear, and in pit-cooking, harvesting and administering medicines, observing animals of all kinds, recounting stories and sharing conversations using all of the specialized vocabulary that goes with the innumerable species encountered and used, teaching children and youth, exchanging products and ideas with neighbours, learning how to care for and manage the land, and all of the other place-based activities that are part of a culture. Furthermore, these seasonal rounds are typical of tribal peoples and all those peoples around the world who depend on seasonal harvesting, from the Highlanders of Scotland and the Saami of Finland and Norway to the Indigenous peoples of Australia (Ommer and Turner 2004; Prober, O'Connor, and Walsh 2011). In this chapter, I discuss the role of seasonal rounds and their relationship to plant harvesting and ethnobotanical knowledge.

Like Edith O'Donaghey, many elders of recent generations remember their seasonal trips to the fish camps along the river, the seaweed and halibut camps on the ocean, and the hunting, berry-picking, and root-harvesting camps in the mountains, as well as their fall fishing and estuarine root garden destinations, with great fondness and nostalgia. Secwepemc elder Mary Thomas (pers. comm., 1994) recalled an experience from her childhood that was quite similar to that of Edith:

> I guess I was quite small yet when I went on one of those trips ... but I can still remember quite a bit of how we camped up there. There was just sort of more or less in the open stars type of camping ... I must have been around about four or five years old ... There'd be quite a few [families] – like my dad, he had brother-in-laws and he had people that he knew – they would take packhorses and go up to collect – and not only that, they would do hunting and drying meats. They had big family units; there'd be quite a few people would go up in one trip ... Around something like [twenty to thirty people], always big family units moved about and working together, which was always fun.

These seasonal rounds would have varied over time, with changes in technologies, such as the introduction of the horse or, later, of motorized vehicles, including automobiles and motorboats. Nevertheless, the essence of the activities and functions has endured – for some people, right up to the present day. For example, the Stl'atl'imx still descend in early summer to the rocky confines of the Fraser Canyon between Lillooet and Pavilion to harvest the spring salmon and early sockeye salmon runs, and many still travel into the mountains later in the summer to pick huckleberries and to hunt. On the northern coast, many members of the Gitga'at Nation of Hartley Bay still move for a few weeks in May to their traditional seaweed-halibut spring harvest camp at K'yel on Princess Royal Island, and then in the late summer, they travel to "Old Town" at Gitgiata Inlet on the north side of the Douglas Channel to harvest and smoke their winter supply of chum, sockeye, and coho salmon and to pick blueberries, crabapples, and highbush cranberries (Turner and Thompson 2006). These trips are now made by fishing boat or speedboat rather than by cedar dugout canoe, and people sometimes preserve their fish, seaweed, and berries in freezers, with the help of electric generators, rather than sun drying or smoking all of it right away. But the basic purpose of the trip remains the same: to acquire the necessities of life and enhance food security, while strengthening social bonds and reinforcing and passing on to younger generations the knowledge and experience needed to survive and thrive in a people's territory.

THE NATURE OF PLANT RESOURCES

Plant resources include not only the different types of food produced by plants but also the various types of material they provide – such as wood for fuel and construction, sheets of bark and fibrous materials for basketry, and tree pitch for glue and caulking – and the medicines they furnish for treating illness and injury and for health maintenance. As recounted in previous chapters, these resources are derived from hundreds of different species, many of which produce different products from the same plant. Thimbleberry, for example, yields edible green shoots in the springtime, sweet edible berries in the summer, and from spring through fall, large soft leaves that can be used as makeshift containers, as a lining for berry baskets and berry-drying racks, and as a wrapping for food when pit-cooking. Year round, insect galls are produced on the stems of thimbleberry, which are burned and used as medicine by some people (Turner, Thompson, et al. 1990). Western hemlock is another example of a multipurpose plant. In late spring, it yields edible cambium and inner bark, and in spring and at other times, it provides soft-needled boughs for harvesting herring eggs, lining food storage pits, ritual scrubbing, and other purposes. Hunters can use its young branch tips, rich in vitamin C, as an emergency hunger suppressant.

Throughout the year, the bark can be harvested and used as a dye and tanning agent and as a medicine for internal bleeding, and the wood can be employed as a fuel, especially for banking fires overnight. Many other plant species are similarly versatile, being used for different purposes at different times of the year. As noted in chapter 7, there is a high degree of congruency between plants used for food and those used for medicine. The same could be said for plant foods and plant materials. In fact, using a plant species in multiple ways is more the rule than the exception.

Species Distributions

Each plant species has its own pattern of distribution and abundance, based on a range of factors related to environmental history, modes and vectors of dissemination, ecological requirements, and tolerances and interactions with other species. Phytogeography is the study of patterns and processes of plant distribution: where plants occur, why they occur there, and in what combinations or vegetation patterns they occur (Klinkenberg 2013). Recognizing these patterns of plant distribution in terms of the species used by First Peoples is essential to understanding patterns of human settlement and movement, as well as patterns of plant use. Many culturally important plants, like lodgepole pine and yarrow, are widespread across the landscape and occur virtually over the entire study area. Others are much more restricted in their distribution, and most have specific ecological requirements that limit their ability to survive in many habitats. Thus, even in regions where they are known to exist (as shown on distribution maps, for example), they tend to occur in patches or in linear zones or to be dispersed sporadically, based on their own particular ecological tolerances and patterns of growth and dissemination.[1] Although a species may occur in a number of different vegetation zones, it usually predominates in just one or two of these. Not even ubiquitous weedy species like dandelion (*Taraxacum officinale*) occur universally on the landscape.

One of the major determinants of overall distributions of plants in northwestern North America today is the area's geological history, particularly the time of the Pleistocene glaciation, when virtually the entire area was scoured, with only a few locations, or refugia, remaining ice-free and able to support plant (and animal) life. Some of these ice-free regions were in the north, in the area of the Beringian land bridge and northeastern Asia, for example. Some were scattered along the Pacific Coast in sites on Haida Gwaii, on the Brooks Peninsula, and in areas now covered by ocean since sea levels are considerably higher today than during the last ice age. There were also some mountain peaks, or nunataks, in the interior not covered by ice, and there was evidently a long, ice-free corridor between the two major continental ice sheets running along the eastern slopes of the Rocky Mountains, at least during the late Pleistocene.

The only other ice-free area in the study region during the last glaciation was south of the Cordilleran Ice Sheet, in parts of Washington, Idaho, and Montana, as well as farther south, in Oregon, California, and the Great Basin (Pielou 1991; see also chapter 2).

All of the trees, shrubs, and other plants that make up the vast forests, the interior grasslands, the wetlands, and the subalpine parklands from Alaska to the southern extent of the ice sheets – and all of the animals these ecosystems support – established themselves in this area originally from one or more of these ice-free regions (Pielou 1991). Some species entered the deglaciated region from the north, some from the west, and many others from the south. Some colonized the area more rapidly and spread farther than others, and some flourished and dominated while others declined over time or became sporadic and infrequent, perhaps less able to compete with more aggressive species. Major changes in climate and geological events over the intervening Holocene period have also determined past and present distributions of species.

Apart from their ability to spread into a given environment and to survive and reproduce within it, plant species vary widely in their growth and reproductive patterns, as well as in the quality and productivity of their fruit, bark, or other useful parts. Fruit-producing species, such as saskatoon berry, black huckleberry, chokecherry, soapberry, and Pacific crabapple, may include varieties or populations that produce an abundance of large, sweet, juicy fruits, whereas other populations or individuals of the same species may produce only small, sparse, bitter-tasting fruits. Some western redcedar trees are bushy, with many branches lower down on the trunk, whereas others grow tall, straight trunks with few branches, ideal for harvesting bark strips. Such differences may be based on either genetic or environmental factors, or both – variable responses to moisture, sunlight, availability of pollinators, competition from other plants, herbivory and insect predation, or seasonal temperatures – playing out on different temporal and geographical scales. Much of this variation is recognized by local Indigenous peoples, and some is linguistically encoded in the names for different types and varieties of plants or in the names for places that produce a particular resource of high quality and repute. The lower portions of Glacier Bay, Alaska, for example, are called Shákʷ Aaní ("Strawberry Land") in Tlingit, in recognition of the quality and productivity of strawberries (*Fragaria chiloensis*) to be found there (Thornton 1999).

Most species and varieties of plants have an optimal range of conditions under which they will thrive, growing and reproducing at stable and predictable rates, even in the face of some level of disturbance or harvest from humans and other animals. If people are able to use the products of these species without depleting them or, conversely, are able to compensate for any depletion by enhancing their growth, distribution, or productivity in some way, this constitutes sustainable resource use.[2] Ultimately, the occurrence, abundance, quality, and productivity of a particular plant resource at a particular place and time, with or without cultural modification, determine its availability. The combinations or complexes of productive, culturally important plant and animal species in a given locality during a given season – along with other environmental factors such as the accessibility of water and a suitable microclimate – are a key to peoples' settlement patterns and seasonal movements.

Variability and Predictability

Aside from the overall presence or absence of particular plant species in a given region based on their overall range and on the preference or tolerance for particular sites or habitats within the range, plants also vary significantly in growth and reproductive capacity from year to year and over cycles of years, as well as due to annual seasonal differences. Berries and fruits in particular seem to fluctuate in productivity from one year to the next; even domesticated apples, cherries, and other orchard crops are known to oscillate in fruit yields.[3] Sometimes the productivity of an entire population of a fruit species will be similarly affected, positively or negatively, or sometimes variation will be limited to a given patch or an individual bush or tree. In any case, berries have been called the quintessential patchy resource, and people's harvesting strategies have to take this patchiness into account (Thornton 1999). Having alternative or backup resource-harvesting sites or fallback species on which to rely when the usual crops fail is part of a group's overall Traditional Ecological Knowledge system.[4]

In contrast to berries, edible "root" species tend to be more predictable and reliable in their availability,[5] and as discussed in chapter 4, they have often served as a backup resource in times of food scarcity. Even edible roots, however, vary in taste and quality through their growing season (Loewen 1998). With a few exceptions, they are usually harvested either when the plant is in its dormant or early growing state and has not yet flowered (e.g., bitterroot and nodding onion) or after it has flowered and gone to seed and the leaves have started to die back (e.g., camas, *Camassia* spp.; and silverweed, *Argentina* spp.). Other resources, such as edible red laver seaweed, eelgrass, and cow-parsnip, may vary in their optimal time for harvest or in the duration of the time that they are considered edible, depending on local weather conditions or on long-term climatic conditions (Turner and Clifton 2009), but they tend to be available and predictable in places where they grow and in their overall harvest season from one year to the next.

One of the most variable and unpredictable of all food substances in the Interior Plateau is Douglas-fir sugar, mainly comprised of a trisaccharide, melezitose, produced by some interior Douglas-firs. Under extremely rare conditions (particularly during stretches of hot, dry summer days) and in specific sites (usually north-facing upland slopes, with high soil moisture), this sugar is exuded from the tips of the fir needles, forming crystalline deposits resembling globs of icing sugar all over the boughs of the tree. A number of Nlaka'pamux elders remembered seeing and/or tasting this sugar, which was never predictable enough to be more than an occasional confection, brought home by hunters and other travellers in the high country, but relished when they had the chance to taste it (Turner 1997a; Turner, Thompson, et al. 1990).[6]

Variation within Species

Almost all culturally used plant species are known to vary in quality from place to place; some localities and habitats are recognized as prime for particular products. For example, Secwepemc elder Mary Thomas (pers. comm., 1994) described a particular site where her family used to harvest *skʷekʷíne* (Eastern dialect) (spring beauty, or mountain potato) at Skwelkwekwelt (formerly Baldy Mountain, now called Sun Peaks), a location that produced remarkably large corms:

> I can barely remember one time; I was with my mother and my dad and a bunch of us. Down below [in the valley bottoms around Chase and Salmon Arm], you had to use a digging stick to dig for the roots. Up there, it was like soft moss covering the ground – places you didn't even need a stick. My mother used to just feel around under the moss and here was these great big potatoes – big as golf balls ... You could just feel it, pushing around under the moss [with your bare feet]. It was just something to pick that and fill a gunnysack ... When they came back, they had bags and bags of the vegetables on the packhorses.

Similar highly productive sites were known for yellow glacier lily, desert parsley (*Lomatium macrocarpum*), tiger lily, balsamroot, saskatoon berries, soapberries, and almost all of the other major food plant species of the Interior Plateau. These special places, like the Tlingit Glacier Bay strawberry site mentioned earlier, are sometimes named for the plants that thrive there. The Secwepemc name for Salmon Arm, Sxʷsemétp,[7] for example, is said literally to mean "soapberry plants"; people came by canoe and boat from all around the lake and beyond to harvest soapberries there (Mary Thomas 2001). In Tsilhqot'in territory is a mountain called Tsinuzchéd (Potato Mountain), famous for the density of its spring beauty, or mountain potatoes, called *súnt'iny* (figure 8-1) (Mellott 2010; Turner, Deur, and Mellott 2011). It is said in Tsilhqot'in origin narratives that the wild "potatoes" were originally planted there by a woman named ʔeniyud, who had separated from her husband, Ts̲'ilʔos̲, and was travelling over toward Tatlayoko Lake; she planted *súnt'iny* on her way across (L.R. Smith 2008b).[8] Many other prime harvesting places have such stories associated with their origins.

8-1 | Spring beauty, or mountain potato (*Claytonia lanceolata*), called *súnt'iny*, from Potato Mountain, in the territory of the Xeni Gwet'in Tsilhqot'in.

On the coast, too, people travelled considerable distances to obtain the best-tasting cow-parsnip shoots in the spring, even though other cow-parsnip plants could be found closer at hand. Some people have noted that cow-parsnip shoots growing in the shade are milder tasting and therefore harvested in preference to those growing in the bright sun.[9] Salal berries, thimbleberries, highbush cranberries, Pacific crabapples, camas, springbank clover, and Pacific silverweed all had their favoured harvesting places in Pacific coast environments, locations where they were known to be prolific and of highest quality. People visited these special harvesting localities year after year, sometimes generation after generation (Bouchard and Kennedy 1990; Kuhnlein and Turner 1991; Kuhnlein, Turner, and Kluckner 1982; Turner and Kuhnlein 1982, 1983).

Not just food plants but other kinds of plant resources as well vary in quality and productivity according to where they are growing and the environmental conditions under which they develop. Many weavers of cedar bark (*Thuja plicata* and *Chamaecyparis nootkatensis*), for example, seek the best bark from trees that are growing somewhat away from salt water, that are well spaced in dense forests on slopes, and that have smooth, unbranching trunks. Cedar roots and spruce roots (*Picea sitchensis*, *P. glauca*) are likewise preferred from trees that have few branches low down and that are growing in sandy (not rocky) soil, meaning the roots will be long and straight, with few branches or kinks (Turner 1992a, 1998, 2004a). Some basket makers seek western redcedar roots from rotten logs because these roots are especially fine and straight; others look for prime, long roots in sandy soil such as alongside a river or in a sand dune area. The best cedar trees for rectangular bark roofing sheets, for planks, and for canoe making were also known to occur at particular sites, where the trees were not only readily accessible, and the right age and size, but also had the right quality of bark or wood: straight-grained, easily split, and not too brittle.

Indian-hemp and stinging nettle also vary considerably in the length and quality of their stem fibres depending on where they are growing: damp, open swales near lakes are considered best for the former and moist shaded forest edges best for the latter. Similarly, cattail, tule (*Schoenoplectus acutus*), basket sedge (*Carex obnupta*), and "three square" (*Schoenoplectus americanus*) have their prime harvesting sites, well known to weavers, and are accessed from these particular places year after year. Medicinal plants, too, are said to vary in quality and therapeutic effectiveness depending upon their location. Sometimes people travelled long distances just to obtain a certain medicine from a special place, even when the same species might be growing closer at hand. As noted in chapter 7, medicinal plants growing at higher elevations were often considered especially strong and potent, compared with populations at lower altitudes (Turner, Deur, and Mellott 2011). In all, people selected the products they needed based on a combination of factors: seasonal and annual weather conditions, general preferred habitat, specific locations known to produce healthy, productive plants, the known quality of particular populations of harvestable plants in terms of desired traits (fewer branches, smoother bark, larger berries), and even sought-after features of one part of an individual plant, whether the fruit, the wood, or the roots. Of course, proprietorship of plant resources and harvesting places was also an important consideration (see chapters 9 and 11).

A good example of how all of these aspects interact is in Mary Thomas's (pers. comm., 1994) description of harvesting paper birch bark for basket making:

> You have to know how to collect that bark, when to get it. You can only gather the birch bark, say, late May, June, and then it starts to stick back. Once it sticks, you can't get it. So you have to get your supply when it's ready. And you can't go and just take bark from any old place; you have to really look for it. And it takes you maybe a whole day of searching to get the right texture. You don't just take any bark at all. You have to test the tree. Some are very brittle; some are nice and spongy. And you have to look for the right kind of bark. The shorter the eyelets [lenticels], the better, and it doesn't crack easily. If you get the long eyelets, it will crack [and] your basket is ruined ... And you don't just go and cut any part of the birch tree, you have to know which side to cut it on ... [the bark] always

8-2 | Mary Thomas with her birch-bark baskets (*Betula papyrifera*).

splits on the north side of the tree. If you go and look at a birch tree, you'll notice the one side has the little fluffs all the way up. And that's the thin part of your bark. So if you cut on the opposite side, you're going to get that thin part right in the middle of your basket and your basket's going to be ruined in no time [figure 8-2].

Mary Thomas (pers. comm., 1994) also noted that the timing for the birch-bark harvest can vary from year to year: "According to the weather the old people would know when to do it [harvest the bark] – it's been so warm this year [1994] the birch bark might be ready around May" (see also chapter 6). Furthermore, aside from the practical knowledge of where, when, and how to harvest the birch bark, for real success in any enterprise, one must also follow the spiritual protocols, approaching the tree or other plant with gratitude and respect, so that the plant will be generous and giving, taking only what one needs, and never harming the plant unnecessarily (see chapters 11 and 13).

Variations in Harvesting

Differences in microclimate and topography within a region or territory influence the growth and maturation rates of species. Sometimes one can go to cooler, shadier locations to find greens, berries, or cedar bark at their prime for harvesting after the same type of resource is finished in other nearby areas. Clan Chief Adam Dick (Kwaxsistalla) described a site on a particular mountain above Kingcome Inlet in Tsawataineuk (Kwakwaka'wakw) territory, for example, where one could always find salmonberries that were still good to pick in August – a month or more after the berries had disappeared in other locations.

Altitudinal differences in ripening times, or times for optimal harvest, are particularly significant in terms of expanding opportunities for harvest. Most culturally important species have relatively wide distributions. Many extend in range from lower elevations to higher altitudes, where their growth and development are correspondingly later (Turner, Deur, and Mellott 2011). For such species, with respect to the timing of ripening, a vertical change of between 500 and 2,000 metres may be the equivalent of a very wide latitudinal change, such as between the locations of the Fraser and the Skeena Rivers along the British Columbia coast. Roots, berries, and other species are generally ready for harvesting much earlier in the valley bottoms than at upper elevations. In some cases, however, since cooler air may settle in the troughs between mountains, plants growing on the hillsides above the valley bottoms may actually ripen earlier than those of the same species on the valley floor. Any berry picker would be familiar with the patterns of ripening related to particular locations within her territory and take these into account when planning her harvest schedule.

McDonald (2003, 54) explains for the Ts'msyen people of the Skeena River, "Just as different environmental and genetic factors contribute to different varieties of wine grapes, mountain berries are affected by their location." Winnie Wesley taught McDonald that berries are sweeter high "up mountain," and she called these "mountain berries." Gitxsan people also gravitate to traditional berry patches halfway up the mountain (Johnson Gottesfeld 1994). For the

Nisg̲a'a, too, some of the prime harvesting sites for berries occur at high elevations, as described by one elder: "The berries they get up the mountains are especially valued because of their sweetness and few seeds: [there is] a black berry, like blueberries, called *maay'im lax̱-sg̲an'ist* ["berry of the mountains," *Vaccinium membranaceum*] in Nisg̲a'a. They don't ripen until late September; they're pretty high up. They are used quite a bit for flavouring the different foods. There are only certain places in the mountains where they are harvested. That's one of the reasons the people live close to the creeks that come down from these sub-alpine areas" (Wilp Wilx̱o'oskwhl Nisg̲a'a 1995, 78–9).

The ability to harvest from a range of altitudes effectively extends the overall harvest season for a given species or resource. When this expansion occurs for a cluster of species, each with a broad altitudinal range, the advantage of seasonal movement is multiplied, and the stage is set for a patterned migration within a group's territory from lower to higher elevations (see Hunn, Selam, and family 1990; and A.G. Marshall 1977). In the eastern part of Secwepemc territory, for example, both *skʷekʷíne* (spring beauty, or mountain potato) and *sxʷixʷ* (yellow glacier lily) grow along the valley bottoms and sidehills around Shuswap Lake above Chase and Salmon Arm. The people of Neskonlith, Adams Lake, and Skwlax (Little Shuswap) start digging these roots there in May and June, just as the plants start to produce fruiting capsules and the leaves die back. This harvest is over by the end of June, but then, in late June, July, and even into August, the same types of roots are ready for harvesting at various nearby upper-elevation sites: Neskonlith Meadows, Skwelkwelkwlt, Tod Mountain, Taft Mountain, Mount Revelstoke, Mount Lolo, and others. The ripening of saskatoon berries follows a similar progression. Thus Secwepemc families travelling to these upland places in the summer are able to continue harvesting these roots and berries, along with hunting, fishing, and gathering their special medicines (Turner, Ignace, and Ignace 2000; Mary Thomas 2001).

For seaweed and some of the green shoots, harvesting itself can also prolong the time period of prime production of the resource. The Gitga'at (Ts'msyen) at their seaweed-halibut camp on Princess Royal Island sometimes picked two "crops" of edible red laver seaweed from the same sites in the same season. The rocks were picked "clean" of the seaweed in early May, and the first harvest was dried for storage and trade. Then a few weeks later, the alga fronds had grown back and could be harvested again, without evident damage to the long-term productivity of the species since this practice is longstanding, with people using the same sites for many generations (Turner 2003a; Turner and Clifton 2006).[10] Similarly, breaking off the young shoots of thimbleberry, salmonberry, and cow-parsnip will trigger the plants' meristematic tissues, stimulating them to produce more shoots within a few weeks, which can also then be harvested, in moderation, just as in picking asparagus from a garden.

Seasons and Phenological Indicators

When the range of variability in animal resources is added to that of plants, the complexity around resource availability is even more pronounced. Shellfish, being more or less stationary, except in their larval stage, may show the same types of variability as plants, but there are times (such as during the red tide season in summer) when it is dangerous to harvest them and other times when they do not taste good (Clan Chief Adam Dick, pers. comm., 2011). Fish, birds, and game all have their own cycles of reproduction and seasonal movement or migration. Because many of these species are often totally absent from an area, and because they are usually at their prime for hunting or fishing only at certain stages in their lifecycles, their availability is even harder to predict than for plants and shellfish. Everywhere, people have developed signs or strategies to assist them in gauging seasonal changes and in predicting the arrival of, or best harvesting times for, fish, birds, or mammals and other key events related to resource availability. Some of these signs are encoded in various languages in the study region in the names of the seasons (table 8-1), which in turn relate generally to the cycles of the moon. The seasonal appearance and relative position of stars and planets also assist in determining resource-harvesting times, as does, of course, the position of the sun and the relative length of days and nights (Clan Chief Adam Dick, pers. comm., 2005; Recalma-Clutesi 2007; Turner, Recalma-Clutesi, and Deur 2013).

The timing of events in the lifecycles of plants, called phenology, is a particularly good indicator of environmentally and seasonally determined events related to resource use. Species whose lifecycle stage or condition helps to estimate the timing of other seasonal events and developments are known as phenological indicators, and "traditional phenological knowledge" is an important component of any Traditional Ecological Knowledge system. In temperate regions, plant lifecycles are relatively predictable, with timing responding to factors such as photoperiod and accumulated heat thresholds in the environment. The generally reliable sequencing of the blooming of flowers and ripening of fruits can thus be used to time other concurrent events and to guide human activities like berry picking or fishing.

A survey of ethnobotanical and ethnographic literature for phenological indicators in over twenty linguistic groups within the study area revealed well over 140 examples, with about half being direct linkages between the flowering time or other phenological response in one species and the onset of a prominent lifecycle stage in a second species, such as a particular bird, mammal, fish, or fruit (Lantz and Turner 2003). One illustration is the Nuu-chah-nulth recognition of the correspondence between the ripening of the salmonberries and the return of sockeye salmon to their freshwater spawning grounds (Bouchard and

Table 8-1 | Examples of seasons named after plants

Approximate month	Language	Name
March	Chilkat Tlingit	*héen taanáx kayaani* 'plants grow underwater'
April	Squamish	*temtsáʔtsqay* 'when the salmonberry shoots ripen' (Bouchard and Turner 1976)
	Stl'atl'imx	*płsxʷixʷm* 'when the avalanche [glacier] lilies start to grow' (Turner 1998; Turner, Bouchard, et al. 1987)
	Okanagan	*sp'itł'mtn* 'bitterroot'; Okanagan First Roots ceremony held when the bitterroot starts to bend over (Turner, Bouchard, and Kennedy 1980); Ktunaxa root-digging season also begins with this plant (Hart 1976)
May	Carcross-Tagish	*kayaani* 'green plants'
	Ts'msyen	*haʔliʔ làx łàʔàsk* 'time for gathering seaweed'
	Straits Salish (Saanich)	*pənexʷəng* 'moon of the camas harvest' (Claxton and Elliott 1993)
	Secwepemc	*pełʔeʔłqten, pesxʷixʷem* 'digging month,' 'time to dig avalanche [glacier] lily bulbs' (Turner, Ignace, and Loewen forthcoming)
June	Gitxsan	*łoxsa maaʔy* 'time for berry picking'
	Ts'msyen	*kiamgam laxmai* 'moon when they pick berries' (Compton 1993b)
	Ditidaht	*tł'ihapax̱pł* 'time of the red huckleberry' (Turner, Thomas, et al. 1983)
	Sechelt	*tem kʷeekʷel* 'salmonberry time' (Maud 1978d)
	Secwepemc	*płltqaitq'atan* 'time of strawberries' (Turner, Ignace, and Loewen forthcoming)
July	Oweekeno	*G̱ùlanx* 'time for picking salmonberries' (Compton 1993b)
	Squamish	*temqʷ'éləm'xʷ* 'time of [trailing] blackberry' (Bouchard and Turner 1976)
	Secwepemc	*pełqʷelqʷ'l't, pełkákaldxmex̱* 'getting ripe' (Turner, Ignace, and Loewen forthcoming) (saskatoon berries are ready to pick) (Teit 1909, 518)
August	Tlingit (Chilkoot)	*quk'eet* 'berry picking' (Thornton 2004)
	Squamish	*tem t'áqaʔ* 'when the salal berries ripen' (Bouchard and Turner 1976)
	Sechelt	*tem t'áqa* 'salal berry time' (Hill-Tout [1910])
September	Tlingit (Chilkoot)	*kaxʷéix* 'highbush cranberries' (*Viburnum edule*) (Thornton 2004)

Table 8-1 | continued

Approximate month	Language	Name
	Ts'msyen	*kiamgam lax tsa'wast* 'when some kind of late [salal] berries ripen' (Compton 1993b)
	Oweekeno	*tsakulstaAm* 'moon when there are no more berries' (Compton 1993b)
October	Nuxalk	*siqaalx̱m* 'time for gathering the rootstocks of spiny wood fern' (Turner 1973)

Note: No plant references were found for winter months (November to February).

Source: Specific references for names are cited in the table. Based on Lantz and Turner (2003).

Kennedy 1990). Other examples are provided in table 8-2. This survey revealed that the ripening of certain berries is one of the most common events to be applied in phenological determinations by Indigenous peoples in the study area. The ripening of a particular berry can both signal other phenological events and be signalled by them, as when the ripening of saskatoon berries is indicated by the blooming of the prickly pear cactus for the Okanagan (Turner, Bouchard, and Kennedy 1980).

One of the most widely recognized examples of phenological knowledge that relates to berry ripening is the relationship between the singing of the Swainson's thrush, or "salmonberry bird" (*Hylocichla ustulata*), and the ripening of salmonberries. The Tlingit, Haida, Haisla, Oweekeno, Squamish, Nuu-chah-nulth, Ditidaht, and Straits Salish all associate the singing of this bird with the ripening of salmonberries (as in the stories and songs of Salmonberry Bird recounted in chapter 12). It is perhaps not surprising that berries should feature so widely in phenological knowledge systems. Berries are a highly valued food yet often require considerable time and effort to harvest, especially because some of the best kinds may occur in distant places. The ability to predict when these berries will be plentiful and at the optimum ripeness can give people a considerable advantage in scheduling their harvest and travelling times.

Considering that wild roses (*Rosa* spp.) (figure 8-3), of all the plant species named by Indigenous peoples in the study region, had the highest frequency of naming (see chapter 3), perhaps it is not surprising that the blooming of the roses in late May and June is a widely recognized occurrence and a primary phenological indicator for a host of different harvesting events, including the ripening of the first wild strawberries (*Fragaria vesca*) and soapberries of the season, as well as an indicator of the ascent up the Fraser River of the second run of spring salmon, called *qel'qásuł* (lit. "rose-hip fish") by the Fraser River

Table 8-2 | Examples of plants as traditional phenological indicators for Indigenous peoples of northwestern North America

Plant species	*Indigenous group*	*Role as phenological indicator*
Green pond slime (*Spirogyra* spp.) and common water moss (*Fontinalis antipyretica*)	Nuu-chah-nulth (Turner and Efrat 1982)	When pond slime and water moss are washed out from the river with the first heavy rains of fall, the coho salmon run is beginning
Bracken fern (*Pteridium aquilinum*)	Lushootseed (Hilbert and Bates 1992)	When the bracken fern starts to sprout and grow, the truly warm days have arrived
Western larch (*Larix occidentalis*)	Okanagan (Ray 1932; Turner, Bouchard, and Kennedy 1980)	When the larch needles turn yellow in fall, the female black bears are entering their dens
Douglas-fir (*Pseudotsuga menziesii*)	Okanagan (Turner, Bouchard, and Kennedy 1980)	When fir cones shed their pollen, the edible inner bark of ponderosa pine is ready to harvest
Bigleaf maple (*Acer macrophyllum*)	Kwakwaka'wakw (Turner and Bell 1973)	Sap running in maples indicates that edible inner bark of western hemlock and other species is ready to harvest
Sitka alder (*Alnus viridis*)	Heiltsuk (Compton 1993b)	Catkin ripening indicates that horse clams are ready to harvest
Saskatoon berry (*Amelanchier alnifolia*)	Nlaka'pamux (Bandringa 1999)	Flowering of saskatoon berry indicates bitterroot roots are ready to harvest
Locoweed (*Astralagus miser*)	Okanagan (Turner, Bouchard, and Kennedy 1980)	Blooming of locoweed indicates that edible cambium of lodgepole pine is ready to harvest
Black hawthorn (*Crataegus douglasii*)	Okanagan (Turner, Bouchard, and Kennedy 1980)	When black hawthorn berries are ripe in the valleys, black mountain huckleberries are ripening in the high country
Fireweed (*Epilobium angustifolium*)	Nlaka'pamux (Turner, Thompson, et al. 1990)	Blooming indicates that it is time to hunt mule deer
Cow-parsnip (*Heracleum maximum*)	Haida (Turner 2004a)	When cow-parsnip is blooming, the seagull eggs are no longer good to eat
Oceanspray (*Holodiscus discolor*)	Comox (Turner 1997b)	When oceanspray is blooming, the butter clams are ready to harvest
Oceanspray (*Holodiscus discolor*)	Saanich (Elsie Claxton, pers. comm., 1998)	When oceanspray is blooming, the sockeye salmon are starting their spawning migration

Table 8-2 | continued

Plant species	*Indigenous group*	*Role as phenological indicator*
Blue lupine (*Lupinus sericeus*)	Okanagan (Turner, Bouchard, and Kennedy 1980)	Blooming of the lupine indicates that it is time to hunt marmots
Prickly pear cactus (*Opuntia fragilis*)	Okanagan (Turner, Bouchard, and Kennedy 1980)	Blooming of the cactus indicates that saskatoon berries are ripening
Mock-orange (*Philadelphus lewisii*)	Okanagan (Turner, Bouchard, and Kennedy 1980)	Blooming of mock-orange indicates that it is time to hunt marmots
Chokecherry (*Prunus virginiana*)	Nlaka'pamux (Bandringa 1999)	Leafing out of chokecherry indicates that bitterroot is ready to harvest
Sagebrush buttercup (*Ranunculus glaberrimus*)	Stl'atl'imx (Fraser River) (Turner 1992b)	Called "spring salmon eye"; blooming of buttercup indicates the first run of spring salmon in the Fraser River
Desert currant (*Ribes cereum*)	Nlaka'pamx (Turner, Thompson, et al. 1990)	Leafing of desert currant in spring indicates beginning of steelhead trout spawning migration
Coastal black gooseberry (*Ribes divaricatum*)	Klallam (Laurence C. Thompson, pers. comm., 1975; L.C. Thompson and Thompson [1975–80])	When the gooseberries are formed, the spring salmon come
Wild rose (*Rosa nutkana*)	Stl'atl'imx (Pemberton) (Turner 1997b)	Blooming of wild rose indicates that western redcedar roots and basket grass (*Phalaris arundinacea*) are ready to harvest
Wild roses (*Rosa* spp.)	Stl'atl'imx (Fraser River) (Turner 1992a, 1992b, 1997b)	Blooming of wild roses indicates the beginning of the second run of spring salmon, as well as the ripening of wild strawberries and soapberries
Thimbleberry (*Rubus parviflorus*)	Nuxalk (Turner 1973)	When thimbleberry shoots are ready to eat, the edible seaweed is ready to harvest
Salmonberry (*Rubus spectabilis*)	Ditidaht (Turner, Thomas, et al. 1983)	Ripening of salmonberries means that oval-leaved blueberries (*Vaccinium ovalifolium*) are ripening
Salmonberry (*Rubus spectabilis*)	Nuu-chah-nulth (Turner, Thomas, et al. 1983)	Ripening of salmonberries means that sockeye salmon runs are beginning

Table 8-2 | continued

Plant species	*Indigenous group*	*Role as phenological indicator*
Red elderberry (*Sambucus racemosa*)	Nuu-chah-nulth (Bouchard and Kennedy 1990)	Blooming of elderberry means that halibut-fishing season has started and that it is time to hunt harbour seals and gray whales
Soapberry (*Shepherdia canadensis*)	Secwepemc (Turner, Ignace, and Loewen forthcoming)	Ripening of soapberries indicates beginning of the sockeye salmon spawning migration
Soapberry (*Shepherdia canadensis*)	Stl'atl'imx (Turner 1997b)	Ripening of soapberries indicates that sockeye salmon smolts are no longer good to eat
Stinging nettle (*Urtica dioica*)	Gitga'at (Ts'msyen) (Turner and Clifton 2009)	Stinging nettle growth parallels that of edible seaweed (*Pyropia abbottiae*), so you can use nettles to gauge when the seaweed is long enough to harvest
Stinging nettle (*Urtica dioica*)	Nuu-chah-nulth (Turner and Efrat 1982)	When the nettle shoots are about 10 centimetres high, Pacific halibut fishing will be good
Evergreen huckleberry (*Vaccinium ovatum*)	Nuu-chah-nulth (Turner and Efrat 1982)	Ripening of berries indicates the beginning of the spawning migration for chum (dog) salmon

Source: Specific references for Indigenous groups are cited in the table. Based on tables in Lantz and Turner (2003).

Stl'atl'imx/St'at'imc. These fish have a deep pink stripe along their sides, the colour of the wild rose blossoms, and are said to be fast swimmers (Art Adolph, pers. comm., 2005; Kennedy and Bouchard 1986).[11] For the Pemberton Stl'atl'imx basket makers, the time when the wild rose blooms is the best time to harvest cedar roots and reed canarygrass (*Phalaris arundinacea*) stems, which are carefully cured and then split open and used along with bitter cherry bark for decorative overlay in their coiled cedar-root basketry (Turner 1992a; see also chapter 6).

The majority of phenological signals that relate to animal resources pertain to fish. In all, the survey identified eighteen indicators for fish lifecycle events, more than half of which are related to the migrating and spawning of salmon. Other indicators are for when birds' eggs are ready, the time for clam harvesting, and the time for hunting bears and marmots (see table 8-2). The examples documented in Lantz and Turner (2003) and in table 8-2 are only a few of literally hundreds of phenological indicators known to and used by individuals and communities in northwestern North America. The immense diversity

8-3 | Wild rose (*Rosa acicularis*). Its blooming around June is a common phenological indicator.

and heterogeneity of the homelands of these peoples – with resources not only separated by great distances over a wide altitudinal range but also varying from year to year in their optimal harvest times and availability as well as in their productivity – have made the development of "proxies" in the form of phenological indicators particularly valuable, potentially saving considerable time and energy, especially when people are seeking less easily seen resources, such as fish.

CULTURAL HABITATS OF HIGH SIGNIFICANCE

Ecologically, geographically, and biologically determined variation across a region produces patchiness, zonality, and variation in resources, ultimately resulting in biological diversity on all scales: genetic, within a species or population, or within a habitat and landscape. This diversity is important for people because it gives more options for resources, in terms of both variety and availability. Habitats where significant numbers of culturally useful species occur together, usually because of similar ecological requirements and tolerances, are particularly sought out, especially when the resources in a place are ready for harvesting at about the same time and when these habitats are relatively easily accessed.

The diverse habitats and special landforms that produce them are featured in the vocabularies of every language throughout the study region.[12] For example, McNeary (1976, 77) recorded distinct Nisg̱a'a names for avalanche track, rockslide, tree blowdown, meadow, swamp, river, stream, lake, cottonwood forest, mud forest, wooden slope, hilly land, mountain, gully, valley, and upstream and downstream areas. Dr Ron Ignace (2008, 5–8) provides a comparable and even more detailed list of Secwepemc place types[13] and emphasizes their importance:

> People can predict what kinds of landforms they will encounter throughout their travels, and what kinds of sources of animal and plant foods, sources of water and shelter, and ecological indicators for all of these they will encounter. For example, as elders explained to us in mapping out these generic landscape terms, you know that a plateau lake will have an outflow, where you usually find trout after break-up in spring. As forest ecologists know, you will find certain plants on the sunny (southwest) side of mountains, others on the moister northeast side. Forested areas in the Plateau will include moist meadows ... that will, in turn, provide pasture for horses, a nearby creek and a good overnight camping location. Along the rivers, back-eddies exist in predictable locations near out-croppings and [these] ... are the locations for salmon fishing.

Some of these habitats – or clusters of adjacent habitats – become special nodes of cultural activity, occupied intensively at times of the year when the greatest numbers of cultural resources are available there. Often these places are situated geographically and ecologically at the intersections, or edges, of two or more different ecological zones – for example, along a coastline or river estuary, at the edge of a wetland or prairie, or at the edge of the treeline in the mountains where one or more creeks are descending through a mosaic of high-elevation woods and montane meadows. By situating themselves in such locations, people gain access to a greater diversity of resources, as well as availing themselves of the best opportunities for travel, shelter, and protection (Turner, Davidson-Hunt, and O'Flaherty 2003).

A good example is, again, the site of Salmon Arm on the southern arm of Shuswap Lake in Secwepemc territory. As noted previously, the Salmon Arm area was known for (and apparently originally named after) its soapberries. But this locale, with its unique combination of wetlands and uplands, also offered a great abundance of other plant resources. In the woodlands, as well as birch, western redcedar, and many other trees useful for shelter and construction, one could find quantities of wild raspberries, saskatoon berries, chokecherries and pin cherries, highbush cranberries (*Viburnum opulus*), and cow-parsnip shoots. At the estuary of the Salmon River were extensive floodplains, with great patches of wapato (*Sagittaria latifolia*), water-parsnip (*Sium suave*), Indian-hemp, and cattail, all of which, as recalled by Mary Thomas (2001), the Secwepemc people would harvest, the first two as root vegetables and the last two as materials. Large cottonwoods along the shoreline provided important animal habitat and prime logs for dugout canoes, as well as shelter and wood for fuel. Salmon River and Shuswap Lake produced abundant fish and waterfowl, including mallards and coots, which were hunted and whose eggs were gathered in season from among the cattail patches at the river mouth. On the benchlands above the lake was once a long slough where people used to pick enormous quantities of trapper's tea and bog cranberries and where the fragrant, little creeping snowberry (*Chiogenes hispidula*) grew in abundance. Secwepemc people used to come from all around the territory to harvest these diverse resources. Unfortunately, most of this highly productive area was changed and destroyed by railway and highway construction, dykes and river diversions, cattle grazing, farming, and urban settlement (Garibaldi 2003; Mary Thomas 2001).

We can call such special diversely productive places "cultural keystone places" because they are so critically important to peoples' lifeways. Every Indigenous territory includes such keystone locales, often incorporating permanent settlement areas or long-term seasonal camps. Collectively, as accessed over the course of a year, these places provided virtually the full spectrum of resources people needed for their survival. Sometimes these cultural keystone places were so productive that more than one group would seek them out, either as shared harvesting sites or as destinations where a particular group hosted visiting neighbours who might be given permission to harvest particular resources under the host's supervision and with some form of reciprocity in place.

Some of these places have been renowned gathering centres for diverse peoples. For example, the area around the mouth of the Nass River in Nisg̱a'a territory on the northern coast of British Columbia is famous for its rich oulachen[14] fishery and oulachen grease production, as well as for other fish, seafood, crabapples, and a host of other animal and plant resources. This locality is said to have been the homeland for the Tlingit-speaking peoples (Thornton 2008), and it features in many Ts'msyen stories as well (McDonald 2003). The Tlingit recognized a few special and exceptionally fruitful places within their own territory that they called "berry land" – places widely known throughout Tlingit country and beyond. Not only were these places owned, defended, and celebrated by Tlingit lineages, but they were even cultivated through eons using "magic" (*heix̲ʷa*) and other methods, such as pruning, that were intended to enhance the productivity of the berries and other resources (Thornton 1999).

Another of these productive resource locales is the lower Fraser Valley, from the Fraser estuary to the Chilliwack region, including an area known as Sumas Prairie, where an immense lake was drained for agricultural purposes in the post–First World War period (K.T. Carlson 2001). Known for its extensive wapato and bog cranberry patches, spawning sockeye, sturgeon, ducks and geese, cedar trees, cattail, and tule, among many other resources, this area attracted people from Vancouver Island and other parts of the coast, as well as from upriver areas. The famous fishing sites of the upper and lower Columbia River were similar magnets for people, not only because they provided fish but also because they were situated within areas of diverse wetland, prairie, and woodland habitats that multiplied available resources.

In the Interior Plateau, there are many other examples. One of these is Botanie Valley in the territory of the Lytton band of the Nlaka'pamux, near the confluence of the Thompson and Fraser Rivers (figure 8-4) (Hanna and Henry 1995; Teit 1900; Turner, Deur, and Mellott 2011; Turner, Thompson, et al. 1990). This upland valley is famous for its productive root and berry resources, as well as for its game and lake fish. Another key harvesting site where people converged was a place called "Many Roots," near the Bridge River (at the intersection of Stl'atl'imx, Secwepemc, and Tsilhqot'in territories), which was known for its productive root resources and accessed by all of these peoples (Teit 1906a, 237, 256). Some people refer to such locations as "one stop shopping" places because they provide so completely for life's necessities. Further examples of such gathering places, and the social outcomes of these areas in terms of sharing knowledge and experience, are described in chapters 9 and 10.

Archaeologically, such "cultural keystone habitats" might be indicated by the presence of clusters of assorted and diverse artifacts, the presence of culturally modified trees, evidence of camping and food processing, and the presence of multiple resource species (berries, edible roots, and important wood and fuel species), as well as fresh water. For example, there are some sites in Kingcome Inlet where archaeologist Jim Stafford (pers. comm., 2009) and his colleagues have noted a strong correlation between intensive bark-harvesting sites indicated by numbers of bark-stripped trees and important berry-picking areas identified by local Tsawataineuk Kwakwaka'wakw.

The presence of petroglyphs (rock carvings) or pictographs (rock paintings) may also be another indication of cultural keystone habitats, possibly associating these places with spiritual aspects of harvesting. An example is the immense number of carved boulders, with images of fish, humans, and apparent supernatural beings, at a site along the shoreline of Douglas Channel, near the fall salmon-fishing camp of the Gitga'at Nation. At this place, two rivers drain into the

8-4 | Botanie Valley, a high-elevation valley in Nlaka'pamux territory, a renowned gathering place and multi-use harvesting site – a "cultural keystone place."

coastline, and the whole area is highly productive, not only for different kinds of salmon but also for clams and crabs, geese and ducks, bear and deer, and blueberries (*Vaccinium ovalifolium*, *V. alaskaense*), highbush cranberries, and different varieties of crabapples (*Malus fusca*), some of which were hereditary property of clan chiefs, matriarchs, and their families. There is a rich oral history centred here, and the petroglyphs (some of which are nearly worn away by the ocean's wave action) provide a considerable time depth for its use and importance as a cultural keystone habitat for the Gitga'at.

These centres for diverse culturally important species can shift over time as environmental conditions and interactions among species change. Natural events such as floods, tsunamis, volcanoes, and wildfires can impact species' distributions, as can human activities such as landscape burning and intensive harvesting. Of course, peoples' lifestyles and preferences may also change, such that once-important resources may be supplanted by others, and places where these resources occurred may become relatively less crucial for survival, whereas others may take on increasing significance. Nevertheless, many culturally important sites are notably enduring. Following the arrival of Europeans, many of these special locales were appropriated as sites for fur-trading posts and settlements, taking advantage of the same conditions that made these such important resource centres for First Peoples.

The routes that people have used to access these resource nodes, or "cultural keystone places," and resource sites in general are significant in themselves. Many of these routes would have been established and determined by the spatial distribution of the keystone habitats, but they would also have provided incidental access to other resources along the way, as well as opportunities for interacting with other groups of people. The routes also served as conduits for dissemination of products, technologies, ideas, names, stories, and other cultural features, and together they allowed an immense network of travel and exchange over the entire study area and beyond. They included canoe routes along coastlines and inland waterways, winter and spring ice routes, land-based foot trails, and later, horse trails and roads. The trails, many probably originally established by migrating animals, traversed mountain passes, reached across islands and peninsulas, threaded along river canyons, cut across prairies, skirted wetlands, and penetrated dense forests. The entire area of northwestern North America was once criss-crossed with these courses, and many still exist. Some branch off from more major throughways; most connect settlements and campsites. Some serve as long-time trade routes, linking communities and nations; others mark boundaries of discrete territories. Some of the cultural keystone places are situated at a convergence of two or more such routes. Many of these travel routes are obvious, even today, from features along them such as campsites and hearth sites, marked trees, stone canoe runs, and worn-down tracks (Blackstock 2001), whereas others are now overlain by roads and highways or are completely overgrown. The "grease trails," leading from locations all along the coast and following some of the major river courses into the interior, are among the best examples of routes connecting different resource nodes and regions (see Birchwater and Ulkatcho and Nuxalk elders 1993). These will be discussed further in chapter 10, on trade and exchange.

THE ETHNOECOLOGICAL CHOREOGRAPHY: SEASONAL MOVEMENTS, TIME, AND SPACE

As already implied, then, virtually all Indigenous peoples of northwestern North America, since time immemorial, have traditionally spent at least a portion of every year travelling and camping in various known sites around their territories, harvesting the variety of foods and other resources they required for survival from the range of habitats encountered. Unlike agrarian societies, where people were relatively sedentary, focusing much of their energy on food production in one area, people who rely on resources from places where they naturally occur have to move to their harvest sites. This movement, however, is not random but cyclical and predictable, within the constraints of weather, tides, seasons, the annual productivity of various resources, and other socially mediated factors.

Cycles of time and productivity are integral to Traditional Ecological Knowledge systems, as explained by Nuu-chah-nulth elder Dr E. Richard Atleo (Chief Umeek): "In this way the process of life is acknowledged to be cyclical and reciprocal. The Creator is One and gives to all, and everything in creation gives, each to another. The Creator sends rain to water the earth, which then blooms to provide food for Nature, which in turn provides food for the human, who eventually dies and returns to the earth, which will bloom again to provide food for Nature, which in turn provides food for the human, again and again, in endless cycles" (Turner and Atleo 1998, 113).

Throughout the study area, these seasonal movements are a significant feature of cultural ecology, influencing virtually every aspect of people's lives, from social organization to spiritual beliefs. In terms of food security, having access to a wider diversity of foods from a broader area helps to mitigate the possibility of food scarcity by lessening a family's or community's reliance on any one type of food or any one locality. Kwakwaka'wakw clan chief Adam Dick (Kwaxsistalla) (pers. comm., 2005, 2006) described the seasonal rounds of his boyhood:

> The people are busy all through the year when a different season comes along. Herring time, cod time, clam time, all different. Keep you busy. How are you going to live through the winter? What you can eat? And what you're going to put away for winter. Has to be smoked heavy or sundried ... If the fish don't come, the clam gardens are a backup. Clams, oulachens, *ts'ats'ayem* [eelgrass], back out [to the coast], up the river ... up to the mountains ... Two or three families on this island, three to four families on that island. Different spots. They keep looking at the stars, Qwam'a and Alujoy, then they move up the inlet, to get the salmon and smoke them. Then, out to clams, then back up to oulachens. After that, they spread out closer to the seaweed picking, then they move to where there's lots of berries to make *t'əqa?* [dried berry cakes]. When I was growing up, there was no time. We go to bed when it gets dark ... Get up in morning and do the same thing.

Often, as described by Adam Dick, Edith O'Donaghey, Mary Thomas, and others who remember these seasonal travels, people would join forces with family members or close friends for harvesting and processing their food. Those who travelled to different sites and had opportunities to obtain different resources would transfer some of their harvest to others through reciprocal sharing and gifting.

In northwestern North America, the usual pattern of the seasonal round – established millennia ago – was for people to reside in permanent dwellings and villages during the coldest part of the year. Then, in early spring, they would leave their houses, usually in small family or clan groups, beginning a cycle of movement that would take them to different parts of their territory – from one cultural keystone place to another – before finishing in the late fall back at their winter homes. Over the past century or so, with participation in the wage economy, and more convenient modes of transportation, people have tended to use their home base increasingly, even during the harvesting times, returning for periods of time between their harvesting trips to bring the processed food back for storage and to reconnect with family and friends. Examples of seasonal round patterns of coastal and interior peoples of the study region are provided in table 8-3.

Springtime on the Northwest Coast

Along the coast, people headed out from their winter homes, usually around March, to start their early fishing for halibut, the first runs of spring salmon, and many other species: red snapper, cod, herring and herring spawn, oulachen, and flounder. They might also dig clams and harvest crabs, mussels, and other shellfish as available. As noted in chapter 5, early spring could be a time of food scarcity for people, especially if there were storms or heavy snows that prevented travel or suppressed the growth of plants or the movement of animals. However, after a long winter of living mostly on dried, stored food, the prospect of fresh greens and other springtime food must have been a great incentive to venture out. The Haida name for the month of March, *xiid gyaas* ("laughing goose month") signifies one of the first signs of the onset of spring, the northward flight of migratory geese. For the Saanich, the chorus of the tree frogs was the sign for people that spring had arrived and that the winter ceremonial season had come to an end; the frog is the "honoured keeper of the sacred seasons" (E. Claxton and Elliott 1993, 27).

For those coastal people harvesting oulachens and rendering "grease" from these small oily fish, the eagerly anticipated oulachen season was heralded by the presence of flocks of gulls and seals following the oulachens to the river mouths, and even well up the rivers. If people killed a seal around this time, it might have oulachens in its stomach, a sure sign that the oulachens were starting to run. As Clan Chief Adam Dick noted, a certain alignment of stars was also one of the signs for people to move to the rivers for the oulachen harvest. Construction of oulachen weirs and traps, catching the oulachens, and then rendering the oil were an immense undertaking that required tremendous preparation,

Table 8-3 | Examples of seasonal round patterns of coastal and interior peoples

Approximate season/moon	*Haida (coastal)*	*Saanich/Straits Salish (coastal)*	*Gitxsan/ Gigeenix (Upriver dialect) (interior)*	*Secwepemc/ Shuswap (interior)*
January/ February	*hlgidguun (łgidguun) kongaas* ('Canada goose moon'); Canada geese come down to the coast	*ngingənə* ('moon of the child') (~ January); snow is waist deep	*kʷ'ithułox̱s* ('ring around the sun/moon'); winter was feast time; families came together; trapping in winter; rabbits snared	*pełtsipwenten* ('cache-pit' month) (~ February); still too cold, and there was no hunting, so the people had to rely on what they had stored in their cache pits
February/ March	*taan kongaas* ('black bear moon'); bears come out of hibernation	*wəx̱əs* ('frog' moon) (~ February); time of the baby moon; the short month	*gutk'ułox̱s k'utk'ułox̱s lasa huʔmal* ('when cottonwood trees snap because of the cold'); stored food depleted; wintering steel-head caught in pools; long arduous trek to Nass estuary for oulachen over the "grease trail"; beaver hunted for meat and pelts	*pełsqepts* ('spring wind') (~ March); snow melts in the valleys; steelhead run in the river, and people fished them with torchlights from boats; people travelled to Loon Lake to fish for cutthroat trout
March	*xiid gyaas* ('laughing [white-fronted] goose moon'); laughing geese fly north	*pəx̱sisəng* (moon of 'opening hands'; 'blos-soming out' moon); frogs start to sing	*'wiiłax̱s* or *'wiiłox̱s* ('when the bears sit around in their den before they come out in spring')	
April	*wiid gyaas* ('salmonberry bird moon'); song of the sal-monberry bird */continued*	*sx̱ʷenəł* ('bullhead' moon); wild geese fly over upon return */continued*	*lasa yaʔa* ('spring salmon month'); still early; medicinal plants and barks for */continued*	*pesłʔewten* ('melting month'); all the snow melts, even in the */continued*

Table 8-3 | continued

Approximate season/moon	*Haida (coastal)*	*Saanich/Straits Salish (coastal)*	*Gitxsan/ Gigeenix (Upriver dialect) (interior)*	*Secwepemc/ Shuswap (interior)*
	(Swainson's thrush) announces that winter is over	from the south	"wood medicine" gathered; later, greens picked (e.g., "lava berries," cow-parsnip stalks)	mountains; people hunted in the mountains and gathered fresh shoots from sprouting plants
April/May	*gansgee 'laa kongaas* ('halibut moon'); also known as the month when berries are forming	*pənex^{w}əng* ('moon of the camas harvest') (~ May); blue camas blooms	*lasa 'yanja* ('leaves come out'); as sap rises, pine and hemlock cambium gathered; cedar bark for basketry; inner bark of willow and maple gathered	
May to early June		*chənthəqi* ('sockeye moon') (~ June); time of warm weather	*lasa maaʔy* ('first berries of the season'); First Salmon ceremony; spring salmon fished; bear hunted when emerge from dens	*pełʔeʔłqten* ('digging month') (~ May); people dug edible roots (*sk^{w}enk^{w}ínem, sx^{w}ix^{w}, tsets'elq*), and gathered inner bark of lodgepole pine; dug basketry roots; went to lakes to fish for rainbow trout
June (late May to early July)	*wa.aay gwaalgee* ('weather is still somewhat cold') (this is the Haida New Year)		*lasa maa'y* ('first berries of the season'); strawberries and soapberries ripen during continued spring salmon fishing	*pełtpantsk* ('midsummer month'); people gathered soapberries, harpooned spring salmon in the river, and trapped spring salmon in weirs in the creek

Table 8-3 | continued

Approximate season/moon	*Haida (coastal)*	*Saanich/Straits Salish (coastal)*	*Gitxsan/ Gigeenix (Upriver dialect) (interior)*	*Secwepemc/ Shuswap (interior)*
June/July	*kong koaans* ('great moon'); because the weather becomes warm and food becomes plentiful; edible inner bark is harvested from hemlocks and spruce	*chənhənən* ('humpback salmon return to earth') (~ July); hottest time of year	*lasa 'wiihun* ('fish are plentiful'); saskatoon berries, huckleberries, lowbush blueberries picked in quantity, right after sockeye run	*pełtqʷelqʷ'eltemx* ('getting-ripe month') (~ July); lots of berries were ripe during this month, e.g. saskatoon berries (*speqpeqʔúw'i*), strawberries (*tqítq'e*), and huckleberries (*wenéx*); people gathered all of them
July/August	*sgaana gyaas* ('killer whale moon'); when cedar bark is stripped from the trees, it sounds like blowing killer whales	*chənthewən* ('coho salmon return to earth') (~ August); mosquitoes are out	*lasa lik'l'insxʷ* ('when the grizzly bears are eating fish'); salmon fishing the mainstay in spring, then sockeye in midsummer; several weeks of intensive berry picking on mountain slopes, where many black huckleberries picked and made into berry cakes for winter	*pesqelqlelten* ('many salmon month') (~ August); people fished for salmon and dried their salmon
August/ September	*k'iijaas* ('belly moon'); animals begin to grow fat and their bellies get big	*chənkʷ'aləx* ('dog/chum salmon return to earth') (~ September); leaves turn colour	*lasa sgangʷiikʷ* ('groundhog getting fat'); Gitxsan hunt them; then fish chum/dog salmon for their roe; late summer mountain goat, */continued*	*pełtemłik't* ('[salmon are] spawned out') (~ September); people hunted and dried their meat for winter

Table 8-3 | continued

Approximate season/moon	*Haida (coastal)*	*Saanich/Straits Salish (coastal)*	*Gitxsan/ Gigeenix (Upriver dialect) (interior)*	*Secwepemc/ Shuswap (interior)*
			caribou, and marmot hunted by men while women pick berries	
September/ October	*k'eed adii* ('in-between month'); the month between summer and winter	*pəq'əlenəxʷ* ('moon that turns the leaves white') (~ October); leaves fall off the tree	*lasa sinlaaxʷ* ('trout are hungry taking to bait, easy fishing'); "early fall"; coho fished; steelhead fished in summer, fall, or late winter	*pesłwelsten* ('abandoning month') (~ October); people were still hunting and drying meat and were also trapping and tanning hides; they fixed up their winter homes and their cache pits
October/ November	*k'algyaa kongaas* ('ice moon'); first ice appears on rainwater in the canoes	*xʷəsəlenəxʷ* ('moon of the shaker of leaves'); everything is put away; too cold to do anything	*lasa sinlaaxʷ* (continued); cranberries and rosehips picked; spiny wood fern gathered in fall (not considered ready for food use until the green fronds had died back); moss gathered; deer, moose, and bear hunted; medicinal herbs gathered; ceremonial purification undertaken; "wood medicine" gathered	*peł?ełxʷ?ułxʷten* ('entering month') (~ November); people entered their pithouses; animals entered their dens to hibernate; people trapped animals at this time
November/ December	*jid kongaas* ('digging month'); bears dig roots to /*continued*	*sch'əlkʷesən* ('moon of putting your paddle away /*continued*	*lasa gʷineek̲xʷ* ('cold'); people prepared for winter; many /*continued*	*pełteteq'em* ('cross-over month') (~ December); /*continued*

Table 8-3 | continued

Approximate season/moon	*Haida (coastal)*	*Saanich/Straits Salish (coastal)*	*Gitxsan/ Gigeenix (Upriver dialect) (interior)*	*Secwepemc/ Shuswap (interior)*
	prepare for hibernation	in the bush'); the ground is shining, or glistening, from frost or ice	activities (e.g., traps, snowshoes, and snares fixed; firewood gathered)	winter solstice month, when the days start getting longer; Shuswap people celebrated and decorated their winter homes
December/ January	*kong gyaangaas* ("cold moon"; lit. 'standing up [to defecate] month'); because the ground is snowy and cold	*sisət* ('elder moon') (~ December); time for winter ceremonials, making and repairing equipment, etc.	*lasa gunkʷ'ats* ('can't go far to outhouse'); winter ceremonies; ceremonial purification undertaken with devil's-club, fasting, and sweathouse use; winter stores eaten	*peɬʔemetmin* ('stay at home month') (~ January); land is frozen during this time of year, and people stayed in their winter homes; a lot of work was done at this time (e.g., making baskets, sewing clothing, and making implements for the harvest season)

Note: The groups for which this information was available were also selected to be representative of peoples from the northern coast (Haida) and southern coast (Saanich) and from the northern interior (Gitxsan, Upriver dialect) and southern interior (Secwepemc).

Sources: For Haida, see de Vries (2012); for Saanich, see Claxton and Elliott (1993); for Gitxsan, see Wale (n.d.), as sourced by Leslie Main Johnson (pers. comm., 2011); and for Secwepemc, see R.E. Ignace (2008).

cooperation, and supervision by knowledgeable people. For the Tsawataineuk community at Kingcome, Clan Chief Kwaxsistalla (the name of Adam Dick and his ancestors) was the recognized guardian of the oulachen.[15] All up and down the coast, especially from the Knight and Kingcome Rivers to the Bella Coola, Kitlope, Kemano, Kitamaat, Skeena, and Nass Rivers, people congregated in the early spring for the oulachen harvest. For many, this was an important social time, when trading and other exchanges took place (Thornton 2008; Thornton et al. 2004).

At the mouth of the Nass River, at Fishery Bay, while people were staying at the oulachen camp and rendering the grease, they also harvested large quantities of the rootstocks of spiny wood fern. Traditionally people pit-cooked these root vegetables or else steamed them on the stovetop until they were soft. Once they were cooked, the "fingers" (leaf bases of previous years' fronds) were peeled and mixed with oulachen grease and sugar. One Nisg̲a'a elder recalled, "We used to get *ax̲* (wood fern roots) too ... You get it right off the ground, and you have to boil that, and you peel the skin off [from the "fingers" and the core] and that's really good. You mix it with grease and sugar ... Get them by the sack. It's used for dessert; mix it with new [oulachen] grease and sugar, that's all. It's sort of a delicacy" (Wilp Wilx̲o'oskwhl Nisg̲a'a 1995, 83).

In relatively quick succession, over the course of the spring, there was more clam digging and other shellfish harvesting, and people also travelled to their estuarine root gardens to dig quantities of silverweed, clover, riceroot, and other root vegetables, before these plants started to sprout and grow their leaves. The succulent shoots of giant horsetail, thimbleberry, salmonberry, and cow-parsnip also appeared at this time and were harvested – in some cases by the armful – to be eaten fresh, with seal or oulachen oil.[16] In the Nass Valley, people also sought the fleshy leaves of "lava berries" (*Sedum divergens*). During these harvests, the clam gardens and root gardens were also tended and repaired; the root gardens were weeded and cleaned and the clam beds cleared of any rocks or driftwood that had been washed up over the winter.

The spawning of the herring was another major event of early spring. Some people relied on the textured fronds of giant kelp[17] as an ideal surface on which the herring could deposit their spawn. Other people anchored boughs, or entire young trees, of western hemlock, western redcedar, or other evergreens in the quiet bays where the spawning herring would accumulate, and within a day or two, these boughs would be densely coated with the tiny translucent eggs. The herring were followed by droves of sea lions, seabirds, and even whales, all feeding on these nutritious eggs. The ocean around where the herring spawn becomes a milky blue from the milt of the male herring. The herring eggs can be cooked and eaten immediately or dried for winter. When kelp fronds are used in the harvest, they are eaten together with the herring eggs, whereas if boughs are used, the eggs are cooked with the smaller branches and then peeled off before being eaten. Today the commercial herring fishery has seriously reduced the numbers of spawning herring; there are some places on the coast where the herring haven't come to spawn for many years. Still, wherever they can get them, people relish herring eggs. Nowadays, like many traditional foods, herring eggs are often preserved by freezing rather than by drying them.

The month of April is called *wiid gyaas* ("salmonberry bird month") in Haida, the time when the song of the salmonberry bird (Swainson's thrush) announces the end of winter. The next moon, at the end of April and beginning of May, is *gansgee ʔlaa [ˈlaa] kongaas* ("month when the berries are forming"), a process assisted by the salmonberry bird's singing. Bears emerged from their winter dens and might be hunted if people were short of meat or fat. However, it is said that a diet of skunk-cabbage in the spring or salmon in the fall made the bear meat poor eating (M. Kat Anderson, pers. comm., 2009). People also trapped fur-bearing animals and hunted small game and birds. The Saanich, Ditidaht, and others hunted ducks with long nets of nettle twine strung across narrows in the ocean. Seals and other marine mammals were hunted for meat and oil. Seagull eggs were gathered from rocky offshore islands and cliffs.

Around May, when the spring tides were lowest early in the morning, many families travelled to their spring harvest camps to pick seaweed, fish, and gather other kinds of seafood. The Gitga'at spring camp at K'yel on Princess Royal Island, previously mentioned in this chapter (and in chapter 5), provides a good example of the intricate requirements of food harvesting, choreographed to the tides, weather, and lifecycles of resource organisms (Turner 2003a; Turner and Clifton 2006; Turner and Thompson 2006). May, or *haʔliʔ làx làʔàsk* ("the time for gathering seaweed"), is the usual time when the Gitga'at move from their home village of Hartley Bay to K'yel.

Generally, the weather is fine in May on the Northwest Coast, with plenty of sunshine. This is important because the optimal harvestable stage of this seaweed coincides with the occurrence of the low spring tides in the morning, as well as a period of winds coming from the north or northwest, which promotes drying of both seaweed and fish. Gitga'at harvesters (mostly women in the old days) can travel in the early morning out from K'yel by canoe or rowboat, or more recently by skiff or punt, to the rocky shoreline along Campania Island and other neighbouring islands, jump from the boats onto the rocks, spend two or three hours picking seaweed while it is exposed by the low tides (figure 8-5), and then return to K'yel around noon, leaving sufficient time for spreading out the seaweed in squares on the sun-heated bedrock or cedar trays. In mid-afternoon, the seaweed squares are shifted and turned over so that they will be thoroughly dried before sundown. This combination of low tide in the early morning, warm sunshine and winds, and optimal seasonal growth of the seaweed itself is essential for this system to work. K'yel is also a halibut and spring salmon fishing camp. The cutting and drying of the fish also depend on good weather, and the work required for both fish and seaweed is so intensive that they cannot be done at the same time. Therefore, if the men are successful in bringing back fish, seaweed harvesting is usually postponed until the halibut are processed.

People accommodated minor variations in the weather by adjusting their days of seaweed harvesting and fishing accordingly. The women would watch the growth of the stinging nettle plants around the K'yel camp as an indicator of the growth stage of the seaweed since it tends to grow at the same rate; if the seaweed season was a little late, this could be determined by the shorter growth

8-5 | From left to right: Rachel, Max, and Morgan Hill of Hartley Bay, picking edible seaweed in May 2011, near their spring harvest camp at K'yel.

of the nettles. Once dried, the seaweed squares are brought back to Hartley Bay, and the women undertake another weather-dependent process for some of their seaweed, namely compressing, chopping, and redrying it, before it is stored away in airtight containers for use as a condiment, soup ingredient, and valued trade good.

All around Ts'msyen territory, too, families and clans had their own resource areas, equivalent to the Gitga'at's K'yel. Different families from Kitsumkalum and Kitselas, for example, had harvest camps for seafood on the lower Skeena River at Port Essington, Ecstall, Gibson Island, Mud Bay, Island Point, Prescott Pass, Eddy Pass, and Baker Inlet. In March people harvested herring spawn at Island Point, and then a little later, there was seaweed picking through Prescott Pass. Crabs, halibut, and salmon were harvested at Gibson Island. Later in the spring (and again in the fall), they would travel upriver to harvest their inland resources (McDonald 2003).[18]

The Kwakwaka'wakw and Nuu-chah-nulth were especially fond of the sweet rhizomes of eelgrass, which they harvested around May.[19] The coordination of tides, season, and weather required for eelgrass harvesting was similar to that required for picking seaweed. Sometimes people walked out to dig up eelgrass in the lowest tide, but as Clan Chief Adam Dick explained, "it's better to go out by canoe and *k'elpax*, or twist it up with a long, thin twisting pole in 'small tide' because you could wash the sand off when the eelgrass was harvested." The fresh rhizomes were greatly appreciated and eaten in quantity at special feasts and family meals; "there's lots of sweet in there, in that [*ts'áts'ayem*] root" (Adam Dick, pers. comm., 2006; see also Cullis-Suzuki 2007).

On southern Vancouver Island and the Gulf Islands, people started to go out in May to their traditional territories in small family groups to harvest camas bulbs and other spring-flowering bulbs; the month of May is called *pənexʷəŋ* ("moon of the camas harvest") in the Saanich language, although these bulbs could also be dug later on in the summer (E. Claxton and Elliott 1993; Turner and Hebda 2012).

Toward the end of spring, as the sap started to rise, coastal people also harvested the edible cambium of western hemlock and other species.[20] It is at this time when the cambium is "fat." People also harvested fibrous barks of cedar and other species for basketry, net making, cordage, and clothing in late spring and early summer.[21] The Gitga'at were always careful, however, to wait until after all of the seaweed and halibut had been harvested and dried before they went out to harvest cedar bark. (It is said that harvesting cedar bark first will cause rain and prevent the seaweed from drying.)

Many medicinal plants, and especially the tree barks used for tonics, were collected in spring and early summer, to be prepared and used immediately or dried for later use. Devil's-club bark, for example, was generally harvested at this time, when the spiny outer bark could be scraped off and the inner bark easily peeled away from the wood (Turner 1973, 2004a; Turner and Thompson 2006).

Springtime in the Interior

In Secwepemc-tsín, the Shuswap language, the fourth and fifth moons are called *peska´pts* or *pełsqepts* ("spring [winds] month"), when there are frequent Chinook winds and the snow starts to disappear, and *pesx˘ü´xem* ("[little] summer [month]") or *pesxʷixʷem* ("has-avalanche-lily month"), when the snow disappears completely in lower areas. This time – for the Secwepemc and other interior peoples – marked the beginning of the harvest season (Teit 1909, 518).[22] Signalled by the blooming of certain early flowers, the songs of particular birds, or the appearance of certain butterflies or other insects, people throughout the Interior Plateau would leave their winter houses in small groups for their camps around their territories. Before the families left their winter homes, Mary Thomas (pers. comm., 2001) recalled, they would clean their houses thoroughly, taking off the turf from the roof to let the interior "air out" and building a big fire beside the house to burn any litter that had accumulated in the house from winter living.

People started to dig a few of the edible roots around late February and March. One of the first roots Mary Thomas (pers. comm., 2001) remembered harvesting in the spring was *$q^{w}eq^{w}$’íle* (desert parsley, *Lomatium macrocarpum*): "I can remember my grandmother, when the robins first come in, they chirp and just chirp and call each other. Then all of a sudden my grandmother would say, 'Listen to the robin!' And you could hear it singing, '*tu weeee sa – te $q^{w}eq^{w}$’ile!*' [Others say it's the meadowlark that sings this song.] They have a little tune and they [people] would say, 'Oh, those root edibles must be ready!' And that's when they'd get together and go up the mountain and start digging for that plant." Mary Thomas also noted how important cow-parsnip and other fresh greens were in the spring after a winter of stored food. Other springtime plants the Secwepemc enjoyed around Shuswap Lake and other wetlands were wapato tubers and water-parsnip roots. Sometimes, too, a few men would venture into the higher country for an early spring hunt, trapping marmots or other small mammals (Teit 1909).

If there was a serious food shortage, people would climb up to the mountains in the spring in search of early roots, greens, or black tree lichen (Turner, Thompson, et al. 1990). Usually, however, the families remained in the valleys and along the benchlands until at least the latter part of spring, collecting the lowland roots before they matured and started to flower: balsamroot and its shoots, nodding onions, bitterroot, yellowbells (*Fritillaria pudica*), mariposa lily, yellow glacier lily, and spring beauty. Sometimes the men would fish near the root-digging grounds. Then, around mid-May, as the snows melted from the mountainsides, people climbed to the higher elevations, following the advancing season and continuing to harvest their favourite root crops: onions, chocolate lily (*Fritillaria affinis*), spring beauty, and yellow glacier lily.[23] Mary Thomas (pers. comm., 1994) explained: "What they did was they collected the avalanche lily [yellow glacier lily] and spring beauty, *$sx^{w}ix^{w}$* and *$sq^{w}eq^{w}$íne*, down in the bottom. When that was completely finished then our people came up to the plateaus. They hunted up here, they picked huckleberries, they gathered more avalanche lilies and spring beauties, and those were brought down to the valley and stored for the winter."

Edible cambium tissues, first of ponderosa pine and then, higher up or farther north, lodgepole pine, whitebark pine, and other species, were harvested in the spring as well (Dilbone 2011). Edith O'Donaghey recalled that her father would return from his hunting trips in the mountains with a bucket of lodgepole pine cambium for the family, and even if he arrived in the middle of the night, everyone would get up and eat it since it was best when eaten fresh (cited in Turner 1992b). Trees that were harvested for edible cambium collection also served later as a source of pitch, to be used as a fire starter during wet weather (Johnson 1997). As described previously, birch bark and other basket materials were also harvested in the late spring, as cedar and other fibrous materials were on the coast. It was a busy time, with harvested edible roots and basket materials needing processing and preparation for storage, to be consumed or worked on later in the fall and winter.

Summer Resources

In summertime, with its more predictable wind patterns and fairer weather, people all over the area, both on the coast and in the interior, continued with their diverse and intensive harvesting activities and their patterned seasonal rounds. There was, generally, an increased ability for safe travel, both on water and on land, and people moved from their spring harvesting camps to other areas, mainly for fishing, hunting, and berry picking. Greens were becoming too tough, woody, and bitter to eat, and tree cambium became thinner and the tissues more woody with drier weather. Some roots, like camas (*Camassia* spp.) and glacier lily, however, were at their prime for harvesting after they had flowered and gone to seed (Hart 1976; Loewen 1998). It was also possible to continue to harvest cedar bark and some other basket materials into the early summer months at higher elevations. Mountain travel was common at this time as the snow melted; from the high points of a people's territory, they could survey their lands, become reacquainted with their sacred places, and reconnect with their ancestors (Turner, Deur, and Mellott 2011). Along the coast, outlying areas and offshore islands were visited during this time as well. Long-distance trips for trade and reconnaissance were also undertaken mainly in the summertime.

Summer on the Northwest Coast

In terms of plant harvesting, berries were critically important for coastal peoples throughout the summer months, and summer travels were often guided by the successive ripening of berries. Starting with salmonberries, wild strawberries, red elderberries, red huckleberries, and blueberries, the berry season progressed to blackcaps and trailing blackberries (both on the southern coast), thimbleberries, currants and gooseberries (*Ribes* spp.), and salal berries.

Skunk-cabbage leaves, commonly used as a surface on which to dry berries, were fully expanded by the end of June and available from wet swampy areas all along the coast. At their campsites, families stayed in simple dwellings, often consisting of a light framework of poles covered with mats. Cooking and food processing were done in the open air. Camps were usually located near a river or lake and with woods nearby, ensuring ready access to water and fuel. Immense quantities of berries would be brought back to the camps, to be cooked up in cedar-wood boxes, or even in small canoes, using red-hot rocks, mashed to jam-like consistency, and spread out to dry, first on one side and then on the other, on overlapping skunk-cabbage leaves set on frames or logs. If the weather was damp, or to keep insects away, the frames were set up on racks over a smoky fire. The dried cakes were then stored away for trading or wintertime use, with usually each kind of berry being kept separate.

The Gitga'at and other seaweed harvesters would often process their dried seaweed cakes from the spring, traditionally in June. Helen Clifton (Turner and Clifton 2009, 184) explains,

> You need to dry it [seaweed] in June. You have to dry it in June, before the grasses really grow long. If the grasses grow long then they retain the dew of the evening. You see, and so the evaporation of that dew is coming ... and you're putting your seaweed close to the ground ... So that takes all day to dry. And ... you're moving that seaweed. About every two hours, you will go and ... move the seaweed so that it's turning over ... so that it all dries ... After the sun starts to set, the seaweed is cooling off now, and before that dew starts again, you gather up the seaweed.

The Saanich and other Coast Salish peoples often harvested their camas bulbs in the summertime after they had flowered and gone to seed. They would steam the bulbs in cooking pits and then flatten and dry them after they had been thoroughly cooked (see chapter 5). When the digging was finished, people usually burned over their camas prairies, timing the burns so that the fire would run quickly over the ground, killing back brush and seedling trees but not harming the grasses and bulbs, which were mostly dormant by this time.

Another major summertime occupation was salmon fishing, as the various runs of sockeye, chum, and other salmon started returning through the ocean channels to their spawning streams in large numbers in the summer. The Saanich reefnet fishery was fully operational by June, around the time when the oceanspray bushes bloomed. At this time people travelled to their hereditary family reefnetting locations around the Gulf and San Juan Islands, remaining there until August.[24] For drying and smoking their salmon, people needed plenty of fuel, as well as wood for salmon spreaders and drying racks. The Saanich held a sacred ceremony to welcome the returning salmon. As soon as the first sockeye salmon was caught in the reefnet, everyone stopped fishing, and a period of prayer and feasting continued for four days to show respect and give thanks to the salmon, who were celebrated as generous relatives. Summer was also a time of family gatherings, visiting between villages, celebrations, and traditional sporting events (E. Claxton and Elliott 1993).

August was the time when cattail, tule, basket sedge leaves, and "three-square" (*Schoenoplectus americanus*) stems were at their prime for weaving. The Nuu-chah-nulth and Ditidaht were renowned for their finely twined trinket baskets of sedge and "three-square," and they and the Salishan peoples also harvested cattail leaves and tule stalks from the edges of lakes and dried them for mats. People gathered many of their medicinal plants in the summer as well, to be dried and stored for winter.

Summer in the Interior

After harvesting roots and other products in the uplands in the late spring, many families in the Interior Plateau would return to the valleys and lowlands for the summer fishing season, heralded by the clicking of the grasshoppers or other local signals. The berry season was also beginning at lower elevations in the month of June.[25] Saskatoon berries, wild strawberries, and soapberries were among the first fruits to ripen, to be followed by gooseberries, blackcaps, raspberries, and thimbleberries, with the berry harvest intensifying into July, August, and September, depending on latitude and elevation. Blue elderberries, black hawthorn berries, highbush cranberries, and chokecherries were among the last of the summer fruits to ripen, and they could generally be harvested well into the fall. Hazelnuts also started ripening in late summer, an event heralded by squirrels; as well as people harvesting them directly, these nuts were sometimes gathered, already cleaned of their encasing husks, from squirrel caches. Otherwise, the nuts had to be buried or roasted for a time to remove the husks, which are covered with sharp, prickly hairs.

Any mountain trips in the summer had to be carefully coordinated with the runs of sockeye and other fish in the river valleys. Usually, people journeyed to high country immediately after the main sockeye run (see Johnson 1997; and Turner 1992b). At their upland camps, people usually spent several weeks intensively harvesting and processing berries, especially black huckleberries.[26] Many baskets of these sweet, juicy, flavourful berries were picked and dried in high-elevation camps. Some elders remember bringing six to eight large pack baskets of dried berries down from the mountains (Johnson 1997; see also chapters 2 and 5).[27] Generally, the women and children picked these berries while the men were hunting deer, caribou, mountain goat, marmot, and other animals and fishing in high-elevation lakes.[28]

As described by Edith O'Donaghey in the quotation that introduces this chapter, people also dug root vegetables at the summertime mountain camps: spring beauty, yellow glacier lily, nodding onion, and tiger lily. Sometimes, too, they found these roots cached in underground stores of voles and other small rodents (Turner 1992b, 1995; Turner, Thompson, et al. 1990). Women would often pit-cook and dry the roots, like the berries, right on site so that they would be lighter and more compact and would keep better for transporting back to the winter villages. Mountain medicines, like false hellebore, mountain valerian, Canby's lovage, and subalpine fir bark and pitch were also harvested and dried for winter use.

At the end of the berry-picking and root-digging season in the mountains, people would often set fire to selected areas in order to maintain clear patches for root species and berries. The burns were undertaken by people who knew just when and how to light the fires, and burning was usually rotated around an entire region over a cycle of several years. The roots – including spring beauty, tiger lily, chocolate lily, and yellow glacier lily – grew much bigger and the berries were bigger and more productive a couple of years after a fire (Johnson 1994, 1997, 1999; Turner 1999).

Fall Resources

Fall was another key fishing time in most parts of northwestern North America. The fall runs of coho salmon and steelhead, and especially of the chum, or dog salmon, were extremely important. Chum flesh is not as oily as that of sockeye and other species, so it dries more readily and can be stored without spoilage. People also processed the roe of chum and other types of salmon, burying them in underground pits lined with leaves, or later in crocks, to ferment them (Johnson 1997). Many of the summer activities continued into the fall; as noted previously, some places were known for late-ripening berries of the kinds normally ready in the summertime. There were also the fruits that normally ripened late to be harvested, and many of the root vegetables were at their prime for digging in the fall. Stinging nettle and Indian-hemp stems were also cut in large quantities in late summer and early fall, after the leaves had started to die back. The stems were split and hung up to dry, to be worked and spun into cordage and crafted into nets over the winter.

Autumn on the Northwest Coast

For coastal peoples, the fall season began with the first frost appearing on the leaves and the first snows settling on the mountaintops. The cottonwoods and bigleaf maples would change to golden yellow, and there was often a period of clear, cold but sunny weather, which was an ideal time for the last harvesting activities of the season. The chum, or dog salmon – the last salmon to be caught and stored for winter – were sought in their spawning streams. People also fished for cod, flounder, and other ocean species.

Deer and elk hunting also increased in the fall, as the animals were in prime condition. Hunters might make a few trips to the outlying islands and into the mountains, and they also hunted geese, ducks, and seals down at the river estuaries. The root gardens were ready for harvesting in the fall, as soon as the leaves started to die back for winter, with the Pacific silverweed leaves often turning bright orange. This was the stage when the roots were said to be good for digging, before the snow set in and the weather got bad. The women would tie the long roots of silverweed and springbank clover in bundles, pit-cook them, and then dry them for storage, as well as eating some fresh. The rootstocks of spiny wood fern were also ready to dig in the fall, or throughout the winter as needed, after their green fronds had started to die back (Johnson 1997; Turner, Johnson Gottesfeld, et al. 1992).

Late-ripening fruits harvested in the fall on the coast included Pacific crab-apples, highbush cranberries, bog cranberries, and lingonberries (*Vaccinium vitis-idaea*). Rather than being dried, these fruits were usually stored under water with a topping of oulachen grease or other oil, or they were stored in boxes buried in the ground or set in a cool part of the house. By the end of the season, people were making serious preparations for winter. Plenty of firewood had to be gathered, the houses weatherproofed, and canoes mended. Medicines, food, basket materials, wood for making implements, and fibre for nets and fishing line were all assembled to be close at hand. The men split cedar logs and roughed out canoes to be ready for winter's work.

Autumn in the Interior

Interior peoples spent the late summer and fall alternating between the salmon rivers and the mountains, where the men hunted for deer, moose, and bear, as well as trapping marmots, rabbits, and other small game. The women were busy cutting up and drying the fish and meat and tanning hides. The peak of the salmon season was around the end of August, and by mid-September a good supply of salmon had usually been dried and cached (Teit 1909). Around October people searched for cottonwood mushrooms along the river bottoms and pine mushrooms in the forests, to be cut up and dried for winter. In the mountains, some peoples harvested armloads of black tree lichen, to be pit-cooked or stored raw for emergency use in the winter.[29] The cones and seeds of whitebark pine were also harvested at this time from the upper treeline in many parts of the Interior Plateau. Edith O'Donaghey has fond memories of these seeds:

> Well [in the fall], there were the pine cones [whitebark pine] up there too. My dad used to bring home sacks full from his hunting trips in the mountains ... They were awfully messy [pitchy]. Granny used to keep a special pan just for roasting them in the oven ... They also dug balsamroots in August and the fall, after the leaves have all died down ... My parents used to go up 'Tit Mountain' with packhorses to hunt. They cooked the deer right there ... And my dad used to go hunting with about six or eight other men and boys ... Sometimes they were gone for a week, or ten days or more. (Cited in Turner 1992b, 418)

Autumn was also when herbalists gathered their winter supply of tree barks and other medicines to last them through the winter and when people undertook cleansing and purification in preparation for the winter ceremonial season.

Many other activities took place, from gathering firewood to making snowshoes and repairing traps and snares.

Winter Resources

Winter was the time when people took advantage of all of their hard work during the rest of the year, living mainly off of their stores during the coldest, most stormy, and darkest days of the year. By November the weather generally becomes unpredictable and treacherous, with strong winds and cold currents, making travel both difficult and dangerous. "Time to quit hunting and fishing, to pull the canoes up and rest ... That's when you light the fires in the big house" (E. Claxton and Elliott 1993, 27). The birds have migrated south, and the bears are now in their winter sleep.

Winter on the Northwest Coast

For many of the Northwest Coast peoples, lighting the fires in the houses signalled the start of ceremonial activities. Winter was considered a spiritual time, when there were potlatches, sacred dances, and initiations, preceded by prayers and ritual bathing at sacred pools and streams. Peoples' connection with the spirit world was celebrated. Winter was also a time of storytelling, of instructing children and youth, and of recounting and reaffirming histories and traditions.

Some harvesting activities were also undertaken even in winter if resources were available or if necessity dictated: fishing locally for cod, halibut, herring, flounder, or trout; hunting for deer or elk; and clam digging and mussel harvesting during the winter low tides. On Vancouver Island and the adjacent mainland south along the coasts of Washington and Oregon, "winter huckleberries" (*Vaccinium ovatum*) could be harvested, sometimes into December, and rose hips and highbush cranberries could also sometimes be found in mid-winter. It was still possible to gather some medicines when needed, and root vegetables could be dug as long as the ground wasn't frozen. Other winter activities were related to preparations for the next harvest season, such as making and repairing a wide variety of tools and containers necessary for survival, including cordage, twine and nets, spears, bows and arrows, fishing gear, baskets and boxes, cooking utensils, clothing, mats, baby cradles, digging sticks, and canoes and paddles. As weather allowed, people travelled from one village to another for visits and to witness sacred ceremonies, initiations, memorials, and naming rituals. It was a time to focus on social and spiritual connections.

Winter in the Interior

For the interior peoples, winter routines were much the same as on the coast, except that there was more snow. The Secwepemc month of November is called *pełʔełxʷʔułxʷten* ("entering month"), and this is when people started to settle in for winter; the Nlaka'pamux fall season extended a little later (Teit 1909). Hunting and trapping continued over this period, when the deer were rutting and easiest to hunt. Beaver were hunted for their meat and pelts, and rabbits were snared. Sometimes people were able to catch wintering steelhead and trout, and in the mid-winter and spring, there was ice fishing on all of the major lakes. It was still winter when the Gitxsan started their long and arduous journey over the "grease trail" to the Nass River estuary in order to be there for the early spring oulachen harvest (Johnson 1997). Other peoples, like the Ulkatcho Dakelh, also journeyed to the coast to spend the harshest months of the year there with their Nuxalk neighbours, where the weather was milder. Snowshoes, toboggans, and sleds of various types were used to carry loads over the snow and ice, sometimes with the help of dogs.

Stored foods – cached in underground pits, in elevated tree caches, or on poles, as well as in baskets and bags or on hanging strings – became immensely important during the winter, and toward the end of the season there was sometimes a period of famine if people had misjudged the required food supply or if stores were damaged or lost. Then people would eat whatever they were able to obtain, including emergency supplies of black tree lichen, rose hips or highbush cranberries frozen on the bushes, and cactus stems, kinnikinnick berries, or blue elderberries and chokecherries buried under the snow. One can imagine how welcome the first signs of spring would be; people would indeed be anxious to get out of their crowded pithouses or wooden lodges and find the first fresh edible shoots (Hayden 1997).

SEASONAL ROUNDS, TERRITORIES, AND TIME DEPTH

It is likely that the patterns of peoples' seasonal movement around their territories have ancient roots. In fact, the territories belonging to different families, clans, communities, and language groups that are known today in many cases probably reflect longstanding connections to different sites and habitats that collectively provided the resources people needed for survival. Once one group established a claim to an area – even a very informal one based on customary use – another group would likely be pressed to set out a claim to other areas. Some of these territorial boundaries are extremely strict; as Nuu-chah-nulth hereditary chief Earl Maquinna George noted, "There's a fine line between two nations" (cited in Turner, Smith, and Jones 2005, 151). Even within a family or clan, there were often strict rules about who had use of and control over the resources of a given place. A group who wished its own territory and harvesting areas to be respected would be careful to respect the boundaries of others. But these boundaries would be maintained by peoples' presence and use of their territories, even if only seasonally. Differences in available resources resulted in different patterns of resource use and different cultural identities. People became known for particular products, which were then traded with neighbouring communities or in some cases farther afield – a practice that archaeological records indicate is ancient.

Differences in resource use and resource locations, even at the household level, were noted from the millennia-old archaeological remains of the Keatley Creek pithouses (Hayden 1997). Trips from this village to several high mountain sites were indicated from different types of stone and other products. Hayden (ibid.) concludes that these artifacts reflect different seasonal round destinations of households and that camps in the surrounding hills and mountains would have been a part of these peoples' lives, serving as places for berry picking, root digging, collecting whitebark pine seeds, hunting deer, and processing meat and hides.

Similarly, on the coast, Moss et al. (2004) document seasonal use of archaeological sites from the Olympic Peninsula to Alaska for at least four millennia, with varying activities, such as hunting sea mammals, fishing for halibut, salmon, herring, and cod, hunting deer and birds, and gathering seabird eggs along the outer shorelines. Harvesting and processing resources at specialized sites, before moving these products to winter village sites, was a well established pattern, and people were constructing their winter houses to allow removal of the siding and roofing planks by canoe to the spring, summer, and fall camps, leaving only the post and beam frames at the winter villages. The harvesting and processing of salmon have been a major impetus for people's seasonal movements along the coast for 3,000 years or more, as indicated by the remains of fish weirs and other evidence (Matson, Coupland, and Mackie 2003).

On the Skeena River and elsewhere, precontact patterns of residential movement have been well studied. People moved in household and extended family groups to interior territories throughout the tributary valleys of the Skeena watershed in the summer and fall, fishing for salmon, hunting land mammals, and picking berries and digging roots. They smoked, dried, or otherwise processed their harvests and then brought them back to coastal villages where they resided over the winter and early spring. Food stores from the interior thus supplied most of the food people consumed during the winter months on the coast (Martindale 2003).

The ability of Northwest Coast peoples to obtain their full year's food supply within only eight or nine months or less has been a key to their cultural development. Arguably, it has been their intensive harvesting and food storage technologies, combined with their capabilities for travelling to and staying at different harvesting sites – their seasonal rounds – that have allowed their immense success, as reflected in their high population densities, monumental architecture and artistic development, economic and social complexity, occupational specialization, regional and local diversity in economy and subsistence, and strong systems of land tenure, trade, and exchange at the local and regional levels and beyond. Slavery and social inequity may have been part of this equation for success as well, as noted by Martindale (2003), Donald (2003), and others.

In terms of historical development of seasonal rounds, Deur (2002b) argues that, although the growth of sedentary villages apparently centred on productive estuarine fishing sites, mainly for spawning salmon, it was an increase in specialized food harvesting and processing that led to the development of small but permanent seasonal resource sites. This development included the expansion of sites for gathering clams, berries, and roots, as well as whale and seal hunting, oulachen grease production, halibut fishing, marine salmon fishing, and so forth. These types of resource specializations are indicated from at least 4,000 years ago. The constellations of outlying harvest and processing sites, together with the primary winter village, thus comprise some of the early evidence for the seasonal round.

The archaeological complex at Cattle Point on the windswept southern end of San Juan Island, mentioned earlier in a note as a probable reefnet location, evidently reflects such resource diversification and seasonal occupancy patterns. Not only reefnet fishing but likely also the harvesting and processing of a range of other spring and summertime resources took place here, judging from the broad sloping prairie, lagoons, offshore fishing bank, and relict beach terraces that comprise the locale. Along with a trench over 20 metres long (edged with stone slabs and with a clay-lined fire pit at one end), which may have been related to salmon processing from reefnetting camps, there were numerous roasting pits dating as far back as 2,750 years, as well as hundreds of stone, bone, and antler tools, many postholes, and immense shell middens, indicating a host of diverse resource activities (Kirk and Daugherty 2007).

Another example of an early multi-use site on the coast is a large midden site situated at the base of a high bluff at West Point, below Seattle's Discovery Park. Dating from about 5,000 to 700 years ago, the site borders on a wetland and reflects peoples' access to a wide range of abundant resources: elk, deer, berries, and roots from nearby lands; and mussels, crabs, barnacles, seals, and dozens of types of fish from the ocean and shoreline. It also shows evidence of cooking and processing clams and other foods on a large scale, possibly for later drying and storage (Kirk and Daugherty 2007). Interestingly, two labrets at the site indicate contact with people farther north as early as 3,000 years ago. Other trade items include pieces of petrified wood, probably from the Columbia Plateau, and obsidian from central Oregon. Woodworking and carving, basket making, and net making are all suggested by the tools found at the site. The site shows a shift – possibly due to environmental factors like storms, tsunamis, landslides, and sea level changes – from an important multiseason base camp in its early occupancy to a seasonal, less important spring-summer camp by about 1,400 years ago (Kirk and Daugherty 2007).

The sophisticated systems of land and resource tenure for which Northwest Coast peoples are well known apparently resulted from an unequal distribution of resource species and particular resource-rich habitats across the landscape and seascape. At the same time, however, these systems have helped to promote the sustainable use of resources and harvesting sites through systematic monitoring, enhancement, and control within clan and community-based governance systems (Deur and Turner 2005). A similar relationship between seasonal rounds and habitat maintenance could be argued for the territories, settlements, and resource-harvesting traditions of the peoples of the Interior Plateau and northern regions. For example, a large huckleberry patch in the northern Cascade Mountains might have served seasonal berry pickers for as long as seventy years or more, but such a tenancy would not have been possible if people had not routinely burned these meadows to reduce brush and maintain young, productive wood that would yield the largest, most abundant huckleberries and blueberries (Kirk and Daugherty 2007; Lepofsky, Hallett, et al. 2005; Mierendorf 1999; Turner 1999; see also chapter 12). These managed areas would also have benefited bears, marmots, mountain goat, deer, and other game, allowed further diversification of resource use, and facilitated access to and from the area, ultimately transforming them into "cultural keystone places." General climatic warming between around 2,400 and 1,200 years ago – at once impacting salmon runs by drying river channels and favouring the development of more open habitats, both the high country subalpine parklands and the lower elevation prairies and savannahs – may also have influenced the development of patterned seasonal rounds (see Lepofsky, Hallett, et al. 2005; Lepofsky, Heyerdahl, et al. 2003; Lepofsky and Lertzman 2008; Lepofsky, Lertzman, et al. 2005; Weiser 2006; and Weiser and Lepofsky 2009; see also chapter 2).

CONCLUSIONS

Keith Basso (1996), who spent many years working with Apache Dene (Athabaskan) peoples of the American Southwest, concluded, "Wisdom sits in places." Indeed, the intersections between space, time, and experience, encoded in language, discourse, and stories, are what make up a people's collective "sense of place." In turn, this sense of place can be said to comprise a major and inextricable portion of their cultural knowledge and identity (ibid.; Feld and Basso 1996; R.E. Ignace 2008; A.D. Palmer 2005; Thornton 2008). Secwepemc cultural specialist Dr Ron Ignace (2008, 164) maintains, "What makes us a people is the common knowledge of living in it [shared habitat], knowing and using its landforms and resources, and talking about it in our language."

The wealth of place names in every language, many of them denoting events that occurred at a particular location (some going back to the myth-times of the Transformers), special resource-harvesting sites, or notable landscape features help to connect geographic locations with memory and history, key activities, knowledge of the environment, and the wisdom needed to both use and care for the resources essential for people's lives. As Dr Ron Ignace (2008) explains, place names "anchor our history in the land" (171). "As our people, over thousands of years, learned to use, know, and find their way in the environment during their seasonal rounds, the names for landscape features evoke memories, evoke sentiment for the contours, the smells, the activities associated with the land, and of course the previous generations who experienced that landscape" (176).

A very focused example of how places, narratives, activities, resources, and history converge is in the Tla'amen (Sliammon, Coast Salish) story told by Rose Mitchell of Mitlenach Island, a small but richly biodiverse island just off of the central east coast of Vancouver Island:

> The people living at the head of Toba Inlet heard that the Transformer was coming, so they packed their belongings. Crow, the leader of the people, told them to gather lots of huckleberries, cranberries, wild crabapples, salal berries and wild blackberries. He wove snakes together as berry baskets. Crow's canoe was changed into a rock, which became Mitlenach Island ... It is also because of the Transformer that there are so many different kinds of berries on Mitlenatch today; these are the berries, some of them from the mountains, that Crow had in his basket woven from snakes. (Cited in Kennedy and Bouchard 1983, 105)

Indeed, peoples' lives and cultures truly are "situated" (Nazarea 1999). They result from a blending of place, practice, and time – cycles of time, where seasons and growth cycles recur, activities are reiterated, and generations are repeated, over and over again. A whole constellation of different factors, occurring on different scales of time and space, come into play in determining the ultimate patterns of how peoples combine their experiences, ingenuity, and creativity to maintain themselves using the diverse resources spread across their territories: the presence, abundance, and productivity of particular species and groups of species in particular places at particular times; weather and other environmental influences that allow the harvest, processing, and sometimes storage of available resources; the availability and productivity of other, potentially competing resources; the degree of competition by animals and birds for the same resources; technologies and innovations that might enhance or intensify the exploitation of certain species and products; the ability of individuals and groups to mediate and manage resource productivity; and any number of combinations of these influences.

A major factor to consider is the social conditions that might either allow or prevent people's ability to access, harvest, and process their resources. Social organization, governance and leadership, and territory, along with the unforeseen impacts of illness, death, war, or other human-based occurrences, are immensely important in determining how people's "situated lives" play out. Social organization is also instrumental in the enterprise of knowledge transmission and acquisition. In this regard, Indigenous peoples' connections to place, including their seasonal movements and use of plant resources, are mediated by the societies and human relationships in which they are embedded, and these factors are ultimately what determine future interactions with place and environment. In the following chapter, I explore the influences of social factors on the application, adaptation, and transmission of ethnobotanical knowledge.

9

Cultural Institutions Related to Ethnobotanical Knowledge and Practice

"Long time ago, whenever people came back on horseback off the mountain after a hunting trip, they would all start singing. It sounded so beautiful, a dozen people or more on horseback breaking into song, coming down the mountain, coming home." Away from the travels, the drumbeat emulates the rhythmic sound of the horse-hoofs through the ravines. They voice the joy of coming home after a long day, or days of riding and camping. (R.E. Ignace 2008, 163, quoting Secwepemc elder Christine Simon)

INTRODUCTION

We humans are social animals, and one of the keys to our success as a species has been our strong ties with one another and our ability to apportion our tasks and work cooperatively for common needs, goals, and interests. Indeed, few other species have social systems as complex and highly developed as ours (Sutton and Anderson 2009). Whereas single individuals can harvest and process enough food, fuel, and other resources to keep themselves alive for many years or even a lifetime, the normal way of living is with others. We thrive on spending time with each other, sharing our work, sharing the pleasure of eating, and celebrating our ceremonies, dances, and songs. We seek to engender a sense of belonging and identity, for it brings us pleasure in living – a common voice of joy, as expressed in the introductory quotation from Secwepemc leader Dr Ron Ignace. Our inclination for social behaviour strengthens our ability to diversify our food systems, provides us with mutual protection in numbers, enhances our survival, and facilitates the passing on of knowledge, practices, and beliefs to future generations. In short, our social systems are the basis of our culture, which can be defined as "the shared ideas, beliefs, values, mores, world views, practices, and information in the minds of human beings" (Ames and Maschner 1999, 30).

Just as our general need for the company and collaboration of others is probably encoded in our genetic makeup, so too may genetics explain our propensity to develop social hierarchies and to recognize and gravitate toward leaders – key dominant individuals who can make decisions, map out plans for the future, and make choices aimed at the beneficial survival of the group. Even in so-called egalitarian societies, some individuals, through their hard work, particular knowledge and experience, acquired skills, and natural abilities, or through the sheer strength of their personalities, emerge as leaders: *primus inter pares*, or first among equals. The processes and protocols for selecting leaders vary from one culture to another, as do the degrees of relative inequity and privilege between leaders and other society members. Nevertheless, it is as "natural" for humans to incorporate leadership into their social structure as it is for wolves, seagulls, and many other species to do so (Pierotti 2008; Pierotti and Annett 1985). The Nuu-chah-nulth (E.R. Atleo 2004) and other Indigenous peoples maintain that their human social system is derived directly from wolves – ancestral totem animals whose "alpha" leaders guide the pack and are responsible for parenting and raising a significant proportion of the group's offspring.

Peoples' narratives of supernatural beings and animal-people over the ages reflect all of the elements of social organization that were – and in part still are – to be found in more recent Indigenous societies around the study region: talented and hardworking individuals who achieve certain goals through the generosity of others and often through some kind of reciprocal sharing; husbands and wives, each with their own special types of work; children who are resourceful and look to future goals, assisted by their parents and grandparents, both human and supernatural; heroes who undergo hardship and privation to assist their people and their families; chiefs and matriarchs, both bad and good ones, who make choices (bad and good) on behalf of others in their clan or village; slaves who are loyal to those who own them and slaves who escape or betray; celebrations and wars between neighbouring communities; and feasts intended to intimidate and insult others, as well as feasts that honour, praise, and reinforce respectful social ties. Thus social relations have historical and spiritual, as well as biological and ecological, dimensions.

The study of how sociocultural complexity has developed in Northwest Coast and Interior Plateau societies of northwestern North America continues to be a major focus of archaeological research (Ames 1995; Arnold 1992; Hayden 1992, 1996; Lepofsky 2004; Matson and Coupland 1995; Moss et al. 2004). It appears that the first migrants entering the region were socially organized mainly in small-scale family groups, travelling over a sparsely populated terrain and setting up single-family or small multifamily camps for periods of time in sheltered, resource-rich sites. Even in the Early Holocene, however, there is evidence of wider social and economic ties between communities, possibly based on networks of trade and communication extending over considerable distances (Ames and Maschner 1999; Kirk and Daugherty 2007; Koppel 2001; Moss et al. 2004). In terms of ethnobotanical knowledge and practice, both the social character of humans and the stratification and diversification of societal roles, including those of leaders, are, and have been, integral to plant resource use and to the processes of knowledge acquisition, transmission, and adaptation. In this chapter, I explore some of the social aspects of plant use, starting with consideration of the smallest social unit, the family, and the varying roles that family members play in relation to plants and ethnobotanical knowledge. I then turn to an examination of social complexity and how it may have influenced plant knowledge and use.

UNITS OF SOCIAL ORGANIZATION

The Family or Household Unit

Most human "families," including those of Indigenous peoples of northwestern North America, consist of a man and woman, their children, and often at least one set of grandparents. In cases of polygamy, which was common in this area particularly with men of greater wealth who might have from two to several wives, the co-spouses would likely have lived and worked together. As well as children, a well-off family in the past in many parts of the study area might have one or more slaves whom they had purchased, traded for, or captured in war. Altogether, such a group might range from ten to thirty people (Suttles 1990a, 1990b). Collectively, these are the people who would spend the greatest amount of time together on a day-to-day basis, usually living in proximity, even within multifamily or extended family dwellings, and moving together to resource-harvesting camps during the seasonal round. Thus such a family group can be considered both the basic social unit of societies in the study area and the basic economic unit.

The fundamental pattern of living, established in the distant past, was that each nuclear family or basic household group resided within an extended territory and moved as a unit from one site to another over the seasons, sometimes with other, allied families, often camping in temporary mat-covered or wood-covered lodges and returning to a permanent winter plank house or pithouse in a centralized village to spend the winter months (see chapter 8). There were many variations of this pattern. In some cases, the territory would be occupied by a broader extended family, clan, or village group or band, or several households would routinely move together or in close coordination to adjacent areas within a broader territory.

Over time, new resource species entered the area or existing ones were enhanced, technologies for food harvesting and processing were improved, populations grew, houses became bigger, house construction and modes of transportation became more sophisticated – and social complexity likely increased simultaneously. Even into the early part of the Late Holocene, winter villages both on the coast and in the interior were very small, with only two or three houses together. On the Northwest Coast, it is suggested that pithouses, or "in ground" houses, preceded cedar-wood plank houses and were replaced by the latter over the course of a few thousand years as western redcedar became more and more common (Ames and Maschner 1999; Lepofsky, Schaepe, et al. 2009). Nevertheless, the same general pattern of families and groups residing in one place in the winter and moving around to different harvest sites over the growing season was evidently maintained into the time of European contact. Until the early 1900s in some cases, a typical family unit often lived with a group of other families in proximity or even all together in large dwellings under a single roof. What changed over time were the overall numbers of families dwelling together and community size, both of which generally increased, leading to greater complexity in social institutions and in social arrangements for communal production.

Between about 6,000 and 3,500 years ago (in the Mid Holocene), immense shell middens – some several metres thick and covering many thousands of square metres – started to appear in the archaeological record on the Northwest Coast, beginning on the northern and central coasts and occurring later in Washington and Oregon. To some researchers at least, these middens suggest that people were expanding their use of intertidal resources, and also of salmon, possibly as a response to more stable sea levels starting around the same time.[1] Salmon weirs may have increased in use over this time since they have been strongly evident on the central coast for at least the past 3,500 to 3,000 years (Matson 2003). During this same time interval, cedar plank houses became common on the coast, and concurrently semi-subterranean pithouses appeared in numbers in the Interior Plateau. Gradually, by the middle part of the Late Holocene, villages became larger as well, with houses on the coast arranged more formally in one or more rows, all usually facing the same direction (toward the water). Furthermore, as woodworking technologies continued to advance, canoes also became larger.

Both the Northwest Coast cedar plank houses and the Interior Plateau pithouses reflected living arrangements that often showed differential social status. On the coast, especially within the larger cedar plank houses, each family would have its own hearth site and sleeping area, with families and individuals of different social status located in different areas of the house. Generally, chiefs and members of the "nobility," along with their immediate families, occupied the back of the house, those of intermediate status resided in the middle positions, and the families of lesser status took the areas nearest the entrance.[2] For the Central Coast Salish, several households totalling 100 or more people might be accommodated in a single shed-roofed house 100 metres or more long (Ames and Maschner 1999; Matson 2003; Suttles 1990a). In the interior, a family might occupy a single, smaller pithouse or share a larger pithouse with other, generally related families. A family's winter stores might be cached within its living quarters or near the house.

The differences in quality and rarity of items recovered archaeologically from houses – both on the coast and in at least some of the pithouses of the Interior Plateau – show differentiation in social status of various families, sometimes within the same house, with the families of chiefs and nobility having higher numbers of rare items like obsidian and nephrite. The antiquity of such differential status in various parts of the study area is not known. It may be relatively recent, possibly developing only within the past 2,000 to 1,000 years (see Matson, Coupland, and Mackie 2003), but some suggest it had beginnings possibly as early as 4,500 years ago, with permanent social inequality – marked by elites and slaves – well established by 2,500 years ago (Lepofsky 2004).

Division of Labour

Ames and Maschner (1999, 137) suggest that the changes in the food economy, starting around the latter half of the Mid Holocene, possibly resulted in changes in gendered social dynamics. People's increasing dependence on clams and other shellfish, as well as on salmon caught in weirs and traps in estuaries and riverine spawning sites, meant that women and children, and older people too, were able to contribute more directly to food procurement. This may have endowed these groups (elders, women, and children) with higher status and possibly shifted the power relationships within families away from the stronger, more dominant male who hunted in groups with other men and toward a more balanced social standing across gender and age groups. However, as emphasized by Moss (2011), women have often undertaken activities, including hunting and fishing, that have typically been cast as male occupations, so suggestions about changes in gendered work remain speculative.

Nevertheless, women, children, and elders have probably always had important and specific roles in the family-based enterprise of survival. Their work not only in harvesting shellfish and easily captured salmon but also in contributing edible greens, root vegetables, berries, and other plant foods to the family diet was a more significant occupation, both on the coast and in the interior, than is often recognized. Women were easily capable of travelling on foot or by canoe, with other women and older children, for berry picking, root digging, clam digging, or fishing. Together the women and older children formed small social groups that had cross-cut households and that facilitated essential tasks undertaken in locations well beyond the main winter dwelling or seasonal camp, as well as within the village context. Men sought out women as wives and bearers of children, and increased opportunities for travel promoted marriage ties across neighbouring communities, or even farther afield by the beginning of the Late Holocene. A wife from a different place would have been a major source of new knowledge and skills related to plant use and other cultural knowledge, bringing her own experiences, and possibly even new products, with her to her new home (see chapters 5, 6, and 10).

In the Nlaka'pamux story of Old-One, the Creator decrees, "Henceforth, men will hunt, fish, and make tools, and women will dig roots, and make baskets and mats" (Teit 1912, 327). Men and women performing different types of work is an almost universal phenomenon in human societies, although the lines dividing particular activities and occupations vary. Not just in Indigenous societies of northwestern North America but also throughout the continent and beyond, in societies where people derived their living from local resources by managing plant and animal populations and habitats in situ, men tended to be the hunters, warriors, and distance travellers, whereas women tended to stay closer to home, harvesting plant foods and shellfish and undertaking household tasks like food preparation and childcare. Although the role of each gender group was vital for survival, archaeologically and ethnographically, men's occupations have been privileged by scholars until recently (Lepofsky 2004). In terms of ethnobotanical knowledge and practice, however, we know that both groups participated in major ways in learning, transmitting, and adapting knowledge about plants and their applications from generation to generation.

Men's Roles

In an elaboration of the role assigned to men in the Nlaka'pamux "Old-One" narratives, after creating four men and a woman, Old-One instructed a man to use a magic knife to cut a piece of saskatoon berry wood (after addressing the bush with, "My friend, I require you"), which Old-One induced to form itself magically into a fish spear. In a similar fashion, Old-One then formed a bow and arrow from yew wood brought to him by the man. Old-One created, in turn, a beaver spear, tools to fell a tree, and a firedrill. He taught the men how to use all of these implements, saying, "Henceforth, men will fell trees, and make fire. They will make tools and weapons, and will be workers, trappers, hunters, and fishermen." Old-One also taught the men how to make nets and to set snares for hunting (Teit 1912, 324). Thus, in a single narrative, the work and skills of the men were mapped out, along with their main tools. These occupations described for Nlaka'pamux men were similar for men throughout northwestern North America, who focused everywhere on hunting, fishing, and working with wood and fire. Men were the specialists in fuel, in different qualities of wood for different purposes, in the manufacture and use of spears, harpoons, bows, and arrows, and in the materials from which they were made. These roles are confirmed in ethnographic reports and even photographs from the late 1800s and early 1900s (Boas 1921; Teit 1900, 1909; Swanton 1905, 1908). They still hold to some extent into the twenty-first century, although the roles are now translated into use of rifles, chainsaws and power tools, large-scale fishing gear, large boats, engines, and mechanical equipment.

Along with men's knowledge of woodworking, hunting, and fishing came an integrated body of cultural knowledge about the spiritual aspects of relationships with trees and forests and with the spirits of animals, about ceremonies around fishing and hunting, and about the importance of the supernatural helpers, spirits, and powers that bring success in these endeavours. Many men, especially those of the nobility, have held important ceremonial and political roles, which are discussed later. As the main long-distance travellers, men were probably the ones who surveyed new territories and identified new dwelling places when it was necessary to move households. They probably also were the first to find and claim potential harvest sites and camps used by families in their seasonal rounds. The men tended to represent their families and communities in relationships with other peoples and probably did so in ancient trading contacts, just as in more recent ones. During the eras of European exploration and trade, men were the main traders and negotiators – the ones who had the most contact with European leaders of the day. As such, they probably led the way in introducing new foods and other new products to their communities, and they would also have conveyed new words, new stories, new technologies, novel artistic innovations, and cultural traits to their own people.

Women's Roles

In the same "Old-One" narrative cited previously (Teit 1912, 324), Old-One made a root digger and showed the woman how to dig roots. In another version (ibid., 321), he "taught the women how to make birch baskets, mats, and lodges, and how to dig roots, gather berries and cure them." As far distant as Haida Gwaii on the northern coast, women's roles were equally distinctive and complementary to those of men, as described by Blackman (1982, 34; see also Norton 1981, 1985): "Although some economic activities, such as collecting shellfish and cooking, were performed by both males and females, in general the Haida division of labor was marked ... clam digging and the implement of procurement, the *gligú* ("digging stick"), were considered part of a woman's domain. The sexual division of labor was summed up for me by one elderly Massett man who offered the following comment on the essential property of the newly married couple: 'Every man's got to have his fishing line and devilfish stick and every woman her digging stick.'"

We know that digging sticks are ancient implements, extending back far earlier than bows and arrows, which were evidently widely adopted only within the past 2,000 years or so (Ames and Maschner 1999). Digging stick tips, for example, were found in the context of a dense wapato meadow at Pitt Polder in the Lower Fraser Valley, dating to around 3,500 years ago (see chapter 2). Many digging sticks were recovered, some as old as 2,500 years, at the Lachane site in Prince Rupert Harbour (R. Inglis 1976; Lepofsky 2004). Evidence of digging sticks extends back into the Mid Holocene in northwestern North America, but it is likely that they have existed in some form from the time humans first entered the New World since they were found in the nearly 15,000-year-old site at Monte Verde, Chile (Dillehay et al. 2008), and it is probable that women were the main users of these implements (see chapters 2 and 5).[3]

In archaeological contexts, most root diggers and digging stick handles, bark peelers, and hide scrapers, as well as baskets and basketry materials, are widely assumed to be connected with women's activities (Bernick 1998b; Hayden 1997; Lepofsky 2004). Digging sticks were often specially made for a specific girl or woman and were kept as important personal possessions for life, sometimes being interred with a girl or woman upon her death. Antler digging stick handles, in particular, were often elaborately decorated and occur more often archaeologically in burial contexts and in house pits of the Interior Plateau than any other woman's tool. Probably the intricate incised designs found on many of these handles signified high status and prestige for an individual, paralleling the status and prestige that went with the occupation of plant gathering in many areas (M.B. Ignace, Ignace, and Nord 2009; Schulting 1994). Other artifacts and features such as hearths, pit-cooking depressions, cache pits, mats and mat-making needles, acorn-leaching pits, and drying racks probably also indicate the past presence and activities of women.

Certainly, women spent an immense portion of their time harvesting and processing foods and materials (see Norton 1985). Much of the processing of both had to be done immediately after harvesting. Weaving materials alone took days of preparation – stripping, splitting, dyeing, bundling, and drying. Then, usually in the winter, these carefully prepared materials were taken out and used in constructing baskets, baby cradles, mats, clothing, blankets, and other woven items – another time-consuming and intensive occupation. Harvesting basketry materials was sometimes done in concert with food harvesting, but both tasks are so intensive that they were usually undertaken sequentially rather than concurrently. Often it was necessary for women to travel for hours, or even days, to gather the resources needed. Sometimes individual women would ascend to the treeline of the mountains to harvest root vegetables, berries, or basket materials, returning in a single day or staying over for two or three days (Dawson 1892; Turner 2003b, 2006a, 2006b). For example, Nuu-chah-nulth elder Stanley Sam (pers. comm., 1994, quoted in Scientific Panel for Sustainable Forest Practices in Clayoquot Sound 1995) recalled that his own grandmother travelled on foot far back into the upland forests of Clayoquot Sound on the west coast of Vancouver

Island to get prime yellow-cedar and western redcedar bark for her weaving; she would be gone for many days – ten or more – at a time on these trips. Edith O'Donaghey, a Stl'atl'imx elder quoted in the previous chapter, recounted that her grandmother sometimes travelled all by herself up to Mission Ridge above Shalalth to dig edible roots. She would leave very early in the morning with a large pack basket on her back and stay for the whole day, coming back down at dusk (cited in Turner 1992b). That said, women tended to travel less than men, except in the context of their family's seasonal rounds. In fact, some women may have resided their entire lives within twenty or thirty kilometres of their home village, perhaps moving to another village when they married. Many may have known or interacted with fewer than a hundred other people in their lifetime, as suggested by Dubin (1999) for Indigenous peoples in the entire region in general. This would have led to a very close and personal familiarity with local environments, mountains, streams, trees, and microhabitats within the sphere of one's own habitat from childhood on.

Haida elder Florence Davidson recalled the considerable time that her mother, Isabella Edenshaw, invested in digging and preparing Sitka spruce roots for her exquisitely crafted twined baskets and hats:[4] "Ever since I can remember, I used to go with my mother in May for spruce roots [*hlii.ng*]. Every fine day we'd go to North Beach early in the morning before sunrise. We'd pack water and food with us and mother would cook our breakfast in the woods ... We'd collect *hlii.ng* all day long ... We'd collect piles and piles of [roots] ... The sun would be going down when we started for home ... Soon as my mother finished eating she'd start splitting the roots" (cited in Blackman 1982, 85–6) (figure 9-1).

In almost all cases, the resources women sought were located in specific, known places often some distance from the camp or village. Critically important aspects of women's specialized knowledge included not only the activities of digging roots, picking berries, and harvesting seaweed and greens, as well as the diverse materials required for weaving, but also knowing the prime areas to go for these products, the best time and season for them, the most efficient techniques for harvesting, which individual roots, cedar stands, berry bushes, seaweed rocks, or plant patches to select, and how to harvest, tend, or care for the plants so that they would remain productive over years and generations. As with men, women also needed to know about the spiritual aspects of their work, including the proper way to approach the plants, the words of thanks and recognition to say, and any taboos that go with harvesting,[5] such as not picking seaweed from under the water but only that exposed on the rocks at low tide (Clan Chief Adam Dick, pers. comm., 2009; Turner 2003a; Turner and Clifton 2006), taking only a portion of cedar bark from each tree, and not cutting too deeply into the inner bark of the birch tree (Turner 2003b, 2009). As illustrated by the complex timing of the seaweed harvest – which requires balancing the times for picking, drying, turning, and bringing in the seaweed with other tasks

9-1 | Agnes Edenshaw-Yeltatzie-Jones, sister of Florence Davidson and daughter of Isabella Edenshaw, Haida, of Massett, preparing *łii.ng* (spruce roots) for weaving, 1971.

like cutting and drying halibut and spring salmon, cooking and rendering seal fat and meat, feeding and caring for children, and tending fires in the smokehouse – women have had to multitask in the extreme (Turner and Clifton 2009).[6]

These tasks and responsibilities generally complement those of men. Just as men's activities contribute irreplaceably to the survival and well-being of the household and community, so too do women's. The story of the Kitkiata hunter, told by Dan Gosnell (Gilodza) to William Beynon in 1954, illustrates how men's and women's roles harmonize. In this narrative, a Kitkiata hunter travelled around until he found the very valley he searched for, where "not only game and salmon were plentiful, but also there was a plenteous supply of berries and fruits and rice[-root]." The hunter took an industrious wife who knew how to weave mountain goat wool. Working together, they were able to put by large quantities of fish and meat and to gather all of the different kinds of berries and "wild rice" (northern riceroot), and the woman wove many dancing robes (cited in Cove and MacDonald 1987, 217–19). It is the cooperation between them that makes men's and women's roles so effective. Even so, most of their tasks (other than women's bearing and nursing children) cannot be said to be exclusively within one or the other domain. Necessity and convenience, as well as diversity in personal preference, skills, talents, and opportunities, have sometimes resulted in a blurring or crossing-over of these roles (R.E. Ignace and Ignace forthcoming; Thornton 1999). Some women hunt; some men pick berries. Some women carve; some men make baskets.[7]

Even archaeologically, there appears to have been some crossing of gender roles. In a detailed study of artifacts associated with burials in the Interior Plateau, Schulting (1994) found, perhaps to his surprise, that there were no cases where an artifact type was exclusively found with bodies of one gender or the other, although males tended to have more celts, scrapers, bone points, pigments, harpoons, antler wedges, stone knives, tubular pipes, and shaft smoothers placed in their burials. In the case of digging stick handles, which one might assume would be exclusively associated with women, of twelve antler digging stick handles found as grave inclusions, five were associated with male burials. Schulting (ibid.) suggests that some of these may actually have been handles of other kinds of tools, but their central hole and their similarity to digging stick handles of the ethnographic period indicate that they were indeed the handles of root digging sticks. Their inclusion with men's interments remains a mystery.

As well, of the stone artifacts, such as projectile points, that archaeologists have assumed reflect men's work and men's occupations, many have actually been used by women. In Schulting's (1994) survey, he found that the burials of eleven females included projectile points and noted that a companion of Meriwether Lewis and William Clark's expedition to the mouth of the Columbia River in the early 1800s observed women manufacturing projectile points. Thus we need to be careful in making assumptions of gendered tool use (Bernick 1998b).

For the Interior Plateau peoples, although women and men traditionally had different roles, Teit (1909, 573) noted that this varied slightly in different families, that many Secwepemc men made most of their own clothes, and that in several families only the men cooked fresh meat. Pit-cooking roots, however, was almost always women's work, and often men were forbidden even to come near while balsamroots, camas bulbs, and other such foods were cooking. On the coast, on the other hand, in some instances pit-cooking was the exclusive activity of men.[8] Pit-cooking often required careful ritual preparation, including sweat bathing and prayers. As Kirk and Daugherty (2007, 66) note, "the skill in roasting roots was more valued than skill in digging them" (see also Peacock 1998; Turner, Bouchard, and Kennedy 1980). Even tasks such as berry picking involved ritual activities, such as the First Fruits ceremony enacted in a number of Interior Plateau cultures. There were also women's dances in connection with their work. For example, Secwepemc and Dakelh women danced with baskets and branches of saskatoon berries, imitating the gathering of berries (Teit 1909, 578).

Gendered roles that may have been more strictly defined in the past have changed with the times, as well. Helen Clifton (Turner and Clifton 2006, 2009) explained that for the Gitga'at and other Ts'msyen, around the turn of the twentieth century and up until about the 1950s, it was solely women who harvested seaweed. The women, with the older children, would take their canoes across to Campania Island from the seaweed camp at K'yel, on Princess Royal Island, early in the morning, leaving Helen, as one of the youngest women, to care for the small children. One of the women – usually the matriarch – would steer the canoe and carefully watch the waves to ensure that everyone jumped safely onto the rocks, picked their seaweed over the course of a couple of hours, and then, when the tide got higher, packed their seaweed and themselves back into the canoes and returned to the camp to dry their harvest. When motorboats, fishing boats, and larger skiffs started to be used in transporting people to and from the seaweed grounds, however, men started to accompany women. Sometimes they would drop the women off and go fishing, or sometimes they would stay and assist in the seaweed harvest. Today, at times, there are more men than women harvesting the seaweed; they might set their halibut nets before picking seaweed and then pull them in and reset them before returning to camp with the seaweed and any halibut caught.

Children's Roles

Children are an ever-present component of households but are often not recognized for the special roles they play in family activities. In some cases, the intended occupations of children would be determined even before birth. A woman of Nuu-chah-nulth or Ditidaht nobility, whose unborn baby boy was intended to be a whale hunter, would undertake special training, preparing her child for the later rigours of life on the ocean, surrounded by and working with spiritual powers for success in capturing whales and other feats that would provide for the community. Then, when the baby was born, if it was indeed a boy, his spine might be rubbed with the charcoal of burned sea palm (*Postelsia palmiformis*), a marine alga of the lower intertidal zone of rocky coastal headlands that is continuously battered by the waves – one of the most flexible, tough, and resilient life forms on earth. This would help to strengthen the infant, right from the start (Turner, Thomas, et al. 1983).

A newborn Secwepemc baby boy was trained to grow straight and tall with the tough withe of a saskatoon berry bush, placed into his tiny gripping hands and used to pull him up, while talking to him and encouraging him to hang onto the stick, "Ohhh, boy! Ohhhh boy!" (Mary Thomas, pers. comm., 2001). Nlaka'pamux babies were washed with different solutions to strengthen them, including a solution of mushrooms. Mabel Joe (pers. comm., 1984; see also Turner, Kuhnlein, and Egger 1985; and Turner, Thompson, et al. 1990) explained that mushrooms, although soft, are very strong, being capable of splitting wood and moving large rocks aside as they grow. Children washed in mushroom juice will therefore become strong and independent. Nlaka'pamux women also washed their babies in solutions of wild rose, desert currant (*Ribes cereum*), and other plants to help them grow strong and healthy (Turner, Thompson, et al. 1990).

Parents and elders carefully observed babies and children from the time they were born to see what talents and inclinations they had so that they could be encouraged in specializations like canoe making, hunting, or basket making for which they showed special aptitudes. Children were also taught discipline from an early age, as described by Mary Thomas (2001, 11):

> Wherever mother went the baby was with her. She packed the baby with a strap across her shoulders with the birchbark cradle on her back. Wherever she went her baby was with her. If she was gathering food out in the bushes or out in the open, like maybe digging wild potatoes, if she got tired she'd take her baby and hang it on the branch and the wind was the baby sitter. The baby slept. If you unbind the baby, it will stretch and kick, just kick away. When it gets sleepy it will not sleep unless it's bound back in the cradle. As soon as you bind the baby, you feed it, it's comfortable, it falls asleep, the mother packs it. That way the child knows it's loved and protected by the parent. That little child automatically grows up to be a loving obedient child.

Clan Chief Adam Dick (Kwaxsistalla), the last of the traditionally trained Kwakwaka'wakw potlatch speakers, described how the elders of his community, when he was only three or four years old, started preparing him to be an orator, dancer, and leader in the spiritual and economic realm of the potlatch (cited in Recalma-Clutesi 2007). Whenever the police boat came to take the children away to residential school, his grandparents took him away, far up the Kingcome River, or over to their small house in a remote part of their territory, at a place called Deep Harbour in the Broughton Archipelago. Here they maintained a *lux̱ʷxiwey* (clam garden), hunted seal, and travelled across to different islands to pick seaweed and fish for halibut in the spring. They also visited their relatives at other locations where there were *t'əkkillak*ʷ (estuarine root gardens) to obtain edible root vegetables – springbank clover, Pacific silverweed, and northern riceroot. Adam Dick (pers. comm., 2009; see also Turner, Recalma-Clutesi, and Deur 2013) recalled that his grandparents had a long dugout canoe that could hold their bedding and food and everything they needed. His grandparents would paddle at either end, and Adam himself had his own paddle and sat in the middle, bundled up against the cold. All the time he was with his grandparents, they were teaching him the skills for acquiring food but also teaching him about correct and proper behaviour, history, stories, and everything he needed to equip him for his later life roles. He remembered, from his childhood, creating his own small clam garden by rolling rocks down to the edge of the low tideline. Similarly, when he was at Kingcome, he and the other children also created their own small estuarine root gardens, in imitation of his grandmother's larger ones.

Children destined to be healers, river guardians, canoe builders, storytellers, historians, or any number of other specialists were watched and groomed for these jobs. For example, when he was still a child, Nuu-chah-nulth hereditary chief Earl Maquinna George was taken by his elders on a journey up the river over which he had hereditary responsibility so that he would come to know it intimately. At this time, he was taught what was needed to keep the river in the best possible state for the salmon that were a precious resource for the community: where to clean out jams and debris, where to leave the logs in the river to create deep pools, and how to transplant salmon eggs from the spawning beds if the river changed its course (George 2003). Like Clan Chief Adam Dick, Chief Earl Maquinna George (pers. comm., 1997) received major training from his grandfather: "A lot of my learning of responsibility for my Chieftainship came from Peter Webster. He was my grandfather. Very intelligent people, more or less told me how I should live, and how I should be, what I am as a hereditary chief, the roles and responsibilities. This was said in outright public instruction of where I belong and how I belong in that position." Dr E. Richard Atleo (Chief Umeek), also of Ahousaht, was taught in a similar way, although, tragically, his grandfather and father passed away when he was still young, so he was left with the principles of knowledge he needed but had to chart his own way in learning the details of a leader's role. In this, he was supported by his grandmother and other family members (E.R. Atleo 2011).

Even at four or five years of age, children begin to contribute in real ways to a community's cultural, social, and economic life. Contemporary elders everywhere remember the tasks and chores they did as children to assist their parents and grandparents: collecting firewood, fetching water, berry picking (figure 9-2), root digging, seaweed harvesting, hunting small animals, clam digging, and selecting salmon from the tidal fish traps. The stereotyped gendered activities were usually not as strongly defined for children. Boys being cared

9-2 | Ethan Dundas of the Gitga'at Nation with wild blueberries he has picked, at Hartley Bay, BC.

for by their grandmothers might go out to pick berries. Years later, these men recall their hands – and mouths – being stained purple from this task. Girls might help their fathers or grandfathers to build dugout canoes, fetching tools or helping to keep the fire going to heat the rocks. Gradually, however, the girls' and boys' tasks became more focused, as reflected in the toys that were made for them. Girls made or were given miniature baskets, whereas boys received small practice bows and arrows. Mary Thomas (pers. comm., 2001) recalled her own experience as a child:

9-3 | Mary Thomas's willow-bark doll, like those her grandmother used to make for her when she was a little girl.

I can remember my brother; Grandpa used to make him toys out of wood. We were made dolls out of natural material [willow bark], by our grandmother [figure 9-3]. And we were taught values through legends. Our legends all had something to learn from them.

I was very young when my grandmother taught me how to make a basket. Because sooner or later, we were going to become part of the food gathering. They would tie a little basket around a little girl's waist. The little girl might fill it with leaves, or dirt, or anything, but she learned that that little basket was a container and it has to be filled. And she'd go along. The bigger they got, the bigger her basket would be. And she was made at a very young age to fill one basket before she would go play. And I can remember, I loved to play, especially swimming when it was hot. And we'd hurry up and hurry up, and our little brother was really slower than us, we'd have to help him fill his basket, before grandma would say, "All right, you go swim."

Clan Chief Adam Dick (pers. comm., 2003) recalled how as a child he and other boys would gather armloads of succulent thimbleberry and salmonberry shoots in the spring, and as they walked through the village at Kingcome Inlet, everyone called out to them, asking for a taste.[9] Being able to share their harvest, to provide food for the community, gave them great pride. Boys were taught to share the first deer, geese, or ducks they killed and the first salmon they caught with one of their elders or with the entire community. And girls were taught – and still are – to give away the berries they picked into their first basket and the first roots they dug, instilling the ethic of generosity and responsibility from an early age (George 2003; Hunn, Selam, and family 1990; Turner 1996).

Children were raised to be self-sufficient. Most of them could handle a canoe from a very young age and had many skills in food harvesting and preparation. Adam Dick and other elders recall as children going out in small groups first thing in the morning. They fished and snared or hunted small game, dug clams and gathered chitons and other shellfish, picked greens or berries and other fruits in season, and even gathered birds' eggs. Then they lit a fire down at the beach and cooked their own food. In this way, they fed themselves for the entire day, not returning home until it got dark.

Young people also participated in a community's ceremonial life. For the Saanich and other Straits Salish peoples, children ceremonially carried the first sockeye to be caught in the reefnet over a pathway strewn with sacred sword fern fronds to a respected elder, who smudged it with the smoke of burning *q'əx̱mín* ("Indian celery," *Lomatium nudicaule*) seeds, cooked it, and divided it up into portions to be served to everyone in the community (E. Claxton and Elliott 1993).

Young people took part in many games and contests that helped to train and prepare them for later roles. For Ditidaht, Nuu-chah-nulth, and some Coast Salish boys, for example, there was the *pilaa-pilaa* contest, in which sword fern fronds – called *pilaapila-apt* ("*pilaa-pilaa* plant") in Ditidaht and similar terms in several other languages – were held and the leaflets removed, one at a time, the contestants pronouncing "pilaa!" (cf. *pála*, "one," a Proto-Central-Coast-Salish term) each time, with as many as possible being pulled off with one breath. The one who pulled off the most – the one capable of holding his breath the longest – was the winner. John Thomas explained that it was important for young men to be able to hold their breath a long time, especially those who were being trained to dive down and harvest the longest stalks of bull kelp for fishing lines and harpoon lines (see chapter 6) and those who, when a whale was harpooned and killed, had to dive underneath the body of the whale to tie its mouth shut so that it would not fill with water and sink. John himself participated in the *pilaa-pilaa* contests, but by his time, in the 1930s, they were using wooden clothes pegs strung up on a line stretched across the big house rather than sword fern fronds (Turner, Thomas, et al. 1983). There were many other races, swimming contests, and other sports that boys took part in.

Girls, too, had their own games and contests that prepared them for their lives as women, such as the Nuxalk *tsayamuus* flower contest (described in chapter 6), which trained girls not to reveal their emotions. Children often made games of their chores, imitating animals and birds, as they participated in harvesting berries or roots. Mary Thomas (pers. comm., 2001) remembered how she and her brother and sister would help their grandmother to gather up wapato tubers and water-parsnip roots along the Salmon River estuary near Shuswap Lake, after her grandmother had waded into the water, dug them up, and then tossed them onto the shore. When they went with their grandmother to dig glacier lily and chocolate lily bulbs and spring beauty corms, their grandmother would use her digging stick to turn over clumps of turf, and then the children would pick out the biggest roots and place them in the baskets. Their grandmother would check over their baskets to make sure they hadn't taken roots that were too small and would remove and replant any small ones. When their grandmother harvested Indian-hemp stems, the children would sling the bundled stems over their backs and gallop with them over to the place where they would be processed, pretending to be horses.

Training at Puberty

Puberty was a special time of learning for both boys and girls. Virtually every group, both on the coast and in the interior, observed particular training or rituals for young people entering adulthood. Boys and young men might undertake long spiritual quests, travelling to a remote location and isolating themselves for periods of weeks, months, or years as they searched for a spiritual helper and supernatural guidance in their future lives as hunters, fishers, or whatever special occupations they sought. With help and advice from one or more elder relatives, they would fast, bathe, or sweat in a sweat lodge to purify themselves, train by running, swimming, and other physical exertions, and pray to the spiritual entities for strength, prowess, and success.

Girls would be sequestered away from their families at the time of their first menstruation and thereafter for a shorter interval at every menstruation. An elder woman, usually a grandmother or aunt, would be their only contact with the outside world. For Interior Salish girls, special huts or lodges, often of Douglas-fir boughs, were made for them, where they would sit and work on intricate tasks like basket making, or even systematically pulling off the needles of fir boughs, one at a time, so that they would learn dexterity and discipline. Teit (1909, 587) reported that Secwepemc girls would leave their secluded lodge in the evening to practise "running, climbing, carrying burdens, digging trenches, the last so that in after years they might be expert at root-digging.[10] Each morning they brought home a small load of fresh fir-brush or fire-wood to their lodge."

Boys at puberty were also often isolated for a period of time. Seeking spiritual powers that would guide them throughout their lives, they retired to remote places where they underwent rigorous training: purification rituals, fasting, bathing in the ocean or in prayer pools, sweat bathing, and scrubbing themselves with tree boughs to toughen their skin. After this period of training, they were considered to be adults and would be expected to marry and take on a man's responsibilities and roles.

Elders' Roles

Older men and women – those who lived through their childrearing years and belonged to the grandparents' generation – had, and still do have, a special and respected position in the household and community. Each stage of life had, and continues to have, its own particular balance of learning and sharing Traditional Ecological Knowledge. The main occupation of the children was learning – through play, combined with imitation, instruction, experience, language acquisition, and storytelling. Then, as described previously, at puberty boys and girls underwent intensive, often very formal, spiritual and practical training to prepare them for their respective adult lives. Some of this type of training continues to this day, although many of these elements of a youth's education were stifled and eradicated over multiple generations by the enforced residential school system and potlatch prohibition (E.R. Atleo 2011). Young adults after marriage spend most of their time raising children and undertaking the more physical activities of food procurement and processing, as well as other day-to-day responsibilities, both practical and ceremonial. Although they never stop learning through this entire period, increasingly they begin to take on the role of teachers, particularly for their own children. Elders, especially, are sought out for their experience and wisdom, and thus their major role is to serve as advisors and teachers to all of those younger than themselves. Although many elders continue active lives of fishing, berry picking, and root digging, as well as processing the foods that others procure, their primary societal position is as guides and mentors, and others increasingly help to provide for their needs.

There has always been a special relationship between children and elders, especially their grandparents. Children were instructed always to be generous with whatever they obtained, particularly to their elders. When a boy caught his first salmon or procured his first deer, he would give it away to an uncle or grandparent, and a girl who had made her first basket would fill it with berries and give it to an older relative (Dr Arvid Charlie [Luschiim], pers. comm., 1999; Mabel Joe, pers. comm., 1995). These teachings of sharing food and provisions with elderly people, whether at feasts, out in the bush, or in their homes, were deeply engrained (see Andre, Welsh, and Turner 2003). Clan Chief Adam Dick (pers. comm., 2001), of the Kwakwaka'wakw Nation, recalled that his grandfather, who was a leader in the community, would gather all of the young people and get them to "Go check on those houses [of the elders]." He explained,

> You know, that's your job to get it [for elders] ... Hunting. If they're starting to go get seal, get three, you know. Give to the elders first. And that's what they mainly go out for. Or, go get a deer. If they get a deer, and they hang it up for a few days and they chop it up for the people that can't go out ... Our elders ... the first of everything always went to the old people ... The first thing we'd do when we're out there fishing ... we'd go out for the widows and the elders. They let them have their winter supply before we'd go out for the rest of the village.

For elders who wanted to continue gathering their own food, special areas might be set aside. Convenient and readily accessed hunting grounds and berry patches would be reserved for those who could not travel in steep terrain. One Nisg̱a'a account exemplifies this: Harry Nyce's great grandmother at Gitwinksihlkw (Canyon City) became blind in her old age, and because she loved to pick berries, everyone in the family saved the berry bushes closest to the house for her to pick and even transplanted some saskatoon, red elderberry, and other berry bushes nearby to give her easy access (Harry Nyce and Emma Nyce, pers. comm., 2013). Sometimes people brought berry-laden branches of soapberry, huckleberry, and other species to the elders so they could pick them off; this had a dual function because the pruning of the bushes was said to enhance the growth and productivity of these plants (Dr Ron Ignace, pers. comm., 2005; Turner and Burton 2010). Helen Clifton said that the Gitga'at always designated the most accessible seaweed-picking rocks for the old ladies – places where they could most easily step off onto the rocks from the canoe or boat with less risk of falling (Turner and Clifton 2006), and Dr Arvid Charlie (Luschiim) (pers. comm., 2011) said that the younger Hul'qumi'num hunters always hunted deer in more distant and hard to access places, leaving the Garry oak prairies, which were easier to walk over, to the older hunters.

In terms of knowledge acquisition and transmission, the elders were, and are, the ones who held the group's collective memory, history, and skills, the ones who had observed any changes, and therefore the ones who guided decisions to be taken. Elders, especially the chiefs and matriarchs, even today, have strong authority – and their wisdom and experience can save lives. For example, they often determine at what point a group should leave camp or travel, depending on the tides, winds, and other environmental factors. Such a decision takes experience and knowledge; if a wrong choice is made, a canoe could capsize and lives could be put in danger. Elders also participated in and officiated at ceremonies, such as the First Salmon and First Fruits ceremonies.[11]

Chief Earl Maquinna George (2003, 33–4) recalls such an elder, named Tushka, from when he was a child at Ahousaht:

> There was another old man called Tushka. He had a cane, and after the dinner ... he got up and banged the cane on the floor, and he was the start of what was going on. He announced what the feast was for, who it was for, and the reason why the feast went on. He started, with the historical stories of where our people came from ... He must have been one of the leading historians of our culture. He knew about the songs, and the dance that goes with each family. There were various types of feasts that went on a lot. The people were proud of the many things that were said through the long talks by Tushka.
>
> This is what Tushka used to talk about, how things came about, ownership of all the beaches and land, resources. Not just only covering forestry. There's a lot to do with fishing grounds, halibut banks, codfish banks, salmon banks ... These expert people talked for hours at a time.

The tremendous changes that have occurred since the coming of Europeans in the recent centuries emphasize the key importance of cultural continuity as a means of building stability and resilience for First Peoples. Maintaining the basic teachings about family roles and responsibilities and about kinship relations with the natural world – as emphasized by many contemporary elders like Dr Mary Thomas, Annie York, Clan Chief Adam Dick (Kwaxsistalla), Dr Daisy Sewid-Smith (Mayanilth), Dr E. Richard Atleo (Chief Umeek), Dr Arvid Charlie (Luschiim), Ida Jones, and John Thomas – has been a key to the amazing capacity of Indigenous peoples to retain the essence of their cultures and, in many cases, their languages, ceremonies, and ability to harvest their own food, carve canoes, weave baskets, and adapt these skills to circumstances of today's world (E.R. Atleo 2011; Turner 2004a, 2005; Turner and Turner 2008), despite all of the immense disruptions people have endured. Comparative studies of Indigenous peoples' knowledge from nearly a century ago to current times (Kay 1993; Turner 2004a; Turner, Thompson, et al. 1990) show that, whereas some names for and knowledge about plants have been lost over intervening generations, a remarkable portion of this information has been retained, largely through the experiences and teachings of elders, both at the family level and more broadly (Recalma-Clutesi 2007).

Slaves

Slavery was a prevalent institution at the time Europeans arrived in northwestern North America (Donald 1997), and many traditional stories mention slaves in some role, sometimes as major actors in narratives. Slaves are often portrayed as assistants and servants of young people and as labourers – for example, cutting firewood. Many took part in food harvesting and processing (Boas 2002). Even Raven was known to have slaves: "Raven finally made salmon and instructed his slaves to build a salmon weir and caught many salmon" (Boas 2002, 385, citing legends of the 'Nakwaxda'xw [Nahwitti]). The extent to which slaves were considered to be part of a family or household varied with circumstances. In all cases, however, slaves would have been important in considerations of harvesting and use of plants and in knowledge transmission. Donald (1997) and Ames (2005) suggest that slaves were indeed a major element in salmon intensification on the Northwest Coast, their labour helping to alleviate the inevitable "bottleneck" in cutting and processing the fish before they spoiled. Slaves were also known to assist with, among other tasks, berry picking and other forms of plant gathering.[12] Pacheedaht chief Charlie Jones (Queesto) recalled that his grandfather used to have slaves whom he assigned to guard his estuarine root gardens and, when they were ready, to dig the silverweed and clover roots for him (cited in Turner, Thomas, et al. 1983). Similarly, Tlingit slaves were assigned to protect berry patches from birds or human intruders, as well as to assist in harvesting (Thornton 1999).

Salish slaves, many of them captives from raids by the peoples of the central and northern coast – Haida, Tlingit, or Kwakwaka'wakw – sometimes ended up nearly 1,000 kilometres from their original homes (Ames and Maschner 1999; Gibson 1992) and must have been major sources of new knowledge and skills that were transferred from one region to another. Slaves were often part of the "goods" that were exchanged at major trading centres such as "The Fountain" (Xaxl'ep) along the Fraser River north of Lillooet, a site where numerous slaves who were captured in war from the Stl'atl'imx/St'at'imc were sold back to their friends and families (Teit 1906a). Farther south, the Chinookan served as middlemen for slaves, along with many other trade goods, between the Columbia Plateau and the Northwest Coast (Ames and Maschner 1999).

One story of a Ditidaht woman from Whyac on the west coast of Vancouver Island, who was taken as a slave by Hul'qumi'num people, illustrates ways that individuals who were captured travelled from place to place. Told by Effie Tate to Kennedy and Bouchard (1994, 46–7), the story is called "The Sasquatch."

> A Ditidaht woman was among a small group of people who were picking salalberries when a canoe load of hostile Cowichan Indians came along. To save her companions, this woman ran towards the Cowichan men,

> giving the others time to escape by canoe. The woman was taken back to the area near the mouth of the Cowichan River [near Duncan on the east coast of Vancouver Island]. Here one of her captors took her as his wife, rather than a slave. When fishing season came, the Cowichans took her with them over to the mouth of the Fraser River. It was here that she had an opportunity to escape. She made her way back to Vancouver Island and began a long walk inland through the forest. Finally the woman got to an area some where along the upper portions of Carmanah Creek.

Eventually, the woman made her way home to Whyac. One can only speculate what she must have seen and learned in her travels and how she might have brought new information with her when she returned. Possibly, she tasted wapato tubers from the Fraser Valley for the first time, and she may have encountered new types of berries, like Canada blueberries (*Vaccinium myrtilloides*), said to have been introduced to the Ditidaht territory from the Fraser Valley at some point (Turner, Thomas, et al. 1983). She probably learned some Salishan names for various plants and may have taught her Hul'qumi'num captors some of her Ditidaht names. This is all speculative but certainly within the realm of possibility. Similar stories could be told for numerous people who were captured, lived in villages far from home, and then were eventually repatriated (e.g., Mary Thomas, pers. comm., 2001; Reid and Sewid-Smith 2004).

Task Groups

Many of the tasks undertaken by women and men – such as creating and using fish weirs, cutting and drying fish, berry picking, root digging, and pit-cooking, and making a large canoe or building a house – were accomplished by groups of people working as a team or in parallel, providing at once social contact and mutual assistance in the tasks at hand. These might be family groups – mother and daughters, aunts and nieces, sisters, brothers, or grandfather and grandsons. Pacheedaht elder Ida Jones (see chapter 5) recalled that working together in a group was highly enjoyable: "They [women] invited each other [to harvest roots], I guess so they won't be lonely or sad, because they are happy, passing on the news" (cited in Turner, Thomas, et al. 1983, 18; see also Turner 2003b). For harvesting, there were efficiencies in transportation when, for example, a group of women went by canoe for berry picking or seaweed gathering. There were also advantages in terms of safety in numbers, efficiency in childcare, and opportunities for younger, less experienced harvesters to learn from the older, more experienced ones (McDonald 2003; Thornton 2008).

The Straits Salish reefnet fishery is an excellent example of such cooperative harvesting. The reefnet site, the reefnet itself, and one or more canoes typically belonged to one individual – part of an entire suite of owned and inherited property that included names, ceremonial regalia, stories, rights to certain ceremonies, and rights to particular resources at particular places. However, the owner could not operate the reefnet alone. It took at least six men, in two canoes, with the willow-bark reefnet, or purse net, positioned between the canoes, with lines running out from the net anchored in place to form an approach and with clusters of beach wildrye grass (*Leymus mollis*) tied to cross-lines to simulate an ocean bottom. All of these materials had to be gathered and equipment had to be made or repaired before the fishing season. Then, as fishing commenced, after observation of a ceremonial "no catch" period (usually four days), following the capture and ritual sharing out of the first salmon, the fishing began in full. When the men spotted an approaching school of sockeye or other salmon, they spread the canoes apart, allowing the salmon to enter the net. When there were sufficient salmon in the net, the men drew the canoes together and individually selected the salmon from the net, to be apportioned to each of the fishers, with the highest share going to the owner of the reefnet (E. Claxton and Elliott 1994; Turner and Berkes 2006; Turner and Hebda 2012).[13] Similar task groups would be operating on shore, where the women would be waiting to cut and smoke the fish that were caught.

This type of teamwork in resource harvesting and processing took place almost everywhere – in duck hunting with nets, oulachen fishing, salmon weir and gillnet fishing, hunting with deer fences in the uplands of the Interior Plateau, and many types of plant harvesting and processing ventures. The seasonal bases, like the Straits Salish reefnet camps, the root-digging and berry-picking camp at Mission Ridge described by Edith O'Donaghey, and the Gitga'at seaweed camp at K'yel on Princess Royal Island, were often attended by groups of families – as many as ten or twelve together – each with its own shelter, boughs for bedding, fireplace, and earth oven. Each family had its own boat or its own horses and horse-grazing areas and was self-sufficient in food and provisions, but all were situated in the same general vicinity, with plenty of visiting and exchanging of labour happening continuously (Turner 1992b; Turner, Thompson, et al. 1990). If one woman filled her berry baskets early, she might help her sister or her niece to fill her baskets. Or if a skilled Gitga'at fish cutter completed cutting all of her halibut into *wooks* ("thin strips") to hang up in the sun, she might cut someone else's halibut for them. These people, in turn, could expect some help some other time when they needed it. Pit-cooking was a frequent communal activity, where the total collective harvest of roots, clams, or other food would be cooked together in a single large pit.[14]

Gitga'at elder Helen Clifton (Turner and Clifton 2006) described the routine at Hartley Bay for task groups that form when a woman is ready to chop her year's supply of edible seaweed. Following the harvesting and drying of the seaweed in squares at the K'yel seaweed camp, the dried squares are transported back to Hartley Bay, and later on in June, a second session of processing ensues.[15]

9-4 | Gitga'at elder Colleen Robinson, an expert at seaweed preparation, demonstrating how to dampen the edible seaweed (*Pyropia abbottiae*) in preparation for pressing it into cedar-wood boxes, one stage in complex seaweed processing.

The seaweed squares are moistened with seawater (figure 9-4), packed in layers in a cedar-wood box, and left for two or three days for the seaweed to "get its flavour." They are then taken out, and using axes and wooden blocks, the compressed seaweed is chopped into small pieces, which are spread out in the sun on sheets to dry, before being stored away for winter or for trade. The chopping of the seaweed is hard and intensive work.

> [I] put the seaweed sheets out, all over the floor, and, put the [yew-wood chopping] blocks on. And, so the women would come to help chop that seaweed ... different women come throughout the village to come and help me chop that seaweed, because I'm married to him [the late Chief Johnny Clifton] I guess. That's why they'll come. He's got many nieces, and ... well somebody will say, "When are you going to chop your

> seaweed?" And I have to send somebody out, "Well, granny's going to be chopping seaweed on such and such a day." I send word throughout the community and so they drift up. Some people have an hour or so, they'll come out and they all help each other ... Some of them have enough daughters or granddaughters to go and help. It works that way in every house, [when] they're chopping seaweed. If you've got an hour to spare, two hours to spare, whatever time you have, you go and help chop seaweed. Especially if you don't have seaweed. You will earn some seaweed; they'll give you some seaweed. You earn it. (Helen Clifton, pers. comm., 2001)

Some task groups are comprised of men and women, or whole families, working together. An example is the harvesting and scraping of the edible inner bark of western hemlock and amabilis fir in years past.[16] Sheets of bark were cut from the tree, sometimes from high up on the trunk, and this was usually done by men, who climbed up the tree with ropes. The men tossed the sheets of bark down to the women, who took them to the beach or some convenient spot and, using a special knife made of mussel shell or iron, scraped off the edible *ksiiw* from the inside of the bark. This intensive work continued sometimes for hours, until enough had been harvested (Curtis 1915; Turner and Thompson 2006). Curtis (1915, 39–40) described the next stage in the process for a Kwakwaka'wakw family, in which the inner bark was pit-cooked:

> The man of the family then takes his place at the side of the pit, while his wife sits at the opposite side and the children are between them. Each has a flat, rough piece of sandstone. The cover [of the cooking pit] is raised, one after another quickly removes a great handful of bast [inner bark], and the cover is dropped. Then each kneads his handful on the sandstone, turning and returning it, tearing it apart and opening it up, until after a while it becomes smooth and pasty. It is then rolled into a ball and laid in one of the boxes, and another handful is taken from the pit.
>
> When after a week or two of this labor the family returns to the permanent wintering village, the woman cuts a mat down to exactly the shape and size of the box into which she is going to place the bast.

Most of the work to be done in connection with food preparation is seasonal, and sometimes the timing for its completion is critically important. For example, without refrigeration, fish like salmon and halibut need to be processed and preserved within a day or two of being caught. Cutting fish is an exacting skill. In the Gitga'at style of cutting thin strips, or *wooks*, from the fish for sun drying, after the skin is removed, each piece has to be cut to the same thickness – usually so one can just see the shadow of the knife through the flesh in cutting halibut – so that it will dry evenly and thoroughly. An expert will be able to predict just how many *wooks* can be taken from any fish of a given size. There are, almost always, particular specialists in any group – women who have the ability to cut the fish quickly and accurately, time after time. Those less skilled will be allowed to practise, but until they gain a certain level of competency, they best serve the communal enterprise by helping to hang and turn the thin strips of fish for the expert cutters, as well as supporting the work by sharpening knives, washing the fish, removing debris from the cutting board, tending to children, and preparing meals. This cooperative approach, with each individual working to her own level of experience and ability, while continuing to practise and learn, allows the greatest efficiency for the overall work. Several such small, focused, and cooperative work groups might exist together at the same camp and would be operating simultaneously in fishing season at comparable locations all up and down the coast or all along the interior waterways.

Many different activities have to be coordinated, similarly, among task groups at any harvest locale over the days and weeks of occupancy. Whether along the coast or in the interior, these activities are always changing slightly, varying with the success of fishing or hunting or with opportunities presented for picking berries or greens or for digging roots, depending on the weather and influenced by any number of other events that happen to occur and must be folded into the routine. Whether at the high-elevation camps of the Interior Plateau, at the seaweed camps along the coast, or at the berry patches of the Nass Valley, groups of women would work together and in parallel, cutting and drying meat and fish, picking and drying berries, and harvesting and processing root vegetables – whatever needed to be done – over a period of days or weeks, all the while looking after children, preparing meals, tending fires, and keeping supplies of water and fuel handy, while men might be away hunting or fishing (Turner 2003b).

Of course, such complexity in the multitude of tasks to be accomplished on different scales of time and space requires strong guidance and leadership. It is not surprising, therefore, that a number of different arrangements for designating responsibility and authority in food production and other types of work, including strengthening relationships with the spiritual domain, were developed, culminating in the complex systems of chieftainships and clans that were encountered by Europeans arriving in northwestern North America in the 1700s and 1800s.

SOCIAL STRATIFICATION AND RESOURCE USE

Over the millennia of occupation of the northwestern North American landscape, Indigenous peoples' technologies and skills developed, sea levels became less chaotic, new species established themselves, some resources became intensified, and human populations expanded. The increased environmental complexity – and in some senses, increased stability – paralleled, and probably led to, increased social complexity. Larger groups required more exacting ways of apportioning and controlling resource use and access to resource localities, differentiation of skills and roles, and more formal means of decision making and planning for the future. Ultimately, social stratification resulted, accompanied by the development and establishment of leaders, who were ascribed high status, and by the role of followers, who looked to these leaders for help in maintaining the necessities of life. There also developed at some point, probably in ancient times, a different dimension of social relationships in the form of clans or lineages (said to be descendants of a common, often supernatural ancestor, as described in the origin narratives) that cut across and complemented direct family and household groupings and provided people with an extra level of social and economic support. Especially in the northern part of the study area, these clans are exogamous, meaning that people could not marry into their own clan, which would be seen as incestuous. In Haida, Ts'msyen, and other matrilineal societies, children assumed the clan of their mother. Different clan membership between family members resulted in more options and opportunities for diverse knowledge systems to develop – ultimately, one could argue, providing greater resilience or adaptability for a given group of people.

The increased populations and development of social complexity, as evidenced especially around the middle of the Late Holocene, some 2,000 years ago (but with roots extending into the Mid Holocene), probably required more exacting claims to particular territories – places on the landscape and seascape where a group could assert exclusive use of resources and control the ways that these resources might be allocated or shared with others (Turner, Smith, and Jones 2005). A group's overall territory would need to encompass sufficient diversity in its resources to provide virtually everything necessary for the group's survival from year to year and generation to generation. Peoples' territories varied over the area in the degree to which others were excluded or in the ways that territories were shared with neighbouring and other peoples. It may be in some cases that the establishment of territory by one group had a "domino effect" on others. If one group prohibited others from using a certain area, those people would, in turn, be forced to turn to other resource sites, and an obvious reaction would be to claim their own exclusivity for such sites. Then neighbouring groups would be forced into a similar action, until no territory was left unclaimed and, by extension, unoccupied or unmanaged. Although there would have been disputed sites with valued resources that would be fought over, boundaries would have been established in many cases: along mountain ridges, around watersheds, or from one island to another. Ultimately, such divisions would have to be respected if a group wanted to keep its own territory intact and unchallenged. Of course, there were also many areas that were shared and many different arrangements for managing shared access to resources between and across territories.

Territories and owned resource sites, usually overseen by chiefs and their delegates, were a key to intergroup social relations and to apportioning resource use (see Turner, Smith, and Jones 2005). Each area or region, depending on its unique features, would have yielded different quantities and qualities of specific resources. In terms of plants, resource sites might be huckleberry patches, crab-apple stands, shorelines producing high-quality seaweed, estuarine gardens of silverweed and clover rhizomes, camas meadows, wapato patches, stands of high-quality cedar trees or yew trees, patches of beargrass (*Xerophyllum tenax*) or cattails (*Typha latifolia*), or the like. Social connections such as trading or reciprocal hosting of harvesting groups allowed people to gain access to different resources – products that might be lacking or of lesser quality in their own territories. Thus, at least in recent times, in places like the territory of the Lytton band of the Nlaka'pamux in Botanie Valley, the Halkomelem Katzie territory in the Fraser Valley, the Straits Salish–Lekwungen lands on southern Vancouver Island, the Chinookan territory at the mouth of the Columbia River, and the Nisga'a territory around the Nass River estuary, people from many communities and nations would congregate annually. Recognizing the rights and authority of the hosting communities, clans, and chiefs, the visiting groups would either trade with the hosts for the resources they sought or be given permission to harvest in a sort of "U-pick" arrangement that was in some way reciprocal (Suttles 1987b; Turner, Smith, and Jones 2005; Turner, Thomas, et al. 1983; Turner, Thompson, et al. 1990).

Affinal Ties and Resources

Intermarriage was one social strategy to assure and legitimize access of one family or community to another group's resources or resource-harvesting areas (Ommer and Turner 2004). Marriage connections were so predominant among the various Central Coast Salish communities that, as pointed out by Kennedy (1995), the tribal designations for individuals set out by the colonial government in the 1800s and 1900s did not at all reflect people's actual status as community members since many people had kin ties and recognized membership in two or more communities, often even across language groups. Rather than a series of isolated villages or tribes, these communities formed networks of relationship that helped to strengthen and diversify their economies and afforded them greater resilience in times of stress (Suttles 1987a, 1987c, 1990a).

Halkomelem (Hul'qumi'num) speakers of Vancouver Island in the vicinity of Duncan, Chemainus, Nanaimo, and the adjacent Gulf Islands, for example, had long standing kinship ties with the Fraser River Halkomelem communities. These communities shared not only a common language but also similar narratives, ritual practices, and general knowledge systems (K.T. Carlson 2001; H.G. Barnett 1955; Suttles 1987a, 1987b, 1987c). The island peoples, through their marriage ties, were able to access the immense salmon runs and wapato and cranberry harvests of the Fraser River peoples, and probably for centuries, if not millennia, hundreds of people routinely travelled from Vancouver Island to visit their relatives and to harvest these resources.[17] In fact, the Cowichan peoples had their own extensive village, Tl'uqtinus, on Lulu Island, along the south arm of the Fraser River, where many families resided for at least some portion of the year for fishing, berry picking, and harvesting cattails and tules for mats, as recounted in many historical documents (Thom 2005). In return, the mainland Halkomelem visited their island relatives to receive camas bulbs, clams, and other desirable products, either by harvesting or through feasts and potlatches.[18] A parallel situation was in place between the Ts'msyen peoples of the Skeena River and those residing along the coast. Each community had its harvest sites within its own territory, but each also had long-recognized histories of harvesting in other's territories. The Kitsumkalum families, for example, harvest seafood in the territories of the Kitkatla and Lax Kw'alaams peoples (McDonald 2003). Similar marriage and socioeconomic ties existed almost everywhere.

Marriage thus often brought more resource rights, accompanied by the establishment of a series of reciprocal and mutually recognized socioeconomic arrangements. Such arrangements are certainly significant in terms of ethnobotanical knowledge transmission. A woman raised and trained in one community, and speaking the language of that community, might move after marriage to her husband's community, where she would be trained by women of her husband's family, learning to speak their language and being instructed concerning all of the knowledge, places, skills, taboos, stories, and protocols for the proper and effective use of plant resources in her new community. In clan-based communities, she would belong to a clan opposite that of her husband and would be embraced by a whole new group of teachers in the form of members of her own clan. At the same time, she would retain the original language and skills she learned when she was growing up, and would to some extent keep her ties with her parents and siblings, so that she would have the capacity to teach her own children about two different territories and knowledge sets. Her sons and daughters, in turn, would grow up bilingual and with a wider set of skills and understandings – and access to a broader set of resources and harvesting sites – than they would if their parents had been raised in the same community.

These kinds of "out-marriages," especially among members of the nobility class, were common between neighbouring villages in the same or related speech communities. Increasingly, in later centuries as opportunities for travel broadened, marriages occurred across wide geographical and cultural spaces. Sometimes men acquired wives through warfare and raids, as was known to have occurred from oral history accounts between the Haida and Heiltsuk, the

Nuxalk and Kwakwaka'wakw, and other peoples. The scale of one community's introduction to another of new knowledge, names, skills, and methods for harvesting and processing plants and other resources due to women changing residence, whatever the circumstances, must have been immense. Numerous traditional narratives tell of such "mixed" marriages, not only between men and women of similar status but also between ordinary human men and women of supernatural status – the daughters of powerful supernatural beings of the sky or undersea worlds like the Haida Goose Woman, for example (see chapter 12). Stories also feature unions between women and supernatural men, such as Devil's-club Man or Stinging Nettle Man, in which the men instruct their wives about the use and preparation of these new medicines and materials. In the ancient myth-times, men and women also married individuals of other species, the most famous examples being the "Dog Children" and "Bear Husband" tales, but there are also accounts of men marrying Mountain Goat wives and entering the communities of the Salmon people, Wolf people, and the like. People even married the Sun or Stars and were taken to live in the sky world, as in the "Star Husband" narratives. In all of these cases, the supernatural or animal being transferred key information to the human spouse, who also acquired resources or other benefits from the marriage, and perhaps all of these episodes are symbolic of the importance of knowledge acquisition through marital alliances.

A woman marrying and moving to a new household might bring important innovations with her in the form of different kinds of baskets, different styles of clothing, or different ways of cutting fish or cooking roots. An example of such an innovation is in the numerous square cedar-wood seaweed-drying trays that Annetta Robinson brought with her to the Gitga'at from the village of Kitkatla when she married Jimmy Robinson and moved to Hartley Bay. These trays (figure 9-5) were a welcome addition to the seaweed-processing equipment of the Gitga'at women, several of whom sometimes borrowed them from Annetta when they needed to dry large quantities of seaweed on the beach at K'yel. These trays, in short, represent an innovation in food processing in the Gitga'at community as a result of Annetta's residency (see chapter 5). One can imagine, over the millennia, hundreds of such introductions and demonstations of new products, devices, innovations, and inventions by women coming into a new community – from digging sticks with removable handles to coiled basketry, different styles of mats and racks for drying berries and roots, better knives for cutting fish, and so forth. Some women also brought living materials to transplant to their new villages: salmon eggs from their favourite runs of salmon; rhizomes of particularly good, straight-growing stinging nettles to plant for their fibre; or food plants such as hazelnut, highbush cranberry, riceroots, and yellow glacier lily bulbs (Dr Arvid Charlie [Luschiim], pers. comm., 1999; Tina Robinson, pers. comm., 2002; Mary Thomas, pers. comm., 1994, 1996, 2001; Compton 1993b; Deur and Turner 2005; McDonald 2003, 2005). As Hul'qumi'num cultural expert

9-5 | Gitga'at elder Helen Clifton of Hartley Bay, BC, holding one of the cedar-wood seaweed-drying trays that Annetta Robinson brought with her when she moved to Hartley Bay. Annetta's mother had reportedly made 400 of these seaweed trays back in Kitkatla.

Dr Arvid Charlie (pers. comm., 1999) explained, "from the stories I used to hear, sometimes, our young ladies would end up somewhere where there was no fish. So when it happened ... they got some salmon eggs and they transplanted them, and they started a new run ... [That goes] Way, way back ... Seseyitsa7, Agnes Ely, told me this ... a great, great grand aunt ... I was already a youth [when she told me], and when I mentioned it to Mother, she confirmed it."

Chiefs and Nobility

The Northwest Coast peoples were, and are, particularly highly organized economically and socially, with various structures and traditions of clans, lineages, chieftainships, and inheritance practices that are validated and perpetuated through feasting, ceremonial crests, dances, songs and narratives, and a complex of systems of exchange, investment, and validation of recognized and assumed roles known generally as the "potlatch" (Trosper 2009). The potlatch system not only acknowledged the status and prestige of chiefs and other nobility but also served as a means to recognize the history of their positions and the responsibilities they bore in caring for the people and the lands and resources under their purview. On the Northwest Coast a chief's status was inherited with his name and position, but he still had to work hard to maintain his place and to retain the loyalty and confidence of his followers. Thus at least part of his prestige was achieved or earned through his wisdom, leadership, and knowledge.[19]

In the Interior Plateau the position of chief was to some extent inherited, but chiefs could also be recognized through particular prowess in hunting, oratory, or some other admired talent. Teit (1909, 582) described the role of chiefs and nobility for the Secwepemc in terms of resource use and proprietorship:

> Fishing-sites belonged to the crest group. All land and whatever grew upon it, all hunting-grounds with the game thereon, and all rivers and lakes with the fish therein, belonged to the nobility; and among the Cañon division all trading-privileges also belonged to this class. The land, with the waters pertaining thereto, was divided among the various bands, although their hunting-grounds were not in all cases well defined. The hunting-territory, root-digging grounds, berrying-resorts, and camping-places in the mountains of each band belonged to the nobility of the band in common, but the trapping-grounds and fishing-places were divided among the crest groups of the nobility of each band.

Chiefs and leaders often relied on supernatural help for their success. The narratives recounting ancestral feats of chiefs and other individuals often stress the importance of such assistance. For example, in the founding of the Ts'msyen community known as Robintown, near the present village of Kitsumkalum, the pioneering ancestor, Nisg̲eel, came to the site of Robintown from a village toward Kitsumkalum Lake and was welcomed into the valley by Robin beings – robins in their human form. The Robin people taught Nisg̲eel and his people how to build fish traps, and then, after awhile, they left. Nisg̲eel established Robintown as his permanent village, where he was joined by Eagle, Wolf, and Killer Whale clan members. They lived on the salmon from their fish traps and on the bounty of berries and other foods in the vicinity of the village (as recounted by Arthur Stevens [Łag̲aax] to Marius Barbeau in 1915, cited in McDonald 2003).[20]

Social Inequity and Burials

Burials and grave goods sometimes show evidence of social stratification and occupational specialization. For example, at archaeological sites along the coast –

Namu, Prince Rupert Harbour, Marpole, and Deep Bay – the presence of labrets, small disc beads ground from stone or shell, beautifully flaked stone knives, and other evidence of wealth indicates not only the higher status of individuals whose bodies were associated with them but also the existence of specialized occupations, including full-time artisans and commissioned art (R.L. Carlson 1976; Bernick 1998b, 1998c). Even the materials used to wrap a deceased person reflected his or her position; in the Interior Plateau, a member of the nobility would be wrapped in a skin robe for burial, whereas a poor person would be wrapped in a tule or cattail mat (Schulting 1994).

Such social stratification usually reflects the ability of a society to generate surpluses of food, which can be amassed and redistributed by its leaders, bringing status and recognition to the elite individuals and, through them, to the entire community (Hayden and Schulting 1997; Schulting 1994). Burials in the Interior Plateau have also included grave goods indicating high status. At the Bell site northeast of Lillooet, the remains of an infant of around eighteen months old buried in the floor of a house pit showed either high individual inherited status or the high status of its family. Dating back around 1,000 years, the burial contained, among other items, a carved antler figurine, a carved steatite pendant, an incised antler comb, a quartz crystal, a siltstone bear figurine, nearly 250 dentalium shell beads, and abundant red ochre. Other burials near Lytton at the confluence of the Fraser and Thompson Rivers contained similar items: dentalium and perforated scallop shells, pendants of abalone and copper, quartz crystals, nephrite celts, carved steatite pipes, carved antler-tine clubs, bone and antler zoomorphic carvings (some of Northwest Coast artistic style), digging stick handles, and numerous other tools (Schulting 1994). As well as evidence of social stratification and societal complexity, these burials reflect the large-scale trade networks and communication linkages between the coast and the interior that would have served as channels for the dissemination of botanical products and information, along with these more tangible and better preserved treasures.

The social differentiation reflected in archaeological contexts is most pronounced from the past 400 to 200 years, but probably extends back several millennia, and shows evidence of intensifying around 2,000 years ago, occurring roughly the same time and over a large area for most or all of the Plateau as well as parts of the Northwest Coast. This transition to more stratified societies is supported by several lines of evidence, including the exotic and prestige goods found in some burials, settlement and house structure data, increasing indications of food storage capacity, and significant elaboration and differentiation of crests, designs, and artistic styles (Schulting 1994). Other than a whole range of rare "status" goods like dentalium and olivella shells especially associated with burials, styles of hats and basketry may be another indicator of status in archaeological contexts. Croes (2003) points out, as noted previously, that hats with distinctive knobs on the top found at both the 3,000-year-old Hoko River site and the more recent Ozette site, may have been worn by noblemen, whereas commoners would have worn hats of a conical, flat-topped design, of a type also found at both sites. At Ozette there were also three spruce-root hats, which were twined up-to-the-left and used an ornamental skip-stitch typical of Haida and Tlingit work – possibly traded from the north and indicative of the high status of those who owned them.

Plants as Crests

Crests and special prerogatives, and the rights to use and display them, are part of a chief's and clan's wealth and prestige. The images associated with elites and their families and/or clans were, and are, shown in a variety of forms: totem poles, dances and masks, blankets, baskets, boxes, spoons, bowls, house fronts, and rock paintings or carvings. They signify both economic and political power in a social group, but they also have deep ceremonial and cultural significance, reflecting layers of history and connections to places, to the deeds of ancestors, and often to particular resources. For example, for the Gitxsan, each "house" belongs to a single village, which has its own leadership and its distinct history, as commemorated in *adaawak*, the "true traditions," which include oral narratives, crests, and songs. Each "house" owns a series of territories, which its members take care of and from which they derive all of the resources they require through hunting, fishing, berry picking, and so forth. Rights and privileges of a "house" group are recounted and validated in front of invited guests as witnesses during feasts and other occasions (Johnson 1997). On the northern coast, among the Haida, Ts'msyen, and Tlingit, crests and other rights and titles are passed down through matrilineal descent, with a man inheriting a chief's position and prerogatives from his maternal uncle, but only after going through a formal process of validating his claims through potlatching before witnesses.

Most crest designs are zoomorphic, being related to animals and supernatural beings, but there are a few crests that reflect people's direct ties and spiritual and historical links to the plant world. One of the major Ts'msyen and Gitxsan clans of the northern coast is the Fireweed, or Gisk̲'aast, clan, whose ancestors were said to have originated on the upper Nass River. Here, as the narrative relates, a Raven group killed all of the members of a rival village, except for a young woman in puberty seclusion and her grandmother. The young woman was rescued and taken as a wife by the son of the Sky God. Her children eventually returned to earth in order to avenge their mother's family. They made war on the Ravens and some coastal Tlingit and eventually settled near Hazelton. The original Fireweed village, Temlax̲'aamit (Prairie Town), was said to exist, possibly as far back as 3,500 years, on the long river flat beside the Skeena River opposite the mouth of what is now Chicago Creek. Then, as recounted in the

Fireweed clan's oral history, Temlax̲'aamit was abandoned in a year of severe and prolonged winter, attributed to supernatural retribution for disrespectful behaviour toward mountain goats by some of the village's hunters. The survivors then founded Gitsegukla and Kispiox, migrated to Git-anyaaw, Gitwingax̲, and Kyah Wiget, and eventually settled at Kitselas (Johnson 1997).

There are other oral histories related to the Fireweed crest and its origins. Another account is from Ken Harris (Chief Hax̲ Bagwootx̲w), in which some people planted *k'ilhaast* ("fireweed") in front of their house and overnight it grew so tall that it pierced the sky: "This was the *gilhast* [*k'ilhaast*] and the beginning of a new clan, the *Gisgahast*" (cited in Johnson 1997, 140). The *k'ilhaast* ("single fireweed") was considered the first totem pole, and Gisk̲'aast is the name of one of the four Gitxsan Clans (ibid.). Another, related oral history recounts the miraculous appearance of a single fireweed through the snow at the southern boundary of a newly claimed hunting territory, which was then called "'Guell haast' [K'ilhaast] – the single fireweed ... To this day Guell haast is the southern boundary of the hunting grounds of the Bear People," and Guell haast "has had its place on the totems to tell of the time of famine and how the salvation of the people was wrought" (recounted in the late 1930s by Walter Wright, a Ts'msyen chief of the Kitselas people, cited in ibid.). The "mountain fern," or spiny wood fern (*Dryopteris expansa*), is another Gitxsan crest, 'Wii Ax̲, which was depicted in stylized form at the base of a totem pole of 'Woosimlaxha at Kispiox village, among other places (Cove and MacDonald 1987; Johnson 1997). For the Haida, yellow-cedar and yellow pond-lily are two known plant crests (Turner 2004a).

Chiefs, Nobility, Resource Use, and Management

Besides the prerogatives of crests, names, and other owned property attributed to designated groups, control over particular territories, harvesting sites, and the resources they yield relates directly to the responsibilities and privileges of chiefs and leaders. These resources include many plants and plant products (Turner, Smith, and Jones 2005). Over the course of millennia in the area, as human populations grew and some resources became more limited, more social controls would have been necessary to prevent overexploitation. Given the seasonality and patchiness associated with most plant resources, having ways to limit or distribute the harvest would have become more and more desirable and necessary. The strength of the controls placed on these resources was aligned with the convenience and perceived quality of the resources, their overall productivity, the potential intensity of use, their storability, the continuity of use, and the amount of labour expended in maintaining and enhancing them. Thus prime and highly desired resource sites like Pacific crabapple stands, highbush cranberry patches, and productive camas meadows near villages or seasonal camps, on which people expended energy through pruning, cleaning, burning, or otherwise tending, were the ones that would be more intensively controlled or "owned" by a chief, clan, family, or village group, depending on the region. Those in control were expected to apply their experience and judgment in order to allow optimal use of the resources and were rewarded accordingly. As Thornton (1999, 39) describes for the berry resources of the Tlingit, "Clan leaders used their knowledge and authority over local berry patches to facilitate others coming to gather when berry conditions were peak and the supply abundant ... enhancing his prestige and 'credit' in exchange for surplus berries, and by responding to the invitation other pickers were, in effect, legitimizing the possessing clan's prerogatives over the territory." Similar protocols were observed, with variations, all along the Northwest Coast, as well as in parts of the Interior Plateau (Boas 1921; Budhwa 2007; McDonald 2003; Teit 1909; Turner, Smith, and Jones 2005).

Just as actual plant resources and harvesting sites were owned, so too were structures made from plant materials: deer fences and deer nets; duck nets and their sites; fishnets, fishing weirs, and platforms; and bridges, shelters, drying racks, and the like at harvest camps (Hayden 1997; Schulting 1994; Teit 1906a, 1909). The various controls established for and around plant resources can be interpreted as a form of management or stewardship and, as such, will be discussed in chapter 11. Stored food, and the capacity for storage, is seen as a necessary prerequisite for the development of socioeconomic inequality, or stratification, in societies (Ames and Marshall 1980; Hayden 1992; Hayden and Schulting 1997; Schulting 1994). Stores – by the very labour that is invested in their production and by the ultimate benefits expected from their use – must have some level of ownership or control. As stated previously, improved technologies allowing effective storage of food went hand in hand with increasing social complexity (Ames and Marshall 1980). As Schulting (1994, 18) comments, when the value of food is increased through processing, "the question of ownership looms large."

Although the storing of dried salmon is usually cited as the primary example of how storage changed the social structure of societies on the Northwest Coast and in the Interior Plateau, plant resources are definitely part of this picture as well. Along the Columbia River, for example, acorn storage pits were individually owned, with acorns (from garry oak) considered a possibly high-status feast food (B. Mathews 2010; Spier and Sapir 1930, 185). Elsewhere, boxes of preserved crabapples and highbush cranberries, baskets of dried roots and dried berries, squares of dried seaweed, cedar-wood boxes of oulachen grease, and many other types of preserved foods were used as potlatch gifts, as marriage gifts, and at feasts given by chiefs, nobility, and their clans, not to mention as trade goods to be exchanged for other products of equal or greater value. However they were used, these stored resources brought additional prestige, power, and status to those who were able to organize and control their use.

Notably, this system of accruing and controlling wealth is different from that assumed by many other societies. In the social system predominating throughout the study area, rather than through conspicuous consumption, the prestige and status of a chief, clan, or community was affirmed and validated through the distribution of wealth. Whereas some surplus food might be kept for individuals or families to consume, much was often given out as a way of gaining stature and esteem. Generosity, in this system, is a virtue; stored food and other items of value, such as canoes or intricately woven baskets created by specialist artisans, were a means of raising the stature not only of an individual but also of an entire clan, lineage, or village group. Especially in the Interior Plateau, chiefs were not always defined or chosen through inheritance but might acquire the designation through repeated generosity or prowess in contributing to the group's stores of food or other valued goods (Hayden and Schulting 1997; Schulting 1994). Teit (1909, 573) described the protocols of the Secwepemc around a leader's role in sharing the proceeds of a hunt, which relate both to prowess in hunting and to leadership:

> Fat and meat were divided equally among the hunting-party by the leader or by the eldest hunter. A man hunting alone had the sole title to whatever game he killed; but when he brought the meat home, he almost always distributed a large part among his neighbors. Even meat of game trapped was generally distributed in like manner. Skins of animals trapped belonged to the person in whose trap they were caught, and skins of animals shot belonged to the hunting-party; but the hunting chief generally allowed the best hunters the greater part of the skins.

A chief or leader typically took charge of distribution of any scarce item or product. For example, among the Secwepemc, who were extremely fond of smoking the native tobacco (*Nicotiana attenuata*), the chief sometimes needed to oversee the harvest. The leaves were dried and pounded with grease before being smoked. Teit (1909, 574–5) explained the chief's role in distribution of this valued product and its successor, the introduced tobacco (*N. tabacum*):

> In some places where tobacco [*N. attenuata*] was scarce, it was gathered under the surveillance of the chief, in order to give every person an equal chance. It grew luxuriantly on some village sites, and some Indians think the seeds must have been planted in these places. In later times the chief of each Shuswap band received annually a present of tobacco [*N. tabacum*] from the Hudson Bay Company's post at Kamloops. The chiefs of distant bands sent young men to Kamloops to receive the annual present ... Upon receiving the gift, the chief called all his men together, gave a feast in

which all smoked and talked for from two to four days, and then he divided the remaining tobacco equally among all present.

Another way for a leader or a community to achieve status that relates in some ways to plant resource use is to gain control over strategic travel routes or passageways. It is likely that elites of Keatley Creek and vicinity were able to acquire wealth and power by this method since it ensured their dominance of the lucrative Fraser River fishery in the region. On the Columbia River, the Wishram were similarly strategically situated to control traffic along the river in both directions, thus serving as the middlemen between trading partners upriver and downriver, which allowed them to control the exchange of baskets of powdered dried salmon, hazelnuts, huckleberries, and acorns, among other goods (Mathews 2010; Spier and Sapir 1930). As Schulting (1994) points out, it is no wonder that it was a modified form of the Wishram language – Chinook – that became the trade jargon used over much of western North America.

Similar trade-controlling communities of high status, with immensely powerful and influential leaders, developed in various regions, especially along the routes between the Northwest Coast and the interior, another prime example being the spring oulachen fishery of the lower Nass and Skeena Rivers controlled in large part, at least for a time in the mid-1800s, by the great Ts'msyen Gispaklo'ots chief Ligeex and his house. Ligeex, through the wealth and power he accrued from his prerogatives, became for a time a paramount chief, not only controlling his own house but also overseeing and controlling other chiefs in the region (Martindale 2003). Likewise, the Canyon Secwepemc elites regulated travel and trade between the eastern Secwepemc and the Tsilhqot'in, who in turn had access to coastal resources through their proximity to Homalthco and Bella Coola Inlets. Individuals in these positions often formed alliances with neighbouring chiefs, thus accruing even more power and prestige. Going far back into history, such individuals must have served as the brokers for the trade of many high-value items, from native copper to marine shells, steatite, nephrite, and obsidian (T. Richards and Rousseau 1987) – and quite possibly various plant products and manufactured objects as well.[21]

Most of the elites were men, but there were many matriarchs – often the wives of hereditary chiefs – who were members of the nobility and were held in equivalent high esteem due to their prestige. Occasionally, a woman would become chief through particular circumstance, either standing in for a son who would grow up to be chief, filling a chief's position if there was no male heir, or simply assuming the position through the force of her personality and leadership skills. Like the male nobility, women of nobility were not only highly respected and influential but also highly knowledgeable, having received training and grooming for their positions from an early age. Lucille Clifton, the Eagle matriarch of the Gitga'at at Hartley Bay, who died in 1962 at the age of eighty-six, was one such revered leader and teacher (Turner, Robinson, et al. 2012). For many decades, she guided the women and children of the community. She recounted stories of Ts'msyen history to the children, taught them about the places they passed in their travels by canoe, settled disputes, delivered babies, and taught the younger people about seaweed harvesting and processing at K'yel and about smoking salmon and preserving crabapples and cranberries at the fall camp at Gitgiata Inlet.

Every fall, after the people had returned to Hartley Bay from their rounds of fishing and other seasonal work, Lucille Clifton and the other Eagle women hosted a feast for the entire community, serving them with all of the food they had harvested and prepared over the year: half-smoked coho (*wüüx*[22] or *ts'a̱al*) and/or dried, soaked humpback salmon (*'wiiyuu stm̱oon*); oulachen grease (*q'əwtsi*) and/or seal oil (*kbaüüla*); toasted, flaked seaweed (*łə'əsk* or *saoolk'a*); inner bark of hemlock and/or amabilis fir (*ksiiw*); salmon egg caviar, cured and smoked (*üüskm laan*); salmon egg cheese (*luayaa*), served with grease and potatoes (*sgusiit*); boiled and drained turnips (*'yanahuu*) and carrots (*q'əʔwts* or *galot*); fried or baked bread with crabapple (*moolks*) jelly or with thick blueberry (*sm̱may*) or salalberry jam; strips of seal flipper, singed or cooked (*xslaxs*); seal meat (*üüla*); venison stew (*samimwan*); Canada goose soup; crabapples (*moolks*) and highbush cranberries (*łaaya*) with whipped oulachen grease; blueberries (*woʔoksil*), salalberries (*dzawes*), and other berries (e.g., *waakyil*, "gray currants"); oranges and apples; Salada tea; and Labrador tea (*k^{w}'ilə'məxs*). These feasts – and the times of working with and learning from Lucille – are still fondly remembered by her grandchildren and their contemporaries, who are the elders of the present generation (Tirone et al. 2007; Turner, Robinson, et al. 2012).

On the Northwest Coast many matriarchs, like many men of high standing, held inherited harvest rights to specific resource sites, including berry patches, crabapple stands, seaweed-harvesting sites, and seaweed-drying rocks. These women served many advisory roles related to the timing of harvests and to the relative order of different tasks. In the Nuxalk village of Bella Coola, McIlwraith (1948, vol. 1, 265) described how "one woman had the prerogative of picking the first berries of a particular kind." When these berries were ready for picking, "she paraded up and down the village with a special decorated picking basket, calling 'Get ready to pick red elderberries (or whatever kind was ripe) tomorrow!'" Matriarchs, like male chiefs, were to some extent "project overseers." They supervised the seasonal movements and food harvesting of their clan or community members, making the decisions about when the roots should be dug or the berries picked, ensuring that enough food stores were put by, and when necessary, arranging for those families who suffered shortages to be given enough to sustain them. These women also often directed management practices such as pruning, weeding, or controlled burning of resource sites. They had the authority to call on others, especially those in their own clan or lineage, to help with food preparation, hosting feasts, and generally supporting the ability of the group to maintain its obligations and responsibilities to the community at large. Matriarchs, like their male chiefly counterparts, were usually trained in their roles from the time they were children, and part of this training was to seek and acquire strong spiritual guidance and help. Girls who were destined to be leaders received particularly intensive spiritual training as well as practical instruction at the time of their first menstruation, during pregnancy, and at other ritually sensitive times of their lives (Reid and Sewid-Smith 2004; Teit 1909).

Many different arrangements existed for ensuring direction and the observance of protocols when harvesting key resources. One model for picking saskatoon berries, or serviceberries, was described for the Secwepemc by Teit (1909, 573), giving a glimpse of the chief's role as a sort of project supervisor:

> Berry-patches were tribal property; but picking was under tribal control. All the large and valuable berrying-spots were looked after by the chief of the band in whose district they were situated. Thus there were several large service-berry patches near Big Bar. The chief there watched the ripening of the berries, and deputed young men to watch and report on the various places. From time to time the watchman brought in branches and showed them to the chief. When the berries were about ripe, he sent out word that on a certain day the berrying would commence at a certain berry-patch. Women would come from as far way as Alkali Lake and Clinton. The first day each woman picked only a little, about enough for herself and her friends to eat fresh during that day and night. After the first day they picked all they could and began to cure them. When they had finished one patch, the chief directed them to the next one which was ripe, and so on until they had finished all.

People who picked the berries without authorization or those who trespassed from outside of the area were fined or driven away by band members acting for the chief. Similar procedures would be followed for other resources, both berries and root crops, as they became ready for harvest. Farther south, the chief would oversee a more formal First Fruits ceremony for saskatoon berries or huckleberries.

Another good example of the chief's role as a project organizer and supervisor – and of the importance of occupational specialization, intervillage connections, and reciprocal obligations – is in a story recounted by Lucille Clifton to William Beynon, "The Narrative of Sarapgyaw" (cited in Cove and MacDonald 1987, 326–31). This story told of a great feast, planned for a group of supernatural beings (who were sea monsters) by a chief named Yaragwenusk, who was a grandson of the Chief of the Sandbar. He wanted to erect a stone totem pole on the Nass River in the next oulachen season. "I must ask the help of every tribe," he said. Thus he requested and received contributions from each tribe, according to the special abilities of each, toward this magnificent effort. He said to the Gitlan people, who traded in wooden dishes, "Your carvers will make a great quantity from yew wood." From the Gispaxloats, who traded with the Hagwilget, he asked for many horn spoons. The Ginaxangik, who traded with the Chilkat, would contribute many food boxes. The Gitsis and Giludzau were to make deep wooden dishes. The Ginadoiks, whose country on the Skeena River had the greatest number of mountain goats, would bring mountain goat fat. The Gitwisabe were to make the dried inner bark of hemlock into cakes. The Kitselas would make cakes of berries, and the Kitsemkalem would produce scores of bundles of dried salmon. The Gitaxta would make shredded cedar-bark robes, and the Gitga'ata would make snuff from tobacco since they had connections with the Haida, who grew this special plant and prepared it by mixing it with finely ground baked clam shells (see Turner and Taylor 1972).

Occupational Specialization and Prestige

Increased occupational specialization, for individuals and entire communities, was both an outcome and a cause of social complexity and stratification. As in any society, people who had specialized talents were often accorded a high degree of admiration and status, especially if the products of their specialty were in great demand and required intensive training to create. The type of specialization on which a person would focus depended on a combination of factors: where their background, inclinations, and talents lay; what access they had to particular resources; and what opportunities they had for training. In the ancient world of northwestern North America, people who became experts at food harvesting – fishing, hunting, root digging, and berry picking – were afforded high respect. Even today, women who can cut halibut and salmon quickly and expertly or who can pick berries more quickly and cleanly than others are greatly admired, as they would have been in the past. It is likely that such highly skilled individuals in the ancient past were the ones who became the leaders and eventual nobility of their communities as populations grew and cultures became more highly organized socially and economically. It is probably not a coincidence, for example, that the main Nuu-chah-nulth, Ditidaht, and Makah whalers were chiefs and that the position of whale hunting was hereditary, with individuals being trained for the role even before birth and all through their infancy and childhood.

Training for many of these specialized roles involves spiritual and ceremonial preparation, praying and purification rituals, as well as practical instruction in hunting, canoe-handling techniques, basketry, and other practical aspects of a given occupation. Seeking supernatural aid, and following the protocols required to gain success through spiritual guidance, took tremendous discipline and practice, no matter what the specialty (E.R. Atleo 2011). Whether people were shamans or herbal healers, expert mountain goat hunters or root harvesters, or canoe makers or mat makers, they had to be connected to the spiritual world, and having such connections gave them high prestige and increased their social standing (Ames and Maschner 1999). One indication of the high social standing and possible spiritual connections of a specialized craftswoman from long ago is in the oldest known wooden carving on the Northwest Coast, from the Hoko River wet site. It is the handle of a mat creaser, which dates to between 2,800 and 2,600 years ago. This remarkable implement, with a painted handle, features two crested birds, probably pileated woodpeckers, touching beaks (Ames and Maschner 1999). A botanical artifact in more than one way, the creaser is a crafting tool for an important technology, cattail and tule mat making, whose products were employed for a multitude of purposes, such as covering shelters, dividing rooms, fashioning mattresses and kneeling pads, drying and serving food, and possibly covering cooking pits (Turner 1998). The woman using this mat-creasing implement was probably also an expert basket maker and maybe, like the individuals who wore the knot-topped hats found at this site, represented a developing group of high-status people who lived at least part of the time at the Hoko River fish camp.

ALLIANCES AND INTERGROUP RELATIONSHIPS

The great Ts'msyen chief Ligeex of the mid-1800s and the ancient chief who organized all of the Ts'msyen villages in hosting the feast for supernatural beings, both mentioned previously, represent a broader plane of social organization and interaction: across community and tribal boundaries. Although such wide-scale interactions must have been in place to some extent since the Early Holocene, they certainly intensified in the past few thousand years, probably reaching their maximum activity in the times of largest settlements and highest populations, toward the end of the Late Holocene, just before Europeans entered the region. The focus for these nodes of social interaction was on the major centres of resource production. Many of them, for example, were associated with rapids in rivers where large numbers of salmon could be caught with dip nets: The Dalles on the Columbia River, Kettle Falls on the Kettle River, Six-Mile near "The Fountain" (Xaxl'ep) on the Fraser River, and the steep-sided canyons and rapids of Kitselas on the Skeena River. These areas – possibly the first places where people started to exploit large numbers of salmon – served as major trading centres, as social centres, and as places where meetings of chiefs and leaders from different communities were held, alliances were formed, and various political organizations were instigated (see Johnson 1997).

Some of the groups as they existed at the time Europeans arrived in the area apparently originated through connections and political unions made at such centres. For example, as reported by Leslie Main Johnson (1997), some of the Gitxsan ancestors were said to have come from the Owikeeno area to the Kitamaat Haisla area long ago and then to have amalgamated with some early Ts'msyens and a group of Haida to become the Eagle clan of the Haisla, which also has connections to peoples on the lower Skeena River and in other areas of the Northwest Coast. Many origin stories trace such complex ancestries. The Ts'msyen also evidently have ancestral connections with Dene (Athabaskan) peoples of the interior.

The Stl'atl'imx/St'at'imc of the Interior Plateau exhibit similar complexity in their social, economic, and political alliances, which developed around Lillooet and "The Fountain" along the Fraser Canyon north of Lytton. These major trading sites on the upper Fraser River are situated at the interior end of a longstanding trade corridor extending south and west through the Coast Mountains to Harrison Lake and down the lower Fraser to the coast. The Scowlitz site, situated at the confluence of the Harrison and Fraser Rivers, is another settlement along this route that has exhibited a high level of social complexity and apparent social stratification within the past 2,400 years, in its burials, in the presence of exotic objects of native copper and other valuables that reflect interior provenance, and in apparent increased house size, external and internal elaboration of houses, and expanded storage capacity, all of which serve as material symbols of wealth and power and suggest increasing social complexity and the elite social status of some (Hayden and Schulting 1997; Morrison 1997; Schulting 1994). The Stl'atl'imx of the Fountain area are also linked to the Tsilhqot'in, Nlaka'pamux, and Secwepemc peoples through intersecting territories like the Hat Creek Valley and "Many Roots" north of the Bridge River.

Access to all of these different places and the different resources they provided would have given key individuals – chiefs and their families – opportunities to gain wealth. Their continued leadership and influence, however, would not have been based on wealth alone. Their personal traits and their skills in oratory and negotiation, as well as in hunting or warfare, would have been just as important. Generosity, in the form of procuring and sharing food, especially in lean times, and evidence of supernatural sanctions through their ceremonial participation would also have been prime factors in the success of these leaders. Hosting feasts for the local community, and for visitors from the entire region, would also have been a hallmark of leadership, serving as a means of redistributing food surpluses and compensating for local shortages (Hunn, Selam, and family 1990; Teit 1909). The more they could manage to provide and control, and the larger their circles of influence, the greater their status and potential for economic and political gain.

Some of the villages encountered by the first Europeans entering the area presented strong evidence of social complexity, the elite status of some individuals and families, and associated large-scale alliances and intervillage connections.

For example, the Wishram village of Nixlúidix (Trading Place), encountered by the Lewis and Clark expedition on 24 October 1805, apparently housed hundreds of people – possibly a thousand or more, not even counting slaves – with at least twenty large cedar plank houses, each occupied by about eight men and twenty or so women and children. The houses measured about 6 by 9 metres and were deeply sunk into the earth, with cedar-bark roofs. The high status of some individuals was reflected in their houses – for example, in doors carved and painted with animal representations. The planks of the houses themselves represented considerable value. Alongside the river at Nixlúidix, William Clark (see M. Lewis and Clark 1805) recorded huge stacks of dried salmon, with an estimated total weight of about 10,000 pounds (about 4,500 kilograms). These people were renowned traders of their salmon, connecting to tribes from the Pacific Coast to the foothills of the Rocky Mountains. During the summer season, thousands of people would have camped in the area in mat lodges, not only trading but also socializing, gambling, and exchanging knowledge, stories, and ideas. Many marriages and alliances would have originated at these gatherings. Slavery was a part of this economic system, with the wealthiest families owning up to ten or more slaves, who would help with fishing, root digging, food preparation, paddling canoes, and child tending (Schulting 1994; Spier and Sapir 1930). Soon, however, the village declined drastically due to disease epidemics. Malaria alone resulted in an 87 per cent population decline between 1830 and 1840 (Boyd 1992). Such major trade centres were particularly susceptible to pandemic diseases, and thus the prestige accrued by powerful chiefs in such places endured in their elaborate funerals, with valuable grave goods and, sometimes, the killing of horses and slaves as a display of their wealth and widespread social status (Schulting 1994).

Warfare, Intergroup Conflict, and Knowledge Exchange

The other side of supracommunity political alliances, with such paramount chiefs as described in the previous section, is intercommunity conflict and warfare, which was also common. The same groups that might have, on some occasions, formed mutually supportive relationships could at other times develop hostile relations that would end in war, with killings and the taking of hostages or slaves. Accounts of such hostilities are relatively common in the traditional narratives and have been historically documented (e.g., Boas 1966, 2002; George 2003; McIlwraith 1948; and Reid and Sewid-Smith 2004).

Some of these conflicts have taken place at the very nodes of common resource use where cultural exchange would have occurred at other times. For example, accounts of hostility between Dene (Athabaskans) and the neighbouring Interior Salish are relatively common in the oral histories. The site called "Many Roots," or "Plenty of Roots" in Tsilhqot'in and Secwepemc-tsín, bordering on

Tsilhqot'in territory on the north and Secwepemc and Stl'atl'imx/St'at'imc territory on the south, is a known place of conflict between these groups, where several battles were waged according to the oral traditions of all of these peoples (Matson and Magne 2007). It is a hill or mound with a lake and wetland nearby, on a tributary of the Bridge River (Teit 1909). Archaeological evidence shows that Salishan peoples once occupied the area where Tsilhqot'in peoples have lived more recently, and these peoples were possibly displaced by past warfare.

Other battlegrounds are described, and cases of warfare, stealing women and children, and slavery intermingle with other oral records of trade, intermarriage, and kinship between neighbouring peoples. Similar situations must have occurred in the Interior Plateau, where the Nicola Valley Athabaskans were eventually replaced or assimilated by the Nlaka'pamux and Okanagan Interior Salish peoples, and on the west coast of Vancouver Island, where Salishan speaking peoples were evidently replaced relatively recently by Nuu-chah-nulth-speaking peoples (Coupland 1989; Keddie 1997; Matson and Magne 2007; Moss and Erlandson 1992). According to Cybulski (2006), skeletal remains and other data on the Northwest Coast indicate that warfare occurred over 4,000 years ago and was a much more serious and aggressive pursuit in the northern regions than in more southern coastal areas.

Warfare in particular may have resulted in certain adaptations in food storage, as indicated in the observation by Hill-Tout ([1910], 29–30) about Sechelt storage practices in reaction to raiding by a Kwakwaka'wakw group from Vancouver Island: "[The Sechelt] ... did not store their winter supplies in their dwellings, but 'cached' them in the woods. Only a few days' supply was ever carried home. This peculiar custom was due to the marauding proclivities of the neighbouring *Yuɛkltas* [Euclataw or Leg^{w}ildax^{w}], who made periodical forays upon their settlement and carried off all they could lay hands upon. It was unsafe, therefore, to keep a large store of food by them."

Drucker (1951, 36–7) also notes the relationship between food shortages and intergroup conflict among the Nuu-chah-nulth: "Family traditions of local groups ... speak of hunger and even starvation that led them to make alliances with or make war on groups who had territories along the inner channels and owned salmon streams." Dewhirst (1978, 21) examines the role of resource scarcity for the Nuu-chah-nulth of Nootka Sound, suggesting that the whale-hunting technologies of the outside groups may have developed in response to "a tight fit between population growth with increased local groups and ownership of limited resource properties." Although these observations and comments focus on fish and marine mammal resources, similar events and circumstances must surely have also involved plant resources, which were, after all, the major materials used in fishing and hunting technologies and cannot be separated from them. Harpoons, weirs, baskets for transport and storage, fuels for cooking and smoking food – all were centred on plant use (see chapter 6) and would have been implicated in any changes that took place as a result of resource use conflicts and economic strategies.

It is difficult to say what other influences warfare and conflict might have had on ethnobotanical knowledge and plant use. The potential role of slaves in transmitting knowledge and practice was discussed earlier. One might assume that in any situations where populations from different communities and language groups intermingled, whether as a result of conflict, assimilation, mutual merging, or affinal arrangements, there would have been some level of knowledge exchange or borrowing related to plant names and specific resource sites, techniques for cooking and processing food, styles of mat making and basket making, techniques for house and canoe construction, and uses of medicinal and ceremonial plants. All of this sharing would have taken place in situ – in the context of a given place or territory that would have been familiar to the original residents and "new ground" for those arriving, whether by invitation and mutual consent or by force. Whatever disturbances and changes that people incurred as a result of social or economic restructuring, the existing plant resources would have assisted transitions and helped to strengthen people's resilience and capacity to adapt.

CONCLUSIONS

Plant resources and the knowledge around their use both affect and are affected by the social roles and interactions of people on multiple scales and through different modes. Furthermore, these interactions with plant resources, like their use and management, are dynamic, occurring simultaneously or transforming, shifting, and replaying at many levels across time and space. The human actors in the acquisition, transmission, and adaptation of ethnobotanical knowledge are one element in the pattern. The other main elements include their practices and actions, the sites at which the actions occur (e.g., campsite, harvesting grounds, or big house), the times – seasonal, annual, and millennial – over which these activities take place, and finally, the plants and other elements of the environment involved at these sites, at these particular times, and in the course of these activities. This framework of ethnobotanical knowledge development, practice, and transmission forms the backdrop for any analysis of such knowledge, regardless of the scale or dimension of human social organization. Every instance of ethnobotanical knowledge practice, belief, and transmission represents an intersection of these factors within the framework.

All of what each of us knows is built to some extent on previous knowledge of others – those who are contemporaries and those who came before. We each create and add to our own knowledge through our individual powers of observation (e.g., by watching and learning from animals and other "kin"), through our abilities for reasoning and applying and testing our ideas, and even through our dreams. Then each of us, in turn, passes our knowledge on to others through demonstration and communication. Knowledge about plants – like other kinds of knowledge – is thus mediated by our relationships with others, our capacity to acquire and apply knowledge from others, and their capacity and willingness to share their knowledge with us through various means. Each individual has a role in both acquiring knowledge and transmitting knowledge. Organized social groupings and relationships, including families, extended households, task groups, and clan- or village-level groups, can all be seen as modes of knowledge transfer – or, conversely, of knowledge exclusion.

The ways and means of disseminating plant knowledge thus seem infinite. In actuality, there are many limitations, from the presence at a particular time and place of a particular individual holding key knowledge to the existence of a given plant or habitat in a particular site at a particular time. By examining these parameters, we can see patterns in the types of knowledge acquired and passed on under different socially mediated situations, at diverse places, and over various time frames.

Social organization has had a major role among the various factors influencing the ways that existing plant populations are known and used by Indigenous peoples and the ways that Indigenous peoples have affected the presence and abundance of plant populations and communities. Social groupings, in turn, have been influenced in major ways by the plant resources themselves – their availability, utility, and productivity. Consider the effects of the availability of western redcedar in allowing the construction of large post and plank houses and oceangoing canoes on the Northwest Coast some 4,000 years ago and how this ability assisted the development of large, diverse, and stratified societies.

Consider, too, the complexity of resource harvesting that requires high levels of direction and organization from project leaders and overseers, often chiefs and matriarchs. Consider the resilience afforded to families and clans from the availability of major resources like saskatoon berries, huckleberries, wapato, camas, and spring beauty, or mountain potatoes – of course, in conjunction with salmon, shellfish, and other animal resources – that have allowed reciprocal resource arrangements and sharing based on differential surpluses. Consider the prestige and esteem accorded to those who had the skills to convert – or to organize others to convert – raw resources into nutritious and storable products, or into items of spiritual and artistic beauty, and then to distribute and share these resources and high-value items with others through feasts and potlatches. Such abilities are translated and cycled into social capital through the designation of certain individuals as leaders: chiefs and matriarchs and others of high standing. Prestige positions, once attained or achieved, can extend into inherited leadership positions through careful guidance and training of one's children or others in one's lineage, enabling them to acquire the skills, abilities, and knowledge to assume similar leadership roles. Continued effectiveness then brings continued success and ongoing acknowledgment of this success in the recognition of leadership.

On the other hand, consider how the abilities of skilled and observant individuals have allowed them to learn about and exploit, process, and gain benefit from particular plant species. Envision how the guidance and wisdom of elders have created a culture of respect and kincentric relationship between humans and other beings that has allowed careful and sustainable use of plants and other resource species and sanctions against careless behaviour toward them. And, finally, reflect on how, often under the leadership of chiefs and nobility, following traditions of numerous generations, certain resource sites are identified, claimed, and accessed seasonally by resource-harvesting groups – and how such intensively used sites and their species are managed, maintained, and enhanced through landscape burning, clearing, weeding, selective harvesting, pruning, tilling, replanting, and transplanting (to be discussed further in chapter 11).

It is, in fact, organized social behaviour of many different sorts and on many different scales – decisions and choices made by individuals, households, clans, and villages and their leaders – that has created, collectively, the diverse patterns of plant use, and the commonalities in relationships with plants and environments, that have played out over time and space, as described and explored in the previous chapters. Ultimately, these choices and actions have affected peoples' abilities for reproduction and thus influenced how human populations have developed and interacted. Not all of these interactions in resource use have been positive for all. The very abilities of individuals and their ancestors that led to the institution of the nobility and leadership class, to population expansion, to resource stewardship and enhancement, and to the creation of beautiful objects of art and spiritual power also created economic and social inequity, resource sequestration, power struggles, warfare, and slavery (see Martindale 2003; and Donald 1997). By and large, however, the system seemed to have been working well for the majority of people, who were living lifestyles that compared favourably with those of contemporary Europeans (E.R. Atleo 2004, 2011). With generally recognized and respected systems of territory and land tenure in place, relationships across neighbouring communities were also evidently mainly positive despite some conflicts.

As increasingly complex societies with specialized occupations emerged in northwestern North America, and as territories with associated resources became well established and defined, people became increasingly identified with the resources under their control and within their capacity to produce in surplus. These resources – especially those considered rare and desirable in other regions – became both the means of enhancing the status of chiefs and nobility and the venue by which their status was affirmed and validated by their neighbours. Ceremonial events and redistribution of wealth through potlatching and feasting were enhanced by a leader's and a community's ability to obtain rare and valued resources from distant sources through commerce. And, as pointed out by Ames and Maschner (1999, 255), "the linkages among the household, social status (including gender) and regional dynamics lie through household production and specialization ... regional economies through activities of chiefs and through trade and exchange."

Trading networks, or "interaction spheres" (Ames and Maschner 1999), encompass the broadest level of social organization and may well be responsible for the remarkable extent to which distinctive cultural traits, economies, and practices – including those involving the knowledge and use of plants – are held in common across broad geographic and linguistic boundaries. Trade and exchange at the regional level and beyond, and their influence on the dissemination of ethnobotanical knowledge, are the focus of the following chapter.

10

Trade and Exchange: Sharing Plant Products and Ethnobotanical Knowledge across Geographical and Cultural Space

INTRODUCTION

Chief Arthur Adolph is a member of the Xaxl'ip (Xaxl'ep) community, of the Upper Stl'atl'imx/St'at'imc (Lillooet) Interior Salish. Xaxl'ep, also known as "The Fountain," is located just north of Lillooet along the Fraser Canyon. In November 2008 Art recounted a story that typifies the kind of plant resource exchange between different Indigenous communities that has been occurring in one form or another probably since the Early Holocene. He explained that the Upper, or Fraser River, Stl'atl'imx have access to certain interior berries, most notably *stsáqʷəm* (saskatoon berries, *Amelanchier alnifolia*) and *sx̲ʷúsum* (soapberries, *Shepherdia canadensis*), which are not generally found in the Lower Stl'atl'imx, or Lil'wat, territory at Mount Currie in the Pemberton Valley. The Mount Currie Lil'wat, on the other hand, have resources like black huckleberries in their Coast Mountain territories that do not occur in the drier Upper Stl'atl'imx territory. Edible berries of all types are highly appreciated, so the people trade with each other for the different kinds. And, to facilitate the exchange, they have developed reciprocal kin ties. Sam Mitchell, Art Adolph's father by customary adoption, had a niece who married into the Lil'wat community at Mount Currie, and this allowed their family access to the productive huckleberry patches and some other resources of the Lower Stl'atl'imx.

When Art was around eight years old, his mother, Susan Mitchell, organized a group of people to go huckleberry picking in the Lil'wat territory.[1] Art was allowed to go on this trip, which was in the summertime. He accompanied Sam and Susan Mitchell, along with other family members, on the train[2] from Xaxl'ep and Ts'kw'aylaxw (Pavilion) to Mount Currie. At the Mount Currie train station,

10-1 | Black huckleberries (*Vaccinium membranaceum*) (darker, shiny) and Cascade bilberries (*V. deliciosum*) (lighter colour), two prime edible berries of the Coast Mountains.

they were joined by a group of their relatives, and they all travelled together to the huckleberry-picking grounds at Alta Lake, toward Whistler, where they spent the weekend picking huckleberries. Art recalled that they stayed in cabins above the train track at Alta Lake. They all worked hard and picked large quantities of huckleberries (figure 10-1), which they brought home to Xaxl'ep. Then, his mother, Susan, organized the sharing of the harvest with the people at home who weren't able to go on the trip.

In reciprocal visits, Sam and Susan's family members from Mount Currie often came to pick *sx̱ʷúsum* (soapberries) around the Lillooet area with the assistance of their Xaxl'ep relatives. Sam would meet them at the train station in Lillooet with his truck, bring them to their house to stay overnight, and then take them the next day up to the open woods of the Fountain Valley to pick the *sx̱ʷúsum* (figure 10-2) (Arthur Adolph, pers. comm., 2008).

This exchange of hosting was typical of the way certain resources like berries were shared within families and across communities. As Thornton (1999) explains, trading and exchange allowed a redistribution of berry resources in space, just as storage technologies allowed a redistribution of their use over

10-2 | Soapberries (*Shepherdia canadensis*), a prime Xaxl'ep resource of the Fountain Valley, BC.

time, and both strategies helped to mitigate issues of supply. Both enhanced a group's capacity for resilience.

All forms of trade and exchange are based on variable occurrence, distribution, and control of desirable resources and are therefore solidly embedded in territorial rights, social relationships, and reciprocity. Within these parameters, there have been many diverse arrangements, enacted on different scales and at different levels of complexity. The situation described above by Arthur Adolph is perhaps the most common and straightforward form of resource sharing. Many elders have recalled similar arrangements with kin residing in different territories (see Suttles 1987a and 1987c for a discussion of such resource exchange among Central Coast Salish). In terms of acquisition and transmission of knowledge about plants and environments, trade and exchange were crucial. It is likely that many of the plant names "borrowed" from one language into another, as discussed in chapter 3, as well as many of the stories related to plants and plant use that are shared across cultural and linguistic boundaries (see chapter 12), are a result of intergroup contact involving trade and exchange. Not only products and materials but also the cultural knowledge and vocabulary that accompany them – as well as, in some cases, more general knowledge, beliefs, and approaches – are communicated between peoples on various scales.

This chapter addresses the role of trade and related economic and social exchange in the transmission of botanical and environmental knowledge within the context of Traditional Ecological Knowledge systems. The historical development of trade and exchange as it relates to the Indigenous peoples of the study area is briefly discussed. Then trade and exchange are defined in cultural terms. The distinctions between gifts and giving, payment for services such as witnessing ceremonies, redistribution of surplus products, and trading of products of equivalent value are blurred. There are explicit and inferred obligations attached to many types of gifts, and these are tied into social organization and differential social roles (Mauss 1990; Turner and Loewen 1998). Following this is a consideration of the types of plants and plant-related information, practices, vocabulary, and other cultural knowledge involved in systems of trade and exchange. What did people bring with them? What were their legacies? How did old and new ideas and practices combine and shift? How did languages reflect the changes? The predominant venues and mechanisms of trade and exchange are described, from feasts and potlatches to major congregation centres or nodes of exchange. Trade networks across geographic regions are then discussed, and the chapter concludes with a consideration of the implications of trade and exchange for how botanical knowledge is disseminated and for the patterns exhibited by this sharing.

HISTORICAL DEVELOPMENT OF TRADE AND EXCHANGE

Natural environments vary in their ability to support human populations, both over space and over time. Few places can provide a complete and consistent array of necessary resources over thousands of years. Late Pleistocene and Early Holocene times must have presented particularly forbidding environments to early humans, whose survival would have depended on flexibility, adaptability, and a capacity to utilize the resources at hand in creative ways. Balancing the limited capability of these environments to support large numbers of people, on the one hand, with the advantages that many individuals with diverse skills and abilities would bring to a group's survival, on the other, must have presented major dilemmas. People's ability to spread out in small groups and take advantage of diversified resources would have been important. Yet the capability to maintain social relationships across these groups, and to exchange both resources and innovations (not to mention opportunities for genetic mixing), would also have been vital. It can be assumed, therefore, that at least some forms of exchange were taking place among the earliest groups of humans in the Americas, allowing them greater flexibility and helping to counter the instabilities in resource availability and accessibility they faced.

Obsidian as a Reflection of Early Trade Networks

This assumption is borne out by archaeological records reflecting extensive ancient trading networks, from which social interactions and knowledge exchanges can be inferred. Obsidian, a naturally occurring volcanic glass highly valued for its cutting qualities, is one of the very early trade goods, easily preserved and detected in archaeological contexts and capable of being traced through x-ray fluorescence "fingerprinting" to the specific volcanic area where it originated (Ames and Maschner 1999). It therefore serves as a possible proxy for other, less durable, and less tangible trade items and ideas – including botanical products and techniques related to plant resource use – that were passed through similar venues from group to group or transmitted across long distances by adventurous individual travellers.

Obsidian has been found at scattered Columbia Plateau archaeological sites dating to at least as early as about 10,000 years ago, with a higher frequency in southern than northern sites, presumably because of closer proximity to volcanic mountain sources (Kirk and Daugherty 2007). Obsidian from the ancient Olcott sites of the Cowlitz River drainage has been traced to five different sources in Oregon (Kirk and Daughtery 2007). At the opposite end of the study area, outside of On Your Knees Cave on the northern tip of Prince of Wales Island, Alaska (where the remains of a young man from about 9,800 years ago were found, as described in chapter 2), obsidian in an archaeological context was traced, at least tentatively, to the vicinity of Mount Edziza near the Stikine River on the mainland about 200 kilometres away, indicating the likelihood of a well-developed coastal trade network even at that early date.

Distribution and archaeological contexts of obsidian and other valued materials may also reflect broader patterns of status, social organization, and household production among these early peoples, determined to some extent through acquisition and ownership of special products such as the artistically and skilfully produced Clovis points at the Wenatchee apple orchard site (Ames and Maschner 1999; Dubin 1999; Kirk and Daugherty 2005; see also chapter 2). On the other hand, the broad exchange of obsidian – and evidently, of the skills and knowledge surrounding its use – may have served to level cultural differences, making it difficult to identify divergent cultural histories and ecological settings of ancient peoples. In any case, we know that obsidian from eastern Oregon reached Puget Sound, the Gulf Islands, the Fraser Delta, and as far north as Namu at the mouth of the Burke Channel on the central coast of British Columbia in present-day Heiltsuk territory. Obsidian from Mount Bes But'a (Anahim Peak; "Obsidian Peak" in Dakelh) reached the Fraser River to the east, the Pacific Coast to the west, and the Skeena River at Kitselas and Hagwilget to the north (Ives 1990). And obsidian from Mount Edziza, as well as evidently reaching the On Your Knees Cave site on Prince of Wales Island, extended as far north as the headwaters of the Yukon and as far south as Namu (R.L. Carlson 1976, 1994). Trade in obsidian was indeed very ancient but was most intensive between about 6,000 and 4,000 years ago.[3]

Marine Shells in Interior Contexts

Another type of trade good that is durable and whose origins are relatively easily traced is marine shell as it occurs in interior archaeological contexts. For example, at Marmes Rockshelter on the Columbia Plateau, olive shells (*Olivella* sp.), most of them drilled for stringing, are present in many strata, including dozens at the 8,000-year level. A piece of California abalone shell, also drilled as a pendant, was found at this site as well (Kirk and Daugherty 2007). Clam shells, pecten shells, and mussel shells have all been identified from numerous interior sites, in the context of pithouses, and often they had been modified – drilled for jewelry or, in the case of mussel shells, ground into adze blades (Hayden 1997; Kirk and Daugherty 2007). Marine shells probably moved with other goods from the coast up the Columbia and Fraser Rivers and their tributaries, where they were most likely exchanged for interior goods of some type (Ames and Maschner 1999).

Among the most popular and widespread marine shells, both in archaeological and more recent contexts, are dentalium (*Dentalium* spp.) shells, or money shells, originating mainly from the west coast of Vancouver Island, where they live in muddy substrate at a depth of up to 15 metres or more. These shells have been found over an immense range in archaeological contexts – north as far as central Alaska, south to The Dalles and Celilo Falls, and eastward as far as the Great Lakes (Byram and Lewis 2001; Kirk and Daugherty 2007). Shells of red abalone, too, have been widely traded, from northern and central California as far north at least as Kiusta on Haida Gwaii (Gibson 1992; Dubin 1999),[4] perhaps at times via Chinookan intermediaries. It is possible that the Northwest Coast tobacco, or Haida tobacco (closely related to *Nicotiana quadrivalvis*, which was cultivated by Native Californians), originated via the same trade route by which red abalone arrived at Haida Gwaii, maybe through long-distance travel and extended residence of Haida among coastal California peoples like the Yurok (see Turner and Taylor 1972; Turner 2004a). Other plant products from coastal environments, both marine and terrestrial, were probably conveyed along similar trade pathways, with interior plant products being exchanged coastward along the same routes.

Although the origins of obsidian, marine shells, and some other products can be determined, the means of their transport to destination sites are not necessarily revealed. Likely, they travelled in diverse states and by diverse means: both as unfinished raw materials and as finished points, knives, or jewelry; carried by single individuals or small groups of travellers conveying them over long distances from their sites of origin to remote settlements; through "middlemen" at gathering places and trading nodes; or by means of a series of links and transmission points from person to person to person or from group to group to group along rivers and other linear routes. We know that people from all over the region were congregating at fishery sites like Kettle Falls and The Dalles from at least 9,000 years ago (Kirk and Daugherty 2007; see also chapter 2), and it seems likely that fishing techniques used at such places and probably also other practices, such as pit-cooking camas and onion bulbs and leaching acorns in pits (Mathews 2010), disseminated from such nodes of convergence and exchange.

Stone Tool Traditions

The growth and development of ethnobotanical knowledge and ethnoecological knowledge systems – and the ways that these traditions have become mixed and blended over time – may well parallel the development of stone tool traditions within the study area. Four distinct and ancient traditions of stone tool making have been identified by Roy Carlson (1976) and others, reflecting the complexities of cultural exchange and intermixing of materials and technologies – including those involving plants – that must have occurred since the very beginnings of human occupation in the Americas: Pebble Tool, Microblade, Fluted Point, and Lind Coulee (or stemmed point). Each tradition represents a slightly different history and different way of life, mainly related to hunting of some form or another. The Pebble Tool Tradition, dating back possibly 12,000 years or more, is characterized by unifacial pebble tools and leaf-shaped bifaces. This tradition was centred in southwestern British Columbia, with early coastal and river sites at major rapids on the Fraser and Columbia Rivers. Apparently related to knowledge and use of marine resources, such as salmon, and possibly to woodworking technologies, pebble tool use extended along the coast at least as far north as Namu in Heiltsuk territory. This tradition interfaced with the Lind Coulee Tradition at Five Mile Rapids on the Columbia River and with the Microblade Tradition at Namu. Carlson (ibid.) suggests an ancestral relationship between the Pebble Tool Tradition and the Proto-Salish and Proto-Wakashan peoples, who possibly migrated north from western Oregon and northern California after deglaciation or arrived via a coastal route from Beringia or the Aleutians.[5]

The Microblade Tradition is based on a system of creating small cutting and piercing tools by striking or applying directed pressure to carefully prepared nodules of stone such as obsidian. The result is the production of numbers of sharp-edged flakes, or microblades, that can be inserted into wooden or bone handles or holders and used as knives or scrapers. Microblade technology had its beginnings in northern China or Siberia some 30,000 years ago, and it appeared in Alaska and Haida Gwaii around 12,000 years ago and then on the Northwest Coast, at Namu and elsewhere, sometime around 10,000 years ago, spreading up the Skeena River to Kitselas, Hagwilget, and other locations (Johnson 1997). Roy Carlson (1976) suggests that the Microblade Tradition is possibly ancestral to the Na-Dené peoples, including the Tlingit and Dene (Athabaskans), and perhaps to the Haida.

The Fluted Point Tradition extends widely from the eastern side of Bering Strait through Alberta and into the continental United States, Mexico, and South America. Although rare, a few of these points have been found on the Northwest Coast (R.L. Carlson 1976). The use of Lind Coulee large-stemmed chipped-stone projectile points centred in the Columbia-Snake drainage of interior Washington and Oregon but extended southward into the Great Basin, eastward and northward into the Kootenays, the upper Thompson River drainage, and possibly the upper Fraser River drainage, and westward to the coast at the mouth of the Columbia River, eventually spreading throughout Puget Sound and the Gulf of Georgia, and as far north as Namu (ibid.).

From these descriptions, it is evident that a general blending and overlapping of these stone tool traditions – which must have come through intergroup contact, intermarriage, or other types of social linkages and exchanges or through simultaneous occupation of territories – took place over the millennia during and following the Pleistocene. By about 5,500 years ago, these traditions, and the cultures in which they were embedded, were becoming less distinct and more integrated. It seems that people of diverse ancestries and histories were becoming more alike, and this convergence was probably due to interactions between them – as reflected in the obsidian trade, which was at its peak over this time – leading to mutual acculturation (R.L. Carlson 1976).

Historical Patterns of Dissemination

Another indication of longstanding continuity among peoples along the coast and up the major rivers is the presence of distinctive rock carvings of similar design over a wide region. These carvings are zoomorphic and anthropomorphic, many with outlined circle faces. They occur most commonly on the northern coast: in southern Alaska, up the Nass River at Gitwinksihlkw, up the Skeena River at Kispiox, along Bella Coola Inlet and up the Bella Coola Valley, up the Fraser River as far inland as the Lytton-Lillooet region, and along the Columbia River from Portland and upstream past The Dalles. They seem to represent communication routes of the past, and these same routes would have served to distribute obsidian and other valuable trade goods, along with cultural traits and cultural knowledge – including ethnobotanical knowledge – within and across cultural and linguistic boundaries (Lundy 1976). These carvings are seldom easily dated, but from the wear of wave action on the coastal carvings, and their widespread consistency of form and design, many would have extended back at least to the Mid Holocene, between about 7,000 and 3,500 years ago (Ames and Maschner 1999; Lundy 1976). The rock carvings are apparently related and show similar traits to the stone sculpture complexes of the Fraser River, Gulf of Georgia, and Columbia River systems, possibly representing an early Northwest Coast art form, with a style also closely related to the later wood sculpture of the Kwakwaka'wakw and other northern tribes (Duff 1956; Lundy 1976). The proliferation of these rock carvings, starting six or more millennia ago, seems to indicate an increasingly spiritual connection to lands and resources, with ceremonial and ritual activities influencing other aspects of life (Lundy 1976).

Stone and shell artifacts, rock carvings, oral histories, and cultural and linguistic traits held in common across divergent groups all stand as evidence of vast networks of communication and interchange – across oceans, along freshwater lakes and rivers, and by land along valley bottoms, river terraces, ridges, and mountain passes. Perhaps the most obvious and enduring orientation for trade was between the Pacific Coast and the interior along the major rivers, some of which penetrated as far eastward as the Rocky Mountains. For thousands of years, people have routinely crossed through the coastal ranges via well-worn trails, known as "grease trails," bringing products such as dentalium and oulachen grease – goods that set the standard for exchange values throughout the region. Oulachen grease, or simply "grease," which was rendered from a small type of smelt (*Thaleichthys pacificus*), was widely recognized as the "cornerstone of a vast northwest trade network" and was certainly among the most important trade items, particularly in British Columbia and southeastern Alaska (Byram and Lewis 2001, 151).[6]

It is likely that the uses of plants and plant products for food, materials, medicines, and ceremonial purposes, along with the knowledge and technologies around plant use, converged upon and were spread by the same routes as the treasured oulachen grease and the durable products like obsidian and dentalia that are more easily found in the archaeological record. Evidence for this begins to show in the transmission of names of plants from one language group to another (chapter 3), in the stylistic similarities and differences in basketry, cooking, and preservation (chapters 5 and 6), in management practices such as landscape burning (chapter 11), in stories (chapter 12), and in the ceremonial use of plants (chapter 13). Of course, these exchanges are mediated by cultural variation and by relative similarities and differences both in skills and technologies and in plant species' abundance, range, and diversity. A given plant resource may occur in one area, but if it is less abundant or less productive, it may also be acquired through trade or through permitted access for harvesting in another people's territory. For example, some of the stands of old-growth western redcedars of Haida Gwaii produce superior dugout canoes, and others, such as particular stands at Naden Harbour, were ideal for large bark roofing sheets, flattened with salmonberry sticks. Haida carvers and bark harvesters from these areas were likewise famous for their skills in constructing products of excellent quality. For this reason, although western redcedar also grows on the mainland coast, the canoes and bark sheets from Haida Gwaii were widely sought by the Ts'msyen, often in preference to their own, and were also traded northward to the Tlingit (Turner 2004a; Turner and Wilson 2008).

Oulachens and dentalium themselves, however, were also inextricably connected to knowledge and use of plants. Oulachens are traditionally harvested using wooden weirs constructed of hemlock poles and boughs, traps of root and wood, stinging nettle nets, and cedar-wood dugout canoes (Stewart 1977; Clan Chief Adam Dick, pers. comm., 2008). Dentalium shells are obtained through the use of an ingenious implement featuring the shrub hardhack (*Spiraea douglasii*) (see chapter 6). Furthermore, these shells would have been carried in baskets or bags or else strung on nettle fibre or cedar-bark twine. Oulachen grease was rendered in wooden boxes and then stored and transported in cured bottles and hollow tubes of bull kelp or in bentwood cedar boxes. There are now-classic photographs taken by Harlan I. Smith and other early ethnographers of men, women, and children carrying these boxes full of grease, by means of a woven tumpline across the forehead, up the steep trail at the headwaters of the Bella Coola River, along the "grease trail" route to the Interior Plateau, through a place called "Kleena Kleene" (after *tł'ina*, the Kwak'wala name for this grease), and far beyond, to the Chilcotin region and the Fraser River. The photographs are relatively recent, but the "grease trails" long preceded the Europeans' arrival on the Northwest Coast (Ames and Maschner 1999; Birchwater and Ulkatcho and Nuxalk elders 1993; Byram and Lewis 2001; Cove and MacDonald 1987; Tepper 1991).

EXCHANGE AND RECIPROCITY

There are many forms of reciprocation or mutual exchange for mutual benefit. "Trade" is one form of such exchange and is often considered a strictly economic transaction, in which goods or products desired by one individual or group are obtained from another in return for other goods or products that are thought to be of corresponding value. In relation to ethnobotanical knowledge and goods, however, this is a rather narrow interpretation that does not cover the full range of activities involved in reciprocal exchange systems. Not only do plants and plant products fit into a broader matrix of "trade goods" that included all kinds of animal products as well as rocks and minerals, but all such tangible objects are also only part of the story of cultural reciprocity that occurs in societies such as those of Indigenous peoples of northwestern North America. Limiting a discussion of exchange systems to the realm of material, physical goods used in "barter" restricts a full understanding of both social and ecological aspects of exchange. A broader perspective is necessary, reflecting less tangible possessions such as names, crests, narratives, songs, ceremonies, and medicinal plant recipes, as well as the role and influences of kinship, political alliances, wealth, status, and obligations of leadership. Also relevant are social organization, seasonal movements, and activities related to control, management, and enhancement of resources. Ultimately – through the medium of people's relationships with the other entities of their universe – worldviews, philosophies, and belief systems are also inextricably tied to reciprocal exchange systems.

Trade and exchange can take place between neighbouring families and communities or over very long distances, where individuals or small groups embark on major expeditions for trade, exploration, or other reasons. Such situations are known from recent times, and it is reasonable to assume that there were long-distance travellers in ancient times, too. Such people may have forged more enduring connections across communities that ultimately resulted in more permanent trade routes and networks. There were many different patterns involving exchange. One example is single individuals or small groups carrying goods or holding specific knowledge or skills who might travel from community to community along a given trajectory, perhaps a coastline or a major river, while stopping for periods of time, even overwintering or residing for a year or two at a community as visitors (Decosse 1980). Such extended stays of visitors also provided important opportunities for individuals of two or more different communities to meet on the common ground of a third community. As described previously, both on the coast and in the interior, there existed major nodes of trade and exchange where different peoples converged at given times, especially around harvests of salmon, oulachens, roots, or berries. Such meetings would have resulted not only in the trade and purchase of diverse resources but also in the forging of social connections, intermarriage, invitations for visiting, and the exchange of stories, songs, and other nontangible "goods." It was through these kinds of meeting places that much learning and borrowing of plant names and disseminating of stories, techniques for processing plants, and so forth likely occurred.

Examples of exchange nodes include upland locales like Indian Heaven, the famous huckleberry-picking grounds of the Mount Adams region (Kirk and Daugherty 2007), Potato Mountain in Tsilhqot'in territory (Minnie Charleyboy, pers. comm., 2003), Green Lake in Secwepemc territory, and Botanie Valley in the territory of the Lytton band of the Nlaka'pamux (Teit 1909; Turner, Deur, and Mellott 2011; Turner, Thompson, et al. 1990).[7] Teit (1900) reported that, as of the late nineteenth century, as many as 1,000 people congregated annually at Botanie, not only from Nlaka'pamux communities but also from Secwepemc, Stl'atl'imx/St'at'imc, and others. Women would harvest roots and pick berries on the slopes of the valley, men would hunt, and everyone would trade and socialize. Nlaka'pamux elder Louie Phillips recalled that women would sometimes bet entire sacks of yellow glacier lily bulbs they had dug on the outcome of horse races in the valley – yet another type of exchange mechanism (cited in Turner, Thompson, et al. 1990). Riverine sites such as Kettle Falls, The Dalles, and Celilo Falls on the Columbia River, the mouths of the Wenatchee and Okanagan Rivers, the confluence of the Thompson and Fraser Rivers at Lytton, "The Fountain" (Xaxl'ep) farther up the Fraser, various sites on the lower Fraser, Namu on the central coast, and the lower Skeena and Nass Rivers are other major sites of longstanding cultural convergence where thousands of people gathered and goods from all of the surrounding areas would be exchanged (Ames and Maschner 1999; Hayden and Schulting 1997; Kirk and Daugherty 2007; Mitchell and Donald 1988).

Chain-style and network-style connections for trading and exchange of goods and knowledge also occurred widely, in which goods and knowledge from one community would pass through an intermediary community to a third community and, eventually, to other communities at higher orders of separation. Some of the major trading centres in northwestern North America were linked to trade routes that extended south to California, east to Montana, north to the Yukon, and ultimately, to the Great Lakes and the east coast of North America. Over the course of its dissemination, especially across language and ecological boundaries, knowledge can become changed or transformed to accommodate different values and perspectives, as well as differences in range and distribution of plants. Plants names, and technologies such as pit-cooking involving plants, for example, might shift to incorporate or apply to more salient species and techniques in different local and cultural situations.

Trading and Gifting

Trading and gifting are not easily separated because gifts invariably carry implications for the receiver as well as the giver, at the very least resulting in the attribution of status to and appreciation for the giver and in recognition of the receiver.[8] There may also be formal or informal expectations that the giver and his or her community will receive backing, assistance, or cooperation from the receiver at some time in the future (Decosse 1980; Mauss 1990). For example, the one giving a gift may receive favours from the recipient by being hosted or provisioned at some other time; as Suttles (1951b, 27) explains in relation to the Katzie as hosts of their Halkomelem kin, "A host at one time and place is potentially a guest at another" (see also Suttles 1987a, 1987b).

Mitchell and Donald (1988) maintain that an early, sophisticated system of trading – in the narrow economic sense – may have existed among Northwest Coast peoples. However, even economically motivated transactions and systems like the *pessa* investment system of the Kwakwaka'wakw (Reid and Sewid-Smith

2007; Recalma-Clutesi 2007) could not operate without the matrix of social relationships, inheritance, and cultural protocols that is well documented for the Northwest Coast potlatch complex (Drucker 1951; L. King 2004; Trosper 2002, 2003). Generosity in both economic and cultural realms had high social value, and a leader who was generous to his people and his neighbours was more likely to maintain enduring internal support and strong social alliances. This affirmation, in turn, not only enhanced his people's status and his own political survival but might also mean the difference between death and continued existence for his people in times of adversity. For example, strengthened social ties between communities might allow reciprocal access to a people's hunting or gathering grounds, leading to enhanced food security. Gifting and generosity thus served as a form of insurance policy.

Sometimes a gift might be in recognition of a past favour, such as a rescue or providing caregiving to a family member; such deeds were important to acknowledge clearly and publicly. Gifts of acknowledgment might be in the form not only of objects but also of names, songs, or rights to resources. For example, Gitga'at Eagle chief Ernie Hill Jr described how his grandfather, Ambrose Robinson, had cared for an elderly Eagle chief who was ill. In appreciation, the chief bequeathed to Ambrose the rights to a certain highbush cranberry patch on Gribble Island, across from Hartley Bay, that he had held. Ambrose Robinson assumed ownership of the patch and was able to bring women from the community over to pick highbush cranberries.[9] When Ambrose passed away, the rights to this patch were returned to the Eagle clan through bequeathment to Chief Ernie Hill himself (Turner and Thompson 2006).

This type of reciprocity was highly complex, extending across clan boundaries and between generations and involving rights to resources (in this case, a highly valued fruit) and territory, social organization, a chief's status and responsibilities, and social cohesion. Furthermore, gifts of food and other valued products were often given for those invited to serve as witnesses to important events such as the transfer of names and titles on the occasion of a feast or potlatch.[10] Gifts might also be given to clan members who assisted by providing food or undertaking tasks requested of them by members of an opposite clan for a feast or ceremony they were hosting.[11]

Similar situations of reciprocity and generosity, translating into social capital, can be seen throughout the entire study area. Another example is from Nisg̲a'a elder Amelia Morven (Niik^{w}'iltlk^{w}) (cited in Wilp Wilx̲o'oskwhl Nisg̲a'a 1995, 48), who describes how the owners and proprietors of specific resource-harvesting areas would give permission for others to harvest in their area once their own needs had been met:

> Whenever it is time to harvest food at one's *ant'aałkw* (place from which any type of edible produce from the ground is harvested) one would first

satisfy his own means, then open it to others. The expression goes, '*Way łaa nii joḵtgat*' meaning: 'one is now residing at one's *ant'aałkw*'; and is an indication that others may go to that place when invited. That is what my father would do. He owned a huge soapberry (*ʔis*) picking ground, and when he himself had taken enough he would then head up to Aiyansh and inform the people that he was now living at his *ant'aałkw*. Many canoes would arrive. I was not ten years old but I remember. Doing this made them very happy, to be able to share their resources with others, as it was not customary for people in those days to hoard their fortune.

In the same Nisg̱a'a volume (ibid., 79), Robert Moore (Niisx̱bakhl) recounts, "There are not too many places where soapberries grow. When the *wilp* ['house'] get enough then the chief goes to the villages and announces that they have enough soapberries and invites others."

Having broader opportunities to access resources was also important at times when people had to forego harvesting certain resources because of scheduling conflicts. For the Stl'atl'imx/St'at'imc communities along the Fraser River, for example, some of the best runs of sockeye and other salmon occurred at the same time as some of the prime berries, like black huckleberries, were ripening at higher elevations. In this case, those families who spent more time fishing down at the river could exchange some of their wind-dried fish for berries picked by those who had spent their time at camps in the mountains picking berries and digging roots (Turner 1992b).[12] Sometimes, too, there were environmentally and geographically mediated differences in resource accessibility between communities that promoted convenient exchange systems. The Ditidaht villages of Clo-oose and Whyac on the west coast of Vancouver Island are a case in point. Fishermen from Clo-oose had access to the best halibut fishing grounds, whereas Whyac was situated beside the Nitinat Narrows at the entrance to Nitinat Lake,[13] where people strung duck nets of stinging nettle and other plant fibres and then, using their canoes, drove the ducks from the lake into the nets. Therefore, halibut from Clo-oose were routinely traded for "black ducks" (scoters) from Whyac (Turner, Thomas, et al. 1983).

Historic rights to access and harvest in others' territories were often widely recognized and longstanding. Hul'qumi'num elder Dr Arvid Charlie (Luschiim) (pers. comm., 1999) recalled that the Quw'utsun' (Cowichan) of Vancouver Island not only had relations whom they visited in the Fraser Valley but, for countless generations, also owned a village site (called Tl'uqtinus) near Canoe Reach and several other sites near New Westminster, where they would go to fish and where they harvested wapato (called *sqáwth*) as well as cattail leaves (*stł'áʔqən*) and tule stems for their mats.[14] The Quw'utsun' also had rights to cranberry (*Vaccinium oxycoccos*) harvesting in Katzie territory (Suttles 1951b). According to Teit (1906a, 232), the Lower Stl'atl'imx/St'at'imc, when they were trading with their Sechelt, Squamish, and Comox neighbours at Jervis Inlet or Howe Sound, "were allowed to pick berries, and to hunt and fish, as much as they liked." Similar access rights were accorded by the Straits and Hul'qumi'num Coast Salish of southern Vancouver Island to the Ditidaht families for harvesting camas in their territories around Victoria (Turner and Kuhnlein 1983; Turner, Thomas, et al. 1983). Likewise, the Hanaksiala of the Kitlope Valley and Haisla of Kitimaat had longstanding ties with the Kitasoo/Xai'xais and Gitga'at Ts'msyen, and for generations they have shared the use of the Kitasoo and Gitga'at seaweed and halibut spring harvesting camps. Conversely, the Ts'msyen peoples have long been welcomed to the oulachen camps of the Haisla and Hanaksiala (Helen Clifton, pers. comm., 2005; Compton 1993b), often bringing dried seaweed and halibut with them as a reciprocal gift. Rights to fish in the major oulachen rivers of Kwakwaka'wakw territory – the Kingcome and Knight Rivers – were also accorded to various tribes from Vancouver Island and the Broughton Archipelago (Curtis 1915), and these rights would be reciprocal, as the Tsawataineuk clans from Kingcome, for example, had clam- and seaweed-harvesting rights in the Broughton Archipelago locations (Clan Chief Adam Dick, pers. comm., 2009; Recalma-Clutesi 2005). As Tagish elder Dora Wedge noted, "Everybody shared together. If somebody picked different kind of berries, and you haven't got it, they'd give you some. And maybe they want something, they haven't get something you got. Well, they just trade around, no fights, nothing, everybody got along" (cited in Greer 1995, 91).

Aggression, Encroachment, Amalgamation, and Acquisition of Knowledge and Resources

The opposite of generosity and amicable sharing – of gifting goods and distribution of wealth, of creating friendships and alliances through trade – was warfare and aggression. These situations could also result in the acquisition of goods, and even in knowledge exchange, but without the social goodwill. Wars may be fought and raids launched for many different reasons, but in at least some cases, wars resulted from a breakdown of peaceful trade relations. Ames and Maschner (1999) note, for example, that a number of intense wars were fought over the spectacular oulachen runs of the Nass River, particularly between the Ts'msyen and Tlingit, and as a result, the Ts'msyen eventually pushed the Tlingit northward (see also Cove and MacDonald 1987; and Matson and Coupland 1995). Slave acquisition was one outcome of raids and wars, frequently described in narratives and undoubtedly a major means of transfer of botanical knowledge along with other types of knowledge (see chapter 9). Sometimes warfare, or the threat of warfare, would ultimately result in a peacefully negotiated settlement, which might – for example, in the Ts'msyen-Tlingit case – include acceded access to resources or territory. Furthermore, slaves taken in raids, and at times transported hundreds of kilometres away from their homes, were sometimes repatriated through payment of ransom goods or through purchase at major trading centres like "The Fountain" near Lillooet (Teit 1909). Thus even warfare was not necessarily a straightforward acquisition of goods on the part of victors but involved some complexity in the ways that knowledge and property were transmitted and acquired. Sometimes, too, latent knowledge or delayed transmission of knowledge might occur in the case of slaves, for example, where it might be necessary to learn another language in order to participate more fully in knowledge exchange. It might even take a full generation or more for information, names, stories, and so forth around plants, food preparation techniques, or medicinal plant use to be passed from one community to another.

Alliances formed through trade might also influence the course of aggression, as in the account from Cowichan elder Simon Charlie and his son Dr Arvid Charlie (Luschiim) (pers. comm., 1999, 2003), who described a long-distance trading relationship that the Hul'qumi'num had long ago with the Secwepemc of the Kamloops area:

> We went up there [around Kamloops] every so often, whatever it was, four years, eight years, for trade. So, one time we got up there, and they had nothing. They apologized. "We don't have nothing. We just got raided. Took all our trading stuff away, and a lot of our people, our ladies and our children." So, overnight, of course, people had a gathering, wondering what to do. They decided to go make war on the enemies of those people. So, next day, they went out, raided that village. Got all the people back, ladies and children, and all the goods that were stolen plus a bunch of others. Yeah. So, that's their story. It was George Manuel [Secwepemc leader] that was telling us. He's gone now, but he learned it from his ... grandfather or a great grandfather, from his elder anyway ... Yeah. Pre-contact. One of the favourite tools, or favourite things, we traded for I understand was the rock from what is now called Alberta, a certain kind of a hard rock.

Another type of situation in which knowledge and rights to goods might be conveyed across linguistic and cultural boundaries would be in cases of encroachment, assimilation, or takeover of one group and its territories and resources by another. This could be a hostile intrusion, as described previously, or a more benign amalgamation, even by invitation of the original group if their numbers were depleted for some reason. Regardless of the cause, there are known elements of cultural knowledge exchange between groups in situations like this. For example, it is suggested that the relatively high number of plant names from Straits and Lushootseed Salish that were borrowed into Ditidaht – as well as a few Ditidaht names in these Salishan languages – might indicate the possibility of Ditidaht expansion into Salish territory at some time in the past (Hess 1987). Similar situations – for example, between peoples speaking Pentlatch, Comox, and Legwildaxw Kwak'wala on the east coast of Vancouver Island, between Salishan peoples and the Nuu-chah-nulth in the Port Alberni area of the west coast of the island, between the Haida and Tlingit on Prince of Wales Island, and between northern Interior Salish peoples and the Tsilhqot'in and other Dene (Athabaskan) groups – may have resulted in the mutual sharing of botanical names, stories, and other cultural information. Some of the situations of occupancy of new territory and resultant acquisition and exchange of knowledge may have taken place at times when the original host population was in decline and therefore did not require as many resources as its territory produced or, conversely, when the original community enjoyed a bounty of resources that enabled it to accommodate more people.[15]

A more recent example of historical movement and settlement is with the migration of Secwepemc chief Kenpesq'et from the North Thompson River, along with fifty or sixty friends and relatives, to form a new community, Kinbasket, at the head of the Columbia River, within the territory of the Ktunaxa. Chief Kenpesq'et and his followers knew the country well, having travelled there previously on hunting trips, mostly by way of the Canoe River. Once in the region, they formed an alliance with the Stoney people of east of the Rocky Mountains, who routinely travelled to the Columbia River in the fall to fish for salmon and pick berries. These two groups supported each other in the event that the Ktunaxa tried to drive them away (Teit 1909, 467). This territorial shift, combined with a transmountain alliance and settlement of an area previously occupied by peoples of yet another culture and language, would have provided many opportunities for exchange and acquisition not only of botanical products and goods but also of plant names, stories, and other types of knowledge. It is likely that the use of both sweetgrass and Canby's lovage, for example, entered the knowledge systems of the Secwepemc and other Interior Salish peoples through this venue (see Turner, Bouchard, and Kennedy 1980; and Turner, Thompson, et al. 1990).[16] The intersection of Secwepemc, Ktunaxa, and Stoney cultures in the Rocky Mountain Trench provides a good example of how knowledge and practice might become similar between two groups of people through mutual association with a third group, with common borrowing – and trading – from the same source.

A similar situation facilitating multicultural, multilanguage exchange evidently occurred within the Bella Coola Valley, in the territory of the Nuxalk Salishan peoples. At some time in the past, Dakelth Ulkatcho families (Dene, or Athabaskan) from the Interior Plateau started travelling seasonally into the Bella Coola Valley to co-reside with their Nuxalk (Salishan) neighbours in the fall and winter (Tepper 1991). Evidently, salmon were so plentiful in the Bella Coola River and its tributaries that the Nuxalk did not mind sharing them with the Ulkatcho. The Ulkatcho were also pleased to enjoy Bella Coola oulachen grease, carried up the valley and onto the Plateau over the "grease trail" in cedar-wood boxes. In terms of botanical products, the Ulkatcho obtained wood and bark of yellow-cedar and western redcedar and boxes of highbush cranberries and crabapples preserved in oil, among other goods, from the Nuxalk. The Ulkatcho also had relationships through marriage to the Tsilhqot'in peoples, and some families travelled annually to Potato Mountain in Tsilhqot'in territory to harvest spring beauty corms, or mountain potatoes, yellow glacier lily bulbs, and other resources (Minnie Charleyboy, pers. comm., 2003; Birchwater and Ulkatcho and Nuxalk elders 1993; Hebda, Turner, et al. 1996; Turner 2004b). In return, the Nuxalk benefited in receiving valuable interior goods brought to the valley either from Ulkatcho territory or through the Ulkatcho homeland from the Tsilhqot'in (Chilcotin): obsidian from Mount Bes But'a (Anahim Peak), beautifully tanned buckskins, dried "roots" of mountain potato and yellow glacier lily, large quantities of dried cakes of soapberries and saskatoon berries (Teit 1909), and wolf lichen (*Letharia* spp.) to use as a yellow dye. The Nuxalk also sometimes used, and learned how to construct, baskets and canoes of birch bark, although they were never as skilled at making these vessels as their Dene (Athabaskan) neighbours (McIlwraith 1948; H.I. Smith 1920–24; Turner 1973).

On the coastward side, the Nuxalk were bounded by Wakashan-speaking peoples (Haisla and Hanaksiala to the north, Heiltsuk to the west, and Kwakwaka'wakw to the south), and they hosted Wakashan visitors from the outer coast, most especially the Heiltsuk, who brought edible red laver seaweed, herring eggs on giant kelp fronds and on western hemlock boughs, clams, mussels, and other marine products to exchange for Nuxalk salmon and some of the interior goods (McIlwraith 1948). Some Heiltsuk women were skilled weavers of mountain goat wool and were hired by the Nuxalk to make blankets, some of them dyed with wolf lichen from the interior (Willie Hans, pers. comm., 1984; H.I. Smith 1920–24). The Nuxalk also travelled to visit, and intermarried with, Hanaksiala from the Kitlope Valley and were said to have introduced springbank clover and Pacific hemlockparsley to that valley (Ken Hall and Simon Hall, pers. comm., 1994, 2007; Compton 1993b; Edwards 1979). Some Nuxalk and Heiltsuk also customarily travelled into the territory of the Qʷiqʷsut'inew tribe of the Kwakwaka'wakw at Haada in the Broughton Archipelago to harvest high-quality rhizomes of springbank clover and other edible roots from their *t'əkkillakʷ* ("estuarine root gardens"), with permission of their host community (Deur 2002b; Deur and Turner 2005; Reid and Sewid-Smith 2004).

As well as numerous different botanical products and other goods, the Nuxalk-Wakashan-Dene network resulted in considerable cultural exchange, linguistic borrowings, and the sharing of narratives, names, and other traditions. The Nuxalk evidently acquired their rich and complex ceremonial traditions, including elaborate winter dances, feasts, and potlatches, at least in part from their Wakashan neighbours (Kennedy and Bouchard 1990). The Nuxalk credit the Ulkatcho and Tsilhqot'in peoples for excellent tracking and hunting skills. Although the Nuxalk traded and intermarried with their various neighbours, they also engaged in war with them on occasion (ibid.). One such time was when a sacred ceremonial object was evidently stolen from one of the Nuxalk women root diggers visiting Haada to obtain springbank clover and silverweed roots. In revenge, the Nuxalk launched an attack on the $Q^{w}iq^{w}$sut'inew, in which men, women, and children were massacred and some taken as captives, including the great great grandmother of Kwakwaka'wakw historian Dr Daisy Sewid-Smith (Mayanilth) (Reid and Sewid-Smith 2004).[17]

Although some intercommunity and internation reciprocal relationships evidently endured over centuries, and possibly even millennia, the diverse occasions of exchange and interaction would certainly have fluctuated over time and space. There may have been pulses of economic and social activity between certain communities or nations at some times, achieved through strong alliances or marital connections, for example. Then these activities might have faded as these relationships broke down, possibly to be replaced by others. In a similar vein, there may have been regional "hot spots" for trade – particular locations coinciding with specific resource peaks – which were then dissembled over time, perhaps being shifted to other sites that became, in turn, more productive. In the following section, the historical development of relationships involving trade and exchange in northwestern North America is discussed.

PLANTS IN TRADE

Starting in the latter part of the Mid Holocene – at a time when there was also evidence of cultures more easily recognizable as ancestral to the First Nations of modern times – trading intensified and started to extend its reach until at least around 1,500 years ago. Already, from the Mid Holocene, culture patterns based on wealth and specialization were becoming evident, with obvious widespread exchange of ideas and innovations (see chapter 9). Populations were growing, food surpluses in the form of stored goods were accruing, and individuals were focusing on particular occupations, many of which involved increasing expertise and knowledge around using plants. Luxury goods – including tools like nephrite adzes[18] that had an obvious connection to woodworking and carpentry as well as to status, wealth, and social stratification – are seen in many archaeological contexts, from southeastern Alaska to Oregon.

By about 3,000 years ago, the development and exchange of perishable items becomes more visible, with the archaeological remains in wet environments revealing fishnets, plaited and twined baskets, bentwood fishhooks, wooden bowls, and wooden wedges that reflect regional centres of development and interchange, such as the lower Columbia River, the lower Fraser Valley and

Gulf of Georgia, the west coast of Vancouver Island and the Olympic Peninsula, Prince Rupert Harbour and the lower Skeena River, and the central coast around Namu. Connections between The Dalles on the Columbia and the Interior Plateau are also suggested in the styles and types of tools from this time (Ames and Maschner 1999; R.L. Carlson 1976; Kirk and Daugherty 2007). Through evidence from sites like these, more clearly recognized are not only the trading of plant products but also the role of plants, botanical products, and access to plant resources as gifts, as tributes to leaders, as payment for services, and in ceremonial exchange. This section focuses on known exchange patterns for plants and ethnobotanical knowledge.

Table 10-1 lists over sixty-five plant species known to have been involved in trade or exchange in some way within the study area. Most of these are foodstuffs or materials used in technology, as well as manufactured goods; a few are medicines. Most trade was likely in preserved or processed products (dried berries, dried roots, bundles of split leaves, or roots for basketry), which were generally lighter and more practically transported than their fresh or unprocessed forms. Everyday commodities, as well as items of high value and rarity, were used in trade, gifting, and exchange. Immense quantities of plant resources were acquired or changed hands through trade and were distributed widely during ceremonial occasions, increasing the wealth and prestige of communities and their leaders (Boas 1921; Deur and Turner 2005; Turner and Loewen 1998). Given their high status value, there would have been strong incentives to enhance and intensify these plant resources, to promote their productivity, and to increase the efficiency of their production through habitat management and technological development. Surplus high-quality botanical resources from managed landscapes – such as springbank clover rhizomes produced at the Haada *t'əkkillakʷ* ("estuarine root gardens") in the Broughton Archipelago – provided opportunities for communities to host harvesters from other regions and to provide quantities of these resources for feasts and gatherings (Deur 2002b; Reid and Sewid-Smith 2004).

Just as in economic systems of the present, peoples of the past often "added value" to the goods they traded, and they standardized the quantities and dimensions to facilitate exchange. Adding value consisted of drying and bundling or carefully stringing foodstuffs like root vegetables in regular sized units, as well as preparing standard-sized containers for cooked, dried camas bulbs, wapato tubers, and seaweed or berry cakes to increase their overall value. On the Northwest Coast not only did the bentwood cedar box serve as a container for storage and transport, but because it could be produced to standardized dimensions and volume, it also facilitated trade. With these boxes, dried seaweed cakes and dried berry cakes could be produced in squares of similar size, and the crabapples, highbush cranberries, and other foods stored in these boxes would be of a standard volume for consistency in value. Even seaweed-drying trays and

Table 10-1 | Plant resources known to have been traded among Indigenous peoples of northwestern North America

Species/material traded	*Notes*
Edible seaweed, red laver (*Pyropia abbottiae*[a] and related spp.; syn. *Porphyra*)	Important item of trade along entire Northwest Coast (especially central and northern) and from the coast inland (e.g., up the Skeena River); sold by Saanich and Nuu-chah-nulth to Oriental buyers from Victoria
Black tree lichen, "tree moss," or *wíla* (*Bryoria fremontii*[b])	Cooked, dried cakes traded from Interior Salish to Coast Salish
Wolf lichen (*Letharia vulpina*[c])	Traded as dyestuff among interior groups and from interior to Northwest Coast (e.g., to Chilkat Tlingit and Nuxalk at Bella Coola), especially for dyeing mountain goat wool
Indian paint fungus (*Echinodontium tinctorium*[b])	Traded from Tlingit to Dene (Athabaskan) groups
Yellow-cedar (*Chamaecyparis nootkatensis*[a])	Bark and bark products widely traded along Northwest Coast and into interior, especially in Chilkat blankets and robes and in Kwakwaka'wakw robes; wood for bows traded from Coast Salish to interior; wood for ceremonial articles traded from northern coast to interior
White spruce (*Picea glauca*[b])	Gum traded from Dene (Athabaskan) groups to the Tlingit for chewing
Sitka spruce (*Picea sitchensis*[a])	Root baskets and hats traded from Tlingit to Dene (Athabaskan) groups and from Haida to Ts'msyen; possibly traded to Ozette and other locations farther south along coast
Whitebark pine (*Pinus albicaulis*[a])	Seeds traded from Upper to Lower Nlaka'pamux and probably from Upper to Lower Stl'atl'imx
Ponderosa pine (*Pinus ponderosa*[b])	Dugout canoes traded among Nlaka'pamux
Pacific yew (*Taxus brevifolia*[c])	Wood and bows commonly traded from Northwest Coast to interior throughout range
Western redcedar (*Thuja plicata*[a])	Cedar-wood dugout canoes widely traded along Northwest Coast (especially from Nuu-chah-nulth and Haida to Makah and Ts'msyen respectively) and into interior; cedar-wood boxes traded along Northwest Coast, generally from north to south and from coast to interior; bark and bark baskets, mats, robes, and ceremonial articles traded widely among Northwest Coast peoples and from coast to interior; roots and split-root baskets traded among Interior Salish and Columbia peoples and from interior to coast
Western hemlock (*Tsuga heterophylla*[c])	Dried cambium inner bark food traded from Ts'msyen to Nisg̲a'a and from Haisla and Hanaksiala to other coastal groups

Table 10-1 | continued

Species/material traded	*Notes*
Vine maple (*Acer circinatum*[b])	Wood for bows traded from Northwest Coast to interior and southern BC
Rocky Mountain maple (*Acer glabrum*[b])	Wood traded from Ts'msyen to Haida; snowshoes traded among Interior Salish peoples
Nodding onion (*Allium cernuum*[b]) and related species	Bulbs traded among Nlaka'pamux; not generally eaten by coastal groups but traded in abundance to Europeans
Saskatoon berry (*Amelanchier alnifolia*[a])	Dried berries and cakes commonly traded among interior peoples and from interior to coast, especially central and southern coast
Indian-hemp (*Apocynum cannabinum*[a])	Fibre, twine, and woven products (bags, mats, fishnets) widely traded among Interior Salish, from Interior Plateau to Northwest Coast, and from Coeur d'Alene to Plains tribes
Kinnikinnick (*Arctostaphylos uva-ursi*[b])	Berries traded from Nuxalk to Heiltsuk and Oweekeno and probably elsewhere in BC; dried leaves as tobacco probably also traded
Silverweed (*Argentina* spp.[c])	*A. anserina* roots traded from Upper to Lower Nlaka'pamx; *A. egedii* roots widely traded among Northwest Coast peoples
Big sagebrush (*Artemisia tridentata*[b])	Bark fibre mats and clothing occasionally exchanged among Interior Plateau peoples
Balsamroot (*Balsamorhiza sagittata*[b])	Dried roots traded among Nlaka'pamux and probably Stl'atl'imx
Paper birch (*Betula papyrifera*[c])	Bark, containers, and wood traded among Dene (Athabaskan) and Interior Salish groups and from interior to coast; birchbark canoes traded among Interior Salish, Ktunaxa, and Dene peoples
Edible camas (*Camassia quamash*,[a] *C. leichtlinii*)	Dried bulbs widely traded from Coast Salish of Vancouver Island to western coast, northern coast, and mainland; also traded among southern Interior Salish, Columbia River, and western Washington peoples, as well as north into Interior Plateau
Slough sedge, or basket sedge (*Carex obnupta*[b])	Split, prepared leaves traded from Nuu-chah-nulth to Sliammon (Comox) Salish basket weavers
Mountain mahogany (*Cercocarpus ledifolius*, *C. intercedens*)	Spokan obtained wood blanks and finished bows from Nez Perce made of these woods; made to "last many generations" (J.A. Ross 2011, 279)
Spring beauty, or mountain potato (*Claytonia lanceolata*[b])	Corms commonly traded (usually dried, sometimes on strings or sticks) within and among interior groups; occasionally traded to coast (e.g., to Nuxalk at Bella Coola)
Hazelnut (*Corylus cornuta*[a])	Nuts widely traded among Salishan groups of BC (e.g., from North Thompson Secwepemc to West Secwepemc and Stl'atl'imx) and among Columbia River peoples; possibly

Table 10-1 | continued

Species/material traded	*Notes*
	traded between Fraser River Salish and Skeena River Ts'msyenic peoples
Silverberry (*Elaeagnus commutata*[b])	Inner bark, mats, and bags traded among Interior Salish and from Interior Salish and possibly Tsilhqot'in to Coast Salish
Yellow glacier lily (*Erythronium grandiflorum*[c])	Dried bulbs traded (loose or on strings or skewers) among Interior Salish and from Tsilhqot'in to Carrier and Nuxalk
Oregon ash (*Fraxinus latifolia*[b])	Bowls (or possibly raw wood) traded to Makah from groups to south and/or east
Chocolate lily, or riceroot (*Fritillaria affinis*[b])	Dried bulbs traded from Upper to Lower Nlaka'pamux
Salal (*Gaultheria shallon*[c])	Large quantities of dried berries and pressed cakes traded among Northwest Coast peoples
Sweetgrass (*Hierochloe hirta, H. odorata*[b])	Nlaka'pamux, Ktunaxa, and other First Nations of BC trade for sweetgrass braids with peoples east of Rocky Mountains
Bitterroot (*Lewisia rediviva*[c])	Dried roots commonly traded within and among Interior Salish groups of BC and Washington
Canby's lovage (*Ligusticum canbyi*[b])	Aromatic roots widely traded as medicine between Interior Salish, Ktunaxa, and other groups
Tiger lily, or Columbia lily (*Lilium columbianum*[b])	Dried bulbs traded from Interior Salish to Coast Salish
Biscuitroot (*Lomatium cous*,[c] and other spp.)	Dried roots traded within Interior Plateau groups, especially from southern to northern plateau (e.g., Okanagan to Secwepemc)
Desert parsley (*Lomatium macrocarpum*[b])	Dried roots traded from Upper to Lower Nlaka'pamux and among Interior Salish peoples
"Wild celery," or barestem lomatium (*Lomatium nudicaule*[c])	Seeds from southeastern Vancouver Island traded to western and northeastern coast of island; greens traded among Nlaka'pamux peoples
Pacific crabapple (*Malus fusca*[a])	Fresh and preserved fruits in water or oulachen grease widely traded along Northwest Coast and from coast to interior
Wild tobacco (*Nicotiana attentuata*[c])	Traded among Interior Salish and Ktunaxa
Haida, or Northwest Coast, tobacco (*Nicotiana quadrivalvis*[c])	Traded from Haida and Tlingit to Ts'msyen, from Metlakatla Ts'msyen to Nisga'a, and from Yuroks to interior peoples
Western sweet cicely (*Osmorhiza occidentalis*[b])	Aromatic roots widely traded as medicine between Interior Salish, Ktunaxa, and other groups

Table 10-1 | continued

Species/material traded	*Notes*
Reed canarygrass (*Phalaris arundinacea*[b]) and possibly common reed grass (*Phragmites australis*[b])	Traded as basket decoration material, and in finished cedar-root baskets, among Interior Salish peoples and from interior to coast in the south
Mock-orange (*Philadelphus lewisii*[b])	Wood for arrows and needles traded by Upper Stl'atl'imx to Secwepemc
Cottonwood, or balsam poplar (*Populus balsamifera*[c])	Dugout canoes traded among Interior Salish peoples; wood traded from Ts'msyen to Haida
Bitter cherry (*Prunus emarginata*[b])	Bark traded from groups of southern coast to central coast; bark as basket decoration exchanged among Interior Salish
Chokecherry (*Prunus virginiana*[c])	Dried cherries traded among Interior Salish and coastward to Fraser Valley
Garry oak (*Quercus garryana*)	Acorns commonly traded along Columbia River; probably an ancient trade product
Labrador tea (*Rhododendron groenlandicum*) and trapper's tea (*R. neoglandulosum*[b])	Dried leaves traded among various interior peoples; probably traded among coastal peoples as well
Currants, wild (*Ribes* sp.[b])	Berries traded (probably dried) among Interior Salish peoples
Cloudberry, or bakeapple (*Rubus chamaemorus*[b])	Berries preserved in water traded from Ts'msyen to Haida
Thimbleberry (*Rubus parviflorus*[b])	Sprouts exchanged locally among Nuu-chah-nulth; taken to halibut fishing grounds for sharing and trading
Salmonberry (*Rubus spectabilis*[c])	Berries traded from Halq'emeylem to Lower Nlaka'pamux and elsewhere along coast; sprouts exchanged locally among Nuu-chah-nulth (e.g., at halibut-fishing grounds)
Wapato (*Sagittaria latifolia*[a])	Tubers widely traded from lower Fraser Valley and lower Columbia River inland to Interior Plateau, to Vancouver Island, and elsewhere along Northwest Coast; also sold to settlers, including Chinese immigrants
Willows (*Salix* spp.[b])	Bark twine traded from Lower Stl'atl'imx to Halq'emeylem
Tule, or hardstem bulrush (*Schoenoplectus acutus*[a])	Tule stems and woven mats widely traded, especially among Interior Salish and between Coast Salish and Nuu-chah-nulth
Three-square, or "sweet-grass" (*Schoenoplectus americanus*[c])	Dried leaves for basketry traded among Vancouver Island peoples (i.e., Nuu-chah-nulth, Ditidaht, and Salish) and to Olympic Peninsula

Table 10-1 | continued

Species/material traded	Notes
Soapberry, or soopolallie (*Shepherdia canadensis*[a])	Berries widely traded as dried cakes, and jarred in water, among many groups in BC, especially from interior to coast and from mainland to Haida Gwaii
Springbank clover (*Trifolium wormskioldii*[b])	Rhizomes traded from Nuxalk to Oweekeno and Hanaksiala; also traded within Kwakwa̱ka̱'wakw communities
Cattail (*Typha latifolia*[a])	Harvested leaves and woven mats traded among Interior Salish and southern coast peoples, from central to northern coast, from coast to interior, and from Fraser Valley to Salish and Nuu-chah-nulth of Vancouver Island
Stinging nettle (*Urtica dioica*[c])	Fibre and twine traded from Hanaksiala and Haisla to Nuxalk and Ts'msyen; living plants transplanted from one village to another
Blueberries and huckleberries (*Vaccinium* spp.[c])	Dried berries traded among Coast Salish, from interior to central coast, and within various interior groups
Black huckleberry (*Vaccinium membranaceum*[a])	Berries widely traded among Salishan peoples of BC and among Columbia River peoples
Bog cranberry (*Vaccinium oxycoccos*[a])	Fresh and preserved fruits in water or oulachen grease widely traded along Northwest Coast and probably among interior peoples
Lowbush cranberry, or lingonberry (*Vaccinium vitis-idaea*[c])	Preserved fruits traded on northern coast among and between Haida and Tlingit
Highbush cranberry (*Viburnum edule*[a])	Fresh and preserved fruits in water or oulachen grease widely traded along Northwest Coast and from coast to interior
Beargrass (*Xerophyllum tenax*[c])	Bundled leaves (dried and sometimes dyed) and finished baskets traded from Quileute and Makah of Olympic Peninsula to Nuu-chah-nulth and Salish basket makers of Vancouver Island
Wild-rice (*Zizania aquatica*[b])	Traded from peoples of Rocky Mountains and Great Plains, especially the Cree, to Nlaka'pamux and Stó:lō

[a] common and widespread trade; [b] limited or restricted trade; [c] moderate levels of trade

Sources: The references cited are restricted to those in which the species traded were identified and Aboriginal groups were specified. Based on the table in Turner and Loewen (1998). Compiled from the following references and citations within Albright (1984), Birchwater and Ulkatcho and Nuxalk elders (1993), Boas (1923), Compton (1993b), Croes (2003), Darby (1996), Dawson (1891), Decosse (1980), Edwards (1979), Elmendorf (1960), French (1965), Gibson (1992), Gunther (1973), Huelsbeck (1988), Hunn, Selam, and family (1990), Kennedy and Bouchard (1983), Kirk and Daugherty (2007), B. Mathews (2010), Mitchell and Donald (1988), Norton et al. (1984), Oberg (1973), Olson (1935, 1954), G. Palmer (1975b), Ray (1938), J.A. Ross (2011), H.I. Smith (1920–24), Sproat (1987), Steedman (1930), Suttles (1951b, 1955, 1987b), Swan (1869), Teit (1900, 1906a, 1906b, 1909, 1930), Turner

(1992b, 1995, 1996, 1997a, 1998, 2003a), Turner and Bell (1973), Turner, Bouchard, and Kennedy (1980); Turner and Efrat (1982), Nancy J. Turner and Marianne B. Ignace (unpub. notes, 1993–97), Turner and Kuhnlein (1983), Turner, Thomas, et al. (1983), and Turner, Thompson, et al. (1990). Also compiled from personal communications with Elsie Claxton (1996), Clan Chief Adam Dick (1994–2009), Daisy Sewid-Smith (1994–2009), Mary Thomas (1993–97), and Annie York (1985–89).

berry-drying racks were constructed with exchange in mind, as they produced finished squares of standard dimensions (Helen Clifton, pers. comm., 2003; Boas 1921). Strings and skewers of dried roots and smoked clams, balls of spun fibre of Indian-hemp or stinging nettle, bundles of basket materials, and finished mats and baskets were all often standardized, both for ease and practicality of handling and use and to facilitate assigning exchange value and equivalency (Ray 1939; Teit 1909; Turner 1996, 1998; Turner and Loewen 1998; Turner, Thompson, et al. 1990; Schlick 1994). With the influx of European goods during the trade and colonial eras, new measures of trade were easily incorporated into the trade of traditional goods: biscuit boxes, wooden kegs, packing boxes, gunnysacks, and cotton flour or sugar bags full of berries, dried salmon, or the like all had their equivalent value and became new measures for a new age. Most recently, plastic totes and zip-closing plastic bags have become a medium of exchange.[19]

A good example of trading equivalencies is from Emmons's (1991, 445) list of the "Value of Native Tlingit Products at Sitka, Alaska," from 1890. For botanical goods, 1 gallon (about 4 litres) of "anything small, round and dried" (i.e., any dried berries) cost sixty cents. Square cakes of either dried seaweed or dried inner hemlock bark were valued at twenty-five cents each. Soapberries cost from twenty-five cents to one dollar per cake, whereas a square cake of dried strawberries (presumably *Fragaria chiloensis*) cost one dollar.[20] These exchange values would have taken into account the perceived quality, desirability, and perhaps rarity of the products, and they reflected the overall effort required to generate them, which is probably why wild strawberries – small, soft, and difficult to pick – were valued more highly than an equivalent amount of most other types of berries. Interior Plateau peoples also had standard equivalencies for trade. The strong fibre of Indian-hemp[21] became a common and valued exchange medium in the late 1890s. Five "packages" of Indian-hemp fibre ("bark") from Upper Nlaka'pamux territory could be exchanged for one large coiled cedar-root basket, two salmon skins full of salmon oil, three sticks of salmon, one large dressed buckskin, one steel trap, or one dugout canoe (Teit 1900). Nlaka'pamux elder Annie York (Turner, Thompson, et al. 1990, 33) noted for her area of the Interior Plateau that bitterroot from the upper regions of Nlaka'pamux territory was "expensive stuff" and that during her childhood in the early twentieth century, a string of dried bitterroot about 1.5 metres long would be worth the equivalent of one salmon. Teit (1900, 261) reported that "ten bundles" of bitterroot had a similar value to one large, dressed buckskin, which would have also

10-3 | Bundle of Indian-hemp fibre.

equated to five packages of Indian-hemp "bark" or ten cakes of dried saskatoon berries. Both on the coast and in the southern Interior Plateau, camas bulbs were a highly valued trade item, sometimes exchanged in the form of loaves of dried, cooked bulbs, some over 4.5 kilograms, which the Nez Perce and Sahaptin traded for horses, although the exact equivalency is not known (Juntunen, Dasch, and Rogers 2005; United States Department of Agriculture 1999).

The most enduring trade networks were established around differences in plant distribution or around variations in the quality of particular plant populations, whether used for food, material, or medicine. For example, Indian-hemp – as described previously, the most important and strongest type of fibre for fishnets, mats, and bags (figure 10-3) – is widespread in the southern Interior Plateau, but the patches are scattered and the plants vary in quality and abundance depending on where they occur.[22] Indian-hemp's patchy distribution, its variability in quality and productivity, and the ease of its rendering into transportable units (balls of twine) made it a good candidate for trade, both within the range of its overall distribution in the Interior Plateau (e.g., from the Okanagan to the Secwepemc) and outside of its range, northward and westward to the coast, as far as Vancouver Island (Kennedy and Bouchard 1983; Turner and Loewen 1998). Nlaka'pamux elder Annie York explains, "Oh, you get them [Indian-hemp] from upcountry. You trade. We have [them] here [at Spuzzum], but not many ... Ours are short ... You get it from upcountry. Some of them are as tall as that [about 1.5 metres], and some of them [stalks] are as thick as my thumb ... The people there, they gather them, bulk like this [large bundled armful], and then they bring that down here" (Turner, Thompson, et al. 1990, 161).

Another material of limited distribution that was widely traded is beargrass (*Xerophyllum tenax*), which is valued for certain types of basketry, especially for the wrapped twined trinket (*pukʷʔuu*) baskets of the Nuu-chah-nulth and neighbouring peoples. Bundles of shiny beargrass leaves, sometimes already dyed, were bought from the Quileute and Makah of the Olympic Peninsula by Vancouver Island Nuu-chah-nulth weavers (who called it "American grass") and by some of the Coast Salish peoples as well (Turner and Efrat 1982; Turner, Thomas, et al. 1983).[23]

Having access to particular high-quality resources enabled some groups to develop exceptional food products or special expertise in some technologies. Their products then became highly sought after, and their skills were emulated by others. For example, the Yakutat Tlingit women were said to excel in production of spruce-root baskets, partly because the Sitka spruce trees around Yakutat Bay grow in sandy soil along the edge of coastal dunes and therefore produce exceptionally long, straight roots (Thomas Thornton, pers. comm., 2009). Yakutat baskets became famous and desired items over the entire region because of this feature. Similarly, as noted for Haida Gwaii, western redcedars at certain locations along the coast are said to make superior canoes, and many different localities are known for the exceptionally high quality of particular foods – soapberries, saskatoon berries, bitterroots, or camas – which then become desired trade goods. Having access to high-quality goods and developing expertise in using them went hand in hand. Availability of high-quality western redcedar roots in the Coast Mountain valleys around Spuzzum, Mount Currie, and Sechelt, for example, allowed the Nlaka'pamux, Lil'wat, and Sechelt women of these communities to become renowned specialists in coiled cedar-root basketry, their baskets widely sought as trade items throughout the region and later by settlers as well.[24]

New techniques for weaving and woodworking as well as other skills were also "products" of exchange, which were learned (and taught) by those who visited or married into communities or were worked out by observing finished products brought in from other places. Functional equivalency, with similar applications of different materials, allowed adaptation to new situations, which was facilitated through trade and exchange. Digging sticks that may have been originally applied for camas or wapato, for example, could be turned to use for yellow glacier lily, or conversely, a digging stick made of one type of wood could be used as a model for a root digger of a different wood if the original wood was not available in a new area. This type of shifting and modification was always taking place, initiated and stimulated by trade (Dubin 1999).

Food Products in Trade and Exchange

Reciprocity in the exchange of food resources was particularly common and widespread. As noted previously, although food items were often used in trading or barter, more often than not, food – and access to food – was exchanged and mediated at a community level, and this exchange was frequently a matter of a chief and his entire clan or community enhancing their status by hosting feasts and potlatches. Family connections invariably resulted in exchange of food; visitors would – and still do – return home loaded with dried fruit, dried salmon, seaweed, or other valued local products. Guests would also bring food for their hosts (Turner, Thompson, et al. 1990). Sometimes these gifts were of foods not common in the host's territory, and thus they helped people to diversify their diets. However, some foods – like black huckleberries – were so highly valued that they were appreciated even if one already had a good supply. As Mary Thomas (2001, 20) explained, "Even if my elder knew that the next-door neighbour had a lot of berries, let's say huckleberries, just the *idea* of giving, the *virtue* in giving ... My grandmother would know that woman had lots of huckleberries, but she would still take a little bag and take it over there as a gift. When she came back, she brought a lot of other types of food; that neighbour would give her something to bring back. It was the idea of sharing and giving."

The origins of some of these edible trade goods, especially those traded into areas where they do not grow naturally, may be reflected in their names. Almost all of the commonly traded plant foods – such as soapberries, edible seaweed, highbush cranberries, camas bulbs, and springbank clover rhizomes – have names that indicate borrowing from one language into another, as discussed in chapter 3 and elsewhere. One particularly intriguing possibility is a trade link for hazelnuts between the Ts'msyenic peoples of the Nass and Skeena Rivers (notably the Nisga'a and Gitxsan) and early Coast Salish peoples.[25] Given the disjunct nature of hazelnut distribution (Klinkenberg 2013), it is possible that these names represent an early introduction of this shrub into the Skeena-Nass region from the lower Fraser Valley.[26]

Plant Technology in Relation to Trade and Exchange

Both raw and processed plant materials were commonly traded, along with finished items made from these materials (Kennedy and Bouchard 1983; Schlick 1994; Turner 1998; Turner and Loewen 1998). Western redcedar – a tree that varies considerably in abundance and quality over the study area – serves as a prime example. As noted previously, sheets of western redcedar bark and cedar planks, used for walls and roofing, along with finished cedar canoes and bentwood boxes were traded from Haida Gwaii to the adjacent mainland. The Haida were thus exporting not only the products of their exceptionally large, straight-grained cedars but also their skills and fine craftsmanship. Similarly, the Nuu-chah-nulth of Vancouver Island traded their cedar house planks and canoes to their Makah neighbours. "The houses of the Makahs are built of boards and planks ... principally made by the Indians of Vancouver Island, and procured by barter with them. There is very little cedar about Cape Flattery, and such

10-4 | Bundles of peeled, split, and dried western redcedar roots, ready for trade, along with dyed cherry bark and basket-making awl and knife, belonging to Margaret Lester, Lil'wat basket maker, Mount Currie, BC.

as is found is small and of inferior quality ... The largest and best canoes are made by the Clyoquots and Nitinats [Clayoquots, or Tlaoquiahts, and Ditidahts] on Vancouver Island; the cedar trees being of a quality greatly superior to that found on or near Cape Flattery" (Swan 1869, 4, 35). In the Interior Plateau western redcedar roots were usually traded already peeled, split, and dried, in small bundles or bags, ready for use in weaving (figure 10-4). Secwepemc elder Mary

Thomas (pers. comm., 1994) recalled, "We [women of Neskonlith on Shuswap Lake] would trade a bag full of *stinestn* ['cedar roots'] to Skeetchestn [west of Kamloops] for something we didn't have."

Even living cedar trees themselves were sometimes exchanged and moved around as gifts. For example, archaeologist Jim Stafford (pers. comm., 2008) was shown a lone western redcedar tree near Blue Earth Lake in Cooks Ferry, Nlaka'pamux territory. He was told that this tree had been planted there and had been presented as a gift from people from lower down the Fraser. The Lower Nlaka'pamux people often presented cedar trees as gifts to the Upper people at meeting places like Botanie Valley, and this tradition continues to the present day.

Because of the tremendous variation in basketry materials, weaving and finishing styles, and ornamentation and design, basketry is a particularly important indicator of connections and innovations among and across different cultural groups and across time.[27] Two baskets recovered at the Hoko River site, for example, were made with a wrap-around plaiting technique, more common at the contemporary (i.e., 3,000-year-old) Musqueam Northeast site in the Fraser Delta. These baskets were utilitarian and were possibly acquired from a basketful of goods traded to the Hoko River site from the lower Fraser region (Croes 2003). At the more recent Ozette site, basketry was recovered of styles known historically from all possible directions of contact: the north (possibly Haida Gwaii or Tlingit territory), the east (Puget Sound and the lower Fraser River), and the south (the Columbia River).[28]

There is modern evidence of learning basketry and weaving styles from others. For example, the Kwakwaka'wakw of the Nimpkish River apparently learned to make cattail mats relatively recently, from their Coast Salish neighbours to the south (H.G. Barnett 1955; Drucker 1951), and in the late 1800s, Sliammon women travelling to Kamloops for Catholic prayer meetings learned the art of cedar-root basketry from Interior Salish women (Kennedy and Bouchard 1983). Haida women apparently learned false embroidery decoration techniques from their Tlingit neighbours about this time as well (Blackman 1982), and Secwepemc and Okanagan birch-bark basket makers may have learned their craft from Dene (Athabaskan), Cree, or even Anishenaabe peoples, whose crafting styles, and associated belief systems, are remarkably similar (Davidson-Hunt 2003; Turner, Davidson-Hunt, and O'Flaherty 2003).

Plant Medicines in Trade and Exchange

Plant medicines, with a few exceptions like Canby's lovage, mentioned earlier, are seldom identified explicitly as trade goods. Some medicines are regarded as private property of the healer and are believed by some to lose their healing powers if others know about them (Compton 1993b; Turner 2004a; Turner,

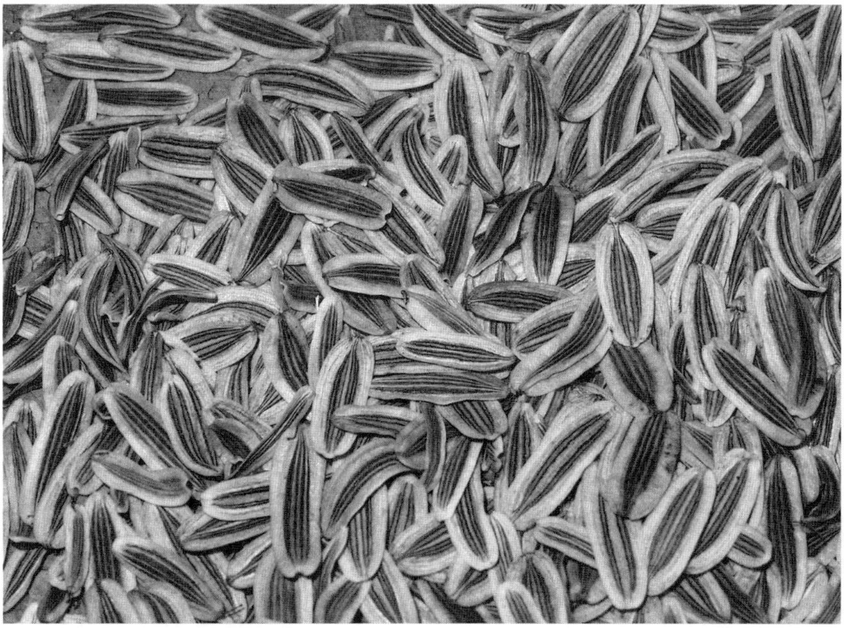

10-5 | Seeds of "Indian celery" (*Lomatium nudicaule*), widely known as *q'əx̱mín* (and variants of this term).

Bouchard, et al. 1987; Turner and Efrat 1982; Turner, Thomas, et al. 1983). Furthermore, individual medicinal recipes tend to be highly specialized and are not likely to be widely shared. Such recipes would be conveyed only within a given family from generation to generation or perhaps between close friends or fellow practitioners. Nevertheless, herbal medicine and its practitioners are an integral part of systems of reciprocity and exchange in many of the area's cultures, as reflected in a substantial congruence of medicinal plant use and practice. For example, more than fifty-one of the eighty-three reported medicinal remedies used by Saanich or Cowichan healers in one study (Turner and Hebda 1990) were used similarly in at least one other Northwest Coast group.[29] Since it is unlikely that all similar medicinal uses of a particular species (or of two or more closely related species) were acquired independently by each group involved, knowledge sharing through one means or another is the obvious conclusion. One can imagine that the properties of particular medicines applied with good results to the treatment of common maladies – use of licorice fern for colds, cascara as a laxative, or devil's-club for arthritis – would quickly become common knowledge. The aromatic and slightly pungent seeds of "Indian celery," or barestem lomatium (*Lomatium nudicaule*) (figure 10-5), were well known as a medicinal trade product of the Northwest Coast.[30] Used in treating sore throats, colds, coughs, and tuberculosis, they were also used ceremonially as a fumigant and for purification at times of illness and death, as well as for burning as an incense during the First Salmon ceremony (see chapters 7 and 12).

Some healers harvested and travelled with their own medicines to treat patients in distant places. This was the case with Annie York's great aunt Josephine, a well-known "Indian doctor," who travelled as a healer from her home near Spences Bridge to many surrounding communities. For example, she travelled downriver to Annie's home in Spuzzum, when Annie had incurred a serious back injury from a bus accident. Before leaving Spences Bridge, Josephine gathered bags of water-hemlock (*Cicuta douglasii*) roots (which are highly toxic and would never be used internally) and brought them with her on the bus to pound and apply as a poultice to Annie's back (Turner, Thompson, et al. 1990).[31]

PATHWAYS OF TRADE AND EXCHANGE

The ecological differences and geographic barriers between the Northwest Coast and the Interior Plateau regions on the eastern side of the Coast and Cascade Mountains made the coast-interior interface a particularly active region for broad-scale trade and exchange. Oulachen grease has already been mentioned several times as a prime coastal product, moved along "grease trails" – multiple riverways and footpaths extending from the lower reaches of the coastal rivers to interior destinations where it was a valued commodity (Byram and Lewis 2001). From Alaska to the Columbia River, from the coast to the Rocky Mountains, and beyond, there were numerous such coast-interior river routes and trails that connected peoples of different cultural and language groups and provided access to entirely different suites of products – including many botanical products – that they would never encounter in their local territories.

The coast-interior social and economic linkages between the Nuxalk and their Ulkatcho and Tsilhqot'in neighbours are particularly well known for facilitating the trade not only of grease but also of many other goods. One of the botanical products that linked these peoples was soapberries, which were, according to a Nuxalk narrative (McIlwraith 1948),[32] originally introduced by Raven himself to the Bella Coola Valley from Ulkatcho territory (see chapter 12). Many other interior peoples, including the Stl'atl'imx/St'at'imc, Nlaka'pamux, Gitxsan, and Tahltan, traded soapberries to their coastward neighbours (Turner and Burton 2010),[33] along with other interior products, from dried saskatoon berry cakes to birch-bark baskets.[34] Even products from east of the Rockies, such as sweetgrass and wild-rice (*Zizania aquatica*), were occasionally traded into the coastal region (Turner, Thompson, et al. 1990).

In exchange, coastal peoples provided – besides the famed oulachen grease and marine shells – such goods as smoked clams and cockles on skewers or strings, dried herring eggs on hemlock boughs, edible seaweed, and many others.[35] These items were also further traded, via the immediate interior and coastal neighbours, to other groups farther inland (e.g., from the Tsilhqot'in to the Canyon Secwepemc and to other Secwepemc peoples) and farther westward (e.g., to the Nuu-chah-nulth, Kwakwaka'wakw, Heiltsuk, Gitga'at, and Haida) (Birchwater and Ulkatcho and Nuxalk elders 1993; Port Simpson Curriculum Committee 1983; Teit 1900, 1906a, 1909; Turner and Efrat 1982; Turner and Loewen 1998).

Even more prevalent than the long-distance routes were the pathways and water routes between villages and families within the same language areas. One example is the system of trails along the Nicola Valley and Thompson and Fraser Rivers, connecting the Upper and Lower Nlaka'pamux, along which the Upper Nlaka'pamux would bring bitterroots, Indian-hemp, silverberry (*Elaeagnus commutata*) bark, and interior saskatoon berries to exchange with the Lower peoples for their famous cedar-root baskets and salmon. Annie York (pers. comm., 1984) described this situation, especially important for women who were widowed: "These women ... have no more husbands to go and gather fish. So that's the way they trade with the people from the interior. They make baskets – round baskets for making soapberries into it, or berry baskets – they trade them. And their mats, too. We use lots of those bulrushes, and the people from here [Spuzzum] likes a silver willow [silverberry-bark fibre] bag. They bring that down here too ... and they trade with that baskets ... That's made out of split cedar."

Alternatively, families might travel over these same routes to harvest their own products – root vegetables, berries, and basket- and mat-making materials – especially if they had relatives there who would host them since they would need permission to harvest in another community's territory (Suttles 1987a, 1987b; Turner, Thomas, et al. 1983). Such was the situation described at the beginning of this chapter, in a modern context, for Arthur Adolph and his family from Xaxl'ep.

Waterways served as some of the most travelled trade routes. In the Kwädąy Dän Ts'ìnchį (Long Ago Person Found) discovery, the young man whose body was found, some 300 years after his death, melting from a mountain glacier at the southeastern end of the Saint Elias Mountains of northwestern British Columbia, was evidently travelling on one of the traditional routes across the mountains from the coast of Alaska to the interior, namely up the Alsek Basin between the Tlingit and Tutchone Dene (Athabaskan) territories. He may have been carrying trade goods with him or may simply have been following a long-used trail to visit relatives or to hunt. He had been on the coast just days before he died, but he showed evidence of having lived in the interior for a good part of his life (Dickson et al. 2004; Mudie et al. 2005; see also chapter 2). He represents a host of ancient travellers, and the Tatshenshini route he followed was just one of several major trade routes in the North between the coast and the interior that connected the coastal Tlingit with inland Tlingit and Dene (Athabaskan) peoples (Bielawski 2007; Greer 1995; Neufeld and Norris 1996). Many of the rivers and lakes that people followed, especially in the northern and interior regions, became ice routes in the winter.

Aside from the major long-distance trade routes from coastal locations to the interior, shorter trails criss-crossed peoples' territories almost everywhere. These often doubled as routes to camping places and harvesting sites, where families from different locations converged. Several trails traversed the Vancouver Island ranges, for example, and a number have been identified that crossed Graham and Moresby Islands on Haida Gwaii. All the smaller islands and headlands along the coast, and major peninsulas like the Brooks and Olympic Peninsulas, had trails across them to expedite travel and trade (Elms 1996; Whelchel 2005).

Testimony (recorded in 1946) from Tlingit elder Paddy Goenette (Guneit') of the Kaagwaantaan clan at Chilkoot village, who lived to about 100 years old (cited in Thornton et al. 2004, 85), explained how he and his family moved around their territory over their seasonal rounds and how their resources were tied in to longstanding networks of trade and exchange, both local along the coast and into the interior:

> I have a cabin on Sullivan Island River ... We used to gather cockles and mussels all the way from Sullivan Island to the mouth of the Chilkat River. We also got ribbon seaweed [probably *Palmaria hecatensis* (Hawkes 1985)] in this area. We got black seaweed [*Pyropia* sp., probably *P. abbottiae*] on the west side of Chilkoot Inlet and around Nukdik Point. Black seaweed was obtained in May and June ... We gather Indian rice [*Fritillaria camschatcensis*] and wild rhubarb [probably *Rumex aquaticus* var. *fenestratus*] at Taiyasanka Harbor and at the Katzehin River.
>
> We obtain crabs and halibut in Lynn Canal. We dry halibut along both shores ... we get flounders in the same way [as crabs] ... We gather hemlock bark and scrape the inner fibers and dry them or eat them fresh ... We get hooligan [oulachen] at Chilkoot Bay, right at the village. We get needlefish at the mouth of the Chilkoot Lake. We dry these in long strings ...
>
> People used to come from other Indian villages to trade. Places like Angoon and Hoonah. They would bring foods not available here – black seaweed with a different flavour, dried cockles, devilfish, etc. – and would trade these with the local people. They would buy dried sockeyes, cranberries preserved in hooligan oil, furs, and Chilkat blankets. If they had connections with our tribe, they might go hunting, if they got permission. When a visitor asks permission to join a hunting party, we generally allow him to go ...
>
> ... The Athabaskans used to come down there to trade for white man's clothes. We sold Chilkat blankets to other people on the coast in the

villages of southeastern Alaska. I used to trade with the Athabaskans myself. I went by foot as far as the Tanana River in Yukon Territory ... The land and waters on which I hunted and fished were also used and occupied from time immemorial. I have been told by elderly people now dead.

Right up to the present generations, Indigenous peoples have been using some of the same routes and trails followed by their ancestors over millennia, both to access resources within their own territories during their seasonal rounds and to link with others for purposes of trade and cultural exchange (see McDonald 2003). However, first horses and then motorized vehicles increasingly have been used for travel and trade.[36] Many of the old foot trails were converted to wagon roads, then to roads for cars and trucks, and finally in some cases, to railways and major highways. Other trails have now been obscured or destroyed altogether with logging, urbanization, or agriculture. The original canoe routes along the lakes and rivers were later, in turn, plied by rowboats, motorboats, steam-driven paddlewheelers, and/or coastal freighters and passenger ferries. Bridges became more common, too, with roads and railways – like the one taken by Art Adolph and his family – commonly being built alongside the waterways on which original routes were based. Throughout all of these changes, First Peoples have continued many of their historic travel patterns – and the trading and gifting of seaweed, cranberries, soapberries, saskatoon berries, and huckleberries continue to the present as well.

KNOWLEDGE TRANSMISSION, TRADE, AND EXCHANGE

Ethnobotanical knowledge, as noted previously, includes not only the actual use of plants but also a whole complex of beliefs and modes of communication about them and their relationships to humans and other entities of the world. The networks created for trade and exchange resulted in a much wider dissemination of such knowledge, along with the plant products and other trade goods – the tangible objects of reciprocity. The venues by which gifts and trade goods were – and are – disseminated provided immense opportunities to convey and demonstrate cultural knowledge, to strengthen ties within and across communities, and to enrich peoples' social as well as economic lives. For example, many goods are traditionally disseminated at potlatches.[37]

For all of the guests, as well as members of the hosting group, including children, potlatches are times for learning and reinforcing this important cultural knowledge. Serving and gifting food and other products at such events is also a way of redistributing resources. In many cases, the food that is served represents surpluses harvested by community, family, and clan members from the group's or chief's recognized territory. On other occasions, and at other places, these same people will be served high-quality food harvested and prepared by other hosts from their own respective territories. Thus reciprocity is achieved, with a redistribution of wealth and resource surpluses. Overall, these occasions help not only to develop and strengthen social ties and alliances but also to reinforce peoples' overall resilience – their capacity to survive and maintain their cultural integrity in the face of uncertainty and resource variability (Ommer and Turner 2004; Trosper 2003).

One could also argue that trade and exchange – as well as conflict and warfare – helped to create and reinforce social inequalities through differential wealth acquisition by elites (Gibson 1992; Hayden 1997; Schulting 1994). As Decosse (1980, 79) explains, "While the desire for trade goods was responsible for initiating social interactions, it was the latter which sustained the material flow." Chiefs and others were able to confirm and increase their social status by obtaining rare and valued items through trade: finely crafted obsidian points, nephrite adzes, beautiful abalone jewelry, elaborately carved or painted boxes, intricately woven hats or baskets, or large, seaworthy canoes. However, they also achieved high status and recognition by *giving away* their accrued wealth to others and by hosting feasts and spectacular theatrical performances that rewarded generosity and leadership talent. Certainly, trade and exchange of both staple and rare items are integral parts of the development of cultural complexity both on the Northwest Coast and in Interior Plateau societies. The need for the members of the nobility to acquire, and then distribute, large quantities of food and other goods would have generated even more trade activity, with botanical products from diverse sources – berries, seaweed, edible roots, baskets, mats, boxes, canoes, and implements – playing a major role.

Interaction Spheres on the Northwest Coast

The Northwest Coast, extending over 15 degrees of latitude and across some 1,920 kilometres of coastline between its northern and southern extremities, shows amazingly strong and coherent cultural links, recognized in its designation as the Northwest Coast culture area (Suttles 1990b). It shares a single ocean's coastline, a similar maritime-oriented resource base, a relatively uniform and mild climate, and a continuous network of waterways that allows relatively easy contact between communities. Nevertheless, its peoples speak many diverse languages, and the area incorporates numerous culturally distinctive sectors, with specialized cultural traits shading more or less imperceptibly with one another at their junctures.

Based on their overview of archaeological findings from the Northwest Coast culture area, Ames and Maschner (1999) suggest two broad "interaction spheres" from the beginning of the Late Holocene: northern coast and southern coast. These regions show broad congruence in features such as social organization, art styles, and ceremonial practices as reflected in archaeological evidence.

Of these two, "the northern coast has remained the most distinctive region of the coast for the last 9,000 years," although it showed an overlapping with the southern coast, most notably in the Mid Holocene, at the beginning of the Early Pacific period around 6,000 years ago (Ames and Maschner 1999, 165). This was a time when there was an apparent expansion of intertidal resource use and greater sedentism, possibly as a result of stabilization of sea levels. At least part of the recognized congruence within these broad spheres, they suggest, is due to trade and other forms of exchange, including intense, long-term interaction, that both supported and were promoted by the development of elites. These individuals and their descendants controlled access to trade and held rights to resources at key places; in turn, they gained status by obtaining special resources from other regions. One could argue an extension of these interaction spheres into the respective northern and southern interiors around this time as well, during the height of obsidian trade and distribution.

Within the southern interaction sphere, Ames and Maschner (1999) delineate five smaller and at times ephemeral interacting subunits: the central coast of British Columbia and adjacent Vancouver Island (roughly Kwakwaka'wakw, Northern Wakashan, and Nuxalk territories); the west coast of Vancouver Island and the Olympic Peninsula (southern Wakashan region); the extended Gulf of Georgia region along the eastern side of Vancouver Island and the southwestern coast of British Columbia to the southern end of Puget South (roughly Central Coast Salish territory); the lower Columbia River region and adjacent coastline; and the coast of southern Oregon and northern California. The boundaries between these subunits were sometimes temporary, shifting substantively over the course of the millennia as contacts developed, as trade ties formed and were broken, and as peoples migrated, amalgamated, or otherwise acquired new territories and developed new relationships with neighbours. In general, however, the peoples within these subareas shared certain characteristic traits and practices that evidently had considerable time depth.

Among the regional similarities are distinctive styles of basketry, woodworking, and mat making, plant food use and preparation, fishing and hunting technologies, and other features that relate directly to ethnobotanical knowledge systems and to their regional exchange. There are also similarities in botanical names and terminology, even across linguistic boundaries. The known borrowings of plant names both between languages within these regions and into their interior counterparts, as described in chapter 3, are congruent with these spheres of interaction. Given the major interaction spheres and subunits of interaction identified by Ames and Maschner (1999) from the archaeological record generally, one expects that the opportunities for botanical knowledge transmission would have been greater and perhaps more frequent within each of these spheres and subunits than across them. In turn, the more intensive interchanges would likely have resulted in greater congruence of ethnobotanical knowledge within the spheres than across them. Although more detailed comparisons are required to demonstrate this definitively, an overview of ethnobotanical traits of different Northwest Coast peoples seems to confirm a major north-south divergence, and within the "southern interaction sphere," a clustering of traits related to plant use and plant knowledge within the subregions of the central coast (Kwakwaka'wakw, Northern Wakashan, and Nuxalk), the west coast of Vancouver Island and the Olympic Peninsula (Southern Wakashan), and the extended Gulf of Georgia region (Central Coast Salish), as delineated above. For each of these Northwest Coast divisions and subdivisions, the interactions also extend inland to the adjacent interior regions, although for the Southern Wakashan, the inland extension may follow the course of the Columbia River.

For example, in terms of shared botanical vocabulary, as noted previously, there is significant overlapping of plant names among Haida, Tlingit, and Ts'msyen, as well as between the coastal and interior peoples of this northern region. There are, likewise, a number of plant names that share common origins among the Nuxalk, the neighbouring Kwakwaka'wakw, the Heiltsuk, and other Wakashan groups, indicating strong cultural linkages and historical channels of trade and communication between them, despite the different linguistic origins of the Nuxalk. In fact, the Nuxalk language has more botanical terms in common with Wakashan languages than with other Salishan languages (Turner 1973). There are also many oral and ethnographic records of relationships – including warfare and slavery as well as trade and intermarriage – between the Nuxalk and the Haisla, Heiltsuk, Ooweykyala, and Kwakwaka'wakw peoples (McIlwraith 1948; Reid and Sewid-Smith 2004), as well as between the Nuxalk and Ulkatcho Dakelh and, through them, to the Tsilhqot'in (Minnie Charleyboy, pers. comm., 2003; Tepper 1991). All of these associations tie into the overall dynamics of ethnobotanical knowledge transmission.

The situation is complex, however, since there are different pathways through which groups can share common knowledge. Two groups evolving from a common ancestral group might retain similar knowledge, such as terms, stories, or ceremonies, from their common past. This is likely the case for a number of the Salishan peoples who share plant names derived from Proto-Salish precursors. To add complication, the meanings of such terms can shift according to cultural and ecological context, making their tracing and the pathways of origin more obscure.[38] Some types of knowledge, or even names, that seem similar may have been derived independently. For example, a name meaning "bear berry" for various types of berries, mostly those inedible for humans, is common in many languages and was likely derived on many occasions through observations of bears and their eating habits. In fact, bears and other creatures like squirrels, ravens, and geese have probably taught similar lessons to groups of people in many different places over a long time period.

Wood-frame shelters, two-ply twine, fishnets, woven baskets, and rush mats are common to many humans around the world and may well have been invented independently any number of times. Similarly, the uses of yew wood (*Taxus* spp.) for heavy implements (Friedman 2005), of scouring rush (*Equisetum* spp.) as an abrasive, and of birch bark (*Betula* spp.) for containers seem to be ubiquitous and are probably related to the obvious physical characteristics of these substances. Some medicines, too (e.g., yarrow, *Achillea* spp.; and wild ginger, *Asarum* spp.), are probably used almost universally, not because of any direct communication across vast areas but because of their innate phytochemical traits, identified and applied over and over again.

Another factor – one that we will usually never be able to determine – is which products, knowledge, information, practices, or beliefs may have been acquired or transmitted on one or more occasions but never became fully adopted or integrated into their new context. Some potential products or techniques may have simply been rejected or might have been adopted, only to be obscured and eventually disappear. Alternatively, some names or knowledge, once adopted, may have become transformed beyond recognition from their original characteristics or form. One could argue that knowledge not acquired is as significant as knowledge that is acquired (French 1981; Johnson 1997), but all we can do is speculate about this. Finally, it must be noted that documentation is still incomplete, based on what was left in the archaeological record or what has been recorded by ethnographers, ethnoecologists, linguists, and/or Indigenous experts past and present. Many voices remain unheard in the story, and much knowledge simply has not been considered. For this reason, too, any conclusions will be incomplete and speculative.

Based on what can be assembled, however, this much is clear: the wide array of similarities and congruencies of plant knowledge and use among all of the various peoples, as reflected in the various languages across the study region and within the different interaction spheres and subunits, presents a convincing picture of multiple exchanges, borrowings, and adaptations of ethnobotanical knowledge, practice, and belief on many geographical and temporal scales. Determining the directions of borrowing and exchange of ethnobotanical traits presents further problems, but these directions can be deduced in some cases by the known distribution of the species involved or by more broadly applying rules similar to those suggested in chapter 3 for determining the direction of borrowings of plant names.

Soapberry: A Case Example in Ethnobotanical Use and Knowledge Transmission

Soapberry, already mentioned many times in this book, is an ancient and widely known species in the study area; thus it can serve as an example of how knowledge and practice surrounding it might spread over time and geographic space. As noted in chapter 3, there is a very high level of congruence of its names, with only nine distinct etymons out of a total of forty-six or more languages for which names are known. The linkages between etymons correlate well with what is known about trading patterns for the berries (e.g., northern Haida with Tlingit; Ts'msyen, Nisga'a, and Skidegate Haida with Gitxsan and Dene [Athabaskan]; Wakashan with Salishan; and Coast Salish with Interior Salish).[39] Possibly, too, there is a link between the Salishan and Dene (Athabaskan) terms, with Dene being derived from Salishan or possibly vice versa (Cecil Brown, pers. comm., 2007; Turner and Burton 2010).

To some extent, we can infer the directions of name borrowings by the overall distribution and abundance of the soapberry plant: generally, from the interior (where it is widespread) to the coast (where it is less common or absent). The situation becomes more blurred, however, when the time factor is considered. All eighteen Salishan names for soapberry are similar, being borrowed or having a common derivation from Proto-Salish *x̱wus* ("to foam, froth"). Although *Shepherdia canadensis* is not presently common or widely productive along the Northwest Coast (including the Proto-Salish homeland) (see Klinkenberg 2013), palaeoecological studies indicate that it used to be very common on parts of the coast in past millennia and was in fact one of the first species to recolonize the area following deglaciation (K.J. Brown and Hebda 2002b; Pielou 1991). What is the time depth, then, of the innovation of the special whipped confection made from soapberries and of the development of more recent trading patterns for the berries? Proto-Salish has a time depth of nearly 5,000 years,[40] at which time soapberry was probably more widespread than it is today and possibly very common in the Proto-Salish homeland of the lower Fraser Valley, where it is now scarce or absent. Over the ensuing millennia, as Salishan peoples started to disperse, one could hypothesize that soapberry became less and less common and more and more patchy in its distribution and availability. People who enjoyed it would increasingly have needed to rely on trade and exchange to obtain it. The shrub's developing rarity – as well as its dioecious (separate male and female flowers) state and the high variability in its fruit production and quality over seasons, years, geographic situations, and ecological circumstances – would have rendered soapberries an increasingly valued status food, right at a time when social stratification was becoming more pronounced.

As a desirable food of high value, traded for oulachen grease, seaweed, and other similar costly products, soapberries needed to be, and were, enhanced – for example, by pruning its bushes and burning over soapberry patches (see chapter 11). People developed a whole suite of innovations to intensify its production and add to its value: tasting the berries before harvesting in quantity to ensure the best quality; generating more efficient methods of harvesting;[41] drying the berries in cakes for large-scale storage and transport; designing special whippers to create soapberry whip most efficiently; designing special soapberry-whipping baskets (including small ones for children); inventing elegant and unique paddle-spoons, possessed by individuals, to enable proper "sipping" of the whip; creating special dishes of yew wood and other high-quality woods from which to eat soapberry whip; developing a special beverage from the berries; creating special songs to accompany the whipping of the berries (Clan Chief Adam Dick, pers. comm., 2008); developing origin stories and other narratives featuring soapberries (see chapter 12); and fostering a general celebratory atmosphere around the consumption of whipped soapberries.[42] Furthermore, not only have soapberries been seen as a high-status feasting food, but they also have a reputation as a nutritious "health food" and tonic, to be administered for influenza, indigestion, and various other ailments (Turner, Thompson, et al. 1990).

A host of terms and vocabulary around all of these aspects of soapberry innovation must have developed in tandem with them: names for soapberry whippers and soapberry spoons, names for the mats and grass used for drying soapberries,[43] and even terms like *'ajaa* (*ʔajaa*) in Skidegate Haida, roughly meaning "smeared with soapberries" (Sapir 1923) and alluding to its use as a festive food, with soapberry contests and soapberry fights ('Ksan 1980; Turner 2004a; Turner and Burton 2010). All of these attributes play into soapberry's role throughout northwestern North America as a highly desired, highly valued food, one that was obviously widely traded and shared and that retains its eminence to the present day.

CONCLUSIONS

Perhaps the most obvious effect of trade and exchange is that they promote a more even distribution of resources across geographic regions and cultural boundaries. They give people access to a wider variety of foods, materials, and other resources than would otherwise be available (Turner and Loewen 1998). Trading and other forms of exchange also help to ensure a group's survival by providing a "safety net" in the event of local or temporary resource scarcity. Trade and exchange often feed into a system of differential social status, with chiefs and nobility asserting control over access to harvesting areas and high-value trade goods but also assuming a higher level of responsibility and obligation regarding the distribution of such goods for the benefit of the entire community, particularly through the ceremonial systems of potlatching and feasting.

However, in the end, it is the associated information, technologies, perspectives, and beliefs accompanying trade and gifted goods that make their exchange so important and relevant to understanding patterns of knowledge acquisition, transmission, and adaptation. Many Indigenous narratives allude, directly or indirectly, to instances of such knowledge exchange regarding plants (see chapter 12). For example, a Nlaka'pamux story about an old woman who lived with her two granddaughters provides an account of dissemination of both goods and expertise in relation to marriage connections. The grandmother taught her granddaughters how to make baskets and how to perform all types of women's work. They were then sent off to marry two brothers, Owl and Eagle, who were "great men, and hunters." The young women started off on their journey to meet their new husbands, each carrying "a large basketful of dried roots and berries" (Teit 1912, 250). This is just one instance of how individuals bringing goods to new locations also convey associated training and knowledge.

The integration and fusion of old knowledge and practice with new have been a continuous process at least over the past 5,000 to 4,500 years and are still ongoing. For example, Swanton (1905, 1908), Acheson (1998), and Boelscher (1988) identify numerous types of blended cultural knowledge and history shared by the Haida and Ts'msyen, including narratives, crests, language elements, and common lineage histories, reflecting a long tradition of contact with and movement of peoples between the mainland and Haida Gwaii. Similar ties are documented between the Haida and Tlingit (Boelscher 1989; Moss 2008a, 2008b).[44] Especially within the past two millennia, there has been a notable cultural congruence between the Haida and Ts'msyen, such that Acheson (1995) suggests that Hecate Strait, which separates these two peoples, should be considered more of a highway than an obstruction between them. It seems that of the two, the Haida have been in their homeland considerably longer, with human occupation of Haida Gwaii predating that of the Prince Rupert Harbour area by about 4,000 years.[45]

The complexities of soapberry and its associated knowledge, discussed in the previous section, provide an illustration of the dynamics of innovation and cultural information in relation to a particular species. Soapberry's differential abundance, patchy distribution, and historic patterns of use give us a good understanding not only of the dissemination of this valued foodstuff and its names but also of the knowledge, practices, and innovations related to its use, which must have developed and accompanied its pathways of diffusion and exchange. Undoubtedly, similar pathways of knowledge exchange, following parallel patterns of trade and sharing of goods, exist for numerous other plants and plant products. Cultural, linguistic, and ecological knowledge about these plants, and the sharing of this knowledge among different peoples, is a key component in overall adaptation and resilience, a factor as true today as ever in the past.

The trade of these plants and the exchange of related knowledge have had significant cultural and ecological implications (i.e., in the form landscape management practices such as burning) that go far beyond simple subsistence (Turner and Loewen 1989). Peoples' actual perceptions of their places, of the species on which they depend, and of the role of humans in the world are also conveyed through cultural means, with major social and ecological outcomes.

The notion that all of the animals, fish, birds, trees, and plants – as well as fire, water, mountains and rocks, and supernatural beings of the spiritual realm – have their own societies, and are relatives of people and have the ability to influence humans profoundly, is a major and overarching belief among Indigenous peoples not only within the study area but also far beyond (Davidson-Hunt 2003; A.M. Miller 2010). This concept of "kincentric ecology" guides peoples' behaviours and actions toward other living beings and environmental entities and is a prime determinant in ethnoecological knowledge and practice. Broad-scale interactions between different peoples through trade and other forms of reciprocity have ultimately led to this commonly held concept, which is a topic of the following chapters. In some cases, as well, the "goods" that were traded or given as gifts were living plants, such as the cedars mentioned earlier, that became established in distant locations. There are many recent examples of transplanting and replanting propagules of plants, as discussed in chapter 11.

11

Management and Sustainability of Plant Resources and Habitats

The late Mrs. Joshua Moody, she used to do that [cultivating springbank clover] in the 1920s, in that little island here at Bella Coola where the tide comes in. She used to choose a place ... where there's soft sand. I hear she's got quite a bit of it because she was looking after it. The more you soften it the more they grow in there. She picked the roots and planted them, early in the spring, like April ... she used that wooden digging stick ... they have to have it where the tide comes in. It's got no flavour when you try to do it inland ... It [the garden] wasn't very big what she used to have [about 8 by 9 metres]. They used to have it fenced up so the people knew that it belonged to somebody. She just had a wooden fence. Nobody would touch it. (Felicity Walkus, cited in Edwards 1979, 6)

INTRODUCTION

The "roots" (rhizomes) of springbank clover (*Trifolium wormskioldii*), cultivated by Nuxalk matriarch Mrs Joshua Moody in the 1920s as described above, were highly valued as a food and also used as a trade good on the Northwest Coast. Archibald Menzies, naturalist and physician with Captain George Vancouver, was evidently the first European to describe its use on the Northwest Coast, in early September 1792 at Tahsis in Nuu-chah-nulth territory:

> In the evening our curiosity was excited in observing a number of Females busily occupied in digging up a part of the Meadow close to us with Sticks, with as much care and assiduity as if it had been a Potato field, in search of a small creeping root about the size of a pack thread. This I found to be the Roots of a new species of Trifolium which they

11-1 | The Kingcome River estuary, BC, site of the Tsawataineuk root gardens of the family of Clan Chief Adam Dick (Kwaxsistalla), showing Adam Dick (centre, holding digging stick) with a group of Tsawataineuk community members and students from the University of Victoria.

> always dig up at this time of year for food ... Wherever this Trifolium abounds the ground is regularly turnd over in quest of its Roots every year. (Cited in Newcombe 1923, 116)

Over 150 years later, ethnologist Philip Drucker (1951, 57) also noted the use of this clover by the Nuu-chah-nulth: "The clover-root patches were owned by the chiefs. In late summer women went to dig them. Each woman carried a burden basket and a root-digging stick. She sat on a cedar-bark mat, the heavy shaft of the digging stick in both hands, gouging and prying the roots up. As she dug them, she shook the dirt off and put them into the basket on her back."

Later still, Clan Chief Adam Dick (Kwaxistalla) (pers. comm., 1996) of the Kwakwaka'wakw Nation described the production of this clover and associated root species in the tidal marshes of the Tsawataineuk territory at Kingcome Inlet: "It was all important. That *tex^{w}sus* [springbank clover], and the *tliksam* [Pacific silverweed], and the *q^{w}'eniy'* [Nootka lupine], and the ... *xuk^{w}em* [northern riceroot]. See, when they go down the flats, they use little pegs. 'This is my area.' You got your own pegs, in the flats. And then you continue on that, digging the soft ground … so it will grow better every year. Well, I guess, fertilizing, cultivating, I guess that's … the word for it. Every family had pegs, owned their little plots in the flats" (figure 11-1).

These and other descriptions of this important root vegetable (e.g., Boas 1921, 186–94; and Turner and Kuhnlein 1982, 423) embrace a wide range of different topics within the overall complex of "knowledge, practice and belief" comprising traditional management systems: seasonal harvesting in specific habitats, tilling, gendered work, associated equipment, resource ownership, control, delineation of plots, soil maintenance, and other aspects of food production systems.

Despite the fact that numerous early explorers, traders, government officials, and ethnographers throughout northwestern North America made such observations pertaining to Indigenous peoples' harvesting and management practices, these practices and their significance have been, until recently, largely – and one might argue, conveniently – overlooked. First, it has been widely assumed that plants were not a particularly important component of people's diets, especially on the Northwest Coast, and this notion has at times perpetuated itself by the frequently minimal efforts applied to document plant resources (Deur and Turner 2005; Deur et al. 2013; Lepofsky 2004; see also chapters 2, 5). Second, any detailed understanding of people's *management* of plant resources has been even more obscured. Even with early records like that from Archibald Menzies or descriptions of huge mounds of clover roots being cooked at Nuu-chah-nulth feasts as noted by Philip Drucker (1951), we have generally remained blind to the overall immensity of Indigenous peoples' capabilities and techniques to sustain and enhance their plant resources, assuming instead that "the plants just grew by themselves" (Deur and Turner 2005; Turner, Deur, and Lepofsky 2013).

In some rare instances from the past when Indigenous peoples' plant resource management practices have been acknowledged, their value has been minimized in comparison to European modes of production. First Peoples' use of fire on southern Vancouver Island to help clear the undergrowth from wooded areas and promote the growth of key food species is a case in point. This practice was noted in grudging and derogatory terms by Captain W.C. Grant in his report to Governor James Douglas (c. 1848), where he described the fires that were "kindled promiscuously by the natives both in wood and prairie between the months of August and October. Their object is to clear away the thick fern and underwood in order that the roots and fruits on which they in a great measure subsist may grow the more freely and be the more easily dug up" (cited in Lutz 1995, 22; see also Turner 1999). To Captain Grant, the First Peoples' burning practices were "abominable" and destructive, even though he seemed to comprehend that their underlying purpose was to enhance local Coast Salish food resources, thus affecting people's very survival. Furthermore, even today, Indigenous peoples of northwestern North America are usually portrayed as "hunter-gatherers." Yet, to characterize their relationships with plants as "gathering" is the equivalent of describing agriculturalists merely as "harvesters"; in both cases, "gathering" or "harvesting" is only a small part of a complex system of applied knowledge developed over time to allow people to survive in a particular place.

Defining Management

Several researchers have suggested that the notion of "management" alone does not do justice to the Indigenous peoples' systems of caring for their lands and for the species on which they rely. "Management," to some, seems too objective and mechanical; instead, terms and practices like "stewardship," "custodianship," "tending," "caring for," "taking care of," and "working with" seem more appropriate descriptions for people's practices and attitudes of relationship with the other species they use (M.K. Anderson 2005; E.R. Atleo 2004; Fowler 2000; Fowler and Lepofsky 2011; Nabhan 1997).

Caring for the land and "giving back" to other species – one's relatives – who contribute their lives or parts of themselves for people's well-being is a deeply spiritual endeavour, far beyond the most stark interpretation of "management" (see chapter 13). However, in its etymology, "management" originates with Italian *maneggiare*, which in turn derives from *mano* ("hand") (cf. Latin *manus*), giving it an original connotation directly connected with people rather than with some mechanical or dehumanizing influence. Furthermore, the common definitions of "manage" include "to handle or direct with a degree of skill," "to treat with care," "to exercise executive, administrative, and supervisory direction of," and "to work upon or try to alter for a purpose,"[1] all of which seem congruent with the actions and approaches people have used to maintain and enhance their lands and resources.

The ultimate outcome of management activities is assumed to be sustainability, and even enhancement, of the managed resources, and therefore Traditional Land and Resource Management (TLRM) could be interpreted as equivalent to, or at least related to, natural resource conservation. Some narrow definitions of conservation, however, restrict this concept or practice to actions that directly prevent or mitigate depletion, with an identifiable *intention* to conserve (E.A. Smith and Wishnie 2000). In contrast, some of TLRM's practices and observances are so deeply embedded in longstanding belief systems (cultural conventions, constraints, and taboos) that they may not be directly attributable to "conservation" in the Western sense of the word (Colding and Folke 2001; Turner and Berkes 2006).

For our purposes, then, Traditional Land and Resource Management can be defined as *the conscious accumulation, application, and adaptation of any combination of techniques and methods drawn from Traditional Ecological Knowledge systems, mediated by particular beliefs and worldviews, that sustain or enhance the availability, abundance, productivity, diversity, and/or quality of a plant or animal population or of an entire resource area or habitat over a period of years or generations* (see also M.K. Anderson 2005; Berkes 2012; Deur and Turner 2005; Gadgil and Berkes 1991; and Menzies 2006b). In such management systems, people engage in reciprocal relationships with their resources and environments, such that they take responsibility for caring for them. As Kwakwaka'wakw cultural specialist Dr Daisy Sewid-Smith (pers. comm., 1996) explained, "People ... think that Nature just grows on its own. But our people felt, to get more harvest, and bigger berries, they did these things [pruning, burning]. Same thing a farmer does ... Traditionalists are very close to their environment, and they know that to survive in this world, you have to have respect for your environment, because without the environment you cannot live. If you destroy your environment, you are destroying yourself." Elders and cultural specialists throughout the entire area – from north to south, from coast to interior, and beyond – have expressed similar sentiments.[2] The idea of people as active and respectful caretakers, not just beneficiaries, of plant and animal resources seems deeply engrained indeed.

The expected outcome of effective TLRM is increased productivity, biodiversity, and heterogeneity across the landscape, with mosaics of diverse, productive plant communities high in structural and compositional variation in different locations. However, for those who have not observed an area in its entirety over an extended period of time, the effects of management may be undetectable or assumed to be "natural." A classic example is the picturesque prairie-studded Garry oak savannah landscape of southern Vancouver Island, in the vicinity of the present site of the city of Victoria. Captain George Vancouver (1798, 227–9), who was deeply impressed by this expansive, scenic landscape with its "delightful meadow[s]," wrote in his journal, "I could not possibly believe any uncultivated country had ever been discovered exhibiting so rich a picture. Stately forests ... pleasingly clothed its eminences and chequered its vallies; presenting in many places, extensive spaces that wore the appearance of having been cleared by art."[3]

Unrecognized by Captain Vancouver was that the beautiful mosaic of oak woodlands interspersed with camas meadows and fern glades was the product of centuries, probably even millennia, of repeated light-impact burning by the local Coast Salish peoples – the very activity observed but completely disparaged by Captain W.C. Grant (quoted previously) some fifty years later. Similarly, for the prairies of western Washington, Wilkes (1845, 336) wrote, "These [camas] prairies have such an air of being artificially kept in order, that they never cease to create surprise, and it is difficult to believe that the hand of taste and refinement has not been at work upon them."

As historian John Lutz (2008, 34) argues, it would have been inconvenient for Europeans to recognize – or if they did, to commend – Indigenous land management practices: "European colonists had to overlook the different agriculture, mariculture, and silviculture practices of indigenous peoples to characterize the non-European world as 'in a state of nature.' They also had to characterize the productive activities of indigenous civilizations as 'not labour' in order to declare America 'unowned' and available for the taking." Certainly, the sophistication and effectiveness of Indigenous peoples' resource production systems were widely disregarded by European explorers, traders, colonists, and religious and government officials, and to some extent, they still are.[4]

In the remainder of this chapter, I first discuss potential responses of plants to various management techniques, including their biological capacity for regeneration. I then discuss the antiquity and possible cultural and historical origins of traditional land and plant resource management practices. I reflect on some of the mechanisms and pathways by which First Peoples of northwestern North America may have developed their management practices and associated beliefs, including impacts and influences on TLRM systems from the European newcomers. Following this is a section delineating the different categories and practices of plant resource management and the nature and scales of their application in the study area – from plant population to entire landscape. I then discuss some of the cultural aspects of plant management, including relationships to social organization, individual and group control of resources, and the connections between plant management and belief systems (a topic further covered in chapter 13). I conclude by emphasizing the diversity and ubiquity of traditional plant management systems as key elements of the dynamic social-ecological systems of Indigenous peoples of the study area.

SUSTAINING PLANT RESOURCES: DISTURBANCE AND REGENERATION

Many of the management methods Indigenous peoples developed to sustain and promote their plant resources are based on the plants' natural regenerative capabilities, which have evolved over thousands or tens of thousands of years in response to various types of natural disturbance, including fire, flood, landslides, insect infestations, and intensive herbivory (Colding, Elmqvist, and Olsson 2003; Hallett et al. 2003; Lepofsky 2009; Lepofsky, Hallett, et al. 2005; Moss, Peteet, and Whitlock 2007). Virtually all plant species harvested and used by Indigenous peoples in northwestern North America are perennials, and most are woody trees, shrubs, or subshrubs. Meristems – tissues comprised of cells capable of rapid growth and differentiation – occur in various parts of plants of all ages, most notably forming banks of regenerative cells in the root and stem tips, nodes, and cambium tissues. These are able, during natural growth or in response to damage, to grow and develop into more specialized tissues. Thus branches or roots, triggered by hormonal changes, can resprout when pruned, cut down, or otherwise injured, and bark can, over time, regrow and heal over a cut or injured area of a trunk or branch.

Herbaceous perennials developing from rhizomes (underground stems) can continue to grow and spread even when some of the shoots and rhizomes are removed. As long as healthy meristematic tissues are retained, and as long as plants have the capacity to absorb and retain sufficient nutrients and water, they can reproduce vegetatively and maintain individuals and populations even with a certain level of harvesting. In fact, many plants are so well adapted to being harvested, browsed, or injured that they actually grow better and faster, and are more productive, after harvesting or disturbance. Furthermore, management strategies such as weeding, clearing, thinning, and burning have the effect of removing or reducing competing vegetation, reducing insect pests, allowing more sunlight and/or moisture to reach the plants that remain, freeing up nutrients, and ultimately promoting growth for the species that are left. Loosening the soil, which accompanies root-digging activities, helps to aerate the soil and allows better root penetration and soil regeneration. Floods and fire also introduce pulses of nutrients that can be easily taken up by the plants as they regrow following the disturbance.

With the exception of cutting down some trees in their entirety for making canoes, house posts, totem poles, or beams, or for purposes of clearing an area of land, plants are generally only partially harvested and are, by intention, not killed in the harvesting process. They are "kept living," as described by Clan Chief Adam Dick (Kwaxsistalla) (pers. comm., 1998, cited in Deur and Turner 2005) in defining the Kwak'wala term *qʷ'aqʷ'alaʔowkʷ* ("keeping it living"). In some cases, for both practical and philosophical reasons, people avoided cutting down the trees they needed for lumber, preferring instead to harvest the trees that were felled by winter snows or that died naturally from disease or windfall (Mary Thomas, pers. comm., 2001).

Digging edible roots might seem to be another harvesting activity that could potentially destroy an entire plant. However, on closer scrutiny, roots, too, are adapted to some level of herbivory and disturbance. When edible taproots like balsamroot, bitterroot, and silverweed are dug out, the tip of the root almost invariably breaks off and remains in the ground, enabling the plant to regenerate itself and continue to grow, eventually to produce flowers and fruits from the same original plant. Sprouts and bulblets of liliaceous species like riceroot (*Fritillaria* spp.), camas, and glacier lily will also readily regrow if planted. Experimental work and observation have borne out this phenomenon.[5]

The same principles of vegetative regeneration are brought into play with the harvesting of spruce roots and western redcedar roots for basketry (figure 11-2); branches of cedar, willow, yew, and saskatoon berry for basketry and implements; inner bark of cedar for weaving and cordage; and bark and branches of

11-2 | Lil'wat basket maker Margaret Lester digging roots of western redcedar for her baskets in 1984 near Mount Currie, BC. She went to special places with sandy soil to get her roots and took only a few from each tree.

various trees and shrubs for medicinal and other purposes (Peacock and Turner 2000; Turner 1998; Turner, Ari, et al. 2009; Turner and Hebda 1990; Turner and Peacock 2005). The plant products in each case are only partially harvested, leaving behind enough to keep the plant living and to allow healing or regeneration over time. Harvesters were taught about this practice from a very early age, and it was widespread and widely applied across the diverse linguistic and cultural regions of northwestern North America: "Take only part of each plant." "Take only what you need."

In the case of some types of bark (e.g., birch, bitter cherry, Douglas-fir, and cottonwood), only the outer layers are taken, leaving the inner bark to protect the growing tissues of the tree. Bark pieces removed to access edible cambium and inner bark tissues, or for bark medicine, are generally cut in relatively small pieces from only one side of the trunk, leaving the remaining bark intact so that the tree will continue to grow. All of these practices leave some tell-tale signs on the trees, mostly in the form of partially grown-over strips or rectangles on the trunks of the trees, forming what are known as "culturally modified trees" (CMTs), which date back three or four centuries in some places and occur by the thousands in some forests – a testament not only to the conservation of trees that were used but also to the management of whole stands by given clans or families (Lepofsky 2004; McDonald 2003; Turner, Ari, et al. 2009; Turner, Deur, and Lepofsky 2013).[6]

PLANT RESOURCE MANAGEMENT OVER TIME: MECHANISMS AND PATHWAYS OF DEVELOPMENT

Although the diverse strategies and approaches to land and resource management discussed in this chapter are based largely on ethnographic documentation, the complexity and sophistication of traditional management techniques, processes, and protocols seem to reflect a considerable time depth. It is reasonable to assume that the systems for sustaining certain resources and creating and maintaining particular habitats and landscapes have been developing ever since people entered the area and possibly well before that (see Turner and Berkes 2006). It is difficult to "tease out" the various elements of these systems as applied to diverse resources and on different scales, but likely there has been a cumulative building of complexity over time.

Basic resource management may begin with focused applications of specific techniques for particular species in certain localities, based on observations and experiences related to the effects of animal and human harvesting, natural disturbance, site comparisons, seasonal development, and other ecological factors on the quality, abundance, and productivity of resources. Increasing pressure from population growth on existing resources probably creates the impetus for developing these techniques. As resources are intensified, in response at least in part to population pressures, through technological innovations – better and more effective harvesting tools, sturdier containers with greater capacity, and new means for processing and storing large quantities, such as with pit-cooking – the need to oversee and manage both the processes and the harvesting of the resources becomes more pressing. This situation may lead to more complex social organization, with diverse roles for individuals: designated leaders as "project managers" to oversee the various activities; experts in determining techniques, optimal timing, and locations; and facilitators of the equitable distribution of resources.

The more different requirements there are, and the more different resources and harvesting areas, the greater the need to develop protocols and widely accepted approaches and practices that govern peoples' actions at a moral or ethical level. This is where the management becomes more abstract and less directed or intentional. Songs, ceremonies, and dances that teach restraint against overharvesting,[7] stories that embody lessons for proper behaviour toward other species, and belief systems like kincentricity (see chapter 13) are not necessarily linked directly to conservation or sustainability but are nonetheless clearly elements of a broader management regime through their general effects on peoples' conduct in relation to their environments. These culturally based philosophies and belief systems may take longer to develop.

Berkes (2012) suggests a progressive layering of and building upon the different elements of management systems. First, people entering a new area quickly learn basic knowledge about the edibility or potential utility of different species they encounter. Through careful observation, they develop practical understandings of the ecological relationships, lifecycles, phenology, and habitats of these species. In time and with growing populations, more complex social structures, organization, and institutions are required to provide the conditions that will support intensive use of an area and its resources, namely a diversified knowledge base, division of labour, control of human activities, management of suites of species and habitats within a territory of a people, and opportunities for knowledge transmission. Finally – or likely concurrently – comes the inception and establishment of worldviews embodying prescriptions for proper and acceptable behaviour in the use of and care for lands and resources. One could argue that it is only with this integrated layer of cultural control that a truly sustainable anthropogenic landscape can be developed and maintained over long periods of time.

The Saanich reefnet fishery, described briefly in chapter 9, is one example of how these layers of management actually work together to conserve resource species. This fishery incorporates all elements of a complex management system: basic practical knowledge about salmon behaviour and migration patterns; development of technology incorporating management of willows, bull kelp, and other plant species; construction of a means of escape for a proportion of the fish captured in the net; traditions of ownership of reefnet sites along with other forms of property; social organization around deploying the nets and exchange of fishing sites and technologies with the neighbouring communities, including the Halkomelem; trade networks; and finally, the embodiment of belief systems that help to sustain the salmon by directing human behaviour at a societal level, with strict requirements for a period of "no fishing," such as reflected in the story about the origin of salmon and the reefnet fishery and in the First Salmon ceremony (E. Claxton and Elliott 1994; Turner and Berkes 2006; Turner and Hebda 2012). There is some archaeological evidence for reefnet fishing in the Gulf and San Juan Islands dating to around 2,000 years or more ago (Kirk and Daughtery 2007). In terms of other types of management systems, particularly of plant resources, what are some indicators of these systems and their antiquity?

Archaeological Evidence of Plant Management

It is often difficult to pick up signals in the archaeological record indicating increased human control of resources and ecosystems (Lepofsky 2004; Lepofsky, Hallett, et al. 2005). In the case of intentional landscape burning, there may be a change in the seasonality, frequency, or extent of fires.[8] For example, the presence of camas in small clearings in the dense coastal temperate rainforests of the Oregon coast may reflect ancient burning practices (Deur 2000, 2005). On the northern coast of British Columbia, an increase in the presence of lodgepole pine after about 2,200 years ago may also indicate past anthropogenic fires (A.S. Gottesfeld, Mathewes, and Johnson Gottesfeld 1991).

It is challenging, however, to distinguish between natural and anthropogenic fires without detailed palaeoclimatic records since there was considerable fluctuation in natural temperature and moisture regimes that would have affected the frequency of lightning-caused fires in any given area. For example, the increase in lodgepole pine noted above may simply reflect the onset of a warmer, dryer period rather than any human influence (Lepofsky 2004). Furthermore, as pointed out by Lepofsky, Hallett, et al. (2005), once a prairie or clearing is established, the fuel load decreases and any fires that do occur will be more difficult to detect because of a corresponding decrease in the amount of detectable charcoal. In this case, other lines of evidence, such as an increase in quantities of grass pollen or grass phytoliths deposited in an area, as reflected in soil cores in wet contexts, can provide an indirect signal for the creation and maintenance of prairies. Such is the case with Ebey's Prairie on Whidbey Island, Washington, in which several different indicators (increased fire frequency; black, organic soils typical of grasslands; and increase in the pollens of grasses and other species of open habitats) align to suggest that people began setting fires to create and maintain this prairie at least within the past 2,300 years or so. Because such fires have been banned since the mid-1800s to early 1900s, with active fire suppression being implemented, this prairie is diminishing due to encroachment of conifers, mostly Douglas-fir (Gedalof, Pellatt, and Smith 2006; Weiser and Lepofsky 2009). Other prairies in the region, like the Ozette Prairies of Olympic National Park, are similarly shrinking (M.K. Anderson 2009).

Weeding or selective propagation may lead to shifts in the frequency and abundance of key species or even to shifts in the genetic makeup of certain populations as reflected in the palaeoecological record, but to date, in the study area, no definitive evidence of this type has been uncovered (on camas, see Tomimatsu, Kephart, and Vellend 2009). Partial harvesting of tree bark or inner bark for food, material, or medicines, in the philosophy of "keeping it living," is reflected, as previously noted, in the presence of culturally modified trees (CMTs). Unfortunately, however, even the oldest CMTs, some of which extend back to the 1500s (Turner and Wilson 2008; Stryd 2001), take us only a relatively short distance into the past, so it is hard to know, beyond that date, when people initiated this strategy or embraced the underlying philosophy that trees and other entities in the environment were sentient beings and should be kept alive whenever possible rather than being killed thoughtlessly (see chapter 13). Nevertheless, the wide occurrence of culturally modified trees of various types and in a range of habitats over the entire area – from the Rocky Mountains to the coast and from Oregon to the Gulf of Alaska, whether at sea level or in montane forests – is a testimony to the importance of this approach and its transcendence over multiple cultures, language groups, and geographical regions.

Indirect archaeological evidence of resource management might be in the form of large pit-cooking depressions, presumably for cooking root vegetables like camas or balsamroot, used over an extended period of time. The shifting frequency and character of earth ovens through time might be indicators of plant-harvesting intensity, and the implementation of different technologies and social controls could signal increases or decreases in overall plant management. In the southern Interior Plateau, the use of earth ovens increased between about 3,500 and 2,300 years ago and again after about 1,500 years ago (Lepofsky 2004; Thoms 1989, 2008a, 2008b, 2009), whereas in the northern Interior Plateau, the use of earth ovens shows an increase starting between about 3,000 and 2,000 years ago, perhaps a reflection of more sophisticated harvesting strategies and/or more developed social structures and specialization. In the northern Interior Plateau, cooking pits diminished in size over the past 2,000 years or so, but this shift, which might indicate fewer roots being cooked and therefore potential depletion of this resource, could equally reflect a change in cooking strategies or in social organization, with one-time use of a given pit rather than repeated use. In fact, it could indicate accelerated preparation and consumption of root foods rather than the reverse (Sandra Peacock and Dana Lepofsky, pers. comm., 2009).

The presence of root-digging sticks in archaeological contexts may also serve as a proxy for root and habitat management, especially in conjunction with other features related to food production. Particularly impressive in this regard are the wooden digging stick remnants from the Pitt Polder wapato wet site in Katzie territory in the lower Fraser Valley (see chapter 2). The combination of digging sticks, seemingly pure and extensive wapato beds in peat, and a cobble and sand underlay (possibly to improve drainage or increase harvesting efficiency), combined with nearby storage and processing features, make this a prime candidate as a managed resource site (Bernick 2007), equivalent to the estuarine clover and silverweed *t'əkkillakʷ* (root gardens) described at the beginning of the chapter.

The potential for transplanting and disseminating species from one place to another as a form of resource management might be found by examining the linguistic record, as well as through genetic studies (which could also refute such a possibility) (see Tomimatsu, Kephart, and Vellend 2009). Hazelnut, with a disjunct distribution along the Skeena River in northwestern British Columbia and with Gitxsan and Nisga'a names related to the Proto-Salish names (see chapter 3), is a candidate for ancient transport from the south. Garry oak, at least from one site at Yale in the Fraser Valley, is another candidate for human transport, although a second oak stand at Sumas was likely a Pleistocene relict (Yousry El-Kassaby, cited in Lepofsky 2004; Ritland et al. 2005).

Ecological and archaeological evidence of plant use may also reflect, at least indirectly, social aspects of plant management systems. For example, long-term or intensive use of particular berry patches, such as the Indian Heaven huckleberry grounds near Mount Adams (see chapter 2) or estuarine root-harvesting sites along the coast (see Deur 2000, 2005; and Lloyd 2011), may represent clan or family ownership, with restricted rights to their use. Stands of culturally modified trees showing repeated use over a long time period may also reflect ownership (Lepofsky and Pegg 1996). The presence of structures such as rock walls or posts or pegs around estuarine gardens, as described by Clan Chief Adam Dick, quoted earlier, would similarly indicate not only cultivation of these sites but also proprietorship and restricted use by individuals or families (Deur 2000, 2005). The existence of discreet stands or patches of high-status foods such as black huckleberries, soapberries, highbush cranberries, or crabapples can also reflect past management as well as ownership of these resources.

Similarly, plant resources known to have been traded[9] may be particular candidates for management and proprietorship since they would require production over and above a family's or community's immediate needs, which might be achieved through some form of intensification. Finally, the presence of plant remains from species that do not occur in a particular locality, and that therefore presumably arrived through trade, may reflect resource management systems that resulted in the establishment of trade links and networks (Croes 2003; Friedman 1978; Gill 1983, 2005a, 2005b; Lepofsky 2004). Trading itself could be considered a form of management, in terms of redistribution of berries, roots, and other resources in geographic space, which would have helped to mitigate supply shortages and to reallocate particularly productive and intensifiable resources. The technologies developed to allow storage of resources, such as smoking and drying, likewise assisted in the temporal and seasonal redistribution of berries and other resources, extending their availability and use over the year (Thornton 1999).

There may well be, as suggested earlier, a direct connection between plant management and the development of social stratification, but much more research would need to be undertaken to verify this and to indicate the direction of cause and effect. For example, digging stick handles, which first appear in archaeological contexts in northwestern North America around 2,400 years ago, at a time of increasing social and political complexity, may reflect simultaneously a convergence of increased harvesting efficiency for prime carbohydrate-rich foods and the enhanced status of the users, as well as, possibly, increased specialization of occupations – particularly women's root digging as an important contribution to a group's overall well-being (Hayden and Schulting 1997; M.B. Ignace, Ignace, and Nord 2009; Schulting 1994; see also chapter 9). An abundance of grinding and pounding implements and the diversification of earth oven hot-rock cooking, which occurred over this same time period (earlier in the southern Interior Plateau), may be other strands of the same overall trend. Combined with growing populations and increasing sedentism, these elements may collectively represent a "carbohydrate revolution," with increasingly sophisticated management of species like camas and land use intensification of sites like prairies and upland meadows (Thoms 2008a, 2008b).

A similar move toward intensification of berry production – also evidently linked to increasingly complex social organization – seems to parallel that of root resources, or "geophytes." Starting around 2,500 years ago, there is evidence from diverse regions of highly productive berry-picking sites. Berries, like some of the key root vegetables, were highly valued in feasting and trade (Thornton 1999). Large quantities of berries in archaeological contexts can signify not only large human populations but also increasing complexity in processing, as well as possible proprietorship and elite control of berry resources. Although the records of focused management practices for berries are somewhat sparse, above and beyond burning over berry patches (Boyd 1999b; Budhwa 2007), there are some compelling descriptions from different parts of the study area that hint at social stratification, large-scale production, and even cultivation in relation to berries from the time of European contact back through the generations (see Budhwa 2007; McDonald 2003, 2005; Thornton 1999; Turner 2005; Turner and Peacock 2005; and Peacock and Turner 2000). One example is the hundreds of thousands of huckleberry and/or blueberry (*Vaccinium* spp.) seeds recovered from hearths and storage areas associated with high-status families at Psacelay, a proto-historic Ts'msyen village on the Skeena River; an entire box of these berries was found, associated with one household chief (Lepofsky 2004; Martindale 1999). This finding not only indicates well-developed technologies for berry processing and storage but also hints at the social contexts for berry intensification and management, with possible intentions for feasting or trade. Similarly, the drying trenches at Indian Heaven in the Mount Adams area, found in association with constellations of bark-stripped trees and remains of temporary shelters, are indications of organized intensification of huckleberry production.[10]

Historical Development of Management Practices in Northwestern North America

Traditional Ecological Knowledge systems are infused with practices and concepts and with modes of teaching, learning, and social conduct that can be related directly and indirectly to resource stewardship and conservation on various scales. Management and resource-conserving approaches, drawn from traditional knowledge systems, have been described for Indigenous and local peoples of many parts of the world and for many different cultures and environments.[11] The origins of these approaches are often obscured in ancient history and embodied in metaphorical teachings. For the Straits Salish reefnet fishery, mentioned previously, tradition holds that it was the Salmon people themselves who taught humans this fishing method, along with the proper protocols for the fishery, including performing the First Salmon ceremony and burning the sacred seeds of *q'əx̱mín* ("Indian celery," *Lomatium nudicauḷe*) along the beach as food for the Salmon spirits (see chapter 12).

It seems likely that a number of different plant – and animal – resource management techniques were originally learned by observing other animals and their effects on the growth and productivity of plants.[12] In an example from north of the study area, in describing bears' harvesting of *masru* ("Indian potato," *Hedysarum alpinum*) as observed by the Kotzebue of Alaska, Anore Jones (1983, 117) writes, "As we dug we came to holes where *masru* had been dug a day ago – there were no tracks on the beach when we came! Strangely enough, not all the roots were taken, just the largest and some were torn off. Then we saw the bear tracks. He too had dug sporadically not taking every plant." Many plants, like *masru*, are adapted to withstand and renew themselves with moderate levels of disturbance, such as caused by animals, including:

- digging, overturning the soil, consumption, transport, and caching of underground storage organs like wapato, camas, spring beauty, and yellow glacier lily by burrowing rodents and by geese, swans, grizzly bears, and other wildlife;

11-3 | Willow (*Salix* sp.) along the Columbia River, BC, with branch cut by beaver, showing the resprouting of long, slender withes just below the cut.

- herbivory of plant leaves and stems by deer, elk, bears, and other animals, including the grazing of eelgrass leaves and rhizomes in the intertidal environment by black brant, Canada geese, and other waterfowl;
- scraping away of patches of tree bark by bears to expose the edible cambium and inner bark tissues (Dilbone 2011);
- sequestration and scattering of fruits, seeds, and nuts by rodents, squirrels, bears, and birds;
- breaking or "pruning" of berry branches and of berries and other propagules by bears, elk, deer, beaver, and other large species (figure 11-3); and
- selection or avoidance of certain parts or growth stages of plants.

Other episodes of natural disturbance – floods, ice jams, storms, exceptionally high tides and tsunamis, windthrow of trees, fire, and even lava flows – also present opportunities for observing the regenerative capacity of plants and were probably a source of "learning through imitation" in the development of management at the habitat level. Lessons from catastrophic depletion from overharvesting or other forms of inappropriate treatment of resource species and habitats may also have factored into the development of management systems. Additionally, one individual or group would probably have learned from the lessons of others. Stories passed from group to group and from generation to generation would serve to spread widely certain lessons learned, without the same mistakes having to be made repeatedly. The social organization necessary for management may also have been learned through observation of social animals such as wolves and of the role of leaders in determining a group's diversified activities (E.R. Atleo 2011; Turner 1997b, 2005; see also chapter 9). An example of how grizzly bears may have informed people about a practice leading to intensification of *k'ám'ts* (yellow glacier lily) was provided by Lil'wat (Stl'atl'imx) elder Alec Peters (pers. comm., 1987) of Mount Currie, who observed grizzlies digging these bulbs in the subalpine parklands:

> You've got to go pretty well up the top of a mountain for it [*k'ám'ts*], the summit. In a certain time of the years they [people] pick it ... The old-timers used to pick it and dry it for winter use. I know the grizzly bears they dig it out too. They use their big claws like that [raking motion], and they just leave it like that in the sun, you know. I guess they must taste good when they're dry. They don't eat it right away. I've watched them. A long time, I've watched the grizzly bear, digging it out. I've seen them *k'ám'ts* laying like that.

Leaving these bulbs to wilt where they were dug, and allowing them to partially dry before cooking them, was a common practice for people, too. It is possible that people actually learned of the edibility of these bulbs by observing grizzlies. Certainly, in traditional narratives of the Interior Plateau, grizzlies are associated with this plant (Loewen 1998; Teit 1898; see also chapter 12). In one example, from a Nlaka'pamux story, Grizzly Bear women harvested glacier lily bulbs and benefited from a burned-over mountainside: "four Grizzly Bear sisters also lived in the mountains [near Botanie Valley] and used to dig roots, wandering about from place to place. At length they came to the spot which had been burned [by four brothers who had burned over a piece of mountainside for their sister so that the *s'ka'mitc* (*sk'ém'ets*) root "should yield a better crop"], and found the *s'ka'mitc* root very plentiful. Here they dug" (Teit 1898, 72). People may have also learned from grizzlies and other animals to harvest the glacier lily bulbs selectively at maturity (when the seed capsules, if brushed, readily release quantities of seeds) and to separate and leave behind the small corms and smaller individual bulbs, as well as to improve their flavour (and possibly nutrient value) through the wilting process (Loewen 1998).[13] Combining these techniques with observations of the plant's responses to burning from lightning-ignited fires in the high country, and of the general correspondence between the size of bulbs and the number of flower heads per stem, would have provided people with a whole suite of practices to test, enabling optimization of this important food of the northern Interior Plateau.

Similar scenarios could be argued for peoples' intensive use and cultivation of northern riceroot. In fact, it seems that in the estuarine tidal marshes of the central and northern Northwest Coast, there was a three-way symbiosis between riceroot, grizzly bears, and humans. In search of nutrient-rich roots and rhizomes of sedges and other plants, as well as the riceroots themselves, grizzlies gouged out and "weeded" multiple small bowl-shaped microhabitats (which they are still creating and can still be found dotted around some tidal flats). Here riceroot plants would thrive and multiply, with the small rice-like bulblets and sprouts generated from the main bulbs being scattered by the disturbance of the digging, being nurtured by the warm, dark soil exposed by the dig, and having plenty of moisture and nutrients. On one occasion, in the Kitlope Valley, I counted the total number of these vegetative propagules from fifteen riceroot bulb sets removed from a single grizzly-created depression: the total number of bulblets and larger sprouting segments was 706 (pers. obs., July 1993; Turner 1994). Each one of these, separated and replanted, would have been capable of growing into a new plant. It seems likely from this evidence that past human harvesting of the grizzly-enhanced riceroot, with disturbance caused by digging sticks and scooping out the bulbs, would invariably have left behind enough smaller propagules to maintain the riceroot populations and that within a few years of harvesting, some of these bulblets would have grown into mature bulbs, big enough to support flowering and fruiting tops. Since, based on its existence and cultural use in northeast Asia and the Aleutians, riceroot may have been one of the most ancient foods along the coast, known to the earliest people entering the New World (see chapter 2, table 2-1), it is conceivable that Asian bears were the ones who originally taught people about its use.

Many people have also noted the close association between black bears and berries; it is easy to recognize the remains of copious amounts of huckleberries, saskatoon berries, chokecherries, soapberries, kinnikinnick berries (also aptly named "bearberries"), crabapples, and other kinds of fruits in their excrement and to see the places where the bears have been feeding, with resulting broken bushes and scattered berries all around. Of particular interest is the presence of bears in areas following fires.[14] The closeness people feel to black bears, because of the similarities between their feeding habits and those of humans, presents special opportunities for learning from them and imitating their actions.

The roles of other animals in traditional management are also hinted at in traditional stories (see chapter 12). In one Haida narrative,[15] Deer (who lived in Tlingit country) was said to keep the skunk-cabbage he ate "just like a garden" (Swanton 1908, 446). Beavers in another story were proprietors of highbush cranberry meadows (Turner 2004a). And the Mountain Chief in Carrier country, featured in a Nuxalk story, grew soapberries on his slopes (McIlwraith 1948). Spiders grew Indian-hemp in the sky country in another story (Turner, Thompson, et al. 1990), and Goose Woman on Haida Gwaii provided edible roots for her husband's people during a famine (Turner 2004a). In Hupacasath (Nuu-chah-nulth) country, at Port Alberni on Vancouver Island, Swan women were apparently regarded as the original diggers of camas bulbs, evidently tending the root patches (called *tł'ayaqak*). They harvested immense quantities of the bulbs by ritual singing, naming their chief camas digging grounds T'ik^{w}atqoowise (Place Where You Dig Roots): "Their heads were bobbing up and down as they used their breasts [to push the digging sticks down]." When they finished digging roots, the Swan women gave a feast, and then they gave about fifty place names to sites in the area (Arima et al. 1991, 190–1; Sapir 1913–14).

Observations about resource depletion and about the positive effects of certain practices would have been readily passed from group to group or from generation to generation by cultural transmission – for example, through oral history and discourse. Lessons gained from "trial and error" on a small scale, or focused on one type of plant or one type of habitat, could be transferred and adapted to other resources and, in some cases, to other habitats. Developing management through such extrapolation must have been one of the major ways that people built up their diverse practices over the entire area. For example, understanding that a tree would eventually heal itself if a piece of bark was removed but that it would die if girdled could be extended to many kinds of management scenarios, including harvesting cedar-bark strips for weaving, gathering inner bark for food, and removing bark for medicinal purposes.

It would have been obvious that certain types of berries tend to respond similarly to fire or to pruning, so once people learned to enhance the productivity of one kind, the lessons would have been transferred to others. Similarly, selectively harvesting bulbs or corms of a certain size while leaving or replanting the smaller ones so that they could continue growing is a practice that would have worked to perpetuate a number of different "root" crops. Even adopting particular worldviews and belief systems (including particular taboos and constraints against harvesting certain resources at certain times) and passing these on through intergenerational learning must have been projected over wide and diverse areas because of their efficacy in conserving key resources or, conversely, because of the damage resulting when these ideas were ignored or bypassed. In other words, sophistication and complexity in management systems probably developed incrementally, starting with observations of specific effects of practices on particular species and building up to suites of practices and overarching approaches embedded in social institutions and worldviews (Turner and Berkes 2006).

Once a particular management technique had been tried and proven effective for a given resource in a specific habitat, the word must have spread quickly, from group to group, via stories, demonstration, and imitation, much in the same way that the potato and its cultivation, once introduced locally, quickly moved from settlement to settlement until, within just a few years, almost everyone was clearing plots and growing potatoes (see chapter 4). In the way of all innovations, each individual learning about a given technique and putting it into practice would have changed it slightly and adapted it to local species, conditions, and circumstances. It is possible, too, that management techniques such as seasonal landscape burning, tilling the soil, pruning berry bushes, or selective harvesting were developed more than once, perhaps multiple times, in more than one region within or beyond northwestern North America. Many of these innovations or their precursors must be relatively old and could also have evolved differently in different regions over time from the same original idea or practice. (This is probably the way that basketry techniques, fishing technologies, and specific house designs emerged.) Some particular practices that were species-specific, such as the partial harvest of redcedar bark, branches, and roots, would have been disseminated together with the spread of that species itself across space and over time.

European Impacts on Indigenous Land and Resource Management

Of the entire range and diversity of methods, strategies, and approaches that people developed to sustain and promote their plant resources and habitats over many millennia, very few were actually recognized as legitimate systems of food production by the European newcomers. A widespread perspective of European officials is summarized in relation to one Ts'msyen group. When the Indian commissioner created the Kitsumkalum people's reserves, he made them a promise: "You will not be confined to your reserves. You can go on the mountain to hunt and gather berries as you have always done. Some think it would be a hardship that the hunting grounds should not be defined, but the government does not see how that could be done, for an Indian goes where he will to hunt, or gather berries. No survey could be made of them" (cited in McDonald 2003, 59–60).

As McDonald (2003, 60) observes, this seemingly generous and straightforward statement is rife with misconceptions. Not only did the commissioner miss the wide range of locations and habitats where people routinely harvested berries and hunted, but he also failed to recognize the diversity of the resources that they used, the fact that all Ts'msyen *laxyuup* ("house lands or territories") and resources, including kelp beds, cedar and crabapple stands, berry patches, and patches or gardens of edible roots, were under the ownership and jurisdiction of Ts'msyen chiefs, and that lineages were subject to strict Ts'msyen laws and to many practices that helped to sustain their productivity. In short, the commissioner trivialized and dismissed the Kitsumkalums' sophisticated plant management strategies. This lack of recognition of peoples' management activities and full occupancy of their territories is reflected in innumerable writings and documents, including treaties, from the colonial era onward. Similar assumptions were made in the imposition of the so-called "Douglas treaties" on Vancouver Island, in which the importance of camas to the Straits Salish and Halkomelem peoples and the practices people used to maintain and promote this important food were completely overlooked. In the wording of the treaties, only "villages" and "enclosed fields" were to be left for people's use, with a stipulation that people could hunt over unoccupied lands and fish as formerly, but there was no mention or recognition of camas cultivation, maintenance of prairies, or the tending of berry patches, all of which were integral parts of peoples' food production systems (British Columbia 1875; see also chapter 4).

Such misconceptions have been far-reaching, and many are still in place. They have, in fact, influenced or biased the directions of research in archaeology and ethnography, perpetuating the stereotype of First Peoples as mere "hunters and gatherers" – or, to use the words of Governor James Douglas, "wandering denizens of the forest" – who were subjected to the whims of nature and had an inferior understanding of their environments.[16]

NATURE AND SCALES OF APPLICATION OF PLANT RESOURCE MANAGEMENT

As described by Peacock (1998),[17] in terms of Indigenous plant management strategies and their impacts on the productivity and availability of plant resources, the use of horticultural methods – including selective harvesting, digging and replanting, tilling and weeding, sowing and transplanting, pruning and coppicing, and burning – was, and is, guided by management activities such as scheduling of seasonal rounds, rotation of harvesting locales, and controlled access (ownership of resource areas), as well as by religious and moral sanctions. Management invariably requires, along with an understanding of the technical aspects of growth promotion and ecological succession, a social context of responsibility for resources, a division of labour, and special expertise about the timing and scheduling of activities, as well as ceremonial recognition and distribution of the products of such activities. Thus "cultivation" of a landscape was thoroughly woven into the social fabric of First Nations communities.

These social and ecological processes collectively regulated the scale, frequency, and intensity of anthropogenic disturbance, and their effects were manifested on different scales.

At the *plant species* or *plant population* level, they resulted in increased productivity of selected species through a combination of altering age structure and longevity, changing density and distribution, increasing the range and yield of a species, and possibly also changing its genetic makeup. At the *plant community* or *habitat* level, these methods and activities may have increased overall species richness and habitat diversity by altering the ecological processes affecting community dynamics, structure, and composition. For example, harvesting and clearing activities by humans can expand the proportion of edges, or ecotones, in areas of homogeneous vegetation and can maintain a higher ratio of certain types of productive plant communities than would otherwise occur (Turner, Davidson-Hunt, and O'Flaherty 2003). Such practices may have, in turn, benefited other animal species, such as bears or geese, and created even more complex and productive habitats. The estuarine root gardens of the Kwakwaka'wakw and other Northwest Coast peoples, the oak savannah camas prairies of the southern coastal and interior peoples, and the subalpine root-harvesting meadows of Interior Plateau peoples are all examples of anthropogenic habitats, formerly maintained by a host of human activities that included tilling, selective harvesting, and creating and maintaining patchworks of early successional stage ecosystems amid wooded or forested areas – in essence, developing and maintaining optimal, or "cultural keystone," habitats for key resource species to grow and thrive (Deur 2000; Deur and Turner 2005; Turner 1999; Turner, Deur, and Mellott 2011).

Ultimately, the application of complexes of management methods, activities, and approaches over time resulted in entire landscapes or territories that embraced all of the species, communities, habitats, and harvesting sites over which a people's influence extended. An anthropogenic landscape could encompass a group's total territory, with all of its interconnecting components, throughout seasons and annual cycles or even over centuries. The development of conventions that related to proprietorship and inherited authority over resources, patterned seasonal rounds and variable harvesting regimes, and culturally mediated prescriptions for humans' relationships to plants and animals were all aspects of anthropogenic landscape development.

People had to pay attention to many ecological and environmental details when harvesting and managing their plant resources. Subtle differences in moisture or sunlight could affect the season and productivity of species. The quality of berries and other resources, for example, can vary considerably with elevation and other environmental factors. Gitxsan elder Winnie Wesley explained that "mountain berries" are sweeter than those growing at lower elevations and that the best berry patches – especially for huckleberries and blueberries –

occur "halfway up the mountain" – in the montane and lower subalpine forest zones dominated by conifers, at about 900 to 1,200 metres (cited in McDonald 2003, 54; and Johnson Gottesfeld 1993, 96). At the Yakima berry grounds at Indian Heaven near Mount Adams, the two kinds of huckleberries people picked – black huckleberries (*Vaccinium membranaceum*) and Cascade blueberries (*V. deliciosum*) – tend to grow at slightly different elevations: the former occur at around 900 to 1,200 metres and the latter at around 1,500 metres. Such berry patches would be used and repeatedly burned over for seventy years or longer (Kirk and Daugherty 2007). Different aspects and slopes also produced different results for management.

Quantities of Resources and Time Required in Harvesting and Processing

Human population densities and corresponding quantities of resource species required to sustain them are obvious questions in the debate around the extent and intensity of food production systems in northwestern North America. There are still diverse views on Indigenous peoples' demographics over time and in different regions. The pre-European contact population of Northwest Coast peoples is generally estimated at 200,000 to 400,000 based on archaeological midden sites and other factors (Boyd 1999b; Hebda and Frederick 1990). A series of smallpox and other disease epidemics, starting in the late 1790s or even earlier, with several major outbreaks in the early to mid-1800s, resulted in a huge mortality, ranging from 20 to 80 per cent of Indigenous populations, depending on the area (Boyd 1999a). In some cases, entire villages were decimated. On the southern tip of Vancouver Island, the pre-epidemic population of the Songhees Straits Salish in the Victoria area alone is suggested at around 11,000, distributed around several villages along the southern coastline (Boyd 1999a). This represents a high density for a population subsisting almost entirely on local resources – and immense quantities of plant resources used on an annual basis.

In terms of camas bulb harvesting, for example, for a Songhees community of 1,000 people, or 100 families (assuming an average of 10 people per family), and given that a single family might collectively harvest around 100 kilograms of camas bulbs in a year,[18] the total harvest for that community alone could be in excess of 1 million bulbs.[19] Among the Tsilhqot'in, elder Minnie Charleyboy (pers. comm., 2003) recalled that each family, working together in small groups, could harvest approximately 100 pounds (over 45 kilograms) of spring beauty corms per day, and that a family needed about 250 pounds (over 110 kilograms) of the corms per year to meet its needs.[20] Unfortunately, there are very few such actual records of quantities harvested or consumed or of harvest rates for plant resources in general. Table 11-1 provides a summary of some rates of harvest for different resources, mostly based on experimental harvesting trials, and table 11-2 presents a compilation of records and estimates from various sources of

Table 11-1 | Measured or estimated rates of harvest for indigenous plant resources of northwestern North America

Species or type of plant resource	*Rates of harvest*	*Reference(s)*
ALGAE		
Red laver seaweed (*Pyropia abbottiae*; syn. *Porphyra abbottiae*): edible alga	~ 350 kg wet seaweed/year/family; 75 lbs (~ 34 kg) wet seaweed/hr/person; 100 squares dried (~ 100 lbs wet, or ~ 45 kg) seaweed/day/family	Gitga'at (Ts'msyen) (Chief Albert Clifton, pers. comm., 2008; Helen Clifton, pers. comm., 2008; Turner, pers. obs., 2008; Turner 2003a)
FERNS AND FERN ALLIES		
Bracken fern (*Pteridium aquilinum*): edible rhizomes	~ 250 mL/15 mins/person (~ 1 L/hr/person)	Nuxalk (Lepofsky, Turner, and Kuhnlein 1985, table 6)
CONIFERS		
Amabilis fir (*Abies amabilis*): edible inner bark	"You can get about two big (10-gallon) buckets [~ 38 litres each] full of this inner bark fairly readily"	Gitga'at (Ts'msyen) (Elizabeth Dundas, cited in Turner and Thompson 2006, 40)
Western hemlock (*Tsuga heterophylla*): edible inner bark	"You can get about two big (10-gallon) buckets [~ 38 litres each] full of this inner bark fairly readily"	Gitga'at (Ts'msyen) (Elizabeth Dundas, cited in Turner and Thompson 2006, 40)
FLOWERING PLANTS		
Saskatoon berry (*Amelanchier alnifolia*): edible fruit	~ 1,000 berries/bush; 250 mL/5 mins/person (~ 1 L/20 mins/person, or ~ 3 L/hr/person)	Nuxalk (Lepofsky, Turner, and Kuhnlein 1985, table 6)
Indian-hemp (*Apocynum cannabinum*): stem fibre	In 2–3 days a woman could harvest enough stems to make five bundles, each 3–4 ft (~1 m) long and 5 in. (~ 12 cm) thick	Interior Plateau (Teit 1930, 68)
Kinnikinnick (*Arctostaphylos uva-ursi*): edible berries	~ 250 mL/20 mins/person (~ 0.75 L/hr/person)	Nuxalk (Lepofsky, Turner, and Kuhnlein 1985, table 6)
Pacific silverweed (*Argentina egedii*): edible roots	~ 250 mL roots dug in 30 mins (500 mL packed roots/hr/person); for Nuxalk, ~ 250 mL/20 mins/person (~ 0.75 L/hr/person), or 0.5–1 kg/hr/person (in a good site)	Kitlope Valley (Haisla) (Turner 1994); Nuxalk (Lepofsky, Turner, and Kuhnlein 1985, table 6; Turner and Kuhnlein 1982)
Camas (*Camassia* spp.): edible bulbs	Estimated 400 bulbs per hour, or 6 bulbs per minute, for *Calochortus amabilis* and other	California (M.K. Anderson 2005, 297); Nez Perce (A.G. Marshall 1977)

/continued

Table 11-1 | continued

Species or type of plant resource	*Rates of harvest*	*Reference(s)*
	bulbs from California (rates of camas bulb harvest are probably similar); 80–90 lbs (36–40 kg) per day for a good digger; 40–50 lbs (18–22 kg) easily per day	
Spring beauty, or mountain potato (*Claytonia lanceolata*): edible corms	Recently (with hardened soil and introduced weeds) ~ 0.5–1 L/hr/person; family of 4 could dig ~ 100 lbs (~ 45 kg) of corms per day	Tsilhqot'in (Minnie Charleyboy and Maryann Solomon, pers. comm., 2003)
Bunchberry (*Cornus canadensis*): edible fruit	Experimental harvesting: ~ 350 berries/m^2; ~ 250 mL/10 mins/person (~ 1 L/40 mins/person, or 1.5 L/hr/person)	Nuxalk (Lepofsky, Turner, and Kuhnlein 1985, table 6)
Black hawthorn (*Crataegus douglasii*): edible fruit	550 fruits/tree; ~ 250 mL/1.5 mins/person (~ 1 L/6 mins/person)	Nuxalk (Lepofsky, Turner, and Kuhnlein 1985, table 6)
Yellow glacier lily (*Erythronium grandiflorum*): edible bulbs	Recently (with hardened soil and introduced weeds) ~ 1 L/1.5 hrs/person; a family of four could harvest 100 lbs (~ 45 kg) per day of these bulbs	Secwepemc (Mary Thomas, pers. comm., 1994); Tsilhqot'in (Minnie Charleyboy, pers. comm., 2003)
Wild strawberry (*Fragaria vesca*): edible fruit	Experimental harvesting: ~ 250 mL/30 mins/person (~ 0.5 L/hr/person)	Nuxalk (Lepofsky, Turner, and Kuhnlein 1985, table 6)
Northern riceroot (*Fritillaria camschatcensis*): edible bulbs	Experimental harvesting: ~ 125 mL (1/2 cup; 7 large bulbs; total of 706 propagules) dug from grizzly hollow in 10 mins (~ 0.75 L/hr/person); for Nuxalk, ~ 250 mL/hr/person	Kitlope Valley (Haisla) (Turner 1994); Nuxalk (Lepofsky, Turner, and Kuhnlein 1985, table 6)
Salal (*Gaultheria shallon*): edible berries	10 berries/branchlet; ~ 250 mL/8 mins/person (~ 1 L/32 mins/person)	Nuxalk (Lepofsky, Turner, and Kuhnlein 1985, table 6)
Bitterroot (*Lewisia rediviva*): edible roots	~ 250 g (1/4 lb)/hr/person to dig and clean fresh bitterroot	Secwepemc (Mary Thomas, pers. comm., 1994)
Biscuitroot (*Lomatium cous* and other spp.): edible roots	~ 300 biscuitroots dug/hr/person; 1 bushel (~ 60 lbs, or ~ 27 kg) dug/day/woman; good digger gathers 50–75 lbs (~ 23–34 kg) of *L. cous* per day	Interior Plateau (Hunn, Selam, and family 1990; journals from the 1840s and 1850s cited in Kirk and Daughtery 2007, 66; A.G. Marshall 1977)

Table 11-1 | continued

Species or type of plant resource	*Rates of harvest*	*Reference(s)*
Nootka lupine (*Lupinus nootkatensis*): edible roots	Experimental harvesting: ~ 250 mL/12 mins/person (~ 1 L/48 mins/person, or ~ 1.2 L/hr/person)	Nuxalk (Lepofsky, Turner, and Kuhnlein 1985, table 6)
Pacific crabapple *Malus fusca*): edible fruits	~ 1,000 crabapples avg./tree; 250 mL fruits/5 mins/person (1 L/20 mins/person, or 3 L/hr/person)	Nuxalk (Lepofsky, Turner, and Kuhnlein 1985, table 6)
Labrador tea (*Rhododendron groenlandicum*): leaves for tea	~ 250 mL/2.5 mins/person (~ 1 L/10 mins/person)	Nuxalk (Lepofsky, Turner, and Kuhnlein 1985, table 6)
Gray currant, or stink currant (*Ribes bracteosum*): edible berries	0–150 berries/bush; ~ 250 mL/5 mins/person (~ 1 L/20 mins/person, or 3 L/hr/person)	Nuxalk (Lepofsky, Turner, and Kuhnlein 1985, table 6)
Coastal black gooseberry (*Ribes divaricatum*): edible berries	350 berries/bush; ~ 250 mL/12.5 mins/person (~ 1 L/50 mins/person)	Nuxalk (Lepofsky, Turner, and Kuhnlein 1985, table 6)
Nootka rose (*Rosa nutkana*): edible rind of fruit	30–50 hips/bush; ~ 250 mL/5 mins/person (~ 1 L/20 mins/person; but need considerable processing for edibility)	Nuxalk (Lepofsky, Turner, and Kuhnlein 1985, table 6)
Wild raspberry (*Rubus idaeus*): edible berries	450–600 berries/bush; ~ 250 mL/30 mins/person (~ 500 mL/hr/person)	Nuxalk (Lepofsky, Turner, and Kuhnlein 1985, table 6)
Blackcap (*Rubus leucodermis*) and trailing blackberry (*R. ursinus*): mixed edible berries	Experimental harvesting: ~ 1-1.5 L/hr/person	Vancouver Island (Turner, pers. obs. picking near Spectacle Lake logged site on southern Vancouver Island, July 2009)
Thimbleberry (*Rubus parviflorus*): edible berries	175 berries/6m^2; ~ 250 mL/10 mins/person (~ 1 L/20 mins/person, or ~ 3 L/hr/person)	Nuxalk (Lepofsky, Turner, and Kuhnlein 1985, table 6)
Salmonberry (*Rubus spectabilis*): edible berries	30 berries/bush; 250 mL/5 mins/person (~ 1 L/20 mins/person, or 3 L/hr/person)	Nuxalk (Lepofsky, Turner, and Kuhnlein 1985, table 6)
Red elderberry (*Sambucus racemosa*): edible berries	Experimental harvesting: ~ 2 L/5 mins/person with stems, or ~ 1 L/hr/person cleaned and destemmed 18 berry clusters per bush; ~ 250 mL/5 mins/person with stems (~ 1 L/20 mins/person, or 3 L/hr/person)	Kitlope Valley (Haisla) (Turner 1994); Nuxalk (Lepofsky, Turner, and Kuhnlein 1985, table 6)

Table 11-1 | continued

Species or type of plant resource	Rates of harvest	Reference(s)
Springbank clover (*Trifolium wormskioldii*): edible rhizomes	~ 500 mL rhizomes dug/30 mins in Kitlope experiment (1 L/hr/person); for Nuxalk, ~ 250 mL/30 mins/person (~ 500 mL/hr/person)	Kitlope Valley (Haisla) (Turner 1994); Nuxalk (Lepofsky, Turner, and Kuhnlein 1985, table 6)
Cattail (*Typha latifolia*): leaves for mats	"Even an experienced mat-maker must have needed 3–4 days to gather materials for a single mat, prepare them, make the string and do the sewing; ~ 6–8 mats needed to cover a shelter of size indicated by postholes from dry part of Hoko site."	Hoko River (Kirk and Daugherty 2007, 98)
Alaska blueberry (*Vaccinium alaskense*): edible berries	Experimental harvesting: 100 berries/bush; ~ 250 mL/10 mins/person (~ 1 L/40 mins/person, or 1.5 L/hr/person)	Nuxalk (Lepofsky, Turner, and Kuhnlein 1985, table 6)
Alaska blueberry (*Vaccinium alaskense*) and red huckleberry (*V. parvifolium*) mixed: edible berries	Experimental harvesting: 1–1.2 L/hr/person, or ~ 5–8 L/day/person	Kitlope Valley (Haisla) (Turner 1994)
Oval-leaved blueberry (*Vaccinium ovalifolium*): edible berries	Experimental harvesting: ~ 350 berries/bush; ~ 250 mL/10 mins/person (~ 1 L/40 mins/person, or 1.5 L/hr/person); "it can take hours to fill a 2.5 gallon bucket"	Nuxalk (Lepofsky, Turner, and Kuhnlein 1985, table 6); Gitga'at (Ts'msyen) (Turner and Thompson 2006, quote at 69)
Red huckleberry (*Vaccinium parvifolium*): edible berries	Experimental harvesting: ~ 300 berries/bush; ~ 250 mL/8 mins/person (~ 1 L/32 mins/person, or ~ 2 L/hr/person)	Nuxalk (Lepofsky, Turner, and Kuhnlein 1985, table 6)
Mountain huckleberries (*Vaccinium membranaceum, V. deliciosum*): edible berries	Family of 7 or 8 people could collect around 500 kg ("half a ton") of fresh and dried berries in a season	Yakima Sahaptin and others at Indian Heaven near Mount Adams (Kirk and Daugherty 2007, 69)
Highbush cranberry (*Viburnum edule*): edible fruits	Experimental harvesting: ~ 100 berries/bush; ~ 250 mL/4 mins/person (~ 1 L/16 mins/person, or ~ 4 L/hr/person)	Nuxalk (Lepofsky, Turner, and Kuhnlein 1985, table 6)

Note: Rates of harvest are based on experimental harvesting or observation.

Table 11-2 | Quantities of plant resources harvested by Indigenous peoples of northwestern North America

Species or type of plant resource	*Quantities harvested*	*Group (references)*
ALGAE		
Red laver seaweed (*Pyropia abbottiae*; syn. *Porphyra abbottiae*): edible alga	~ 350 kg wet seaweed/year/family; 100 squares dried (~ 100 lbs wet, or ~ 45 kg) seaweed/day/family; Annetta Robinson's mother made 400 seaweed trays back in Kitkatla for her family to use (Helen Clifton)	Gitga'at (Helen Clifton, pers. comm., 2008; Turner, pers. obs., 2008; Turner 2003a)
FERNS AND FERN ALLIES		
Spiny wood fern (*Dryopteris expansa*): edible rootstocks	Rootstocks of this fern dug up "by the sack" and pit-cooked by the Nisg̲a'a, among other peoples	Nisg̲a'a (Nisg̲a'a Tribal Council 1995, 83)
"Fern roots" (probably referring to bracken fern, *Pteridium aquilinum*): edible rhizomes	Immense piles cooked for feasts (apparently higher than a person's height) by steaming (water had to be poured from the roof of the house)	Nuu-chah-nulth (Drucker 1951, 62)
CONIFERS		
Subalpine fir (*Abies lasiocarpa*), white pine (*Pinus monticola*), western redcedar (*Thuja plicata*), and other spp.: bark sheets for dwellings, canoes, etc.	Dozens of bark sheets needed for each summer lodge or dwelling; 3–6 large sheets per canoe; 5–6 sheets per cache pit lining; possibly 25–30 rectangular bark sheets, each 1 × 2 m, per family per year; Fort Langley bought enough cedar-bark sheets to roof the buildings within two years	General (Turner, pers. obs. from photographs and canoes)
Western larch (*Larix occidentalis*): sweet edible sap	A productive tree could yield at least 4–5 L of sap twice a year	Spokan (J.A. Ross 2011)
Whitebark pine (*Pinus albicaulis*): edible seeds or "nuts"	"Considerable quantities" of seeds eaten by Upper Stl'atl'imx; Edith O'Donaghey's (pers. comm., 1985) father used to bring the cones home "by the sackful"	Stl'atl'imx (Teit 1906a, 222; Turner 1992b)
Western redcedar (*Thuja plicata*): wood for canoes, house posts and planks, boxes; bark sheets for siding and roofing	Single large (multifamily) house required 500,000 to 1 million board feet of lumber to maintain (might stand for 400 years) [single family modern house requires 10,000–12,000 board feet]; cedar-bark sheets for building Fort Langley	Meier archaeological site near Portland, Oregon (Ames and Maschner 1999, 167)

Table 11-2 | continued

Species or type of plant resource	*Quantities harvested*	*Group (references)*
Western hemlock (*Tsuga heterophylla*): edible cambium or "sap"	"Many hundred score of dried cakes of hemlock sap"	Peter O'Reilly, Indian reserve commissioner (Department of Indian Affairs 1900, 83)
FLOWERING PLANTS		
Saskatoon berry (*Amelanchier alnifolia*): edible fruit	Each family in any Interior Plateau community would have dried and stored several large basketfuls of these sweet fruits; probably totalled ~ 10 gallons (~ 38 litres) of dried berries every summer	Interior Plateau and Nlaka'pamux (Turner, Thompson, et al. 1990)
Indian-hemp (*Apocynum cannabinum*): stem fibre for basketry, mats, cordage, and nets	"Vast quantities" needed; ~ 5 stalks to produce 1 ft of cordage (~ 15 stalks/m of cordage); a 40-ft deer net requires ~ 7,000 ft of cordage (~ 35,000 plant stalks)	Interior Plateau peoples generally and Native Californians (Blackburn and Anderson 1993; M.K. Anderson 2005, 231; Stevens and Anderson 2006)
Edible camas (*Camassia* spp.): edible bulbs	"canoes laden with camas" brought by Cowichan to Fort Langley; ~ 100 camas bulbs per litre (conservatively estimated); ~ 100–260 kg/year/family (~ 10,000–26,000 bulbs harvested/year/family depending on size of bulbs); ~ at least 1 million bulbs per community of 1,000 annually (based on ~ 30 L/bushel, or 3,000 bulbs/bushel), or 2.6 million using figures calculated by Beckwith; 4 bushels per family equals ~ 12,000 bulbs per family; ~ 200,000 bulbs per family harvested every year from the interior camas (*C. quamash*) grounds of eastern Washington, Idaho, and western Montana (Thoms)	Central Coast Salish (Christopher Paul, pers. comm. to M. Babcock, 1968; Christopher Paul, pers. comm., 1969; Beckwith 2004, 101-2; Deur and Turner 2005; Maclachlan 1998, 40); Interior Salish (Thoms 1989)
Spring beauty, or mountain potato (*Claytonia lanceolata*): edible corms	~ 50–120 kg corms/year/family, or ~ 250 lbs/year/family	Nlaka'pamux (Turner, Thompson, et al. 1990); Tsilhqot'in (Minnie Charleyboy, pers. comm., 2003)

Table 11-2 | continued

Species or type of plant resource	*Quantities harvested*	*Group (references)*
Yellow glacier lily (*Erythronium grandiflorum*): edible bulbs	Households would harvest 500 lbs (~ 225 kg) of bulbs annually; family of 4 needed 250 lbs (~ 115 kg)	Secwepemc (Ike Willard, cited in G. Palmer 1975b); Nlaka'pamux (Turner, Thompson, et al. 1990); Tsilhqot'in (Minnie Charleyboy, pers. comm., 2003)
Seaside strawberry (*Fragaria chiloensis*): edible berries	Recently, ~ 4–5 gallons (~ 15–20 L) per year/family	Haida (Turner, pers. obs.; Turner 1995)
Salal (*Gaultheria shallon*): edible berries	~ 2,000 dried berries per litre; ~ 40,000 dried berries per 20-litre storage basket; 4 baskets per family equals ~ 160,000 berries/year/family; 10–20 million berries/year/community of 1,000 people; salal cakes 10–15 lbs (~ 4.5–7 kg) each (Gunther)	General, Northwest Coast, based on berry counts (Deur and Turner 2005; Gunther 1973)
Cow-parsnip (*Heracleum maximum*): edible green shoots	Stems of "wild rhubarb" (*Heracleum maximum*) were "very much used" by Secwepemc peoples	Teit (1909, 515)
Biscuitroot (*Lomatium cous* and other spp.): edible roots	300 biscuitroots/hr/person, or ~ 30 kg (~ 66 lbs)/day/person	Sahaptin (Hunn, Selam, and family 1990; journals from the 1840s and 1850s cited in Kirk and Daugherty 2007)
Pacific crabapple (*Malus fusca*): edible fruits	Lucille Clifton (cited in Turner, Robinson, et al. 2012, 349) used to put by ~ 4–5 barrels, each 5 gallons (~ 20 L), packed with crabapples and highbush cranberries every winter	Gitga'at (Ts'msyen) (Turner, Robinson, et al. 2012; Turner and Thompson 2006)
Prickly pear cactus (*Opuntia* spp.): edible succulent stems	"much used" by the southern bands of Secwepemc	Secwepemc (Teit 1909, 515)
Thimbleberry (*Rubus parviflorus*): edible sprouts	Harvested by the "armload" and canoe-load in the spring	Nuu-chah-nulth (George 2003); Kwakwaka'wakw (Adam Dick, pers. comm., 2004; Turner and Turner 2008)
Salmonberry (*Rubus spectabilis*): edible berries	Edible sprouts gathered by the "armload" in the spring	Kwakwaka'wakw (Adam Dick, pers. comm., 2004; Turner and Turner 2008)

Table 11-2 | continued

Species or type of plant resource	*Quantities harvested*	*Group (references)*
Wapato (*Sagittaria latifolia*): edible tubers	Large quantities dug around Fort Langley; "highly esteemed by the [Columbia River] natives, who collect vast quantities of it for their own use and for barter."	Halkomelem and Chinookan (Cox 1957, 79; Darby 1996; Garibaldi 2003)
Pacific willow and other willows (*Salix lucida* ssp., *lasiandra* and other *Salix* spp.): inner bark for reefnets and cordage	For a reefnet of ~ 10 × 10.3 m (~ 30 × 34 ft), 750–1,688 m of willow shoots (1–3 cm diameter) would be required; each family constructed a new net each season	Straits Salish (Townsend 2009)
Tule, or hardstem bulrush (*Schoenoplectus acutus*): stems for mat making	~ 1,000 tule stalks per large sewn tule mat (~ 1 × 4 m); ~ 10 mats or more per family, especially for the large lodges used as dwellings in Interior Plateau, requiring a dozen or more large mats, totalling 12,000 tule stems per mat lodge; for Spokan, 20 woven tule mats (1.1 × 3 m) for each lodge	Straits Salish and Interior Plateau peoples, based on counts of tule mat elements in museum collections (Deur and Turner 2005; J.A. Ross 2011)
Springbank clover (*Trifolium wormskioldii*): edible rhizomes	Immense piles cooked for feasts (apparently higher than a person's height) by steaming (water had to be poured from the roof of the house)	Nuu-chah-nulth (Drucker 1951, 62)
Mountain huckleberries (*Vaccinium membranaceum*, *V. deliciousum*): edible berries	"Large quantities" dried on racks over fire hearths; well over 200 berry-drying features in upland meadows (archaeological); 100 gallons of huckleberries per acre (~ 380 L/0.4 ha)	Indian Heaven (southwest of Mount Adams), Yakima, and others (Kirk and Daugherty 2007, 167; Lepofsky 2004; Mack 1992; Mack and McClure 1996, 2002)
Stinging nettle (*Urtica dioica*): stems for fibre and cordage	3.5 m length of nettle twine from 6 1.5-m stalks; net 10 m long and 1 m deep (10 m^2) can require ~ 30 m twine (50–60 nettle stalks) and take 20–25 hours to construct; cattail or tule mat 1.2 × 2.4 m (4 × 8 ft) required ~ 30 m (~ 100 ft) of string	General, calculated from personal experience (see also Price, Samford, and Steponaitis 2001; and Kirk and Daugherty 2007, 98)

Table 11-2 | continued

Species or type of plant resource	Quantities harvested	Group (references)
Highbush cranberry (*Viburnum edule*): edible fruits	Lucille Clifton (Turner, Robinson, et al. 2012, 349) used to put by ~ 4–5 barrels, each 5 gallons (~ 20 L), packed with crabapples and highbush cranberries every winter	Gitga'at (Ts'msyen) (Turner, Robinson, et al. 2012; Turner and Thompson 2006)

known quantities of particular plant resources harvested or used by a family or given group. The ultimate quantities of plant resources that can be harvested and processed relate in part to the productivity and abundance of the resource being harvested (which in turn relates to effectiveness of management), in part to the rates and efficiency of the harvest, in part to the time available to devote to harvesting the resource in light of what other resources need to be harvested at the same time by the same individuals, and/or in part to how long it takes to process a given resource into a storable condition before it spoils. Ultimately, too, restricted availability of the resource itself – as well as external factors like weather conditions, tides, presence of bears in the area, or other factors – might present the ultimate constraints on the amounts one is able to harvest. The information in tables 11-1 and 11-2, limited and fragmented as it is, indicates that immense quantities of many diverse resources were needed for survival and trade and that correspondingly immense amounts of time were collectively devoted to acquiring and processing the resources. Often, family groups, or small groups of women, might spend half of each day gathering a particular resource and the rest of the day processing it, sometimes day after day until sufficient quantities were obtained.

The time required to harvest and process seaweed presents a good example. The general practice was, and still is, for groups of people to go out by boat from a spring harvest camp – like the Gitga'at camp at K'yel on Princess Royal Island – early in the morning when the tide was low. They spread out along the shoreline of the islands in the vicinity and picked seaweed for two to four hours, until the tide started to rise. Then they returned to the camp, around noon or one o'clock in the afternoon. Once harvested, the seaweed had to be dried almost immediately,[21] a task and process that took at least a full sunny afternoon. The seaweed was spread out on flat bedrock areas, usually in squares or shapes conforming to the shape of the rocks, or on cedar-wood trays. The quantities harvested daily could translate into 100 squares (of about 60 by 60 centimetres) or more per family group. The squares had to be turned and shifted after a couple of hours, and then, by sundown, fully dried, they were brought indoors so that they wouldn't become damp in the evening dew. Once dried, the harvested seaweed was stored indoors until the next stage of processing, usually undertaken back at the main village. In this stage, the dried cakes were rewetted with salt water, pressed into cedar-wood boxes, weighted down, and left for a number of days, before being taken out, chopped into fine pieces, redried, and then stored in airtight containers. Following these different steps, a family might harvest and process its year's supply of seaweed – around 300 kilograms of fresh seaweed, for Helen Clifton's (pers. comm., 2001; see also Turner and Clifton 2006) family at Hartley Bay – in a total of about five full days, not counting the time when it is in the cedar-wood boxes or the time travelling to and from the seaweed camp.[22]

The time required to weave the baskets and make the trays and boxes needed for harvesting, processing, and storing plant resources also must be taken into account when estimates of time and resource-harvesting rates are considered. In the seaweed example, openwork cedar-withe pack baskets would have been needed, in the past, to hold the fresh seaweed, as well as many cedar-bark baskets to transport it and several wooden boxes with lids to store it. The time taken to maintain and repair these containers would also have been considerable, although with care and upkeep, most could be reused over several years.[23] The construction of canoes, paddles, anchor lines, and mats, travel to the harvest camps and seaweed-picking grounds, and the time and materials required to build the small lodges where people stayed at their camps are other investments of time and energy that factor into the overall equation. These items and structures would generally be quite durable, however, and like the baskets, trays, and boxes, can be expected to have been functional over generations, with planks and posts replaced in the lodges as required.

Right after seaweed harvesting, at least for the Gitga'at, women would begin the work of harvesting and processing cedar bark for mats, hats, clothing, and baskets – a task that could be accomplished with ease only while the bark was still fresh and pliable.[24] Once cleaned and pulled apart, the layers of bark could be dried and stored in whole sheets for winter, a time when many of the baskets and other equipment were actually made. Some weavers would split the freshly harvested and cleaned bark sheets into strips of various widths and lengths, ready for use, and then bundle and dry these strips separately. Before the bark could actually be woven, it was soaked in water until pliable, and it was continuously wetted during the weaving process. Thus a woman might devote several days to the harvesting and initial processing of cedar bark each year and then many days in the wintertime working with this material to make her family's requirements of clothing, mats, and basketry, as well as perhaps some items to be traded for other goods.

Obviously, the efficiency of any harvest is dependent on high abundance, productivity, and predictability of the resource, as well as on easy accessibility and availability. The higher the overall quality of the resource, and the more concentrated and productive it is – whether a seaweed-picking area, a cedar stand, a berry patch, or a root-digging ground – the less the time and energy that need to be expended in its harvest. Knowing that one can count on harvesting valued staple resources at the same sites from year to year and generation after generation is critically important for a people's survival and well-being. This is where land and resource management in all its different forms and on multiple scales comes into play.

Furthermore, to what extent did peoples' ability to augment the natural availability of their plant resources, or to intensify the production of plant foods and other resources required for survival, enable their own populations to grow and thrive? Conversely, one might ask, to what extent did the need to meet increasing levels of human consumption from population growth and expansion help to motivate the development of management practices and approaches? These questions may never be adequately answered, but they should be kept in mind in the discussions that follow.

Management Techniques and Outcomes

The diverse approaches and practices that relate to Traditional Land and Resource Management in northwestern North American cultural areas, some of which have already been mentioned in the previous section, are summarized in table 11-3. Many of these are interrelated, with a range of different outcomes reflected on different temporal and spatial scales.[25] As noted previously, these practices also apply to different groups of species, such as fruiting shrubs, root vegetables, or closed forest communities. Table 11-4 lists individual plant, algal, and fungal species for which some evidence of management in the form of selective or partial harvesting, habitat manipulation, ownership of stands or patches, or other ways of maintaining and enhancing their production has been documented. Over 100 species or closely related groups of species are identified as having had some form of management associated with their maintenance or productivity within one or more cultural areas in northwestern North America.

The practices and techniques applied to maintain and enhance the productivity of key resources and prime harvesting sites are both cumulative and dynamic, building upon observations and experiences of earlier generations and adapting to new technological and socioeconomic changes. Many of them are quite specialized and are relegated to skilled practitioners who then teach others through processes of observation and experiential learning.[26] Significantly, the management methods tend to work in concert with natural processes to which plants and other species are already adapted: soil development, nutrient cycling, photosynthesis, genetic divergence, decomposition, light and moisture sequestration, seasonality, phenology and reproductive cycles, tidal cycles, vegetative reproduction, responses to fire, flood, and other disturbance,

Table 11-3 | Techniques and approaches applied by Indigenous peoples of northwestern North America to maintain and enhance plant resources

Management practice	*Notes (examples)*
SPECIES- OR POPULATION-LEVEL MANAGEMENT	
Gardens of annual species cleared and sown, mulched and fertilized	Northwest Coast tobacco (*Nicotiana* sp.) and native tobacco (*N. attenuata*)
Partial harvesting of roots for fibre, implements, etc.	Only a few roots taken from each tree (western redcedar, Sitka spruce, yellow-cedar)
Partial harvesting of edible root vegetables	Portion of root remains in ground to regrow (balsamroot, bitterroot)
Partial and selective harvesting of medicinal plants and roots	Harvested from remote places; spot-harvested; plants often regenerate from fragments left in the ground (like a pulled-up dandelion in one's lawn) (Canby's lovage, mountain valerian, false hellebore, yarrow)
Partial harvesting of bark as material	Fibrous tree bark pulled or cut in slabs from living trees (western redcedar, paper birch); bark for dye (red alder)
Partial harvesting of bark for medicinal use	Rectangular pieces cut from living trees (red alder, cascara, Pacific crabapple, Pacific yew)
Partial harvesting of edible cambium and inner bark	Rectangular pieces cut from living trees (lodgepole pine, amabilis fir, western hemlock)
Partial harvesting of planks for construction, etc.	Wooden planks split from standing, living trees (western redcedar)
Partial harvesting of woody withes and branches for basketry, rope, fish traps	Pruned from growing trees or bushes (saskatoon berry, hazelnut, Rocky Mountain juniper, Pacific crabapple, western redcedar, willows, Pacific yew)
Partial harvesting of fibrous stems and leaves of herbaceous perennials for mats, cordage, baskets	Cut from perennial plants at end of growing season; often only vegetative plants taken (cattail, tule, stinging nettle, Indian-hemp, slough sedge); spot-harvesting from many different plants (sword fern)
Harvesting edible stems and leaves from living herbaceous perennials	Vegetables (cow-parsnip, fireweed, giant horsetail, Indian celery)
Partial harvesting of edible seaweed	Fronds of *Pyropia abbottiae* and related species pulled from the rocks by hand, allowing rapid basal regeneration; giant kelp fronds coated with herring spawn harvested from living plants
Harvesting pitch or gum from living trees for material or medicinal purposes	Collected from natural or human-made wounds in trees or from pitch blisters (western hemlock, lodgepole pine, Sitka spruce, subalpine fir, and other conifers)

Table 11-3 | continued

Management practice	*Notes (examples)*
Pruning and coppicing of berry and nut bushes	Bushes pruned or cut down to the base to stimulate new growth and larger, higher-quality fruits (saskatoon berry, salmonberry, highbush cranberry, salal, hazelnut, black huckleberries and blueberries, soapberries)
Pruning and coppicing of arrow shaft, basketry, and cordage materials	Bushes pruned or cut down to the base to stimulate new, straight, vigourous shoots for basket rims (saskatoon berry), cordage (hazelnut, sandbar willow), basketry (Rocky Mountain maple), nets (willows), or arrows (saskatoon berry, oceanspray)
Burning of individual fruiting shrubs or trees	Individual bushes sometimes burned down to produce vigourous young shoots and eventually increase fruit production (saskatoon berry, hazelnut); individual or clumps of trees burned to create fuel (subalpine fir)
Burning of populations of basketry materials	Cattail and tule patches burned over after harvesting to clean them for the following year; beargrass patches similarly burned over
Selection by size, age, or reproductive stage of plant or organism	For several edible and material species, only vegetative plants harvested (desert parsley, *Lomatium macrocarpum*; slough sedge, or basket sedge, *Carex obnupta*); others harvested only at certain stage of maturity (yellow glacier lily, balsamroot, spring beauty, edible seaweed – spore-bearing plants avoided)
Digging and tilling before, during, or after harvesting	Edible root vegetables said to grow faster and more productively after soil disturbance (springbank clover, Pacific silverweed, northern riceroot, blue camas, yellow glacier lily, spring beauty)
Replanting propagules during or after harvesting	Fragments of plants returned to the ground (springbank clover, Pacific silverweed, northern riceroot, yellow glacier lily)
Protection of growing tissues (e.g., covering over soil after harvesting)	Mushroom stalks cut (mature specimens only) and soil covered over afterward (pine mushroom); turf replaced after roots dug (yellow glacier lily, blue camas, spring beauty)
Transplanting	Individual propagules (bulbs, corms, rhizomes), seedlings, or rooted shoots transplanted from one location to another (camas, *Camassia quamash*; stinging nettle; highbush cranberry; soapberry; cattail; wapato); seeds or nuts transplanted (hazelnut, garry oak); trailing blackberry vines transplanted
Scattering seeds, fruits, corms, bulbs	Berries or seed heads scattered during or after harvesting (spring beauty, blueberries and huckleberries, "wild celery"); corms of spring beauty, roots of silverweed

/continued

Table 11-3 | continued

Management practice	*Notes (examples)*
	(*Argentina anserina*), and bulbs of tiger lily and yellow glacier lily ritually scattered after harvesting (Tsilhqot'in)
Insect control	Tent caterpillar nests systematically removed/burned (chokecherry); control of insect pests by landscape burning (J.A. Ross 2011)
Diversification as management	Developing and using alternative resources can ease harvesting pressure on prime resources (e.g., using clams instead of, or along with, salmon, onions instead of camas bulbs, or clover rhizomes instead of silverweed may have allowed continuing sustainability of the original resources while allowing populations to survive and thrive)
PLANT COMMUNITY- OR HABITAT-LEVEL MANAGEMENT	
Site selection for prime populations	Berry gardens created on terraces beside waterfalls as ideal moisture regime (Heiltsuk); root gardens located at particular tidal zones; barriers placed to take advantage of flood tides and river floods (Nuu-chah-nulth, Kwakwaka'wakw); prime productive berry patches identified and owned (Tlingit, Ts'msyen)
Altering successional stages by burning over meadows or woods	Patches intentionally burned over periodically and routinely (oak savannahs, boreal forests, montane meadows) to remove underbrush, create clearings, increase size and productivity of berry bushes or root vegetables, and create forage for deer and other game; provides dead snags and branches for fuel
Weeding, tilling, and clearing root "gardens"	Competing woody shrubs, grasses, and other vegetation are removed by hand before, during, or following harvesting (blue camas meadows, tiger lily patches, strawberry patches, tidal estuarine gardens) (Coast Salish, Secwepemc, Tlingit, Kwakwaka'wakw)
Clearing for settlements	Trees are felled around houses or village sites; stumps and cleared ground provide habitat for certain berry bushes; also provides wood for construction and protection from surprise attacks by enemies
Moving rocks, erecting fencing, installing posts or pegs, etc.	Rocks are moved and piled up to clear root-digging grounds and to delineate owned berry patches or root "gardens"; posts or pegs set around patches or individual trees to indicate ownership
Terracing or use of terraces	Terraces constructed or natural terraces used for house placement but also provide areas for plant growth, as with the orchard gardens of Robintown (Ts'msyen) and the terraces at Nimpkish River estuary (Kwakwaka'wakw); berry garden terraces constructed (Heiltsuk)

Table 11-3 | continued

Management practice	Notes (examples)
Enhancing nutrient regimes and fertilizing	Adding ashes, clamshells, or animal and fish remains to berry patches, root gardens, or around individual trees (e.g., Heiltsuk berry gardens); ceremonially "feeding" strawberries (*Fragaria chiloensis*) and nagoonberries (*Rubus arcticus*) with salmon eggs (Tlingit) (Thornton 1999)
Preferential use of dead trees, downed limbs, and other self-produced plant materials	Retains standing forests; harvesting wind-thrown or snow-downed logs for construction (western redcedar, trembling aspen); harvesting wood from downed logs or driftwood as fuel; using tree knots from rotten logs for fishhooks (western hemlock and others); driftwood gathering; harvesting tree roots for basketry from already exposed root systems
LANDSCAPE-LEVEL OR TERRITORIAL MANAGEMENT	
Ownership, proprietorship, or caretaking responsibility	Social controls of various types enacted over resource species, habitats, and territories; harvesting usually overseen by chiefs as part of inherited responsibility (see Blackman 1990; McIlwraith 1948; Suttles 1987b, 2005; Thornton 1999; and Turner, Smith, and Jones 2005; see also table 11-4 for examples)
Kincentric belief system	"Resource" species also regarded as sentient relatives of humans, requiring respect and reciprocity with use; constraints against overharvesting or wastage (Thornton 1999; Turner and Berkes 2006)
Ceremonial harvesting, dissemination practices, and constraints against harvesting	First Fruits and First Roots ceremonies, diverse puberty or bereavement ceremonies, and taboos against certain types of harvesting by certain people at certain times effect constraints on and controls over resource use (Turner and Berkes 2006)

Sources: Based on compilations of Christine Joseph (Watta) (pers. comm., 1999), Daisy Sewid-Smith (pers. comm., 1994), Mary Thomas (pers. comm., 1997), M.K. Anderson (2005), Bandringa (1999), Blackman (1990), Boas (1921, 1930, 1966), Bouchard and Kennedy (1990), Boyd (1999b), Collins (1974, 175), Compton (1993a, 1993b), Curtis (1915), A. Davis, Wilson, and Compton (1995), Deur (1998, 2000), Deur and Turner (2005), Drucker (1951), Gill (2005a, 2005b), Haeberlin and Gunther (1930), Hunn, Selam, and family (1990), Johnson (1997), Johnson Gottesfeld (1993), Kennedy and Bouchard (1983), Kimmerer and Lake (2001), Kuhnlein and Turner (1987, 1991), Lepofsky and Lertzman (2008), Lepofsky, Lertzman, et al. (2005), Mack and McClure (2002), McDonald (2003), Minore (1972), Norton (1979a), Peacock (1998), Peacock and Turner (2000), Richardson (1982), J.A. Ross (2011), Shebitz, Reichard, and Dunwiddy (2009), L.R. Smith (2008b), Stewart (1984), Stryd (1997), Suttles (1951a, 1951b, 1955, 1987b, 2005), Teit (1900, 1909), Thornton (1999, 2008); Turner (1987, 1995, 1997a, 1998, 1999, 2003a, 2004a, 2005), Turner and Berkes (2006), Turner, Bouchard, and Kennedy (1980), Turner and Brown (2004), Turner and Clifton (2006), Turner, Deur, and Lepofsky (2013), Turner and Efrat (1982), Turner and Hebda (1990), Turner, Ignace, and Ignace (2000), Turner and Kuhnlein (1982, 1983), Turner, Kuhnlein, and Egger (1985), Turner and Peacock (2005), Turner, Smith, and Jones (2005), Turner and Taylor (1972), Turner, Thomas, et al. (1983), and Turner, Thompson, et al. (1990).

Table 11-4 | Plant resources managed and sustained through various means by Indigenous peoples of northwestern North America

Plant resource	*Management methods applied*
ALGAE	
Giant kelp (*Macrocystis pyrifera*)	Fronds harvested with herring spawn; picked from living plants; same kelp beds used generation after generation (Haida, Heiltsuk, other Northwest Coast peoples)
Bull kelp (*Nereocystis luetkeana*)	Special beds accessed for prime fishing lines; certain kelp beds owned (general, Haida, Ts'msyen, Ditidaht, Northwest Coast)
Red laver edible seaweed (*Pyropia abbottiae* and related spp.)	Pulled off rocks when young, allowing base to regenerate; said to grow better with harvesting; two crops taken from same spots each season; patches used generation after generation (Haida, Ts'msyen, Kwakwaka'wakw, other Northwest Coast peoples)
FUNGI AND LICHENS	
Black tree lichen, or *wila* (*Bryoria fremontii*)	Harvested only from certain species of conifers, at certain elevations; pulled off living trees with twisting poles, allowing base to regenerate (Interior Plateau peoples)
Pine mushroom (*Tricholoma magnivelare*), cottonwood mushroom (*T. populinum*), and other spp.	Mature individuals cut at base; soil carefully replaced to protect those still growing (Stl'atl'imx, Nlaka'pamux, Secwepemc, other Interior Plateau peoples)
FERNS AND FERN ALLIES	
Spiny wood fern (*Dryopteris expansa*)	Partially harvested so remaining rootstock grows into new plant (widely used along Northwest Coast and inland along the Coast and Cascade Mountains)
Licorice fern (*Polypodium glycyrrhiza*)	Rhizomes partially harvested; the rest left to grow (general, Northwest Coast)
Sword fern (*Polystichum munitum*)	Fronds for pit-cooking, matting, and ceremonial purposes harvested only a few from each plant to maintain plants (general, Northwest Coast)
Bracken fern (*Pteridium aquilinum*)	Patches maintained by burning and clearing; some patches owned and inherited (general, Northwest Coast, especially western Washington prairies)
CONIFERS	
Amabilis fir (*Abies amabilis*)	Boughs, bark, and pitch harvested from living trees (general and widespread); edible inner bark partially removed so as not to kill tree (Kwakwaka'wakw, Ts'msyen, and others)
Grand fir (*Abies grandis*)	Boughs, bark, and pitch harvested from living trees; pitch harvested over generations from the same tree (Kwakwaka'wakw, Coast Salish)

Table 11-4 | continued

Plant resource	*Management methods applied*
Subalpine fir (*Abies lasiocarpa*)	Boughs, bark, and pitch harvested from living trees (Interior Plateau, northern interior)
Yellow-cedar (*Chamaecyparis nootkatensis*)	Bark harvested for weaving material; only partially harvested so as not to kill tree (widely, Northwest Coast)
Rocky Mountain juniper (*Juniperus scopulorum*)	Bow staves possibly cut from standing, living trees in Interior Plateau, as in Great Basin area and California; boughs cut from living trees for medicine, smudging, etc.; branches from living trees for bows (Interior Plateau)
Engelmann spruce (*Picea engelmannii*) and white spruce (*P. glauca*)	Sheets of bark and roots harvested from living trees; only partially harvested so as not to kill tree; roots and pitch harvested from living trees (Interior Plateau, northern interior)
Sitka spruce (*Picea sitchensis*)	Roots harvested for basketry; only a few taken from a single tree in any one year; limbs and sapling trunks used for fish weirs; inner bark (food) and pitch (medicine, adhesive) harvested from individual living trees over generations (Haida, Tlingit, Ts'msyen, Kwakwaka'wakw, other Northwest Coast peoples)
Lodgepole pine (*Pinus contorta*)	Partial harvesting of inner bark (food) and pitch to keep tree alive (Haida, Gitxsan, other Northwest Coast peoples for pitch; Interior Plateau and northern interior peoples for food and pitch)
Western white pine (*Pinus monticola*)	Sheets of bark harvested for baskets and canoes from living trees (Ktunaxa, Lakes, Secwepemc)
Ponderosa pine (*Pinus ponderosa*)	Pitch and edible inner bark harvested from standing, living trees (Interior Plateau peoples); fallen needles harvested for pit-cooking and insulation
Douglas-fir (*Pseudotsuga menziesii*)	Bark for fuel cut from standing, living trees; self-pruned limbs from living trees used extensively for fuel; branches cut from living trees for pit-cooking, bedding, floor covering, and sweathouses (widespread, Vancouver Island, southern coast, Interior Plateau)
Pacific yew (*Taxus brevifolia*)	Bark removed for medicine from branches or partial harvesting from trunks of living trees; wood cut from standing living trees for bows and other implements; called "begging from" the tree (Kwakwaka'wakw, Halkomelem, other Northwest Coast peoples)
Western redcedar (*Thuja plicata*)	Bark sheets, inner bark, and planks taken from standing, living trees; also branches and roots for rope and basketry; limbs and sapling trunks used for fish weirs; "groves" managed over generations for canoes and bark production; stands "owned" by families/lineages (general, Northwest

/continued

Table 11-4 | continued

Plant resource	*Management methods applied*
	Coast, Interior Plateau); seedlings said to have been given as gifts from coastal to interior Nlaka'pamux at gatherings, like in Botanie Valley (Jim Stafford, pers. comm., 2008)
Western hemlock (*Tsuga heterophylla*)	Bark harvested for food (inner bark) from standing, living trees; branches and sapling trunks cut for herring egg harvesting, lining acorn-leaching pits, and reinforcing weirs; poles cut for weirs and twisting eelgrass (general, Northwest Coast)
FLOWERING PLANTS	
Vine maple (*Acer circinatum*)	Branches and wood harvested from living trees, for snowshoes and other implements (mainland Halkomelem, Lower Stl'atl'imx, among others)
Rocky Mountain maple (*Acer glabrum*)	Branches, wood, and bark harvested from living trees for weaving, snowshoes, and other implements (Interior Plateau)
Bigleaf maple (*Acer macrophyllum*)	Bark and wood harvested from living trees for weaving, paddles, spindle whorls, and other implements, as well as for fuel (Coast Salish)
Yarrow (*Achillea millefolium*)	Stems and leaves harvested for medicine, tea, and food storage from perennial plants (widespread, coast and interior)
Sweetflag (*Acorus americanus*)	Evidently transplanted (e.g., to Salmon River estuary at Salmon Arm)
Nodding onion (*Allium cernuum*)	Patches maintained by burning and partial harvesting; possibly transplanted (Interior Plateau, Tsilhqot'in)
Mountain alder, green alder (*Alnus incana, A. viridis*)	Wood and bark harvested from living trees for use as fuel, tools, and dye (widespread, Interior Plateau, and northern Interior)
Red alder (*Alnus rubra*)	Limbs and sapling trunks used for fish weirs; wood and bark harvested from living trees for use as fuel, tools, and dye (Straits Salish, Kwakwaka'wakw, other Northwest Coast peoples)
Saskatoon berry (*Amelanchier alnifolia*)	Individual bushes pruned and burned; maintained by landscape burning; occasionally transplanted (Nlaka'pamux, Nisga'a); many varieties recognized (Secwepemc, Nlaka'pamux, other Interior Plateau peoples)
Indian-hemp (*Apocynum cannabinum*)	Patches maintained by continual harvesting; said to grow taller and straighter in tended patches; transplanted (Interior Plateau, Tsilhqot'in)
Arbutus, or Pacific madrone (*Arbutus menziesii*)	Wood spared as fuel and for any other use by Saanich for spiritual reasons (because of its service to people during the Great Flood); leaves and bark harvested for medicine from living trees (Straits Salish)

Table 11-4 | continued

Plant resource	*Management methods applied*
Kinnikinnick (*Arctostaphylos uva-ursi*)	Leaves for smoking and medicine harvested from living plants (widespread, Interior Plateau, recently Northwest Coast); transplanted (Nisga'a) (Emma Nyce, pers. comm., 2013)
Silverweed (*Argentina anserina*)	Roots spread around ritually following harvest (Tsilhqot'in)
Pacific silverweed (*Argentina egedii*)	One of the estuarine tidal marsh garden species; tended, weeded, roots selectively harvested, and propagules replanted; patches "owned" by chiefs and families or clans (Kwakwaka'wakw, Nuu-chah-nulth, Haida); plants transplanted (e.g., Knight Inlet) (general, Northwest Coast)
Balsamroot (*Balsamorhiza sagittata*)	Selectively harvested by size; "mother plants" and very young plants spared, only daughter plants harvested; root tips regrow into new plants; achenes scattered during harvesting for food (Secwepemc and other Interior Plateau peoples)
Paper birch (*Betula papyrifera*)	Bark for baskets, lining cache pits, burials, etc. removed from standing, living trees without killing the tree (general, Tsilhqot'in, Ulkatcho Dakelth, Dene [Athabaskan], Secwepemc, Okanagan, Ktunaxa)
Edible camas (*Camassia leichtlinii*, *C. quamash*)	Camas prairies maintained by burning (Vancouver Island, western Washington, Willamette Valley); patches cleared and weeded; bulbs selectively harvested by size; ground tilled; turf replaced after digging; patches "owned" by families (Coast Salish, southern Interior Plateau); occasionally bulbs transplanted (Nuu-chah-nulth); often harvested in summer season when seeds are ripe and easily scattered
Slough sedge, or basket sedge (*Carex obnupta*)	Leaves for basketry taken from same site over and over; said to grow better with "cleaning" and routine harvesting; only leaves from vegetative plants taken (Nuu-chah-nulth, Ditidaht, Makah)
Spring beauty, or mountain potato (*Claytonia lanceolata*)	Patches maintained by fall landscape burning (Interior Plateau, Tsilhqot'in); corms selectively harvested; turf replaced after digging (general, interior); mature stems and handfuls of corms ritually scattered around areas without the plant to extend their range (Tsilhqot'in)
Pacific hemlockparsley (*Conioselinum gmelinii*)	Roots selectively harvested and smaller roots replanted; patches "owned" by chiefs and families or clans (Kwakwaka'wakw)
Red-osier dogwood (*Cornus sericea*)	Branches "pruned off" for use in basketry and sweat lodge construction; inner bark harvested for medicine, from living bushes (mainly Interior Plateau)
Hazelnut (*Corylus cornuta*)	Individual bushes pruned and burned for nut production and use of withes for rope, arrows, etc.; growth improved

/continued

Table 11-4 | continued

Plant resource	Management methods applied
	by landscape burning and clearing (Kalapuya, Nlaka'pamux, other Interior Plateau peoples); patches "owned" by chiefs and families or clans; occasionally transplanted (Ts'msyen)
Black hawthorn (*Crataegus douglasii*)	Transplanted to garden in late 1800s (Nisg̱a'a) (Emma Nyce, pers. comm., 2013)
Silverberry, wolf willow (*Elaeagnus commutata*)	Inner bark harvested in quantity from living bushes in particular designated patches (Nlaka'pamux, other Interior peoples)
Fireweed (*Epilobium angustifolium*)	Shoots for eating and stems for fibre harvested routinely from same patches; patches "owned" by chiefs and families or clans (Haida, Ts'msyen, Stl'atl'imx, among others)
Yellow glacier lily (*Erythronium grandiflorum*)	Patches maintained by landscape burning (Interior Plateau); bulbs selectively harvested by size (indicated by multiple flowers on a stalk); harvested in season when seeds are ripe and easily scattered; propagules from bulbs replanted; turf replaced after digging (Secwepemc)
Seaside strawberry (*Fragaria chiloensis*)	Strawberry patches owned and weeded; spirits of the berries ritually "fed" with salmon eggs (Tlingit)
Wild strawberries (*Fragaria vesca, F. virginiana*)	Patches maintained by landscape burning and clearing (Coast Salish, Vancouver Island, Gulf Islands)
Cascara (*Frangula purshiana*)	Medicinal bark harvested in strips so as not to kill the tree (Northwest Coast, Interior Plateau); quite a few other trees were also harvested like this (Turner and Hebda 1990)
Chocolate lily (*Fritillaria affinis*)	Bulbs selectively harvested; bulblets scattered with tilling; patches maintained by landscape burning (Coast Salish, Vancouver Island, Gulf Islands; Interior Plateau); handfuls of the bulbs ritually scattered around after harvesting (Tsilhqot'in)
Northern rice-root (*Fritillaria camschatcensis*)	One of the estuarine tidal marsh garden species; tended and weeded, bulbs selectively harvested, and propagules replanted; patches "owned" by chiefs and families or clans (Kwakwaka'wakw, Ts'msyen, other Northwest Coast peoples); transplanted to new areas (e.g., Knight Inlet, Robintown)
Salalberry (*Gaultheria shallon*)	Patches tended by burning (Makah, Nuxalk, other Northwest Coast peoples); harvesting berries by their stems said to increase future productivity; patches "owned" by chiefs and families or clans (general, Northwest Coast); possibly transplanted (Heiltsuk)
Sea milkwort (*Glaux maritima*)	Prime patches "owned" by chiefs (Kwakwaka'wakw)
Rattlesnake plantain (*Goodyera oblongifolia*)	Plants for medicine selectively harvested to leave smaller plantlets growing (Quw'utsun' Halkomelem)

Table 11-4 | continued

Plant resource	Management methods applied
Cow-parsnip (*Heracleum maximum*)	Edible shoots selectively harvested from particular patches; picking season prolonged by harvesting; patches used over generations (widespread, Northwest Coast, northern and southern interior)
Roundleaf alumroot (*Heuchera cylindrica*)	Roots for medicine selectively harvested from patches to leave smaller plantlets growing (Secwepemc)
Oceanspray (*Holodiscus discolor*)	Shoots harvested from living, growing shrubs; pruning said to produce straighter stems for arrows, needles, digging sticks, etc. (Straits, Squamish and neighbouring Coast Salish, Lower Stl'atl'imx)
Bitterroot (*Lewisia rediviva*)	Roots selectively harvested and propagules replanted; roots transplanted (Secwepemc, Nlaka'pamux, Ktunaxa, other Interior Plateau peoples)
Giant wild rye (*Leymus cinereus*)	Leaves and stems cut from perennial rootstocks in particular patches that are used over years (Okanagan, Nlaka'pamux)
Canby's lovage, or licorice root (*Ligusticum canbyi*)	Roots harvested selectively, leaving small segment in ground to regrow (Secwepemc)
Tiger lily, or Columbia lily (*Lilium columbianum*)	Size and quality of bulbs maintained by landscape burning; selectively harvested from known, delineated patches; weeded and possibly transplanted; prime patches owned (Skagit Valley, Secwepemc, Tsilhqot'in); handfuls of the bulbs ritually scattered around after harvesting (Tsilhqot'in)
Canby's biscuitroot (*Lomatium canbyi*)	Various species of *Lomatium* known from Washington State and Oregon; edible roots selectively harvested (Hunn, Selam, and family 1990; Hunn and French 1981)
Desert parsley (*Lomatium macrocarpum*)	Edible roots selectively harvested; only "female" vegetative plants harvested for their roots (Okanagan, Interior Plateau)
Indian celery, or barestem lomatium (*Lomatium nudicaule*)	Green leaves picked selectively in spring from patches; plants perennial and soon regenerate (e.g., like asparagus); seeds harvested with care, leaving enough to fall to the ground and reseed; patches maintained for generations (Tsilhqot'in, Stl'atl'imx, Secwepemc, and others for greens; Coast Salish, Kwakwaka'wakw, and others for seeds)
Nootka lupine (*Lupinus nootkatensis*)	One of the estuarine tidal marsh garden species; tended and weeded; roots selectively harvested from living perennial plants; patches "owned" by chiefs and families or clans (Kwakwaka'wakw, Nuxalk); transplanted (e.g., Knight Inlet)
Skunk-cabbage (*Lysichiton americanus*)	Leaves for berry drying, pit-cooking, and food preparation routinely harvested from living plants and patches (general, Northwest Coast, southern and eastern Interior Plateau); Deer said to keep his skunk-cabbage "just like a garden" in Haida story (Swanton 1908, 446)

Table 11-4 | continued

Plant resource	Management methods applied
False lily-of-the-valley (*Maianthemum dilatatum*)	Plants known to have been transplanted for medicinal use (Tlingit); leaves for medicinal purposes harvested from perennial rhizomes (general, Northwest Coast)
Pacific crabapple (*Malus fusca*)	Trees tended, pruned, lopped, and transplanted; crabapple stands "owned" by chiefs and families (Kwakwaka'wakw, Haisla, Ts'msyen, Nisg̱a'a, and other Northwest Coast peoples)
Wild tobacco, or coyote tobacco (*Nicotiana attenuata*)	Evidently cultivated from seed in Interior Plateau; grown in prehistoric times (Nlaka'pamux, Okanagan, Ktunaxa)
Northwest Coast tobacco, or Haida tobacco (*Nicotiana quadrivalvis* var.)	Cultivated from seed capsules on northern coast; grown in gardens in prehistoric times (Haida, Tlingit, Ts'msyen); apparently obtained centuries ago from Native Californians, perhaps via Columbia River trading
Devil's-club (*Oplopanax horridus*)	Stems harvested for medicine (inner bark) by pruning from "daughter" shoots; mother shoots left to regrow; cut shoots replanted (Halkomelem and other Northwest Coast peoples); constraints against harvesting flowering or fruiting stems or more than four stems (Ts'msyen, Haida)
Wild caraway, or yampah (*Perideridia gairdneri*)	Apparently selectively harvested; one of the prairie root species improved by landscape burning (Interior Plateau, Vancouver Island, western Washington)
Reed canarygrass (*Phalaris arundinacea*), reed grass (*Phragmites australis*), and other perennial grass spp.	Stems harvested from perennial plants in given patches for basketry decoration (Coast Salish) and mats (Interior Plateau)
Mock-orange (*Philadelphus lewisii*)	Bushes trimmed to produce long, straight shoots for arrows and mat needles (Straits and neighbouring Coast Salish)
Cottonwood (*Populus balsamifera*)	Inner bark harvested for food from standing, living trees (Nuxalk, Stl'atl'imx); bark sheets cut from living trees for lining cache pits, buckets, etc.; outer bark cut from living trees for high-quality fuel for smoking salmon (Gitxsan); self-pruned limbs from living trees used extensively for fuel; seedlings transplanted (Heiltsuk)
Bitter cherry (*Prunus emarginata*) and pin cherry (*P. pensylvanica*)	Outer bark harvested in strips from standing, living trees for basketry decoration, cordage, etc. (Coast Salish, Interior Plateau)
Chokecherry (*Prunus virginiana*)	Bushes tended, pruned, and cleaned of tent caterpillars; different varieties recognized (Secwepemc, other Interior Plateau peoples)
Garry oak, or Oregon white oak (*Quercus garryana*)	Oak savannahs maintained by landscape burning (Coast Salish, Kalapuya); possibly trees tended in the past for acorn

/continued

Table 11-4 | continued

Plant resource	*Management methods applied*
	production (as with California oaks) (M.K. Anderson 2005); trees possibly relocated, planted, or transplanted (Lepofsky 2004; Norton 1979a)
Labrador tea (*Rhododendron groenlandicum*)	Leaves and shoots selectively harvested from living plants for beverage tea and medicine; plants weeded by some (Ts'msyen, Makah, other Northwest Coast peoples)
Trapper's tea (*Rhododendron neoglandulosum*)	Leaves and shoots selectively harvested from living plants for beverage tea and medicine (Interior Plateau)
Wild gooseberry (*Ribes* sp.)	Patches cleaned and pruned (and possibly burned) to make them more productive (Ts'msyen)
Gray currant, or stink currant (*Ribes bracteosum*)	Bushes pruned back after harvest; productive patches "owned" by chiefs and families (Kwakwaka'wakw, Ts'msyen, other Northwest Coast peoples)
Northern black currant (*Ribes hudsonianum*)	Patches cleaned and pruned (and possibly burned) to make them more productive (Ts'msyen)
Trailing black currant (*Ribes laxiflorum*)	Productive patches "owned" by chiefs and families; possibly maintained by fire (Kwakwaka'wakw, other Northwest Coast peoples)
Nagoonberry (*Rubus arcticus*)	Berry patches owned; spirits of the berries ritually "fed" with salmon eggs (Tlingit)
Wild raspberry, or American red raspberry (*Rubus idaeus*)	Grows more productively after landscape burning (Stl'atl'imx, Interior Plateau, possibly northern interior)
Blackcap (*Rubus leucodermis*)	Patches maintained by landscape burning (Straits Salish, Halkomelem, Stl'atl'imx, Interior Plateau)
Thimbleberry (*Rubus parviflorus*)	Green shoots cut from plants and regenerate later (widespread, Northwest Coast, Interior Plateau); grows more productively after landscape burning
Salmonberry (*Rubus spectabilis*)	Green shoots cut from plants and regenerate later; berry bushes pruned after harvesting to enhance production (Halkomelem, Kwakwaka'wakw, other Northwest Coast peoples); patches "owned" by chiefs and families (Kwakwaka'wakw and others)
Trailing blackberry (*Rubus ursinus*)	Patches maintained by landscape burning (Straits Salish, Halkomelem)
Wapato (*Sagittaria latifolia*)	Patches owned, tended, weeded, and cleared; tubers transplanted into new areas; possibly grown over cobble base to keep tubers shallow (Halkomelem, Fraser Valley, Columbia River Valley)
Sandbar willow (*Salix exigua*)	Bark and stems cut from living plants; regenerate in following years (Stl'atl'imx, other Interior Plateau peoples)

Table 11-4 | continued

Plant resource	*Management methods applied*
Willows (*Salix lucida* ssp. *lasiandra* and other spp.)	Bark and stems cut from living plants for cordage and netting; regenerate in following years (Straits Salish and others)
Red elderberry (*Sambucus racemosa*)	Bark and stems harvested from living trees; shoots occasionally transplanted (Ts'msyen); prime patches "owned" by clans and chiefs (Kwakwaka'wakw)
Tule, or hardstem bulrush (*Schoeno-plectus acutus*)	Stems routinely harvested from given patches used over generations; possibly transplanted (Coast Salish, Interior Plateau)
Olney's "three-square" (*Schoenoplectus americanus*)	Plants routinely harvested from given patches; known to have been transplanted (Coast Salish, Nuu-chah-nulth, Ditidaht, among others)
Soapberry (*Shepherdia canadensis*)	Patches maintained by landscape burning; bushes pruned; berries scattered; occasionally transplanted (Tlingit, Ts'msyen, Coast Salish, Interior Plateau)
Springbank clover (*Trifolium wormskioldii*)	One of the estuarine tidal marsh root garden species; tended and weeded, rhizomes selectively harvested, and propagules replanted; patches "owned" by chiefs and families or clans (Kwakwaka'wakw, Nuu-chah-nulth, other Northwest Coast peoples); transplanted to new areas (Nuxalk: Kimsquit to Kitlope)
Cattail (*Typha latifolia*)	Leaves and stalks routinely harvested from particular patches; rhizomes sometimes transplanted to new areas (e.g., Hesquiaht Nuu-chah-nulth); patches possibly burned, as they were in the boreal forest; seed fluff harvested for stuffing and diapers (widespread, Coast and Interior Salish, Ditidaht, among others)
Stinging nettle (*Urtica dioica*)	Stems for fibre routinely harvested from particular patches; rhizomes known to have been transplanted to new areas; always found at ancient village sites (widespread, Northwest Coast); constraints against harvesting until fruiting is finished (Ts'msyen)
Alaska blueberry (*Vaccinium alaskaense*)	Productivity improved by burning; bushes pruned, fertilized, and sometimes transplanted; patches "owned" by chiefs and families or clans; transplanted to new areas (Northwest Coast)
Dwarf bilberry, or mountain blueberry (*Vaccinium caespitosum*)	Productivity improved by burning; patches "owned" by chiefs and families or clans; transplanted to new areas (Interior Plateau, Northwest Coast, Olympic Peninsula)
Black mountain huckleberry (*Vaccinium membranaceum*)	Productivity improved by burning; bushes pruned, fertilized, and sometimes transplanted; patches "owned" by chiefs and families or clans; transplanted to new areas (Northwest Coast, Interior Plateau)

Table 11-4 | continued

Plant resource	*Management methods applied*
Velvetleaf blueberry, or Canada blueberry (*Vaccinium myrtilloides*)	Productivity improved by burning; bushes pruned and transplanted to new areas (e.g., Ditidaht, Nuu-chah-nulth, Sechelt, Sliammon)
Oval-leaf blueberry (*Vaccinium ovalifolium*)	Productivity improved by burning; bushes pruned, fertilized, and sometimes transplanted; berries "scattered"; patches "owned" by chiefs and families or clans (Northwest Coast, Interior Plateau, Ts'msyen)
Evergreen huckleberry (*Vaccinium ovatum*)	Maintained by burning and pruning (Makah, possibly other Northwest Coast peoples)
Bog cranberry (*Vaccinium oxycoccos*)	Cranberry patches "owned" and harvesting controlled (Katzie, Haida, others); productivity improved by burning (Olympic Peninsula prairies, Ts'msyen) (Gill 2005a, 2005b; M.K. Anderson 2009)
Red huckleberry (*Vaccinium parvifolium*)	Productivity improved by burning; bushes pruned, fertilized, and sometimes transplanted; patches "owned" by chiefs and families or clans (widespread, Kwakwaka'wakw, Northwest Coast)
Bog blueberry (*Vaccinium uliginosum*)	Productivity improved by burning; patches "owned" by chiefs and families or clans; occasionally transplanted (Northwest Coast – e.g., Sliammon, Makah)
Mountain valerian (*Valeriana sitchensis*)	Medicinal roots selectively harvested; often regenerated from fragments left in the ground (Secwepemc and other Interior Plateau peoples)
False hellebore (*Veratrum viride*)	Medicinal roots selectively harvested; often regenerated from fragments left in the ground (widespread, coast and interior)
Highbush cranberry (*Viburnum edule*)	Sometimes transplanted to new areas and "gardens"; bushes pruned to renew growth; patches "owned" by chiefs and families or clans (Kwakwaka'wakw, Haida, Ts'msyen, other Northwest Coast peoples)
Beargrass (*Xerophyllum tenax*)	Leaves routinely harvested from particular patches; burned over periodically to enhance productivity and improve quality of leaves for basketry (Makah, Quileute, Coast Salish of Washington)
Eelgrass (*Zostera marina*)	Harvesting rhizomes through moderate disturbance said to enhance growth and palatability (Kwakwaka'wakw, other Northwest Coast peoples)

Note: For species found growing in the Robintown "orchard garden" near Kitsumkalum, see below in this chapter.

Sources: See table 11-3.

herbivory, and intra- and interspecies competition and symbiosis. These processes are incorporated into the following discussions of different aspects of management, as are the associated culturally mediated processes of narrative and ceremony, settlement and seasonal movement, food-processing methods, social stratification, governance, division of labour, trade, and affinal exchange, among others.

Species- or Population-Level Management

Annual Species

The only annual species that has been documented as being grown in plots similar to the European style of gardening – and therefore clearly recognized as "gardening" by the European newcomers – is Northwest Coast, or Haida, tobacco.[27] Evidently extinct since the 1880s, this tobacco had large, globular capsules and relatively small leaves, and it has been suggested to be a distinct species or variety that was grown, used, and traded by the Haida and Tlingit long before contact with Europeans. A number of European explorers attest to its presence under cultivation, both in Haida territory, such as at Cumshewa and Skedans, and in Tlingit lands, such as around Yakutat Bay (Curtis 1916; Emmons 1991; Newcombe 1897–1916; Turner 2004a; Turner and Taylor 1972). It was also known to Ts'msyen peoples but was apparently obtained by trade.

Women were the usual cultivators of Northwest Coast tobacco. They used several recognized "gardening" practices in its production (Emmons 1991; Turner 2004a):

- *ownership* of the gardens (designated only for growing tobacco) by individual women, with the size of the garden in proportion to the owner's rank;
- *clearing and weeding* of square patches of ground for the gardens;
- *fencing* the gardens with stakes of crabapple lashed with cedar withes and other such barriers, for security against both people and dogs;
- *sowing* "seeds" (actually, apparently the large, spherical fruiting capsules) usually around the end of April, with each capsule being placed on a small hill of earth;
- *mulching and fertilizing* with rotten wood (which was sometimes mixed with the capsules when planted); and
- *harvesting the plants all together* when they had reached the right stage and then drying and curing the plants carefully on cedar frames over small fires of alder.

Notably, July, known as *us whook disi* ("native tobacco drying moon"), was the first month of the Chilkat Tlingit calendar (Emmons 1991, 462). For the Haida, however, the plants were apparently harvested at the beginning of September, and at this time, rotten wood was put into the ground to enrich the soil for the next year. These techniques are similar to those used by Native Californians for growing their tobacco (Donn Todt, pers. comm., 2003) and, significantly, are quite analogous to the practices used in the early contact period for growing potatoes, which the Haida in particular became famous for producing and trading (Turner and Taylor 1972). Possibly the tobacco cultivation techniques – and the original "seeds" of Northwest Coast tobacco itself – were derived in prehistoric times through trade at the mouth of the Columbia River, at The Dalles, or as far south as California and were brought, either via a coastal route or northward through the interior, to Haida and Tlingit territory (Meilleur 1979). Some of the origin stories for this species place its origin with the "Stick" Indians, or Dene (Athabaskans), up in the interior (Turner and Taylor 1972). A second species of tobacco (*N. attenuata*), used by the Nlaka'pamux, Secwepemc, Okanagan, Ktunaxa, and other peoples of the Interior Plateau, being cured and smoked in pipes of antler, steatite, or wood, may also have been cultivated, but its status as a cultivated plant is rather tenuous.[28]

With the possible exception of chenopods (*Chenopodium* spp.) – annuals whose seeds are found in copious quantities at diverse archaeological sites of the Interior Plateau (Lepofsky 2004)[29] – most of the other plant species that can be considered to have been cultivated, tended, or managed in some way within the study region are perennials, and the associated management activities would have taken into account the previously discussed abilities of these plants to regenerate their tissues or to regrow entire plants from vegetative propagules – mainly root fragments or underground stems – as well as from seed.

Partial Harvesting: Perennials and "Keeping It Living"

As previously described, many of the foods and materials people obtained from plants were partially harvested from the plant, leaving most of the plant intact, or at least with sufficient meristematic tissues to enable it to heal or regrow. The most widely recognized form of partial harvesting is in the removal of bark from various types of trees, as already described in this chapter and in chapters 5, 6, and 7 (see also Turner, Ari, et al. 2009). Species whose "dead" outer bark was taken off, still leaving the viable inner bark to protect the tree, include paper birch, bitter cherry, and pin cherry. In addition, the outer bark of old-growth Douglas-fir (Mathews and Dady 2008) and cottonwood (Mary Hanson, pers. comm., 2009) was removed from living trees for use as hot-burning fuel.[30]

Species whose inner as well as outer bark was required and therefore only partially harvested, so as not to kill the tree, include fibrous resource trees like western redcedar and yellow-cedar, edible cambium species like lodgepole pine and western hemlock, and dye- or medicinal-barked species like red alder, Pacific crabapple, cascara, Pacific yew, and subalpine fir. Additionally, pitch-producing

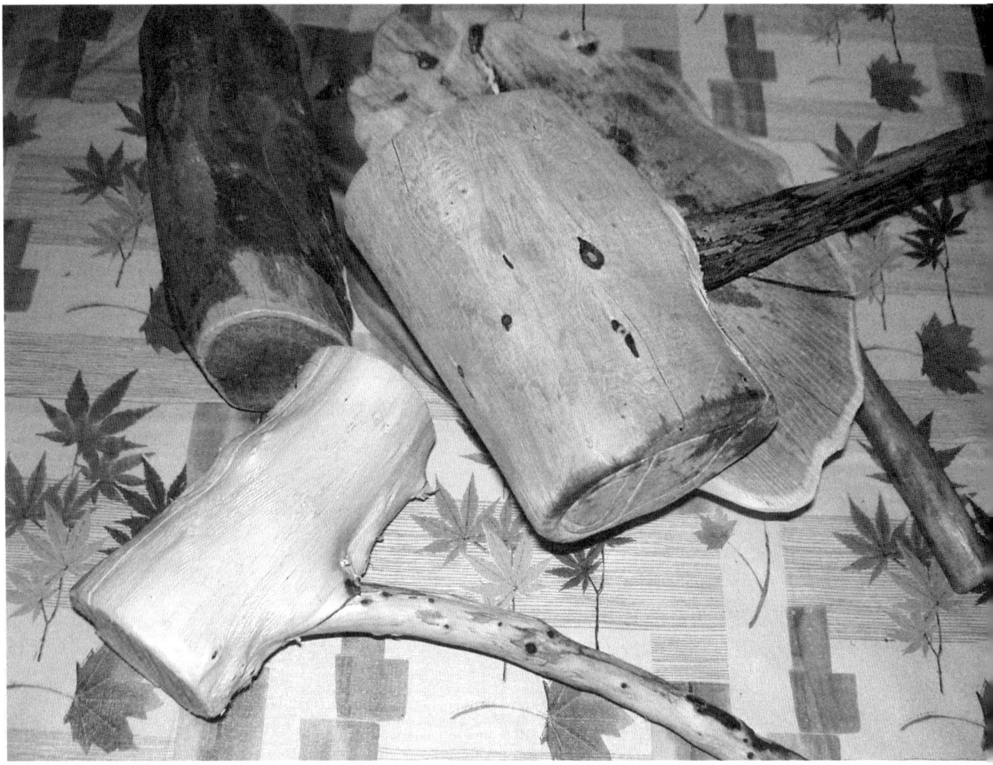

11-4 | Mallets made of yew wood (*Taxus brevifolia*). Since yew has the capacity to regenerate from cut stumps, branches and even trunk sections could be harvested without killing the parent tree, as long as it was done with care. These Gitga'at mallets are used for pounding dried halibut.

trees like spruce, hemlock, and grand fir were sometimes cut or burned in one spot to activate pitch production. Some of these pitch trees were employed continuously for such purposes over many generations, in a similar way to how the eastern sugar maples (*Acer saccharum* and other spp.) serve as an annual source of sap for maple syrup. The pitch was variously chewed for pleasure, used as waterproofing or adhesive for implements and containers, or prepared for medicinal use. Individual Sitka spruce and grand fir trees that have been used for eighty years or more, and are still used today, are known from certain camps and villages.

Only on occasions when logs were required for construction or as fuel, or when an area was being cleared for settlement or some other purpose, was an entire tree felled.[31] Whole planks were cut from standing western redcedar trees without killing them, and large limbs, branches, or roots were also routinely harvested from living trees and shrubs for basketry, cordage, nets, or other purposes, always leaving a sufficient part of the plant behind to enable it to survive and continue growing (figure 11-4).

Edible green shoots from plants like fireweed, thimbleberry, and salmonberry, as well as cordage, basketry, and mat materials from cattail, tule, Indian-hemp, stinging nettle, beargrass, and slough sedge (*Carex obnupta*), can be harvested sustainably from perennial rootstocks, either at the beginning of the growing season for edible plant parts or when the season's growth is mature for the material plants. In all cases, the plants are often said to grow better and to be of higher quality in the years following harvesting than if left unharvested to die back naturally. Edible seaweed (*Pyropia abbottiae*), too, was said to flourish with harvesting.[32] Similarly, partial harvesting of edible rhizomes like springbank clover and eelgrass, and of medicinal "roots" like licorice fern and showy aster (*Eurybia conspicuua*) rhizomes, allowed these species to regenerate readily, as long as the harvesters removed only limited quantities and did not take too much at any one time (Cullis-Suzuki 2007).

Pruning and Coppicing

Another form of partial harvesting, but one combined with a process that intentionally stimulated regrowth and production of flowers and fruits, was the practice of breaking off the branches – essentially pruning – of berry bushes. This technique was widespread and was often undertaken by women during and after harvesting berries or fruits. As Clan Chief Adam Dick (Kwaxsistalla) (pers. comm., 1997) commented, "A lot of people think we never touched the wild ... berries. But we did. We cultivated it. We pruned it ... Especially that *g̱ʷadems* [red huckleberries], when they finished picking the *g̱ʷadems*, you know, they pruned them. They break the tops off. Salmonberries too." Breaking off berry-laden branches of soapberry, red huckleberry, and other berries had a dual function. Not only did it increase the succeeding years' growth and berry production, but it was also a means of ensuring that elders could share in the harvest. Once removed, the branches could be carried by the armload to a convenient location where the grandmothers (and very young children, too), unable to manoeuvre over the rough ground and undergrowth, could sit and pick off the berries. Other species for which there has been specific oral evidence of pruning include saskatoon berry, Pacific crabapple, gray currant (*Ribes bracteosum*), blueberries, and highbush cranberry (Elsie Claxton, pers. comm., 1997; M.K. Anderson 2005; Compton 1993b; Johnson Gottesfeld 1993; Peacock and Turner 2000; Turner and Peacock 2005; see also table 11-4).

M. Kat Anderson (1993a, 1993b, 1996, 2005) has demonstrated, working with Native Californians, that pruning and coppicing enhance the production of long, straight, flexible shoots (withes) for basketry, cordage, fish traps, and other purposes. Although there is little current evidence of this practice for First

Peoples in northwestern North America, it is likely that it was once widespread. Stl'atl'imx elder Sam Mitchell (pers. comm., 1974) demonstrated how the long stems of "rope willow" (*Salix exigua*) were twisted into rope while still growing and then cut off at the base (see chapter 6). This would have had the effect of coppicing in the same manner as Native Californians did for this species, producing immense quantities of high quality withes for basketry. In the Secwepemc area of the Interior Plateau, Mary Thomas (pers. comm., 1996) remembered her mother looking at tall, bushy saskatoon berry bushes and saying, "Yes, it's time to cut it down." Then she would cut or burn the bush right back to the ground. In the year following, long, straight, flexible sucker shoots would arise from the rootstock. These were ideal for crafting the rims of birch-bark baskets, making arrows, and other purposes. However, only a few of these young, straight shoots were cut from any one bush. The rest were left to grow and develop, and within just a couple of years, the coppiced bushes would have regrown and would be covered with large, plump, juicy berries. After perhaps eight or ten years, they would again become too large and bushy, and berry production would decline; then they would be cut back again (ibid.). Through this technique of always creating young, vigorous growth, these bushes could be kept at a relatively young, productive stage for decades (M.K. Anderson 2005).

Other types of shrubs were also pruned back for material production. Some implements, like digging sticks, arrows, and mat-making needles from oceanspray, were considered best if made from older stems; the first-year shoots were too thin and soft. In this case, the bushes might be pruned but the wood used in multiyear cycles (Dr Arvid Charlie [Luschiim], pers. comm., 1999). Hazelnut, mock-orange, willows, and Rocky Mountain maple are other species whose shoots, splints, and bark were used for basketry and implements and whose quality and productivity were probably improved through coppicing or pruning, as was the case for their equivalents tended by Native Californians (M.K. Anderson 2005; Deur and Turner 2005).

Burning Individual Plants

Berry bushes were sometimes managed by burning them,[33] as indicated in the Kwakwakaw'akw "prayer" or address to berry bushes as recorded in Boas (1930, 203, emphasis added):

> I have come, Supernatural Ones, you, Long-Life-Makers, that I may take you, for that is the reason why you have come, brought by your creator, that you may come and satisfy me; you Supernatural Ones; *and this, that you do not blame me for what I do to you when I set fire to you the way it is done by my root (ancestor) who set fire to you in his manner when you get old on the ground that you may bear much fruit*. Look! I come now dressed with my

large basket and my small basket that you may go into it, Healing-Women; you Supernatural Ones. I mean this, that you may not be evilly disposed towards me, friends. That you may only treat me well.

In this case, the type of berry bush referred to is unspecified, but it is apparent that the words are from a female berry picker, and she speaks to the supernatural character attributed to this resource, addressing the bushes as "Supernatural Ones" and "Long-Life-Makers" and the berries as "Healing-Women," "Supernatural Ones," and "friends." Hazelnut bushes were also burned back periodically, according to Nlaka'pamux plant specialist Annie York (Turner, Thompson, et al. 1990), as they also were by the Kalapuya in the Willamette Valley (Boyd 1999b). After just a few years, these treated bushes would produce an abundance of nuts, with the enhanced productivity continuing for several more years, until the bushes were deemed ready to be burned back once again.

Selection by Size, Age, or Reproductive Stage

Selective harvesting of plants by their status, state, or size was an important technique related to a plant's reproduction abilities and may potentially lead to genetic selection of traits such as early fruiting, larger bulbs and roots, sweeter, larger berries, and other desirable attributes. For example, it was common to harvest root vegetables such as edible camas, yellow glacier lily, spring beauty, springbank clover, Pacific silverweed, northern riceroot, and Nootka lupine at their fruiting or postfruiting stage, when the leaves had started to fade and die back. In these situations, harvesting tended to result in scattering of ripe seeds since most of these species, having dry capsules, are well adapted to "shattering" or releasing their seeds when the capsules are brushed or bumped, or even when blown by the wind. Thus, at the same time as one generation of plants was being removed, the ground was being prepared for new growth through digging, tilling, and weeding, and the seeds for the next generations were being scattered or dropped into ideal niches for germination and growth.

Alan Marshall (1977, 161) emphasizes the importance of timing of the harvest for the Nez Perce in terms of regeneration of the resource species:

> The conservation of these plants was affected by the timing of their collection. Many of the root foods, especially important ones as */q'eq'í.t/* [*q'eq'éet*] [Canby's biscuitroot, *Lomatium canbyi*], */'qámsit/* [kouse, *Lomatium cous*], and */qém'es/* [camas, *Camassia quamash*], were not collected in quantity until their seeds had ripened. This was important in maintaining the stands of these plants which thrive in disturbed soils. By digging after the seeds ripened, the people ensured that plant seeds would be distributed in "prepared" seedbeds. This may have been especially critical for camas, which is restricted to habitats with poorly aerated soils. Increased

> aeration resulting from digging probably increased nutrient cycling rates. Furthermore, not all */qém'es/* were collected; small bulbs were left in the soil. Some people distinguished between "male" and "female" bulbs, and utilized only female bulbs. The depth of disturbance was about 12 to 18 inches [30 to 45 centimetres]: the largest, most prized bulbs being found deepest. Disturbance was extensive. Women used digging sticks – */tuqes/* – to loosen the soil, then sorted through the clods, breaking them up by hand, to gather the appropriate bulbs.

By harvesting after the seeds were ripe, the people were sure of getting tasty roots. This timing was important since stored nutrients are somewhat depleted through the fruiting process. Prior to seed maturity, the root tissue can be soft, somewhat spongy, and flavourless. Once the seed has set, however, the root gains a finer texture and flavour and probably also has a higher nutritional value (see Loewen 1998). In situations when the ground was burned over after the harvest, the seeds would already have been shed and many would have been lying protected from the fire under the soil or settled into crevices that would then have caught any nutrient-rich ash washed into the soil by succeeding rains (Beckwith 2004).

Root size was also a factor in management strategies. Notably, for arrow-leaved balsamroot in particular, the medium-sized taproots were the ones selected for harvest. The gigantic "mother" roots – which can be as large as one's arm and produce as many as forty to fifty flower heads – were left in the ground, and only the middle-sized, or carrot-sized, "daughter" roots were dug for pit-cooking and eating (Aimee August, pers. comm., 1991; Peacock 1998). In this way and for this species, the most productive plants (in terms of flower and seed production) continued to grow and generate large quantities of achenes. Similarly, the largest bulbs of camas (*Camassia* spp.) were said not to taste as good as those of slightly smaller size, so they were left behind in the harvest.[34] The same was said by some for the largest corms of spring beauty, or mountain potato; Patrick Lulua (pers. comm., 2003) observed that if the corms become really big, they go hollow inside and then are not good to eat any more. Instead, people selected the corms that were between about 2 and 3 centimetres across, leaving the largest ones behind. The smallest, youngest roots, bulbs, and corms of balsamroot, camas, and mountain potato respectively were also either left in the ground during digging or replanted (as is done with most of the edible "root" species) so that a continuous crop would be produced as the younger individuals matured and themselves became ready to harvest after a few years. This type of multiyear rotation of harvest, coupled with age selection, was an effective way to sustain a root resource.

For the edible root *masru* ("Indian potato," *Hedysarum alpinum*), the Kotzebue Inuit, just north of the study area in Alaska, harvested the plants based on their age as a way of perpetuating the populations; fast-growing, medium- to small-sized clumps usually had the best roots – white, crisp, sweet, tender, and of good size, which is 2.2 centimetres in diameter and 30 to 45 centimetres long: "That first year we had easy digging in the loose, sandy soil of an old river channel. Only *masru* plants grew there, in clumps of all ages and sizes. We didn't dig every plant. Only a third of these were the proper age. This left older plants to re-seed the garden" (A. Jones 1983, 117).

In terms of vegetative reproduction, given the common practice of replanting propagules that were separated from the bulbs as they were dug, in the case of species such as riceroot and yellow glacier lily, and given that plants were selected for large bulbs (i.e., multiflowered glacier lilies and large, tall riceroots), replanting the same genetic stock might tend to result in increasingly large and vigorous plants being perpetuated (see Reedy-Maschner and Maschner 2012) – a proposition admittedly hard to test.

For berries, intensive picking almost invariably results in a scattering of some of the berries under the bushes, particularly with species like soapberries and huckleberries – which were picked by hitting the branches or raking the fingers through the branches – but also with saskatoon berries, raspberries, thimbleberries, salmonberries, and most other fruits. If berries from bushes that produce the largest, sweetest-tasting berries are the ones selected for picking, the picking process itself – which leaves some of these behind on the ground – might ensure that the desired high-quality strains are perpetuated.

Digging, Tilling, and Weeding

Digging up the soil in the process of root harvesting (either for edible "roots" or for roots as materials for weaving or cordage) results in a loosening and aeration of the soil, which in the immediate and succeeding years allows for better regrowth of the roots and propagules left behind in the soil. Digging and tilling also allows easy removal of unwanted grasses or other "weedy" plants, which gives a further advantage to the desired species. The Tlingit of Glacier Bay weeded their seaside strawberry patches of unwanted brush that might grow up and advance the successional stage to shrub or forest cover, when strawberries would be shaded out (Thornton 1999). Camas prairies of southern Vancouver Island were cleared of bushes and tree seedlings for similar purposes: to maintain the open areas most conducive to camas production and to prevent these areas from being taken over by woods (Christopher Paul, cited in Babcock, 1967; Beckwith 2004; Turner and Hebda 2012).

Sometimes, weeding of a root-harvesting area was undertaken during different occasions and seasons than the actual harvest, such as in the early spring, giving a competitive advantage to the camas, springbank clover, riceroot, or wapato over the growing season, before harvesting in the summer or fall (McDonald 2003; Turner and Peacock 2005).[35] Digging and tilling of the soil – and production of the edible roots themselves – were easiest if the soil was sandy and loamy, without too much clay or too many rocks. Therefore, these practices were usually combined with the selection of sites for enhancement that had the right kind of soils and a good moisture regime.

Elders like Tsilhqot'in plant harvester Minnie Charleyboy (pers. comm., 2003) have said that picking and digging spring beauty corms, or mountain potatoes, makes them grow better. Long ago, she said, when everyone was digging them, they were much larger than you can find them today. She suspected that overgrazing by cattle had had an impact on their size, but the lack of digging and harvesting in recent years was also a factor. She had been coming to Potato Mountain to harvest these corms for over fifty years. Her family would start harvesting these "potatoes" at lower elevations in early June and then move up into the mountains in early July to continue the harvest. The Tsilhqot'in also burned over areas of the mountainsides from time to time to increase the productivity and size of mountain potatoes and other roots (Mabel Solomon, pers. comm., 2003).

Replanting Propagules

As well as partial harvesting of rhizomes and other plant resources (a practice that allows the remaining parts to continue to grow and regenerate), people also systematically and purposefully replanted parts of the plants as they were harvesting or after the harvest was finished. Clan Chief Adam Dick (Kwaxsistalla) (pers. comm., 1996) and Dr Mary Thomas (pers. comm., 1996) described this process for northern riceroot and yellow glacier lily respectively. Both of these elders had the opportunity of harvesting plants with their grandmothers, back in the 1920s to 1940s, and they learned many plant management approaches and practices from them. It was Adam Dick's job – as a boy of around eight or nine years working with his grandmother and mother at their *t'əkkillakʷ* (estuarine root garden) at Kingcome Inlet – to separate off the small sprouts, called *g̱ág̱emp* ("grandfather"), growing from the bottom of large riceroot bulbs, and to replant them in the garden (see also Deur and Turner 2005) (figure 11-5). Mary Thomas (pers. comm., 1996) remembered well her grandmother and mother breaking off the small appendages on the lower part of the glacier lily bulbs they were digging (actually small corms with rootlets growing from them), which they called the "whiskers," and either planting them directly back in the ground or, when they were cleaning the bulbs in the evening at the camp, putting these "whiskers" aside in a little pile and then bringing them back up to the root-harvesting grounds the following day for replanting (see also Loewen 1998; and Peacock and Turner 2000). She also remembered replanting small bulbs of

11-5 | Small sprouts, called *g̲á̲gemp* ("grandfather"), produced by the bulbs of northern riceroot (*Fritillaria camschatcensis*), which were replanted during the riceroot harvest by Clan Chief Adam Dick (Kwaxsistalla) when he was a boy. The sprouts are surrounded by the rice-like bulblets that give this plant its name. Each is capable of producing a new plant.

glacier lily and chocolate lily (*Fritillaria affinis*) as well as pieces of the crown of Canby's lovage (*Ligusticum canbyi*) root so that the plant would resprout from this part.

Other evidence for replanting includes planting the "navel" (embryonic plant) part at the top of the bitterroot, which was said to regrow the following year (Bandringa 1999), and replanting small pieces of "wild carrot" (*Conioselinum gmelinii*) roots and springbank clover rhizomes (Compton 1993a), as well as Pacific silverweed roots, small individual bulbs of camas, corms of spring beauty, and roots of balsamroot (Peacock and Turner 2000; Turner and Peacock 2005). This was obviously a widespread and longstanding practice, especially related to "root" harvesting and production, which likely led easily and obviously to transplanting and seed-planting practices.

Transplanting and Scattering Seeds, Fruits, and Propagules

There is little way of proving the extent to which humans have knowingly established new plant populations in places where they were not growing in past centuries and millennia in northwestern North America by transplanting seedlings, roots, or other propagules or by scattering seeds and fruits in new locations.[36] However, evidence suggests not only that such practices occurred but also that in some cases they were widespread and longstanding. The translocation of salmon eggs and young salmon, oulachen, herring, trout, young deer, and other animal species is also part of people's oral histories in various places (Dr Arvid Charlie [Luschiim], pers. comm., 2000; Clan Chief Adam Dick, pers. comm., 2009; Linda Smith, pers. comm., 1986; Pauline Waterfall, pers. comm., 2002; George 2003). Table 11-4 incorporates notations about culturally important plant species having been brought from one area and planted in another. Examples include:

- transplanting of one (tall) variety of saskatoon berry from the Merritt area to Spuzzum (Annie York; see Turner, Thompson, et al. 1990);
- transplanting of camas (*Camassia quamash*) bulbs from the Victoria area to the meadow behind Hesquiat Village in Clayoquot Sound (Alice Paul, cited in Turner and Efrat 1982);
- transplanting of hazelnut bushes from the Kitselas Canyon to around Kitsumkalum (McDonald 2005) and from the Pemberton area to Lillooet (Edith O'Donaghey, pers. comm., 1987);
- transplanting northern riceroot from Port Essington to Robintown (Chief Don Roberts, pers. comm., 2008);
- transplanting wild lily-of-the-valley (*Maianthemum dilatatum*) as a medicine into gardens at Yakutat (Moss 2005);
- transplanting wapato from the Fraser Valley to lakes in the Gulf Islands and between other locations (Dr Arvid Charlie [Luschiim], pers. comm., 1999; Haeberlin and Gunther 1930);
- transplanting soapberries from interior Alaska (Chilkat) to sites on the coast and from southeastern Alaska to northern sites (de Laguna 1972, 409);
- introducing young western redcedar trees to Botanie Valley and interior Nlaka'pamux territory as gifts from the downriver Nlaka'pamux (Jim Stafford, pers. comm., 2008);

- transplanting springbank clover from Kimsquit to the Kitlope Valley (Ken Hall and Simon Hall, pers. comm., 1993, 2007) and around Nuu-chah-nulth territory (Craig and Smith 1997);
- relocating cattail to Hesquiat Lake (Alice Paul, cited in Turner and Efrat 1982);
- relocating prime stinging nettle plants for cordage from village to village along the Northwest Coast (Tina Robinson, pers. comm., 2003);
- transplanting Canada blueberry (*Vaccinium myrtilloides*) from the Fraser Valley to Vancouver Island (Turner, Thomas, et al. 1983) and the Sechelt Peninsula (see Bouchard and Kennedy 1977a; Turner, Timmers, and Bouchard 1972); and
- transplanting highbush cranberry from meadows up the Kingcome River to gardens at Kingcome (Clan Chief Adam Dick, cited in Deur and Turner 2005).

Most of these examples are from recollections of contemporary elders, who either were told about these plants being brought or participated themselves in transplanting them. John Ross (2011, 297) reports a rather unusual type of transplanting by the Spokan, who used trailing blackberry plants to create pathways for animals leading into traps: "When available, the strategically placed long branches of trailing blackberry branches served effectively to direct animals into traps and snares ... women would carefully remove the root system with digging sticks, wrap the roots in moist buckskin, and transport them to an intended site for replanting (*heł qʷ'ołłq*)."

Another example of probable transplanting, likely occurring many generations ago, is suggested by the discovery, in 2012, of an extensive population of sweetflag (*Acorus americanus*) growing in the vicinity of a traditional Neskonlith (Secwepemc) campsite in the Salmon River estuary of Salmon Arm. This is a blue-listed species (i.e., vulnerable in its habitat) found in only a handful of locations mostly in the northern part of the province. Its presence as an isolated population in an estuary at the edge of Shuswap Lake – at a site where the Neskonlith people have gone for many generations to hunt ducks, dig wapato and water-parsnip, and harvest various edible berries, Indian-hemp, and cattail (described in chapter 8) – probably reflects a human-assisted extension of its natural range, as part of multifaceted traditional resource management. As noted in chapter 4, possibly it was brought in by Cree or Sekani, who used sweetflag extensively and had ventured into Secwepemc territory in the past (Teit 1909).

There are also numerous examples of the contemporary transplanting of introduced berry bushes, rhubarb, or garden flowers from one garden to another. It will take focused genetic survey work to determine the antiquity of such practices, as well as the sources and interrelationships of specific populations of native plants.

Intentional, and often ceremonial, scattering of seeds, fruits, and other propagules was also a known practice, as described by contemporary Indigenous consultants. In at least one case – the ritual scattering of the ripe stems of *súnt'iny* (spring beauty, or mountain potato) by recently bereaved widows of the Xeni Gwet'in Tsilhqot'in – dissemination by people was directly attributed by the elders to the wide range and density of spring beauty in the Potato Mountain area above Chilco Lake. In Tsilhqot'in tradition, especially if a recently widowed woman was having trouble recovering from her husband's death, she was urged by the other women in her community to go up on the mountain, pull up the mature stems of mountain potatoes from a place where they grew thickly, and spread them onto the bare areas of a place where they were not growing. These stems, which remain fleshy and greenish after flowering, contain fruiting capsules that will ripen and produce viable seed even after they are pulled up from the ground. Gilbert Solomon explained that a recently widowed woman had a special energy, which was why she was asked to do the work to make more *súnt'iny* grow (Mabel Solomon and Gilbert Solomon, pers. comm., 2003; see also L.R. Smith 2008b).

Individuals from the Secwepemc and Spokan Nations, during the course of annual berry picking, have also practised the routine scattering of handfuls of blueberries and huckleberries in places where the bushes were not growing (Dolores Bebbington, pers. comm., 2006; J.A. Ross 2011), and this was conceivably quite a widespread practice.[37] There are other instances of humans distributing seeds and other propagules for ceremonial and conservation purposes. For example, Kim Recalma-Clutesi (pers. comm., 1994) explained that Kwakwaka'wakw harvesters of *q'əx̱mín* ("Indian celery") seed for ceremonial and medicinal purposes intentionally waited for the stage when many of the seeds break loose and drop to the ground before or during the harvest, thereby ensuring continued populations of this highly valued plant.

Some of the traditional narratives from different nations also hint at the idea of individuals or groups spreading plants and influencing their distributions and densities. These include the well-known Nuxalk story of the introduction of soapberries to the Bella Coola Valley (McIlwraith 1948, vol. 1, 88), the Tlingit narrative "The Boy Who Shot the Star," involving the translocation and transformation of berries through magical means (Swanton 1909, 210; Thornton 1999), the Sinixt (Lakes Okanagan) story of four Mountain Goat brothers bringing a gift of black huckleberry bushes to Eagle and her grandmothers (Nancy Wynecoop, pers. comm. to William Elmendorf, 1935-36, cited in Turner, Bouchard, and Kennedy 1980, 103), and the Ktunaxa story of a culture hero, Yaaukuéikaam, who steals "serviceberry" (*Amelanchier alnifolia*) from the grizzly bear and spreads it all over the country so that the people can have arrow wood (Boas 1918, 93) (see chapter 13). Additionally, in a story explaining why the Nlaka'pamux site in Botanie Valley is a prime location for digging many kinds of edible roots, a powerful woman who lived at Lytton was taken away to marry a great chief, possibly the Sun. Before she left, she dropped edible roots at Botanie as provisions for her people, saying, "Roots will grow in abundance in this place" (Steedman 1930, 477). These, along with other origin stories – for tobacco, bitterroot, camas, cactus, highbush cranberry, Indian-hemp, and many other species (Schaeffer 1940; Spier 1938; Swanton 1905; Teit 1912) – may symbolically portray ancient but actual routes by which certain important plants travelled to new destinations along with their human agents, just as there are stories today of modern introductions, such as the importation of Himalayan knotweed (*Persicaria wallichii*; syn. *Polygonum polystachyum*, called *halax̱saʔnax̱ʔnóx̱*, "spirit plant") to Hartley Bay gardens (Turner and Thompson 2006).

Other Measures That Protect Growing Species

There are other, diverse practices associated with harvesting or producing particular foods and other products that were aimed at protection and conservation of the individual populations and that would therefore have been part of an overall management system at the species and population levels. One example is the convention of carefully covering over areas from which mushrooms, such as pine mushrooms (*Tricholoma magnivelare*), or particular "roots" like spring beauty corms were harvested so as to protect the underlying soil and "root" systems from exposure to sun and heat and to keep them moist and viable (see also R.T. Richards 1997). Another is the cutting or burning of tent caterpillar nests from chokecherry and other fruiting bushes, as recalled by Secwepemc elder Mary Thomas.[38] This type of individual care matches the pruning, selective harvesting, and other focused practices that led to enhanced health and productivity of the species on which people relied. The combination of these practices that played out for individual species and populations, taken together and across many species, would have had significant cumulative – and, in terms of resource enhancement, positive – effects over entire localities and specific habitats.

Diversification as Management

Having "fall-backs" in the event of depletion of a given resource (from any number of causes) is an important strategy in resource management. An example is from the late 1700s when the Kwakwaka'wakw people of the Nimpkish Valley transferred their major food source from salmon to clams at a time when the salmon failed to come up the river to spawn. The Kwakwak'awkaw chiefs decided to move the villagers from the valley to the estuary of the Nimpkish River in order to enable the people to live on clams and other seafood until the salmon returned. This is akin to allowing a resource area to lie fallow so that it can rejuvenate without the pressure of human harvesting. There is evidence that people even longer ago than this episode were turning to alternative food resources at sites like Namu on the central coast. Here, and elsewhere, archaeologists have observed fluctuations over the past 6,000 years or so in people's reliance on major resources, such as when salmon and herring use has been alternated with more diversified exploitation of other fish, shellfish, marine mammals, deer, and small land mammals (Ames and Maschner 1999).

In terms of plant resources, people everywhere in the region practised such alternative harvesting as necessary for many different applications. For example, fuel – a critically important resource – can be drawn from a wide variety of woods of trees and shrubs. The archaeological record shows many places where variations existed in fuel use over millennia, apparently related to the changing availability of certain species. It could be that particular fuel woods were depleted through overharvesting, necessitating a shift to other species, or that changes in climate resulted in vegetation shifts over time, reducing opportunities to use certain woods and increasing the availability of others (see Lepofsky 2004). In any case, the ability to utilize and adapt to a range of different fuels has helped people to cope with resource fluctuations and has possibly directed a trend toward sustainable fuel production from living trees, as with the selective harvesting of the outer barks of Douglas-fir and cottonwood, mentioned previously.

Berries often have undulating cycles of productivity, contingent on such factors as weather conditions. Unusually cold winters can kill back the flower buds, heavy spring rains can suppress pollinator activity or encourage fungal infestation, and cool, damp, or excessively dry summers can prevent optimal ripening of the fruit. Each species, and even different varieties within a species, have different environmental requirements and vary in tolerance to stress from weather conditions, insect pests, and other factors. Thus having knowledge of and access to a whole range of usable species strengthens a group's resilience and its ability to withstand variations in particular resources. Similarly, having access to – and being able to modify or create – diverse habitats at different elevations and in various geographical areas also increases available resource diversity. Rotation of burns at different sites across the landscape is one example of how people diversified their available resources by creating sites of different successional stages and therefore of different species combinations with different levels of availability and productivity (Lepofsky, Hallett, et al. 2005; Turner 1999).

Habitat-Level Management

As described in chapter 8, Indigenous people did not confine themselves to only one place but moved over their landscapes and territories in patterned seasonal rounds that were choreographed to key productive resource sites in different ecosystems and that were mediated by growing cycles, weather, and other environmental factors. Virtually every type of prime productive habitat, from coastal beaches to coniferous forests and montane meadows – and every successional stage of vegetation – was known to and influenced by Indigenous managers. Within these general habitat types were favoured locales, known widely for producing the largest, sweetest berries, the most succulent greens, or the biggest and best-tasting roots. Many of these sites are places where people have been harvesting intensively since time immemorial and that bear the evidence of human management, such as periodic clearing and burning, with associated archaeological features including hearth sites, pit-cooking and pithouse depressions, camping areas, altered and disturbed soil, culturally modified trees, rock alignments, rock carvings or paintings, fish weirs, and areas with evidence of drying and processing food. Two good examples of such places are the Indian Heaven huckleberry-picking grounds near Mount Adams and the Tsilhqot'in areas for harvesting spring beauty corms and blueberries on Potato Mountain, but these are only two of hundreds of focal cultural sites. The exceptional quality of the berries, roots, or other resources identified with such places could well be due, in part, to generations of occupation, use, and tending (Peacock and Turner 2000).

All of the practices referred to in the previous section have implications for habitats in general. Weeding and clearing, transplanting, tilling and pruning, and other management activities, when enacted toward groups of individuals or collectively on different species of roots or berries growing in a given plant community, affect entire suites of species. Sometimes, for example, Straits Salish people cleared their camas beds of large rocks and brush by pulling shrubs and seedlings out by hand. Removing competing species and converting rocky ground to level soil patches would have improved the growing conditions not only for camas but also for onions (*Allium* spp.), wild caraway (*Perideridia gairdneri*), false onion (*Triteleia* spp.), chocolate lily, bracken fern, and other root vegetable species, as well as possibly for wild strawberries, trailing blackberry, and other low-growing fruits. It also would have enhanced habitat for deer and elk by providing prime forage grounds.

Physical alterations in specific places – clearing land and building terraces for houses, moving large rocks (such as for clam garden construction or other purposes), and installing walls and fences – would have affected water and nutrient regimes (e.g., nitrogen and phosphorus levels), soils and sediments, and ultimately, the assemblages of plants and animals and the habitats themselves (Deur 2000, 2002b, 2005). These practices are therefore part of an overall habitat management regime. For example, weeding and clearing undertaken in wetlands for wapato production (Suttles 1955) would have affected an entire habitat and community of species, influencing the growth and production of willow, cattails, tule, cranberries, and cedar, as well as of fish, waterfowl, muskrat and beaver, in various ways. In the tidal estuarine gardens such as at Kingcome Inlet, removal of rocks and large clumps of grasses and sedges would have enhanced the growth of all of the traditional root garden species (Clan Chief Adam Dick, pers. comm., 2004; see also Turner and Peacock 2005). Weeding peat bogs to improve the production of Labrador tea and clearing riverside areas like at the Ecstall River of northern British Columbia of excess vegetation to promote riceroot and bracken fern production (McDonald 2003, 57–8) would similarly have affected the entire ecosystem. The ultimate result of these activities would have been the creation of "patches within patches," where specific growing areas were situated within a matrix of culturally important sites that had been cleared or thinned of certain undesirable species in order to improve the production of others.

Each household, clan, or village group had leaders and individuals who were skilled in identifying and overseeing exceptional resource locations, who were knowledgeable in the techniques, such as clearing and burning, that would produce higher concentrations of resources – patches of the largest roots or the tastiest berries – and who were able to achieve easier access through trail building and land modification (McDonald 2003; Turner 1999). Prime locations, such as berry patches growing near camps, might be specially selected for enhancement through burning and other practices.

One example of such a site is a "berry garden" at Roscoe Inlet in Heiltsuk territory, as described by the late Cyril Carpenter[39] and by Pauline Waterfall (both pers. comm., 2002; see also Turner 2005), both of whom had learned about this site – a wide, bushy ledge beside a waterfall above a campsite – from their respective grandmothers, Bessie Brown and Beatrice Brown, who were related. In this and similar sites, people identified areas that had optimal exposure to sunlight at certain hours of the day but were protected from the extreme prevailing winds and located beside waterfalls, where there was always sufficient moisture from the mists. Here, they fertilized and mulched existing blueberry, huckleberry, and salal bushes with herring eggs and with remains from cleaning and dressing their salmon, deer, and other game by digging holes around the bushes and burying the organic material near their roots. People also scattered ashes around the berry bushes from the fireplaces of the long houses, as well as adding clamshells. They also used to transplant berry bushes from surrounding areas to these productive sites, and when they did, they added crushed clamshells to the holes. Cyril Carpenter recalled being told that the berries from these bushes were healthier, bigger, more productive, and tastier than untended berries; one could harvest them from the branches "by the handful." Such berry gardens were apparently created and maintained in various places along the coast but were seldom noticed or recognized as "gardens" by anyone beyond the peoples who created and cared for them.

Forests and woods, too, can become specially managed habitats, depending on the extent and intensity of human interactions within a given area. Even the practice described by Mary Thomas (2001) for the Secwepemc of preferentially selecting dead and downed trees to remove and use, rather than cutting down living trees, would have affected entire forest habitats by helping to clear and thin woodland areas, thus making them more accessible for humans and deer and more productive for berries and the trees that remained. Cedar stands where bark strips, roofing sheets, or planks are harvested similarly bear the mark of human interactions, sometimes over centuries. Many are, in fact, managed habitats that are owned and monitored by individuals or lineages. Not only the trees themselves but also the undergrowth, berry bushes, ferns, and other species are all affected by humans' continued presence within the entire area. Even when whole plants were removed, as in cutting trees, it was done with care and purpose. The trees cut down around villages to be used for house construction, canoes, or totem poles, for example, offered habitats on their stumps for salal, trailing currant, red huckleberry, and blueberries, and thus the area became a berry garden for the people of a village (Captain Gold, pers. comm., 1996; Turner 2004a).

Estuarine root gardens serve as another key example of managed habitats, being carefully selected, established, and maintained over generations. Known in Kwak'wala as *t'əkkillakʷ* ("place of manufactured soil") (Clan Chief Adam Dick and Dr Daisy Sewid-Smith, pers. comm., 1997; Deur and Turner 2005; Lloyd 2011) and in Nuu-chah-nulth as *tł'ayaqak* ("place for roots or berries") (Arima et al. 1991; Sapir 1913–14), these productive gardens once extended over many estuarine habitats of the Northwest Coast but are perhaps most remarkable, or at least best documented, from the territories of the Kwakwaka'wakw, Nuu-chah-nulth, and Nuxalk.[40] Situated in the upper intertidal zone, these plots were, and in some cases still are, owned and inherited by individual family or clan units, with the entire area being part of a community's territory. Their boundaries were marked out by natural features or by posts, logs, or rocks placed by the owners. The root vegetables produced in these "gardens" included springbank clover, as described at the beginning of this chapter, Pacific silverweed, northern riceroot, and other species.[41] Soft sandy-loam soil was preferred for growing these root vegetables because they were easier to dig and, especially, because the roots tended to grow long and straight in this environment, making them culturally more desirable (Deur 2000). As previously noted, the plots were often weeded of grasses and other competing vegetation in the springtime, with the edible roots being harvested mainly in the fall but also to some extent in the winter and early spring months before the plants sprouted up again.

Sometimes these root gardens were actually built up with rock and log walls, which functioned to trap floating marine algae and other tidal detritus brought in by the highest tides, as well as silt and organic materials washed down the rivers and deposited during flooding, thus contributing mulch and enriching the soil (Deur 2000, 2005). The nutritious roots of the tidal garden patches also

11-6 | Crabapple tree from Robintown "orchard garden" showing partially cut top.

attracted ducks, geese, and swans and were sometimes deliberately used as bait for these waterfowl (Edwards 1979). Clan Chief Adam Dick (Kwaxsistalla) (pers. comm., 2009) noted that bears and even seals also eat the clover and other edible roots from the *t'əkkillak*ʷ (estuarine root gardens) of Kingcome. Generally, too, Pacific crabapple trees and sometimes highbush cranberries, blueberries and huckleberries, and salal line the moist shorelines just a little higher than the root gardens, thus extending the array of food available for harvest in the fall. As well as weeding these plots, management techniques applied to enhance the productivity of the root foods, as described in the references cited previously, included tilling, partial and selective harvesting, and replanting propagules. These highly productive root gardens were extensive enough to provide food for over 500 people of the Tsawataineuk clans living at Gwa'yee on the lower Kingcome River, as well as to provide a certain quantity of the roots for trade.

Adam Dick (pers. comm., 1999) said that at one time "there wasn't *one square inch*" of that tidal flat at the Kingcome River estuary that was not under care and ownership of the people of Gwa'yee. Each family had long rectangular plots, bordering the river on both sides, that were marked out with pegs and carefully tended, especially by the women and children.[42]

Another, similarly complex managed habitat can be seen at Robintown, a now-vacated settlement, which was occupied by the Kitsumkalum Ts'm̓syen until the mid to late 1800s (Downs 2006; McDonald 2003). This village site, which can be termed an "orchard garden" because of its mix of trees, shrubs, and herbaceous plant populations, features several large terraces still supporting an array of habitats and many culturally important species that would have been owned and tended when people were living there. As well as dozens of Pacific crabapple trees, the site sustains many other key plant resources. The crabapple trees still show the effects of pruning and cutting at the tops, which causes them to partially fall over and then regrow closer to the ground, allowing easier access to the fruit (figure 11-6). Another prime species of the "orchard garden" is hazelnut, which may have originally been transplanted as saplings to this site, as they were known to have been distributed from the Kitselas Canyon to other sites, including the town site at the mouth of the Kitsumkalum River. As McDonald (2003, 57) notes, "people now consider the presence of [hazel]nut trees and crabapple trees to be a sure sign of an old residential community or camp site." Northern riceroot is also still common along the terraces and was said to have been introduced to Robintown from Port Essington (Chief Don Roberts, pers. comm., 2008).

Over fifty other plant species of known cultural significance can be found in the Robintown orchard gardens, including:

- material species: western redcedar and skunk-cabbage;
- medicinal species: licorice fern, wild ginger, wild lily-of-the-valley, false Solomon's-seal, sarsaparilla (*Aralia nudicaulis*), devil's-club, black snakeroot (*Sanicula marilandica*), and stinging nettle;
- edible fruits: saskatoon berry, bunchberry (*Cornus canadensis* and/or *C. unalaschensis*), black hawthorn (*Crataegus douglasii*), wild strawberries (*Fragaria vesca, F. virginiana*), gray currant (*Ribes bracteosum*), Nootka rose, wild raspberry, thimbleberry, salmonberry, red elderberry, blueberries (*Vaccinium alaskaense, V. ovalifolium*), black huckleberry, red huckleberry, and highbush cranberry; and
- other edible species: fireweed, northern riceroot (said to have been transplanted there), and cow-parsnip.

Few sites anywhere can compare in the number of important plant resource species within a limited area – a tribute to the ingenuity of the people who lived there and who used and apparently cared for these species (Downs 2006; Mackin 2009; McDonald 2003; Turner, Deur, and Lepofsky 2013). To the north, evidence of managed "gardens" of culturally important plants has been found for the Aleuts (Reedy-Maschner and Maschner 2012), as well as for the Tlingit and other Alaskan peoples (Thornton 1999, 2008).

Other major evidence for tended crabapple gardens comes from a petition written by three Kitamaat Haisla leaders to the chief commissioner of lands and works of British Columbia on 10 November 1897 (Kitamaat Indians 1897). This letter was a plea to the commissioner to reinstate their precious "Crab Apple Gardens" as part of their reserve lands. Their letter speaks for itself:

> Sir –
>
> We the undersigned Indians of Kit-a-maat humbly petition you and ask your kind and just intercession with assistance, in securing our ancient, and, as we think, just right to our old Crab Apple Gardens, planted by Nature in the old time for the use of Indians for both food and medicine.
>
> These Crab Apple flats held for ages, equal to all past time, by our forefathers, from generation to generation, and frequently bought and sold by Indian families; are in our part of the Country, situated in the valley of the Kit-a-maat River in seasonal small patches of low land aggregating in all about ___ acres in the River Valley, and about ___ acres in a small Inlet near our village named Kil-dalla.
>
> These Apple Gardens always jealously guarded by their Indian owners, were overlooked by the Honourable Peter O'Reilly, Reserve Commissioner, and the few Indians who were at home when he came here years ago to reserve some land for us. Unfortunately at that time but few of our people (old men) were at home to meet the Commissioner, and they did not understand fully what was being done, and never for a moment suspecting that the Indians might lose their consistent right to use the fruit patches always heretofore held by us Indians and found so necessary for our health and strength failed to ask, or secure their reservation. These few patches of fruit trees producing apples not valued by white men, are now in danger of being included in Land Claims made by strangers to us, and if allowed by you and our Governor to be taken from us and destroyed by clearing of the land will greatly inure us and destroy our faith and trust in the Justice and Friendship of our government.
>
> We hopefully ask you to protect our Old Rights to those small parts of our valley here so useful to us.
>
> We enclose herein a rough sketch plan of the different spots where the Crab Apples always grow – and we fully trust you to take proper steps to secure to such of our Families as can show ancient right to those Gardens, protection against their loss.

And your petitioners will ever pray:
Signed: –
Chief Jsosie [?] Morrison
Chief Wm Young
Charlie Adams

Landscape Burning

Fire has been used as a tool for vegetation management in many different ecosystems across the study area, from the boreal forest (Johnson 1997, 1999; H.T. Lewis 1982; H.T. Lewis and Ferguson 1988) to Douglas-fir woodlands, camas prairies, and subalpine parklands (Boyd 1999b; J.A. Ross 2011; Turner 1999). Forestry and government regulations from the early 1900s, in both the United States and Canada, discouraged and then banned intentionally lit fires, so there are very few people in the southern part of the area who remember the intricacies of landscape burning. Even lightning-ignited fires, which would have formerly burned over some areas and from which people would have learned about fire behaviour, have been actively suppressed in the past century, so contemporary people often know less about the effects of fire in different environments than did their ancestors.

One of the most detailed accounts of the practice of landscape burning, from the Lil'wat (Lower Stl'atl'imx) of the Pemberton-Lillooet area, is in a narrative told by Baptiste Ritchie of Mount Currie, called "Burning Mountainsides for Better Crops" (recounted first in the Stl'atl'imx language to linguist Leon Swoboda in the 1960s and then translated by Baptiste Ritchie, cited in Turner 1999). Baptiste Ritchie described the circumstances when people used to light fire to a given mountainside: "When it get's too bushy." He and his friends, Charlie Mack Seymour of Mount Currie and Sam Mitchell of Xaxl'ep, both also Stl'atl'imx, confirmed that fires were generally lit in the fall, immediately after the huckleberries and other high-elevation berries were harvested (as confirmed by other elders like Nlaka'pamux plant specialists Annie York and Hilda Austin), and that they "rarely burnt the big trees" (ibid., 189). A couple of years following a burn, root vegetables like nodding onion, spring beauty, yellow glacier lily, and tiger lily were known to grow very productively, and berries like blackcaps, soapberries, and blueberries and huckleberries fruited abundantly. Around six or seven years later, when the bushes started to become too dense, the people would burn the area over again. This practice was carried out in multiyear cycles, with mountainsides being burned in alternating years so that there were always good berry-picking sites available to people at any one time (L.R. Smith 2008b; Turner 1999; Turner, Thompson, et al. 1990).

The timing of burns often depended on the type of habitat; generally, burning might take place in the spring at lower elevations and in the fall at higher elevations. Huckleberry areas might be burned over in the late spring, when the normal soil moisture was higher and heat penetration shallower (J.A. Ross 2011). For prairie burns, the grass might be ignited early in the morning, and those overseeing the burns would be careful to watch the winds in order to ensure the fire did not become too hot. On the coast, even watching the tides was important because winds are influenced by rising and falling tides (Dr Arvid Charlie [Luschiim], pers. comm., 2011).

Farther north, along the Skeena and Nass Rivers, for example, burning was usually done in the early spring, when low air temperatures and high humidity created shorter flame lengths and when lower fuel temperatures made fires easier to control (Johnson 1999). Dry grasses and dead twigs provided a limited fuel load, enough to carry the fire but not so much that it would blaze uncontrollably. Streams and snowbanks served as barriers to the spread of the fires. The burned-over slopes above the villages often became prime habitat for grazing and browsing deer, elk, and moose, as well as for the production of nodding onions, berry bushes (huckleberries and blueberries, saskatoon berries, currants, and gooseberries), and hazelnuts. Keeping the areas around settlements and camps clear of thick brush had the added advantage of protecting the sites against accidental fires in the summertime. This practice also kept open trails and campsites for better access and allowed easier lookouts against enemy attacks. The burning of slopes and hillsides in the Skeena region continued longer than in the southern part of the study area (Deanna Nyce, pers. comm., 2004). The burning was said to make the berries "fat and plump" (Pat Namox, quoted by A. Mills, cited in McDonald 2003, 55).[43]

All along the Northwest Coast, from Oregon to Alaska, even within the dense temperate rainforest, in which few natural fires occurred over centuries or sometimes multiple millennia (Gavin, Brubaker, and Lertzman 2003), people formerly burned over particular berry patches to enhance their growth and created meadows and clearings among the big trees, diversifying the habitats and thereby creating a higher diversity of resources.[44] Here, burning to create openings in otherwise dense forests created desirable habitat for deer, elk, and bear, for ground-nesting birds, and in forest sloughs and wetlands, for geese and ducks (M.K. Anderson 2009).

The grassy plains and savannahs of southeastern Vancouver Island (such as those seen by Captain George Vancouver as so pleasing to the eye), the wide prairies of the Olympic Peninsula (such as the Ozette Prairies), the Gulf and San Juan Islands, the Puget Sound region, the Skagit Valley, and the Willamette Valley also represented extensive human-modified landscapes, cleared by periodic burning and maintained by other cultural practices such as tilling, brush clearing, and pruning. Managed camas prairies also extended eastward to Idaho and Montana and south into Oregon and California. Depending on soil and moisture regimes associated with these areas, they were maintained as well-drained prairies, wet meadows, or even peat bogs, which were spotted around the country and interspersed with woodlands of oaks and other broad-leaved trees and with conifer forests – a rich and diverse mosaic of vegetation types, many of which were identified by European newcomers as highly promising in terms of potential as agricultural sites (M.K. Anderson 2009; Beckwith 2004; Boyd 1999b; Gill 2005b; Turner 1999; Weiser and Lepofsky 2009).

Burning and clearing created the prime habitat for camas (*Camassia* spp.) and other spring-flowering roots, including chocolate lily, taper-tip onion (*Allium acuminatum*), bitterroot (in the interior), wild caraway, or yampah, and bracken fern, whose rhizomes were once a prized food (Norton 1979a, 1979b). It also enhanced the productivity of the berries, especially the trailing blackberries, wild strawberries, and blackcaps. Burning in these environments was undertaken mostly in the summertime after the spring-blooming species like camas had gone to seed and the strawberries, trailing blackberries, and other fruits had been picked. People knew just where and how to burn, and the fires tended to be quick and light, such that one could actually run through the line of flames as it moved across the prairie (Beckwith 2004; Gill 1983, 2005b; Norton 1979a; Turner 1999). Without repeated fires every few years, the meadows, open wetlands, and forest glades would gradually be taken over by tree saplings and eventually revert to second-growth forests with closed canopy and relatively low productivity of food species.[45] Furthermore, fires helped to reduce the numbers of weeds, wasps, and other undesirable species (J.A. Ross 2011).

Landscape burning is an art, requiring knowledge of fuel loads, expected patterns of ecological succession, responses to fire of individual species at various lifecycle stages, variability and cycles of productivity and regeneration of plants and plant communities, effects of different weather and wind conditions, soil moisture regimes, and so forth. Camas meadows might be burned over every year or two. Upland berry patches and coastal prairies within wet forests might be burned over in five- or six-year cycles (Turner 1999), with cranberry bogs burned over in ten-year cycles (M.K. Anderson 2009). Burning was usually overseen by experts, such as clan or house chiefs, who were often trained from a young age in the skills and practices they needed to take on responsibility for managing resource-harvesting grounds (Boyd 1999b; Johnson 1999; McDonald 2003).[46] For example, for the Gitxsan, "When it is the right time, he [the chief] burns the berry patches so the berries are fat and plump. If he didn't do that the berry patches would become old and overgrown and there would be berries but they would just be small. But he knows when to burn so that it cleans up just the berry patch and doesn't spread to the trees" (Pat Namox, quoted by A. Mills, cited in McDonald 2003, 55–6).

Sometimes it was women who were responsible for burning and carrying out other management practices. For example, one woman of the Straits Salish of Vancouver Island told anthropologist Wayne Suttles (1974, 60) that she was the one responsible for burning the camas area that belonged to her grandmother, a practice she undertook once they had finished harvesting for the season. Other people were allowed to harvest from this locality, but responsibility for managing it through landscape burning was clearly delegated to ensure continued productivity of the camas. For the Spokan, elder women and sometimes "two-spirited" men (those with female qualities) were designated to oversee the burning of tule and cattail patches following the harvest of these plants (J.A. Ross 2011).

Selective burning of resource areas was called *p'ʕapúleʔxʷ* in the Spokan language, and the people who acted as burners were called *sxʷp'ʕápm* ("one responsible for setting a forest fire") (J.A. Ross 2011, 267, 271). The designated burners were afforded complete control over decisions regarding the frequency and strategies for burning.[47] Different animals, however, were attributed special powers for burning, and it would be advantageous to have their assistance in this task. The black bear, for example, could give a neophyte – male or female – multiple powers, such as the ability to locate huckleberries and cure burns. Those having the guardian spirit power of the blue grouse, chipmunk, western meadowlark, or various woodpeckers would be assisted by these powers in prescribed burning (ibid., 271).

Anthropogenic Landscapes

Taken in total, all of the practices and approaches for sustaining and enhancing plant and animal populations, species, and habitats and special ecosystems over the entirety of a people's territory combine to produce *anthropogenic landscapes*: mosaics of geographic and ecological regions and sites where people participate in diverse ecological processes while still maintaining the overall biodiversity and productivity of the region. Apart from the practical strategies people apply to their resources and harvesting areas as a means to utilize resources yet sustain them, their belief systems, seasonal rounds, social organization, and systems of ownership and proprietorship for particular sites and territories apply on a broad scale to the sum total of all resources, which are interconnected in a seamless web of natural and cultural processes that extends across generations and millennia.

Landscapes as distant as the Potato Mountain area of the Chilcotin region and the high meadows of Botanie Valley in the interior, as well as the seaside bluffs and camas meadows of Vancouver Island and the Kingcome River estuary and surrounding areas, can be considered anthropogenic, at least to some extent, having assumed their recent ecological status through a degree of human influence and management that potentially included the introduction and promotion of culturally important species.

One such anthropogenic landscape is in Tlingit territory, around Glacier Bay, Alaska: the "berried landscape" described by Thornton (1999). Berries were said to "hold" or represent the landscape on which they occurred, and thus they symbolized the Tlingits' material, social, and spiritual ties to their lands. These ties have had both practical and symbolic value. As inherited property, the berry patches – like the Ts'msyen highbush cranberry and crabapple places – were linked to the matrilineal clan and house system. Those who controlled the berry production through various means were enabled to participate in the ceremonial networks, including potlatching and feasting. A few special places in Tlingit country were called "berry land" and were widely known throughout and beyond Tlingit territory. One place on the lower portions of Glacier Bay was called Shákʷ Aaní (Strawberry Land), after the seaside strawberries growing there in profusion. Not only were these particular landscapes owned, defended, and celebrated, but they were also nurtured by means of *heix̱ʷa* (spiritual "magic").

It was essential in maintaining such a system that the land and the berries themselves – regarded along with other plants and animals as nonhuman persons – be attributed appropriate respect. If the plants were not properly treated, they would cease to make themselves available for harvesting and might even cause harm to those who did not follow the correct protocols. One basic tenet held by the Tlingit and other peoples was that berries "*need* to be picked"; not harvesting and appreciating them reflected disrespect and might cause them to disappear, or the plants might fail to bear fruit for a period of time. Thus it can be argued that the active use of berry resources was necessary for their renewal, an idea that has also often been tendered by Indigenous peoples for resource and land use in general. If people stop going to special places and using them as intended, the places will become less productive and more grown over (Deur and Turner 2005; Thornton 1999; Turner, Ignace, and Ignace 2000). Sometimes people showed respect to their berries through ritual feeding of their spirits.[48] The Huna Tlingit similarly "fed" dog salmon eggs to their wild strawberries and nagoonberries (*Rubus arcticus*) to enhance the productivity of the berries in succeeding years. This practice could be termed ritual application of fertilizer, but it was instead looked on as an offering to the spirits of the berries, not as a directly practical measure (Thornton 1999). In such cases, environmental manipulation is so embedded in beliefs and customs that it is impossible to pull them apart; in combination, they conserve and perpetuate key resources.

Other anthropogenic landscapes can be identified across the subalpine parkland environments of the Coast Mountains and Interior Plateau, such as those focused in places like Botanie Valley and Potato Mountain that include mosaics of plants and habitats managed in different ways and on different scales (Turner, Deur, and Mellott 2011). As noted previously, people travelled to these areas in the late spring through summer to harvest and process a rich array of root vegetables and upland berries, among other resources. Like the camas prairies, these montane meadows were maintained largely by controlled burns in multiyear cycles and were also affected by complexes of practices related to belief systems – such as First Roots and First Fruits ceremonies and the ritual scattering of seeds – with knowledge about them being passed on through narratives and oral history. Upland root- and berry-harvesting sites, managed by fire, tilling, weeding, selective harvesting, seasonal use, and rotating digging sites over multiyear cycles, would have been connected with:

- nearby creeks, lakes, and wetlands that provided drinking water, fish, waterfowl, tule (*Schoenoplectus* spp.), cattail (*Typha latifolia*), willow (*Salix* spp.) and other resources;
- upland forested areas where needed wood and bark for fuel and construction were harvested;
- forest edges and openings where berries and roots were often exceptionally highly productive; and
- grassy meadows where deer and other game came to graze and where marmots could be found, with the remains of weirs and traps, rock walls or deer fences, and other evidence of hunting and fishing now possibly present in the vicinity.

Family campsites (identified by proximity to creeks, by hearths with charcoal and fire-cracked rock, and by remains of lodges and dwellings), sacred sites (possibly indicated by rock paintings and carvings), and burial sites might also occur within anthropogenic landscape complexes. Trails and boundary lines may be marked in some way by culturally modified trees, rocks, or simple evidence of wear from trampling. There may also be evidence of social stratification, social institutions developed to assist in management of resources, and division of labour – for example, with the presence of high-value goods or tools like digging stick handles generally used by women. Evidence of social activities, such as playing lehal and other games, may also reveal periodic occupancy. Oral histories reveal multiple place names for such diversified areas (Hunn 1996; L.R. Smith 2008b), as well as stories connecting them to ancestral use.

CULTURAL APPROACHES TO PLANT MANAGEMENT

Responsibilities for Management

As described by McDonald (2003, 50–1) for the Kitsumkalum Ts'msyen of the Skeena River, "Prior to the settlement of Terrace, the Kitsumkalum Valley and nearby lands were under the land management of the people of Kitsumkalum for the benefit and prosperity of each *Waap* ['House'] and, ultimately the *Galts'ap*.

Each family took care of its *laxyuup* ['lands'] with the store of local ecological knowledge that they and their ancestors had accumulated over generations through experiment and observation ... They had to ensure the conservation of the resources for future generations and they did so with management practices that enhanced the productivity of the *laxyuup*." McNeary (1976, 61) described a similar situation for the Nisga'a: "To the Niska [Nisga'a], the Nass valley is far from a wilderness. It is a collection of familiar localities, each with its own particular resources. The ownership of each place is known and many old village sites and fish camps dot the valley. There is a richness of historical and supernatural associations which make the landscape virtually a textbook of Niska history and religion."

Within the territory of each major group, people tended to undertake their seasonal rounds as smaller family units or clans, spreading out over the land to different salmon streams, clam beds, river valleys, or mountainsides. This scattering of the population helped to reduce pressure on individual resource locales, ensuring enough resources for everyone. As noted for the Sahaptin, in the spring, "The women climbed to the ridges above to dig bitterroot and Lomatiums ... when the local supplies were exhausted ... the people moved camp further upstream to the next tributary canyon" (Hunn, Selam, and family 1990, 123). Furthermore, the social ties and the kinship and clan linkages were always maintained to provide backup systems in the event of failure or diminishment of resources in one territory or another – in other words, to maintain overall adaptability of a group (A.G. Marshall 1977; Suttles 1987a). Even when groups of families travelled together to the same general area, they often used separate campsites and harvesting sites but visited back and forth to socialize in the evenings and to help each other as required in order to ensure that everyone was able to harvest sufficient roots, berries, fish, or game for each family to survive over the winter (Turner 1992b). Thus, while people were dispersed over their territory, the "social safety net," woven of kinship and other kinds of relationships, continued in place, strengthening resilience and assisting the collective management of their lands and resources.

Everywhere, too, people held an overall philosophy of stewardship over, or caretaking responsibility for, the resources and places they relied on for survival. Sanctions against waste predominated, and children were taught from an early age not to overharvest or take more than they needed or more than they were able to process and preserve at any one time, whether it was food, medicine, basketry materials, or other resources. Respect and care for the species – the "relations" – that provided for people and the importance of sharing with others were teachings that led to an ethic of conservation or resource management at a broad level (M.K. Anderson 2005; Cruikshank 2005; Thornton 1999; Turner 2005; Turner, Ignace, and Ignace 2000; see also chapter 13). As described by Nuu-chah-nulth elder Roy Haiyupis,

> The idea and practices of over-exploitation are deplorable to our people ... Respect is the very core of our traditions, culture and existence ... Respect for nature requires a healthy state of stewardship with a healthy attitude. It is wise to respect nature. Respect the Spiritual ... it is not human to waste food. It is inhuman to over-exploit. "Protect and conserve" are key values in respect of nature and natural food resources. Never harm or kill for sport. It is degrading to your honour ... it challenges your integrity and accountability. Nature has that shield or protective barrier [that], once broken will hit back at you. (Cited in Scientific Panel for Sustainable Forest Practices in Clayoquot Sound 1995, 6–7)

Along the same lines, people were widely encouraged to use the resources that made themselves available to them, such as driftwood logs, dead snags and fallen trees in the forest, and roots and bark for basketry from trees that had fallen over or whose roots were already exposed (Mary Thomas, pers. comm., 2001; Lepofsky, Heyerdahl, et al. 2003; Lepofsky, Moss, and Lyons 2001; McDonald 2003). Self-pruned fallen limbs of trees, particularly Douglas-fir and black cottonwood, were preferred fuels, routinely harvested from the forest floor (Lepofsky 2004).

Many of the culturally special places – artifacts of many generations of focused resource use and manipulation – were given specific names in recognition of these practices. For example, in Clayoquot Sound on the west coast of Vancouver Island, there is a site in Hesquiaht territory named *shishp'iqa* ("cultivated"), so called because of the way people used to look after the beds of Pacific silverweed there, as noted by Alice Paul (cited in Bouchard and Kennedy 1990, 43). Another place in the vicinity is called *t'iik^{w}uwis* ("root-digging beach") (Stanley Sam, cited in ibid., 335). In the interior, there is a flat area in Botanie Valley (a famous root-harvesting locale in Nlaka'pamux territory) called *k'em'k'em'ats-útsiyem'tsw* after the yellow glacier lily, *sk'ém'ets*, which grows there in abundance and over which people would have assumed a caretaking role (Turner, Thompson, et al. 1990). These are only a few of hundreds of such examples (Johnson 2010; Johnson and Hunn 2010; Kari 2007).

Rights, Privileges, and Controls in Traditional Management Systems

The lands and territories of First Peoples in northwestern North America cannot be considered open "commons," with resources "free for the taking." Although the strategies for control over lands and resources varied from group to group, some form of proprietorship was always present to some extent. For the peoples of the central and northern Northwest Coast, the territories and resource sites belonged to specific houses or clans, and their use and occupancy were overseen by the hereditary chiefs and their designates (Richardson 1982). The boundaries of each chief's specific resource areas were known and often marked out or signified by well-recognized natural features. The rights and privileges of a Nuu-chah-nulth chief that relate to rivers, fishing and hunting areas, and plant-harvesting sites, for example, are bound up in a concept known as *ha̠h̠uułi*. In the past, it would be unthinkable to violate a chief's *ha̠h̠uułi*.

A chief of the Nuu-chah-nulth or some other Northwest Coast group might bequeath to an individual or family specific rights to, for example, goose hunting over one half of a lake, with the other half being apportioned to another, separated by known and recognized markers such as a creek estuary, a distinctive rockslide or cliff, or an artificial marker such as a line of posts or rocks. Others might be given the rights to pick bog cranberries or highbush cranberries, to harvest cedar bark, or to fish for sockeye around the same lake. All kinds of prime harvesting sites – from root gardens in the tidal flats to clam beds, fishing locations, berry patches, and crabapple stands – were known as belonging to a family, clan, or individual. Table 11-4 identifies particular plant resources that were known to have been specifically "owned" (usually along with other forms of management) in one area or another in northwestern North America.

The owner of a resource – often a chief, matriarch, or designate – must be asked for permission to use these areas. Rights to certain resources are determined by people's ancestral origins as well as by membership in a community, clan, or family, social status, marriage ties, and gifts and other formal exchanges. Nuxalk tribal members, for example, trace their descent from the original ancestral families, as determined by Ałquntam, the Creator. It is by having names from, or being a member of, these ancestral families that a person can claim the rights to use certain territories (McIlwraith 1948). Similar affiliations between ancestral origin narratives and rights to lands and resources are seen with other Northwest Coast groups, such as the Gitxsan: "For us, the ownership of territory is a marriage of the Chief and the land. Each Chief has an ancestor who encountered and acknowledged the life of the land. From such encounters come power. The land, the plants, the animals and the people all have spirit – they all must be shown respect. That is the basis of our law" (Gisday Wa and Delgam Uukw 1989, 7–8).

Although Coast Salish peoples like the Straits and Halkomelem recognized private ownership of key harvesting locations such as reefnet sites, duck- and deer-hunting sites, and wapato and camas beds (Babcock 1967; E. Claxton and Elliott 1994; Suttles 1955), they also included many areas that were open to general harvesting. Even people from other communities or other nations like the Ditidaht, especially if they had marriage connections, were allowed to come and harvest salmon and camas at some places (Turner, Thomas, et al. 1983).

In the Interior Plateau similar protocols were observed, enacted by the chief of a community or band, as described for the Secwepemc by Teit (1909, 573) in a passage quoted in full in chapter 9: "Berry-patches were tribal property; but picking was under tribal control. All the large and valuable berrying-spots were looked after by the chief of the band in whose district they were situated." Concerning Secwepemc hunting and fishing grounds and resource sites in general, Teit (ibid., 582–3) observed, "All land and whatever grew upon it ... belonged to the nobility." Elsewhere, however, Teit (ibid., 573) noted, "Root-digging grounds were all common tribal property. Some people of the Northern Fraser River bands laid a claim on the root-digging grounds of Quesnel Lake, where very large [tiger] lily-roots grow, but their claims were not recognized by the rest of the tribe." Teit (ibid., 583) also explained that "In a great many places ... the more distant hunting-grounds and root-digging and berrying places were looked upon as tribal property; and the nobility of the bands either did not claim them, or, being unable to enforce their claims, had decided to leave them as common property." As Teit suggested, it is often the case that areas more distant from settlements and campsites might be more generally available to people from a given group or community. The situation, however, varies not only from group to group but also over time as social inequities become more engrained or societies become more egalitarian.

In general, in both interior and coastal locations, it was the highly productive, highly valued, and well-localized resources that were "owned" or recognized as belonging to certain individuals, families, lineages, or communities. For such places, one was obliged to seek permission from the owner or a designate to harvest high-quality berries or other resources. Outsiders without specific rights to an area could normally obtain such resources only through trade. It would be a grave offence to harvest in some places without permission, with transgressions potentially punishable by confiscation of the harvest, destruction of one's canoe, or even death. In circumstances where property rights were strictly enforced, one would need to seek permission just to pass through another group's or a chief's territories (Johnson Gottesfeld 1994).[49] Less prestigious, less productive, or more widely dispersed species or populations in general tended not to be as carefully controlled.

Ownership with Social and Resource Management Responsibilities

There was more to ownership than simply receiving the benefits and the prestige of hosting a feast, gifting, or accruing surplus goods for trade and exchange. Chiefs and clan leaders applied their knowledge of and authority over local berry patches and other resource areas to organize, facilitate, and supervise groups of harvesters and to ensure that they adhered to the necessary protocols and harvested the resources only at the prime stage of ripeness and only if the supply was abundant enough to withstand intensive harvesting. They were also charged with resource stewardship and support of community members. As Nuu-chah-nulth cultural specialist Roy Haiyupis explains, regarding *h̲ah̲uułi*, the Nuu-chah-nulth system of hereditary leaders, "*Ha hoolthe* [*h̲ah̲uułi*] ... indicates ... that the hereditary chiefs have the responsibility to take care of the forests, the land, and the sea within his *ha hoolthe*, and a responsibility to look after his *mus chum* or tribal members" (cited in Scientific Panel for Sustainable Forest Practices in Clayoquot Sound 1995, 9).

Ownership thus carried with it obligations. A person who "owned" productive patches of Pacific silverweed and springbank clover or other resources would indeed receive a prime portion of each harvest but was, in return, expected to share these food resources with his village on behalf of his family or clan. In some cases, he was obliged to host feasts for particular groups or societies within the community (Dr Daisy Sewid-Smith, pers. comm., 1997; Blackman 1990; Boas 1921; Drucker 1951; Johnson Gottesfeld 1994; Thornton 1999; Turner and Kuhnlein 1983; Turner, Smith, and Jones 2005; Turner, Thomas, et al. 1983). For example, Boas (1921, 560) reported that among the Kwakwaka'wakw (Southern Kwakiutl), "When the people have a winter dance, the owner of lily-bulbs [*Fritillaria camschatcensis*] promises a lily-bulb-feast to the Sparrow Society."

The situation is described for the Gitga'at (Ts'msyen) as follows:

> An owner of a berry patch would ask or invite several people to pick berries in this patch. The owner would provide the transportation and food while picking berries. The first basket, we say "*uusil*," that an invited guest picks goes to the owner of the patch. The owner of the patch may end up with several barrels of berries. Especially if the owner insists that the people he has invited must pick until he is satisfied that it is enough. The understanding is that the owner is obligated to give a feast, a "*lulgit*," usually during the winter at which time the berries will be served. (Hartley Bay School 1997, 7)

For their part, those who responded to a chief's invitation to pick berries or dig roots in his sites, and who conformed to the requirements and codes of behaviour, effectively legitimized the leader's prerogatives to control and oversee a territory of the community or clan group. Harvesters of plant resources who took on this role might include relatives of the chief himself, mainly but not exclusively women, as well as nonkin slaves. When large quantities were needed of high-value, highly productive crops for which there was a limited harvest time, entire families, including men and children, might also participate in the labour (Thornton 1999). Because of the high social status they imbue, productive berry patches are situated in the midst of institutions of status and ownership, requiring the application of particular protocols, as explained by Thornton (ibid., 30):

> Traditionally among the most celebrated of gifts were heaping quantities of berries, especially bearberries, blueberries, huckleberries, nagoonberries, salmonberries, soapberries and strawberries ... that had been preserved (typically through drying) and were served with fish oil or seal grease in huge feast dishes ... Ritual protocol calls for the hosts to honor prominent guests by providing them with a large feast dish brimming with fruit and then formally inviting them to eat it *du xwaa x'utín* (... "with pals and buddies"). In this way the honoree and his family and friends are "recognized" and compensated for their role in the ritual cycle.

Clans, houses, and their chiefs who owned rights to certain valued resources were able to capitalize on those resources in terms of trade.[50]

Belief Systems and Management of Plant Resources

The ritual "feeding" of berry plants by Tlingit berry harvesters, described previously (Thornton 1999), is one example of the close interdependence between beliefs and sustaining harvests of plant resources. As discussed further in chapter 13, spiritual beliefs can be regarded in many instances as providing all-inclusive codes of behaviour that pertain to any situation and thus constitute elements of a "conserving" religion or ethic.[51] They connect inextricably with the recognition of all species and environmental entities as sentient nonhuman persons or beings with powers to affect human affairs positively or negatively depending on how they are treated. The beliefs require a general respect for all species but are particularly focused on prime "cultural keystone" species such as camas, balsamroot, and huckleberries (Garibaldi and Turner 2004; Turner and Atleo 1998).

In many cases, the practices these beliefs engendered were undertaken by women, as the main harvesters of plants, and paralleled the practices of men when they embarked on hunting or fishing missions. Just as hunters followed, and still do, special protocols and rituals – prayers, bathing or using the sweat lodge, sexual abstinence, and cleansing of their gear with devil's-club or other purifying solutions – women often undertook ceremonial preparations prior to, and following, berry picking or root digging.[52] Young women undertaking puberty rituals often prepared themselves spiritually – through prayer and special tasks such as digging trenches – for finding, digging, and preparing roots and for berry picking and other food harvesting (Teit 1909). Ritualized acts like the ceremonial scattering of berries or the careful timing of harvests of species like "wild celery" (*Lomatium nudicaule*) to ensure dissemination of some of the fruits during harvest, as mentioned previously, show direct and obvious connections to management systems.

Linda Smith (2008b, 136–7) describes some of the ritual practices of Tsilhqot'in people in relation to the care and perpetuation of *súnt'iny* (spring beauty) and other edible roots:[53]

> After harvesting *súnt'iny* 'spring beauty corms', *èsghùnsh* 'yellow glacier lily', *sáchén* 'tiger lily', and *chìns̲dàd* 'silverweed roots' [*Argentina anserina*], *nímính* individuals [ritually sensitive or powerful people, such as twins, parents of twins, or recently bereaved widows] are to gather the plants and return a handful of the harvest back to the fields. The ones still attached to the plants are replanted and covered over with soil. The remaining plants are strewn about with their seeds intact, and are told, "*Łàn tsínsh nàghùndláx*" 'become plentiful.' These observances are called *nàts'ègùts'àd* 'the area is treated with a ritual' and this ensures continuous plant yields. If these procedures are not followed by harvesters who are *nímính*, the plants will not grow well. There would be decreased yields and smaller bulbs, corms, and roots in the years following the harvest.

Parallel practices were observed by the Spokan, as described by John Ross (2011, 346): "Before departing the harvest area, each women would *give some of the berries back to nature* (*x̲áq'l'eʔxʷi*); not to do so was considered gluttonous and disrespectful, a rude affront (*chsłtsut*) to the plant, which may not produce again."

Cedar-bark harvesters took care not to girdle cedar trees when harvesting their bark so that the tree would continue to live and, with time, heal over itself.[54] There is a similar prohibition against harvesting the "mother" stalk of devil's-club for medicine; many people maintain that only the side branches of devil's-club should be cut, leaving the main stalk intact so that it will soon sprout more branches (Lantz, Swerhun, and Turner 2004; Turner 2004a). The Gitga'at and others always limited their harvest of devil's-club at any one time to just four stalks, out of respect and consideration for the plant and its conservation (Turner and Thompson 2006). Leaving the "mother" balsamroot plant when harvesting balsamroots for food is also a recognition both of the value of this plant and its need to continue its life and of its spiritual power (Peacock 1998). The taboos against picking certain flowers that influence the weather or against ruthless destruction of "flowers, plants and grass" as "the covering or blanket of the earth" (see chapter 13) are also a part of a religion of conservation. "Keeping it living" thus represents a philosophical approach to plant management, as well as a practical one (Deur and Turner 2005).

CONCLUSIONS

Traditional Land and Resource Management for any group of people can be viewed as a subset of a system of Traditional Ecological Knowledge and, as such, is more than just a suite of practices based on empirical knowledge (Berkes 2012). Rather, as shown in this chapter, activities that relate to resource use and production are deeply embedded in belief systems and are modulated by webs of interrelated social, cultural, and ecological influences. Spiritual values, along with human population fluctuations, governance and social organization, seasonal rounds, anticipated ecological cycles, and unforeseen environmental change are all factors in the development and enactment of resource use strategies over time and across geographic space. Even seemingly mundane and straightforward activities, such as picking berries, digging roots, and harvesting cedar bark, have intensely spiritual and social components that are seen to be critical to their short- and long-term success. For the well-being of both the harvesters and the harvested resources, it has been stressed, these activities must be undertaken with care, must be guided by longstanding conventions, and must adhere to generally held cultural practices.

The practices, effects, and outcomes of Traditional Land and Resource Management also come into play at different ecological and social levels and on different temporal and spatial scales. Their extent ranged, for example, from one-time pruning, tending or weeding around, or transplanting of an individual berry bush to the development of routine and systematic burning regimes over particular patches of prairie or meadow in multiyear cycles, sometimes over many generations in a given territory. Aspects of belief systems, such as the concept of kincentricity – the "kinship" relationship between humans and all other living things and environmental entities (see chapter 13) – have even more far-reaching and long-lasting influences, which ultimately affect the composition, productivity, and sustainability of entire suites of resources and living landscapes. Enacted by a given cultural group over its whole territory, these approaches of kinship and reciprocity collectively result in long-term maintenance and perpetuation of standing forests, productive wetlands, and entire marine and terrestrial food webs, in concert with human occupation.

Indigenous management systems – based on locally referenced knowledge, practice, and belief and guided by culturally embedded institutions – came into conflict with perspectives and management systems imposed first by missionaries, settlers, and colonizers and later by provincial and national governments and by corporations holding tenure rights assigned by those governments. In some cases, people adapted their Indigenous management approaches to accommodate the new situations. Thus it is probably no coincidence that, when the European style of gardening was adopted in the Ts'msyen, Nuxalk, Straits Salish, and other First Nations territories around the Northwest Coast in the late 1700s, these new types of gardens were often established on the same sites where people had originally tended their camas beds and root gardens, and the original practices of Indigenous plant cultivation were transferred to the successful growing of potatoes and other crops (McDonald 2003; Moss 2005; Suttles 2005).

The more that is uncovered in the archaeological record, the more we reassess and re-evaluate the information about plant use already documented in journals and ethnographies, and the more we observe and listen – carefully – to what elders like Dr Mary Thomas and Clan Chief Adam Dick (Kwaxsistalla) have described and demonstrated from their own personal experiences, the clearer it becomes that the practices and perspectives of Indigenous peoples that relate to resource management, far from being rare and unique, were as ubiquitous as they were complex. Every one of the practices documented by M.K. Anderson (2005) as having been applied by Native Californians in managing their plant resources (i.e., burning, pruning and coppicing, cleaning, selective harvesting, tilling, transplanting/replanting, and ceremonial management) has been known to at least some of the First Nations of northwestern North America (Turner, Deur, and Lepofsky 2013). Furthermore, these diverse modes of resource use and management have been documented over a wide spectrum of ecosystems. Virtually every habitat, from coastal beaches and tidal marshes to coniferous forests, montane meadows, and all successional stages of vegetation, was known to and influenced by Indigenous resource managers. Management practices that matched the rhythm of the growing and lifecycles of plants and animals were applied to many different species and were mediated by fluctuations in climate and resource productivity as well as in human populations and overall lifeways. Many of these practices have been little recognized beyond the knowledge of the peoples who carried them out, yet collectively, they have been key elements within the dynamic social-ecological systems of Indigenous peoples of the study area and have significantly influenced not only the people themselves but also the structure, function, and composition of ecosystems across the entire region, converting them in many cases into truly "cultural landscapes."

PART FOUR

Underlying Philosophy

12

Narratives in Transmission of Ethnobotanical Knowledge

And my ancestor *ts'eqamey* [*c'eqamey*] was told that this flood was coming, and he was told how to prepare for it. And he was told to go to look for a large cedar tree, and to hollow it out ... And so that's what he did. He prepared the tree and he put all the food and all the things he would need in the tree, and then he sealed himself and his family in this tree ... When he entered the cedar tree, his name was *hawilk^{w}ala*, which means "cedar tree," but when he came out, he was told, "you will no longer be *hawilk^{w}ala*; your name will be *ts'eqamey*, which means that you will be the first person to ever perform the cedar-bark ceremony" (Dr Daisy Sewid-Smith, pers. comm., 1994; see also Sewid-Smith and Dick 1998).

INTRODUCTION

Dr Daisy Sewid-Smith (Mayanilth), Kwakwaka'wakw historian, culture, and language specialist,[1] shared her family's origin story, about how her ancestor from ancient times, *hawilk^{w}ala* ("cedar tree man"; cf. *wəlk^{w}*, "a full-grown cedar tree"), renamed *ts'eqamey*, survived the Great Flood through supernatural help and guidance by taking refuge inside a colossal western redcedar tree. Only *hawilk^{w}ala*, his wife, his daughter, and his four sons were directed to enter the cedar tree; of his whole tribe, only they were meant to survive. During the time they were sealed inside the tree,[2] *hawilk^{w}ala* was taught about the sacredness of the cedar and about the ceremonies to be enacted from that time forward to honour and respect the cedar. Assuming the new name *ts'eqamey*, he was the first one ever to perform the Sacred Cedar Bark ceremony.

As a child, Mayanilth (pers. comm., 1994) was told this story many, many times by her grandmother, Q^{w}iq^{w}asutinux^{w} noblewoman Agnes Alfred (ʔA<u>x</u>uu). Mayanilth explained, "Every tribe has their origin, what they call their sacred origin, and they talk about how they originated, and how their ancestors came down. And that's exactly what they call it, *gelgalis* – how they originated, and how they survived after the flood." Since that day back in time immemorial, when *ts'eqamey* came out of the cedar tree after the floodwaters had receded, he performed the Cedar Bark ceremony according to instruction, and through him the ceremony, called *ts'eqa*, spread among all the Kwakwaka'wakw people. It is always performed slowly and regally, with great dignity; every stage of its preparation and performance is done with solemnity, befitting the sacredness of *ts'eqamey*'s teachings. Only certain individuals are entitled to harvest the cedar bark for the potlatch or perform the sacred winter rituals that comprise the Cedar Bark ceremony. Mayanilth's family's story of the origins of the Kwakwaka'wakw Sacred Cedar Bark ceremony exemplifies the inextricable ties between narrative, belief, ceremony, and the plant world that infuse Indigenous cultures in northwestern North America, as in many other parts of the world (see Reed and Calman 2008). The discussions in this chapter focus on the ancient narratives, as Mayanilth (pers. comm., 1994) would say, "from the beginning of time," or from what is sometimes called the "myth world" or the "spirit world." The roles of plants and knowledge about plants in the context of these stories are the focus of this chapter.

It is important to acknowledge at the outset that, as my friend Nuu-chah-nulth hereditary chief and scholar Dr E. Richard Atleo (Umeek) has emphasized, these stories are *true*. They reflect different *dimensions* of truth, serving as parables and ways of remembering the past, encoding memories, lessons, and approaches, and passing them on to future generations in an effective and meaningful manner. Storytelling is about conveying truths at a metalevel and is therefore highly important as a means of transmitting critically important ideas and information in a given cultural context.

In the ancient times that are portrayed in the origin stories, the world was a much different place. The spiritual side of existence was more evident; mountains, rivers, animals, and trees and other plants had obvious human persona and societies like those of humans of today. There were different worlds that were relatively easily transcended – a sky world, an under-ocean world, and a place where ghosts and dead people resided, living lives that paralleled those of people on the earth. Many different supernatural beings, some of them benevolent, some fearsome monsters, but all very powerful, wandered the earth, moving through the different worlds and using their mystical powers to shape and change the world into its present form. There were notable culture heroes, who had the ability to transform themselves from their supernatural or spiritual existence into human form. Conversely, human ancestors, with careful training and supernatural help, often crossed into the spirit world, interacting with animal societies like the Bears, the Wolves, or the Mountain Goat people and venturing into the sky country and the country under the ocean. These connections with the spirit world and with long-ago times are recognized and acknowledged today not only through stories but also through special places or sacred sites, through ceremonies and dances, through clan, house, and family crests as exhibited on masks and dance regalia, totem poles, and many other items, and through names that are passed by way of inheritance from generation to generation.

Right up to the present time, many people, like Mayanilth, tell and retell these special time-honoured stories to their children and grandchildren. Such narratives are usually the property of a clan, house, or lineage and must be acknowledged as such. Not just anyone has the right to tell many of these stories. The narratives, together with the rituals and ceremonies they engender, form the foundation of a people's sense of their place and belonging and of their relationships with each other and the natural world. The stories place humans – together with plants and animals, fish, birds, rivers, and mountains – firmly as active participants within the environment, not apart from or above it.

There is an immense body of published texts and narratives that bring these ancient "myth-times" to life in oral histories.[3] The stories embody teachings and reinforce beliefs that govern human behaviour toward the environment as well as fostering an appreciation of the gifts brought to humans through great effort and sacrifice – even of the lives – of ancestral beings (R.E. Ignace 2008; Turner and Berkes 2006). The stories are the basis and a reflection of the "kincentric" approach that Indigenous peoples often embrace: all of the other species – plants and animals – are our relations, or kin, and their welfare affects the well-being of humans in countless and interconnected ways (Cajete 1994; Martinez 2008; Salmón 2000; Turner 2005; Turner and Atleo 1998).

It must be understood at the outset that, as well as being the intellectual property of particular individuals and their lineages, the narratives were originally told, and ceremonies enacted, in their own Indigenous languages. As described by Linda Smith (2008b), the names, terms, and phrases used in these narratives often do not translate readily or accurately into English, no matter how much effort is put into the translation. The stories referred to here fall into this category, and in some cases, such as those of Boas (2002), they represent not just one translation but sometimes two or three.[4] Some of the original meaning is bound to be lost under such circumstances.

Furthermore, as emphasized by Dr Ron Ignace (2008), oral histories cannot "stand alone." They cannot really be separated from their historically and culturally constituted meanings and experiences. The moral and social messages they embody need to be carefully explained, and they may be contextual, varying both with the storyteller and the audience. Narrative "provides a framework for experiencing the material world and how local stories intersect with larger social, historical and political processes" (ibid., 28). The traditional stories integrate factual and historical information with lessons and guidance. They provide details of past geological and social events, of ecological relationships among plants and animals, and of specific resources and environments that are intertwined with both human and environmental history. At the same time, however, they offer broader and deeper interpretations of the spiritual qualities and metaphysical powers of the world, imparting not only practical but also moral and social messages and metaphors that can guide people's decisions and approaches in their lives (R.E. Ignace 2008; see also Cruikshank 1998, 2005; H. Robinson and Wickwire 1994; and Turner and Berkes 2006).

Plants play a broad range of roles in these Indigenous narratives, including as items of everyday cultural use, as entities possessing or embodying magical or supernatural properties, and as active persona of stories, with their own human or superhuman characters and characteristics. References to plants used as food, materials, or medicines can convey some ideas about their relative importance in more recent times, as well as the possible antiquity of various cultural roles. Supernatural or magical properties attributed to plants, and positions of plants as actors in narratives and stories, can also serve as indicators of their cultural significance. In general, those species featured in traditional narratives also tend to have other attributes that identify them as "cultural keystone" species (Garibaldi and Turner 2004). The nature and frequency of these various roles reflect their relevance and cultural salience in a way that parallels plant names, as discussed in chapter 3.

Belief systems and worldviews in Indigenous cultures are further addressed in chapter 13 as a central component of systems of Traditional Ecological Knowledge, inextricably linked and interwoven with practices and strategies for sustainable resource use and with the modes of communicating both this practical knowledge and the ethical and moral principles that underlie it (Berkes 1999; Turner, Ignace, and Ignace 2000). The present chapter focuses on how traditional narratives – those stories drawn from ancient times – are linked to plant knowledge and use and how the patterns of narratives relate to the acquisition and transmission of ethnobotanical knowledge, practices, and ideas across time and space. Narratives are both a major venue for this development and exchange and a major reflection of the past connections between and among peoples over countless generations. Reaching back into the mists of time, many of the stories from northwestern North America have themes and episodes that are common to narratives in other parts of North America, and even to those of Asia and Europe, yet, as will be seen, they are often locally situated and compellingly relevant to today's world (see E.R. Atleo 2011).

Widespread as these narratives are, they also show a great deal of local consistency and conformity in many ways. Storytellers are trained, often from a very young age, to carry forward the important oral histories of their families, clans, and communities.[5] It is no wonder, therefore, that certain themes would have held true across countless generations, cultures, and languages, with just the embellishments shifting over time to match the places and cultural norms of the storytellers.

OVERARCHING THEMES INVOLVING ETHNOBOTANICAL ELEMENTS IN INDIGENOUS NARRATIVES

Indigenous narratives from northwestern North America and beyond are full of references to plant use in everyday life: women going out to pick berries or cooking roots, men using cedar roots and branches to make rope, women making mats, men cutting down trees for firewood, healers and shamans using plants for medicine, and hunters and those in training bathing and rubbing themselves with evergreen boughs. Such references, although commonplace, are informative because they reflect the cultural embodiment of ethnobotanical knowledge and practice in all facets of peoples' lives. Table 12-1 provides a few examples – by no means exhaustive – of plants or plant products employed in "everyday" activities and applications, as referred to in various sources, most notably the diverse collection of Indigenous narratives recorded by Boas (2002). As shown in table 12-1, three key themes involving plants and plant products appear over and over in narratives from throughout the region: fire, food, and famine.

Fire

It is no wonder that the gift of fire dominates many Indigenous narratives. Fire has been such an important element for all of humanity; the ability to control and apply fire and to cook food is a key to human and cultural development (Lévi-Strauss 1975; Pyne 1995, 2001). Stories about how fire was captured for people by cultural heroes, or sometimes provided by the Creator, are found in almost every cultural tradition across the region. References to fire, tinder, and firewood, and to roasting and boiling food, are almost universal. According to stories from northwestern North America, in the very early days of the world all of humanity was suffering from a lack of fire, and it was up to one of the animals or other beings with supernatural qualities to "steal" fire by some ruse and bring it to people. Firedrills, fire-sticks, torches, and glowing coals are all featured in stories about how fire is given to humans for comfort, warmth, and cooking. A Comox story typifies this time and event. First, the daughter of a supernatural chief was able to obtain a fire-stick for her father from a huge whirlpool (the ocean's navel) in which rubbing-sticks for fire making were swirling around. Then Deer wanted to steal fire from this chief. Deer split resinous wood and put the slivers in his hair. Then he offered to entertain the chief by dancing and singing for him. He danced so close to the fire that the resinous splints in his hair ignited. At this point, he leapt out of the house and ran away, thus bringing fire to the people (Boas 2002, 203).[6]

Table 12-1 | Examples of plants or plant products employed in "everyday" activities and applications, as referred to in Indigenous narratives of northwestern North America

Example	*Details of reference*	*Group (reference)*
Berries, as food	Four women left Coyote to watch the house while they were picking berries; Raven and his sisters, Bluejay, Crow, and Snail, went to gather berries; many women went into the forest to gather berries; "We will give people everything they need -- berries, shellfish and fish"; "Once two women wanted to bring a box of berries to a friend in a distant village; made all kinds of berries and gave them to people. They dried and ate them and their hearts were glad"; Masmasalániq (Son of Raven) showed people where to get berries; chief's daughter used to go with the other girls into the forest to pick berries; Mink's wife, Grizzly Bear, ate berries; Raven scared his sisters away and then ate all of the berries; women gathering berries; berries and berry cakes as feast food and everyday food; Old-One taught the women to gather berries and cure them	Nlaka'pamux (Boas 2002, 88) Comox (Boas 2002, 198, 201) Nuu-chah-nulth (Boas 2002, 246, 259) Nahwitti (Boas 2002, 385, 388) Oowekeeno (Boas 2002, 452) Ts'msyen (Boas 2002, 581) Comox (Boas 2002, 191; Kennedy and Bouchard 1983) Nuxalk (Boas 2002, 505, 510) Haida (Swanton 1905, 32, 300) Ts'msyen (Cove and MacDonald 1987, 315) Nlaka'pamux (Teit 1912, 250)
Bow and arrows	Many episodes of story heroes shooting with bow and arrows; Raven (as a baby) cried for a bow and arrows from the chief who owned the sun	Comox (Boas 2002, 189, 203) Nuu-chah-nulth (Boas 2002, 268, 272) Oowekeeno (Boas 2002, 444–5) Ts'msyen (Cove and MacDonald 1987, 307)
Canoes, paddles	Many references to travelling by canoe, paddling; canoe capsizing; transformer showed a man and a woman how to hold canoe paddles and paddle their canoe; Raven (as a baby) cried for a canoe and paddle from the chief who owned the sun; Masmasalániq (Son of Raven) made canoes and paddles for the people he had carved from cedar wood; Raven taught people how to construct canoes; people had canoes at the time of the Great Flood	Comox (Boas 2002, 198, 211, 213) Kwakiutl (Boas 2002, 361) Oowekeeno (Boas 2002, 444–5, 452) Heiltsuk, Nuxalk (Boas 2002, 483, 490, 504) Ditidaht (Joe Edgar, cited in Kennedy and Bouchard 1994, 44–5)

Table 12-1 | continued

Example	Details of reference	Group (reference)
Digging stick	Mother of the dog children hung her cedar-bark cape on her digging stick (to fool her children); Old-One made a digging stick and showed women how to dig roots	Nuxalk (Boas 2002, 532, 34) Nlaka'pamux (Teit 1912, 324)
Firedrill	Masmasalániq (Son of Raven) made a firedrill so that people would have fire; Wren used his firedrill to make a fire inside Grizzly Bear; Mink stole a firedrill from the Wolves; Beaver showed people how to use a firedrill; Old-One made a firedrill and showed men how to use it	Oowekeeno (Boas 2002, 452) Nuxalk (Boas 2002, 522) Ts'msyen (Cove and MacDonald 1987, 126) Ditidaht (Kennedy and Bouchard 1994, 18–24) Nlaka'pamux (Maud 1978c, 102; Teit 1912, 324)
Firewood	Going into the forest to fetch wood; log of driftwood to split up; a chief owned twelve slaves who always had to cut firewood for him; Bear women collected good dry firewood; big tree for firewood; old man with a big belly splitting a tree with wedge; different woods tested for their burning capacity	Comox (Boas 2002, 211) Nuu-chah-nulth (Boas 2002, 272) Oowekeeno (Boas 2002, 472) Ts'msyen (Boas 2002, 581, 585) Nuxalk (Boas 2002, 527) Ts'msyen (Cove and MacDonald 1987, 307)
Fish spear/ harpoon	Man was catching salmon with a harpoon; Old-One created a fish spear from saskatoon berry wood	Lower Fraser (Boas 2002, 100) Nlaka'pamux (Teit 1912, 324)
Healing herbs	Provided to people by Transformer	Nuu-cah-nulth (Boas 2002, 246)
Net (fishing)	Raven made a net of cedar bark; many references to nets from nettle fibre and other materials; often made by Spider; Old-One taught the men how to make nets; old woman in sky country (Spider) makes a net of Indian-hemp	Oowekeeno (Boas 2002, 446, 452) Ts'msyen (Cove and MacDonald 1987, 84–8) Nlaka'pamux (Teit 1912) Halkomelem (Maud 1978a, 141, 146)
Pitch	A man chewing pitch (in the form of a dog); going into the forest to get pitch; chewing pitch to spit on eyes of blind women; pitch chewed to form the shape of whales, sharks, sea lions; octopus chewing pitch; used by Coyote	Sliammon (Boas 2002, 227) Nuu-chah-nulth (Boas 2002, 291) Nahwitti (Boas 2002, 430) Comox (Boas 2002, 184, 185)

/continued

Table 12-1 | continued

Example	Details of reference	Group (reference)
	to glue Grouse's eyes shut; a woman always got her pitch from a particular tree; used by Cannibal Woman to seal the eyes of children she steals	Secwepemc (Kuipers 1989) Ts'msyen (Cove and MacDonald 1987, 492) Many groups (Boas 2002)
Roots, edible	A man went out with some women to collect roots; Crow went with his sisters to dig roots; four blind girls sitting around a fire, in which they were heating a stone in order to cook four roots in a wooden pot; Masmasalániq (Son of Raven) made edible roots; Mink's wife, Grizzly Bear, wanted to feed him only roots; Old-One taught the women how to dig roots; Crow was determined to try to get some roots from the people; Sandhill Crane gives a man roots to eat	Kwakwaka'wakw (Boas 2002, 361) Nahwitti (Boas 2002, 412, 430) Oowekeeno (Boas 2002, 452) Comox (Boas 2002, 191) Nlaka'pamux (Teit 1912, 250, 266, 326–7) Halkomelem (Maud 1978c, 139)
Tinder, for fire	Chewed fern roots as tinder; shredded inner cedar bark inside a mussel shell or clamshell; Masmasalániq (Son of Raven) beat inner cedar bark for tinder; Wren used ball of pulverized wood as tinder to light a fire inside Grizzly Bear	Nuu-chah-nulth (Boas 2002, 268) Oowekeeno (Boas 2002, 452) Ts'msyen (Cove and MacDonald 1987, 126)
Wood, for carving	Women made a bowl out of a piece of driftwood (who was Coyote in disguise); Raven carved four fishes from cedar wood; Masmasalániq (Son of Raven) went into the forest to seek a cedar tree, made a stone axe to cut down the cedar, made a box to store dried salmon	Nlaka'pamux (Boas 2002, 88) Oowekeeno (Boas 2002, 446, 452)
Wood, for salmon weir	Two girls building a salmon weir near a strong rapid and driving poles into the cracks and holes of the rocky shore to brace their weir; young people built themselves a house and a salmon weir; man and his wife and sons had a salmon weir; Wren taught people to make salmon weirs	Nuu-chah-nulth (Boas 2002, 285) Nimpkish (Boas 2002, 317) Kwiksootainuk (Boas 2002, 343) Halkomelem (Maud 1978a, 56)
ALGAE, MOSSES, AND FERNS		
Algae, marine	Various kinds mentioned in stories	Haida (Turner 2004a)
Bull kelp (*Nereocystis luetkeana)*	Raven and his brother were found as babies in a kelp patch; kelp used as halibut fishing line	Ts'msyen (Cove and MacDonald 1987, 1) Nahwitti (Boas 2002, 422)

Table 12-1 | continued

Example	Details of reference	Group (reference)
"Raven's mustache" (unidentified seaweed)	Poor material for harvesting herring eggs in one story	Haida (Walter McGregor, cited in Swanton 1905, 135)
Bryophytes (moss)	Abandoned daughter cried and cleaned her nose with moss	Comox (Boas 2002, 207)
"Fern" roots (probably wood fern, *Dryopteris expansa*)	Mountain goat hunter eating fern roots; women gathering fern roots for winter; during a famine, people lived on fern roots	Comox (Boas 2002, 238) Lekwiltok (Boas 2002, 294) Nuxalk (Boas 1916, 166)
"Fern" (probably bracken, *Pteridium aquilinum*)	Chief's sons secretly looking for fern roots and eating them	Comox (Boas 2002, 181)
CONIFERS		
Yellow-cedar (*Chamaecyparis nootkatensis*)	A root-digging stick of the side branch of a yellow-cedar was chosen as the best after "all sorts of sticks," including western redcedar, had been rejected; mentioned in other Haida stories; spear for sea lions from the wood	Haida (Swanton 1905, 181; Turner 2004a) Northern coast (M.L. Beck 1989)
Sitka spruce (*Picea sitchensis*)	Used by dancers; mentioned in a variety of roles in Raven stories	Haida (Turner 2004a)
Douglas-fir (*Pseudotsuga menziesii*)	Two heroes bathed in a lake every day and then washed with fir boughs until they had become purified and had no human smell about them any more (see also under *Thuja* and *Tsuga*); bark broken off by Bear used as fuel	Pentlatch (Boas 2002, 239) Hesquiaht (Nuu-chah-nulth) (Turner and Efrat 1982, 20)
Pacific yew (*Taxus brevifolia*)	Club of yew wood; spear of yew wood; harpoon of yew wood; yew tree with deep roots; yew tree uprooted as a test of strength; yew wood used to make a baby's swing; yew wood made into a bow by Old-One	Lekwungen (Boas 2002, 170) Nuu-chah-nulth (Boas 2002, 247) Kwakwaka'wakw (Boas 2002, 310, 372) Ditidaht (Kennedy and Bouchard 1994, 18–24) Nlaka'pamux (Teit 1912, 324)
Western redcedar (*Thuja plicata*)	Wa'walis took a rope of cedar branches to lower himself to the bottom of the sea; Magpie's wife said she was going to get inner cedar bark; woman went to peel inner cedar bark; three beautiful /*continued*	Nuxalk (Boas 2002, 534, 527) Nlaka'pamux (Boas 2002, 97)

Table 12-1 | continued

Example	Details of reference	Group (reference)
	daughters who used to sit in front of the house plaiting mats; Crow went with sisters to forest to gather cedar bark; Raven sent four women into the forest to fetch cedar bark; Raven had his sisters make mats on which salmon were served; cape of inner cedar bark worn by mother of dog children; young woman who wants to marry the Sun and her slave purify themselves on their journey by swimming in lakes and rubbing themselves with cedar branches; Eagle and Woodpecker sent their youngest sons into the forest to bathe and rub themselves down with cedar boughs in order to become strong; son of chief bathed and rubbed himself with cedar branches to gain strength and power; Raven made a net of cedar bark; Raven had his sisters make mats on which salmon were served; chief tied up with ropes of inner cedar bark by his slaves; cedar-withe rope used during the Great Flood	Nuu-chah-nulth (Boas 2002, 279) Lekwiltok (Boas 2002, 298) Nahwitti (Boas 2002, 412) Oowekeeno (Boas 2002, 444, 450) Nlaka'pamux (Boas 2002, 83) Lower Fraser (Boas 2002, 114) Nuu-chah-nulth (Boas 2002, 263) Oowekeeno (Boas 2002, 446, 450, 472) Ditidaht (Joe Edgar, cited in Kennedy and Bouchard 1994, 44–5) Ts'msyen (Cove and MacDonald 1987, 308; Boas 1916, 251)
Western hemlock (*Tsuga heterophylla*)	Young man purified himself with hemlock branches (see also under *Pseudotsuga* and *Thuja*); inner bark eaten; hemlock boughs used for collecting herring eggs	Nimpkish (Boas 2002, 338) Ts'msyen (Boas 1916, 205; Boas 2002, 586) Haida (Turner 2004a; Walter McGregor, cited in Swanton 1905, 135)
FLOWERING PLANTS		
Rocky Mountain maple (*Acer glabrum*)	Canoe material in story (Haida); found to be the longest burning wood in test of different firewoods	Haida (Swanton 1905; Turner 2004a) Ts'msyen (Cove and MacDonald 1987, 307)
Nodding onion (*Allium cernuum*)	Eaten by the woman in the story "Sès Ts'íqí Ghàghìndá" (The Bear Who Married a Woman)	Tsilhqot'in (L.R. Smith 2008a, 42)
Saskatoon berry (*Amelanchier alnifolia*)	Woman who married a Bear was eating saskatoon berries; wood made into a fish spear by Old-One	Tsilhqot'in (Smith 2008b, 42) Nlaka'pamux (Teit 1912, 324)

Table 12-1 | continued

Example	*Details of reference*	*Group (reference)*
Indian-hemp (*Apocynum cannabinum*)	Old-One showed men Indian-hemp and how to make nets; nets made by old woman in the sky	Nlaka'pamux (Teit 1912) Halkomelem (Maud 1978c, 141, 143)
Silverweed (*Argentina anserina*)	The woman who married a Bear was harvesting silverweed roots (*chìnsdàd*); Bear will only eat uncooked silverweed roots	Tsilhqot'in (L.R. Smith 2008b, 42)
Pacific silverweed (*Argentina egedii*)	Four blind women, who are Ducks, cooking *tłeqsεm* with hot stones; food of Goose Woman	Nimpkish (Boas 2002, 308) Haida (Walter McGregor, cited in Swanton 1905, 264)
Paper birch (*Betula papyrifera*)	Old-One taught the women how to make birch-bark baskets	Nlaka'pamux (Teit 1912, 324, 326–7)
Pinegrass, or "timbergrass" (*Calamagrostis rubescens*)	Bedding used by the Bear in the story "Sès Ts'íqí Ghàghìndá" (The Bear Who Married a Woman)	Tsilhqot'in (L.R. Smith 2008b, 42)
Camas (*Camassia* spp.)	Women out camping and digging camas ("Star Husband" tales); camas bulbs dug by daughters of Swan in Hoopach'asʔath country, used to give a feast	Saanich (Violet Williams, pers. comm., 1990; Jenness n.d.) Nuu-chah-nulth (Arima et al. 1991, 233–5)
Spring beauty, or mountain potato (*Claytonia lanceolata*)	Woman in the "Dog Children" stories went out to dig "mountain potato"; food of grizzlies and black bears	Tsilhqot'in (L.R. Smith 2008b) Nlaka'pamux (Teit 1912, 321, 331)
Yellow glacier lily (*Erythronium grandiflorum*)	Eaten by the woman who married a Bear; food of grizzlies and black bears	Tsilhqot'in (L.R. Smith 2008b, 42) Nlaka'pamux (Teit 1912, 321, 331)
Northern rice-root (*Fritillaria camschatcensis*)	A Kitkiata (Gitga'ata) hunter found a valley where "not only game and salmon were plentiful, but also there was a plenteous supply of berries and fruits and wild rice."	Ts'msyen (Cove and MacDonald 1987, 217–19)
Sea milkwort (*Glaux maritima*)	Roots as food for Duck women	Comox (Boas 2002, 182)
"Grass"	Frequently mentioned in stories (e.g., eaten by mountain goats)	Haida (Turner 2004a) Ts'msyen (Cove and MacDonald 1987, 246)

Table 12-1 | continued

Example	*Details of reference*	*Group (reference)*
Bitterroot (*Lewisia rediviva*)	Three owl sisters (three different owl species) were out digging bitterroot	Nlaka'pamux (Teit 1912, 347)
Indian celery, or barestem lomatium (*Lomatium nudicaule*)	*Q'əx̱min* seeds sprinkled into the fire as an incense	Klahoose (Boas 2002, 223) Saanich (Jenness n.d.)
Seaside lupine (*Lupinus littoralis*)	Roots mentioned as a feast food	Haida (Swanton 1905, 300)
Skunk-cabbage (*Lysichiton americanus*)	Skunk-cabbage as food of Deer; during a famine, people baited their halibut hooks with skunk-cabbage roots; used to cover food in cooking pit	Haida (Swanton 1908, 446, 574) Ts'msyen (Boas 1916, 68)
Pacific crabapple (*Malus fusca*)	Mentioned as a food and feast food in a number of stories	Haida (Swanton 1905, 300) Ts'msyen (Boas 1916, 240)
Single delight (*Moneses uniflora*)	Eaten as medicine, along with devil's-club, to bring luck in gambling	Haida (Swanton, 1908, 689–700)
Coyote tobacco (*Nicotiana attenuata*)	Old-One showed the men tobacco and taught them how to smoke	Nlaka'pamux (Teit 1912)
Devil's-club (*Oplopanax horridus*)	Devil's-club eaten as medicine to bring luck in gambling	Haida (Swanton, 1908, 689–700)
Cottonwood (*Populus balsamifera* spp. *trichocarpa*)	Best trees for starting fires; Beaver throws fire on them	Nlaka'pamux (Maud 1978a, 102)
Sagebrush buttercup (*Ranunculus glaberrimus*)	Wolf taught hunters how to poison their arrows with the flower, *nkukaxemús* (*nkʷəkʷax̱m'-ús*)	Nlaka'pamux (Teit 1912, 378)
Wild rose (*Rosa* sp.)	Bushes were used by the Bear for bedding in the story "Sès Ts'íqí Ghàghìndá" (The Bear Who Married a Woman)	Tsilhqot'in (L.R. Smith 2008b, 42)
Wild currants (*Ribes* sp.)	Favourite food of bears	Haida (Walter McGregor, cited in Swanton 1905, 264)
Salmonberries (*Rubus spectabilis*)	Reference to salmonberry bush in relation to Swainson's Thrush; reference to salmonberry blossom in Raven story; many references in Haida stories	Comox (Boas 2002, 205) Nahwitti (Boas 2002, 382) Haida (Turner 2004a)

Table 12-1 | continued

Example	*Details of reference*	*Group (reference)*
Willows (*Salix* spp.)	Abandoned boy makes a bow and arrows, with a bowstring from willow bark; the Salmon people taught the Saanich how to make a reefnet (of willow)	Nlaka'pamux (Boas 2002, 94) Saanich (Jenness n.d.)
Red elderberry (*Sambucus racemosa*)	Used to anchor canoes at mountaintop in the Great Flood; red elderberries seen carried by Raven as a sign of spring	Ditidaht (Joe Edgar, cited in Kennedy and Bouchard 1994, 44–5) Ts'msyen (Cove and MacDonald 1987, 255)
Soapberry (*Shepherdia canadensis*)	Mentioned in various narratives as a food and feast food (e.g., berries served by a boy's grandmother to a supernatural guest); feast food in Nuxalk origin story	Haida (Swanton 1905, 32, 301) Ts'msyen (Cove and MacDonald 1987, 315) Nuxalk (McIlwraith 1948)
Springbank clover (*Trifolium wormskjolidii*)	Rhizomes mentioned as a food; Black Bear goes out to dig clover roots; roasted on stones by Slug Women; food of Goose Woman	Haida (Walter McGregor, cited in Swanton 1905, 264; Turner 2004a) Nuu-chah-nulth (Boas 2002, 272) Nuxalk (Boas 1898, 111; Boas 2002, 530)
Stinging nettle (*Urtica dioica*)	Used to make nets; mentioned in Haida stories; used by Coyote to kill Willow Grouse (by stinging)	Ts'msyen (Cove and MacDonald 1987, 84–8; Johnson 1997, 138–9) Haida (Turner 2004a) Secwepemc (Bouchard and Kennedy 1979, 15–18)
Blueberries (*Vaccinium alaskaense, V. ovalifolium*)	Raven and his sisters went to pick blueberries	Nuxalk (Boas 2002, 505)
False hellebore (*Veratrum viride*)	Medicine in stories, used by shamans for protection	Haida (Turner 2004a)
Highbush cranberries (*Viburnum edule*)	Mentioned in various narratives as a food and feast food (e.g., berries served by a boy's grandmother to a supernatural guest)	Comox (Boas 2002, 198) Haida (Swanton 1905, 32, 300) Ts'msyen (Cove and MacDonald 1987, 315)
Eelgrass (*Zostera marina*)	Shoots eaten by Goose Woman	Haida (Swanton 1905, 264)

Note: The Indigenous groups are listed more or less in the order of the episodes cited, and their names are those given in the sources. Nahwitti, Nimpkish, and Kwiksootainuk are Kwakwaka'wakw subgroups cited by Boas (2002).

In a Fraser River Halkomelem story, "The Origin of Salmon and Fire," it was Beaver who stole fire from the Salmon people (see Boas 2002, 128–9). In other stories, it was Mink who brought fire. In a Snuneymuxw (Nanaimo) story, for example, Mink went to fetch fire with his grandmother. He stole the child of the supernatural chief who was keeping fire, and he used the child as ransom to exchange for a firedrill; "Thus people obtained Fire" (ibid., 152). A similar story is told in the Ditidaht oral tradition, in which Mink ransomed the baby he stole from the Wolves in exchange for fire (Kennedy and Bouchard 1994, 18–24). In another Fraser River Halkomelem narrative, Beaver gave fire to the Ghosts, and Mink went to get it. He cut off the head of the Ghost chief and then used the head as a ransom to get a firedrill (ibid., 130). In a Quw'utsun' (Cowichan) story, the Creator, Qäls (X̲e'ls), after performing many different transformations, encountered a group of Lekwungen people near Victoria who were catching flounders but didn't know how to cook them. Qäls gave them a firedrill and taught them how to make fire and roast their fish (ibid., 138).

Another common episode occurs after fire is initially acquired, but it also demonstrates how essential fire is for survival. In this episode, a young person's family and village abandon him or her as punishment for a transgression. Often this happens to a young woman who has become pregnant through deceit on the part of another and is consequently abandoned by her father and all of her family. As part of this abandonment, the father extinguishes all of the fires before leaving, with the assumption that the girl he is punishing will not be able to survive. This theme is prevalent in a widely known and very ancient cycle of stories referred to as "The Dog Children," to be discussed later. Another, similar, widely occurring theme is of a boy – often the son of a chief – who shames his family by stealing or begging for food and who is therefore abandoned and left to perish. In both cases – of the shamed girl and the gluttonous boy – the abandoned one is aided, usually by a grandmother, who takes pity on him or her and leaves a "slow match" hidden somewhere, often together with a bit of food.[7] Usually with some kind of supernatural help, the glowing coal is retrieved and used to rekindle a fire. An example of the "shamed girl" theme is the Secwepemc narrative "Lynx and the Girl" (Boas 2002, 73), paraphrased and abridged as follows:

> A girl who had many suitors lived in an underground house. Lynx, who was an old man, used trickery to sleep with her and make her pregnant. Her father made her marry Lynx, and then he and the girl's entire family abandoned the woman, Lynx and their child. "They extinguished all the fires and left. But the girl's grandmother, Magpie, pitied her. She placed a glowing coal in a mussel shell, added some food and hid it." (Later, the shell called to her.) "She took the coal and the food and lit a fire ..." With

> fire and a little food, she and her child were able to survive and, eventually to thrive. (See also the Nuxalk version, in ibid., 532)

A version of the "abandoned boy" story is from the lower Fraser River Halkomelem:

> A boy was always begging, and was abandoned by his family. "Only the abandoned boy's old and blind grandmother took pity on him." She took some chewed fern-roots [probably bracken fern rhizomes], wrapped a glowing ember in them and put them in a shell, which she hid under a board. She told the dog to scratch the board so he'd find the fire. The boy wept when he found he had been abandoned, but then he found the fire. He made a bow and arrows, with a bowstring of willow bark, and he used this to shoot many birds. With the skins he made a cape so beautiful that the Sun came and asked to swap his supernatural mountain goat skin cape with the boy's bird cape. The Sun's cape would immediately fill with schools of herring when the corner of the cape was immersed in the river. Using this magical cape, the boy was able to catch and dry many herring, and eventually became rich. (Ibid., 93–5)

In a Nlaka'pamux story (Charles Hill-Tout, in Maud 1978c, 101–2), not only does Beaver bring fire to the people and show them how to kindle fire with a drill, but he also shows them how to carry fire as a "slow match," comprised of glowing coals encased in a rope (a story recounted later in this chapter).

Seeking good fuel wood is also a prevalent topic in traditional narratives. For example, in the Ts'msyen story "Tsag.atilâ'o [Tsagatiláo]" (The Woman Who Married a Bear), the Bear women always went out to collect "good dry wood" so that they could have a bright fire ready for their husbands when they arrived back from fishing. These women collected great bundles of firewood and propped them up against a tree (Boas 2002, 581).

Another persistent element in relation to fire and firewood is the episode of a young man who marries the daughter of a powerful supernatural chief, usually in the sky country or under the ocean, and who is then threatened and challenged by his father-in-law. The chief tries to kill him by a number of different means. One of these is by sending him out to split a tree (for firewood), held apart by wedges, which the chief then knocks out, assuming that the split will close up and kill his son-in-law (or prospective son-in-law). But the young man, with the aid and advice of supernatural helpers, thwarts the plan by jamming a hammer into the tree so that it won't press together.[8] This episode occurs in stories of the Tlingit and Ts'msyen, the Kwakwaka'wakw and Nuu-chah-nulth, the Nuxalk, Comox, and Halkomelem, and even the Chinookans. Boas (2002, 641) suggested that the story originated on the central coast and that its components gradually disintegrated toward the north.[9]

Food

Food and the quest for food are also predominant subjects of narratives in the study area, often associated with fire, as well as with famine, when food is not forthcoming. Many of the ancient stories refer to specific types of food and activities related to food – for example, picking berries and cooking roots. Animals are often associated with particular foods in these narratives, reflecting ancient and enduring knowledge about the habits of certain animals and birds. Stories about Salmonberry Bird (Swainson's Thrush), with his magical abilities to ripen salmonberries, for example, are found all along the coast (see chapter 8). Another episode that occurs over and over, with many variations, marks the association of mallard ducks and Canada geese with estuarine roots. Mink or some other culture hero travels to the sky country, where he encounters two or more blind women sitting by a fire and cooking root vegetables – usually Pacific silverweed or springbank clover – by roasting them or using red-hot rocks to steam them. The hero gives the women their sight and then transforms them into ducks or geese. One example of this episode is the Nuxalk story "The Visit in the Sky" (Boas 2002, 531), paraphrased and abridged as follows:

> A young man, "*Mōk.oa'nts*" [Mok'oánts], shot many birds and covered himself with bird skins. He ended up being blown up to the sky country in the shape of a big feather. When he arrived in the sky, he found a path and followed it. After some time he saw a number of blind women sitting around a cooking box in which they were cooking roots (*tqsos*) [*t'x̱wsus*, "springbank clover"]. They were Ducks. He took one of the roots away from them and tasted it. Thereupon the women shouted, "I can smell *Mōk.oa'nts*, he must be here." The root made the saliva collect in his mouth. He spat into the eyes of one of the women and she could see again ... He then brought the sight back to the others. Then he rid them of their pungent smell. Then he threw them down to the earth and said, "Henceforth you shall serve as food for mankind."

This theme occurs not only in several versions in the Nuxalk area (see also McIlwraith 1948, vol. 2, 472) but also in various forms in the Kwakwaka'wakw, Nuu-chah-nulth, Comox, Fraser River Halkomelem, and Secwepemc traditions (Boas 2002). A comparable story is the Haida narrative "Goose Woman." This woman ate only silverweed roots, wild clover rhizomes, and eelgrass. When the people in her husband's village were starving, she transformed herself into a goose and brought the people a large amount of these roots. The woman later fled when someone in the town said, "They think a great deal of goose food" (told by Walter McGregor, cited in Swanton 1905, 264).

Berry picking is another activity often mentioned in ancient stories. For example, the stories of the "Bear Husband," widely told in different languages and cultures of the study region, almost always begin with a young woman of high standing going to pick berries with a group of women. The Ts'msyen narrative "Tsag.atilâ'o [Tsag'atiláo]," mentioned earlier in connection with the importance of firewood, begins, "Once upon a time there was a chief who had a beautiful daughter. In autumn, when the time to gather berries had arrived, she used to go with the other girls into the forest to pick berries." She ends up being taken as a wife by a Bear and goes to live with the Bears (Boas 2002, 576; see also McDonald 2003, 95–100; and L.R. Smith 2008b).

Another theme, involving both cedar-bark gathering and the origin of edible berries, relates to a jealous husband who strands his wife (by tying or impaling her) at the top of a cedar tree to kill her, after which her blood (from her wounds or from menstruation) runs down the tree and transforms into berry plants. The woman's brothers hear her death song, try in vain to rescue her, and then one of them, masquerading as their sister, takes revenge on her husband. This story, known variously as "The Revenge of the Brothers" and "The Jealous Husband," is widespread on the coast (Boas 2002), but the detail of the creation of berries from the woman's blood is found only in a few versions. In the Lower Nlaka'pamux, Lushootseed, and some other versions, it is trailing blackberries or blackcaps that are created from blood (Fleisher 1980, 203; Teit 1912, 286), whereas in the Nuu-chah-nulth version, it is bunchberries (*Cornus canadensis*). Alice Paul of Hesquiaht, who shared her version of the story, told about how the woman, near death, decreed that her own blood that was running down the cedar's trunk would become bunchberries. She said to her brother when he saw her blood turning into bunchberries, "Don't be afraid! You can eat this. I made it. It is good to eat. It will always grow around that [cedar] tree." Alice Paul noted that today one always finds bunchberries growing at the base of cedar trees (cited in Turner and Efrat 1982, 64) (figure 12-1).[10]

Bears – both black bears and grizzlies – are often associated with eating berries (Turner 1997b), as well as with digging and eating roots. For example, Teit (1912) recorded a Nlaka'pamux narrative, "The Four Black Bears," in which a hunter living near the famous root-digging grounds of Botanie (Beta'ni) Valley had two wives, Grizzly Bear and Black Bear. Grizzly became jealous of the man's attention to his Black Bear wife: "One day she said to her husband, 'I am going to dig roots. Come along with me, so I may not feel lonely! You can do shooting at the same time.'" She wanted to kill Black Bear's children, so she told her own children to entice the young Black Bears by saying, "Let us play at feasting!" and then setting a basketful of "*nqáux*" (*ngex*^{w}), a "pudding" of roots and berries,[11] in front of them and inducing them to eat heavily so that they would

12-1 | Bunchberries (*Cornus canadensis*) at the base of a cedar snag.

lose their strength (Teit 1912, 218–21; see also 321, 331). In another Nlaka'pamux story, Skunk asked his wife to cook *ngex*ʷ and fish for his guests (all of the other animals) so that they would have plenty to eat (ibid., 233). The Tsilhqot'in version of the "Bear Husband" tale, "Sès Ts'íqí Ghàghìndá" (The Bear Who Married a Woman) (recorded and interpreted by L.R. Smith 2008a), contains important references to foods that are known to both people and bears: *chìnsdàd* (silverweed roots, *Argentina anserina*), saskatoon berries, yellow glacier lily bulbs, spring beauty corms, tiger lily bulbs, and nodding onion bulbs. The Bear husband liked only raw silverweed roots, not cooked ones. Black bears also like wild currants (*Ribes* sp.), and in one Haida story, they went to war with the Land Otter people over these berries.[12]

Another example of human food that animals and birds also eat is in a Secwepemc story (Boas 2002, 71) where Coyote encounters Foolhen's children

(Ruffed Grouse, Prairie Chicken, and Blue Grouse), and they explain that their mother is "in the forest, searching for berries." In another Secwepemc story, Magpie collects *wíle* (black lichen, *Bryoria fremontii*) from pine trees and roasts it, before giving it to her children along with deer fat (ibid., 74).[13] In yet another story, Mountain Goats make a large basketful of soup of "black and white moss" (the lichens *Bryoria fremontii* and *Alectoria sarmentosa*, the first of which is edible by people) (Teit 1912, 260).

Fish and fishing – especially for salmon – are themes alluded to many times in traditional narratives of the region (Boas 2002). Plants and plant products (as the materials needed in fishing technologies) feature strongly in episodes related to fishing. The origin and construction of fishing weirs, fishing nets and spears, canoes and paddles, and implements for drying and cooking fish are topics central to many stories that incorporate plants (e.g., Boas 2002; Cove and MacDonald 1987; Maud 1978a, 1978c; and Swanton 1905). Significantly, too, a number of narratives on the Northwest Coast (e.g., Nuxalk, Heiltsuk, and Kwakwaka'wakw) describe how salmon were actually created from cedar wood, cedar bark, alder bark, or some other plant material (see Boas 2002, 503, 536). In one Nimpkish story, cedar-bough tips were transformed into herring (ibid., 310). In a story from the Kwagiulth, the Transformer turned four "pine needles" into four kinds of fish: silver salmon (coho), dog salmon, trout, and humpback (pink) salmon (ibid., 358). A 'Nak^{w}axda'x^{w} (Nahwitti) story describes how Raven stole the sun because the world was dark and how he also created all of the lakes and rivers. Then he carved a wooden fish and ordered it to swim up the river because he wanted it to be a salmon. But it soon turned around and became a halibut. Then he took a salmonberry blossom and made a salmon from it. But it became a red snapper. Raven finally made salmon and instructed his slaves to build a salmon weir, and he caught many salmon (ibid., 385).

In some of the Haida and Tlingit stories, as recounted by Mary Beck (1989), it was Fog Woman[14] who called the salmon back each year from the open ocean to the streams of their birth. They followed her call in such abundance that the people could take all they needed for food, and still there were salmon. At one time, it is said, there were no salmon, and fishermen had to try for cod, halibut, or sculpin. Fog Woman, who was married to Raven, captured all of the fog in Raven's spruce-root hat. Then she used that same hat to capture all of the different kinds of salmon and showed Raven how to pit-cook salmon, wrapping it in skunk-cabbage leaves, and how to smoke it.

Another example of how plants feature in stories that relate to fishing is the Saanich story "Origin of Salmon," as recorded by ethnographer Diamond Jenness (n.d., 94):

> Once there were no seals and the people were starving; they lived on elk and whatever other game they could kill. Two brave youths said to

each other, "Let us go and see if we can find any salmon." They embarked in their canoe and headed out to sea, not caring in what direction they travelled. They journeyed for three and a half months. Then they came to a strange country. When they reached the shore a man came out and welcomed them, saying, "You have arrived." "We have arrived," the youths answered, though they did not know where they were. They were given food to eat, and after they had eaten their host led them outside the house and said, "Look around and see what you can see." They looked around and saw smoke from *q'əx̱mín*[15] that the steelhead, sockeye, spring and other varieties of salmon were burning, each for itself, in their houses.

The youths stayed in the place about a month. Their hosts then said to them, "You must go home tomorrow. Everything is arranged for you. The salmon that you were looking for will muster at your home and start off on their journey. You must follow them." So the two youths followed the salmon; for three and a half months they travelled, day and night, with the fish. Every night they took *q'əx̱mín* and burned it that the salmon might feed on its smoke and sustain themselves. Finally they reached Discovery Island (*kttses*),[16] where they burned *q'əx̱mín* all along the beach; for their hosts had said to them, "Burn *q'əx̱mín* along the beach when you reach land, to feed the salmon that travel with you. Then, if you treat the salmon well, you will always have them in abundance."

Now that they had plenty of salmon at Discovery Island they let them go to other places – to the Fraser River, Nanaimo, etc. Because their journey took them three and a half months, salmon are now absent on the coast for that period.

The coho said to the other salmon, "You can go ahead of us, for we have not yet got what we wanted from the lakes." That is why the coho is always the last of the salmon.

The young men now had salmon, but no good way of catching them. The leaders of the salmon, a real man and woman, taught them how to make *sx̱wələʔ* (reefnets),[17] and how to use *q'əx̱mín*. They also told the young men how their people should dress when they caught the salmon, and that they should start to use their purse net in July, when the berries were ripe. So today, when the Indians dry their salmon they always burn some *q'əx̱mín* on the fire (or on top of the stove); and they put a little in the fish when they cook it. Also, when they cut up the salmon, before inserting the knife they pray to the salmon, that they may always be plentiful.

This story not only explains the origin of the different kinds of salmon but also identifies the important ceremonial role of *q'əx̱mín* ("Indian celery," *Lomatium nudicaule*), whose seeds are widely used in the region as a protective incense as well as a medicine for colds and respiratory diseases (see chapter 7). Finally, it alludes to the ingenious fishing technology of the Saanich reefnet, which is constructed from Pacific willow bark and various other plant materials and is unique to the Straits Salish peoples (E. Claxton and Elliott 1994).

Another key plant in relation to fishing technologies is stinging nettle, which is the single most important source of fibre for nets and short fishing lines on the Northwest Coast.[18] There are multiple narratives that relate to the origin of nets, particularly nettle nets. A good example is the Gitxsan story of the origin of nettle fibre, "Prince of Spiders."[19] This story is set in early spring on the coast near the present site of Metlakatla. There was a general famine at the time. The people tried many times to catch salmon with stone traps on the sea coast, but the spots for traps were few and were often controlled by only a few families. Also, there had been many days of storms and cold weather, making food even harder to get. Many people were starving and were moving around to try to find food.

Two people, an old woman and her daughter, were left behind when the others left their village. They began to gather roots and wild rose hips, which still remained on the bushes through the winter. They made a shelter at the foot of a large spruce tree and managed to get enough food to keep themselves alive. They saw many salmon but had no way of catching them, and they were both near death from starvation.

Then the young woman dreamed that there was someone sleeping with her. It turned out to be a handsome young man. He said to her, "Today we will gather much nettle and clean and dry them, then I will show you how to catch the spring salmon." So they gathered a huge pile of stinging nettles. He told her, "Take off the leaves and the outer skin ... Now take this stone and shred the stalks into these fibres and then hang them where the hot sun will dry them. When you have them nearly dry, sprinkle saltwater on them from the sea and then dry them on the racks I have made over the fire place."[20]

The man took all of the nettle fibre and began to make a net, saying to the young woman, "This you will call net." He showed her how to fasten the net across the creek to catch the spring salmon. Every day from that point, they caught salmon in these nets, and they had a plentiful supply of food. Soon a baby boy was born to them, and through his father's supernatural abilities, he grew extremely quickly. His father showed him how to hunt and snare animals and how to make nets to catch fish. One day, after their son was grown, the father told the mother he was going to leave her, saying, "My son shall become a famous chief among his people, who had scorned you." Before daybreak, the woman awoke and saw her husband leaving through the doorway. But although she arose quickly and followed him outside, she saw no one there, only a large spider climbing up into the sky. Then she knew that it was a supernatural spider whom she had married and who had fathered her child.

The mother, son, and grandmother returned to their village with all of the boxes of food that the son had caught, which the mother had prepared and dried. Their canoe was filled with all of the different kinds of food – dried salmon, dried halibut, smoked seal meat, smoked deer meat – as well as many robes of different animals, and "above all they had a large net." The people who had originally deserted the woman and her mother were in the throes of another famine, so the three shared their food with the people. They went back and retrieved more food that they had left behind, and the chief, who was the woman's maternal uncle, prepared a great feast with this food for all of the Ts'msyen tribes, all of whom were suffering from the famine.

The chief told the people the story of the supernatural spider, saying, "The spider showed my niece many methods of gathering food and among these was the net. No longer need we wait till the salmon goes into the river. We will now be able to gather it while it is in the saltwater, and so for the other kinds of fish as well. You will be able to use this net and make it, as my grandson will show you."

The grandmother spoke: "The great chief of the heavens had pity on us, so he sent down one of his sons to marry my daughter and thus teach us how to gather many different foods which we have been unable to get before ... he was the Prince of Spiders. My daughter saw him going into the heavens as a giant spider." All of the guests at the feast were presented with gifts of food. The young woman then began to show the method of making nettle fibre as her husband had taught her, as well as how to make fibre from the fireweed. The story ends, "When all of the people learned this method of making nets, they began catching much salmon, and soon the Ts'msyen became traders in this food to many of the other tribes."

This narrative is a pivotal one, as it details a crucial innovation in peoples' collective history – the processing of stinging nettle and other types of fibrous materials and the technology of net making. This discovery possibly extends back to the earliest times on the coast, with at least indirect archaeological evidence for nets during the Early Holocene (see Ames and Maschner 1999, 251; and also chapter 2). This special knowledge of making twine, rope, and nets, so important for the survival of the ancestors, is almost universally associated with spiders (figure 12-2).[21]

The interior equivalent of the stinging nettle and fireweed as fibrous materials on the coast is Indian-hemp, widely known as *sp'éts'n* in Interior Salish languages (see chapter 6). Stories from the Interior Plateau that connect plant fibre to spiders and fishing nets invariably allude to *sp'éts'n* rather than to nettle. For example, there is a Nlaka'pamux narrative recorded by Charles Hill-Tout (in Maud 1978c, 141, 146), with other versions recorded by Teit (1912), of Coyote's son travelling to the sky country and encountering an elderly couple, who turned out to be the Spider people. They were busy rolling *sp'éts'n* on their legs

12-2 | Orb spider weaving its web, said to have been the inspiration for the original fishing nets, according to a number of origin stories.

to make twine, but they were having difficulty because the only Indian-hemp plants nearby were short and scrubby. They were kind to Coyote's son, so he pulled out hairs from his own abdomen and threw them on the ground. "Immediately three or four acres of the land adjoining the stream became covered with fine *sp'éts'n*" (Maude 1978c, 90–1).[22]

Famine

As in the Saanich story "Origin of Salmon," previously recounted, food shortages and famines are referred to many times in stories from northwestern North America (Cove and MacDonald 1987; Boas 2002) and must have occurred regularly, far back into cultural memory, especially in late winter and early spring. Two examples from the Boas narratives are provided here:

> There once was a severe winter. It stormed, rained and snowed incessantly so that the people were unable to catch fish. They dug for shellfish and cooked them, but were unable to find enough and many starved to death. (Nuu-chah-nulth legend, in Boas 2002, 269)

> Once there was a famine in a village and all the supplies had been exhausted except for a box of dried fish eggs. The people lived on fern roots which they had gathered laboriously in the forest and on which they survived wretchedly. (Lekwiltok legend, in ibid., 294)

These passages detail the use of shellfish and "fern roots"[23] as famine foods. Plant foods (along with clams and other shellfish) have had a particular role as famine foods because their locations are predictable. Thus, especially for roots and other underground storage organs of plants, if one knows where to look, they are available year round and are generally most nutritious in their dormant period, from fall through early spring until they start to sprout (see Turner and Davis 1993). Springtime greens like fireweed, cow-parsnip, and prickly pear cactus are also known famine foods (e.g., Turner, Thompson, et al. 1990; see also chapter 5). Fireweed features in a Ts'msyen account, for example, in which a fireweed plant appeared miraculously through the snow at the southern boundary of a newly claimed hunting territory, and "from that day Guell Haast [K'il-haast], the single fireweed, has had its place on the totems to tell of the time of famine and how the salvation of the people was wrought" (recounted in the late 1930s by Walter Wright, a Ts'msyen chief of the Kitselas people, cited in Johnson 1997, 140; see also chapter 9).

Famines and food shortages, deeply embedded in the ancient narratives and origin stories, are also known from more recent times, within the memories of contemporary elders. It is likely that the stories themselves, by naming the plants that people can seek out at such times – roots, greens, and winter fruits – serve as a "safety net," helping to remind people of what to do when they face a food shortage. In a similar vein, the Haida story "The Famine of Ti'An"[24] provides possible advice for others facing famine: "Since they had nothing with which to catch fish, they baited their hooks with skunk-cabbage roots," and they caught lots of halibut with this method.

Origin Stories

Origin stories, or creation stories, are often long and complex. A single narrative may be recounted over several days and/or nights by a skilled and trained orator. In their entirety, these chronicles encompass and explain multiple aspects of cultural practices of a people. Children, and the entire audience, are expected to listen carefully, for there are many lessons woven into the accounts, embedded instructions necessary for survival and proper cultural behaviour. Two of these stories, one from Nlaka'pamux traditions and one from Oowekeeno, illustrate how ethnobotanical knowledge is firmly embedded within such fundamental narratives.

"Old-One and the Earth, Sun, and People" (Nlaka'pamux, or Thompson)

A long time ago, before the world was formed, there lived a number of people together. They were the Stars, Moon, Sun, and Earth. The latter was a woman, and her husband was the Sun. The Earth-woman always found fault with her husband, and was disagreeable with him, saying he was nasty, ugly, and too hot. They had several children. At last the Sun felt annoyed at her grumbling, and deserted her. The Moon and Stars, who were relatives of the Sun, also left her, and moved over to where the Sun had taken up his abode. When the Earth-woman saw that her husband and his friends had all deserted her, she became very sorrowful, and wept much. Now Old-One appeared, and transformed Sun, Moon, and Stars into those we see in the sky at the present day, and placed them all so that they should look on the Earth-woman, and she could look at them. He said, "Henceforth you shall not desert people, nor hide yourselves, but shall remain where you can always be seen at night or by day. Henceforth you will look down on the Earth." Then he transformed the woman into the present earth. Her hair became the trees and grass; her flesh, the clay; her bones, the rocks; and her blood, the springs of water. Old-One said, "Henceforth you will be the earth, and people will live on you, and trample on your belly. You will be as their mother, for from you bodies will spring, and to you they will go back. People will live as in your bosom, and sleep on your lap. They will derive nourishment from you, for you are fat; and they will utilize all parts of your body. You will no more weep when you see your children."

After this the earth gave birth to people, who were very similar in form to ourselves: but they knew nothing, and required neither food nor drink. They had no appetites, desires, knowledge, or thoughts. Then Old-One travelled over the world and among the people, giving them appetites and desires, and causing all kinds of birds and fish to appear, to which he gave names, and ascribed to them certain positions and functions. He said to the people, "Where you see fish jump, there you will find water to drink ... It will quench your thirst, and keep you alive." He taught the women how to make birch baskets, mats, and lodges, and how to dig roots, gather berries and cure them. He taught the men how to make fire, catch fish, shoot, snare, trap, and spear game. He taught them how to make nets, beaver-spears, and snares.

He showed them the spátsan-tree [*sp'éts'n*, "Indian-hemp"], telling them the bark from it was the best for making thread and rope. He taught them how to make dead-falls for marten, and showed them the white and the black arrow-stone, telling them it was best for making knives, spear-points, and arrow-heads. He taught them how to snare grouse, and use the feathers on arrows so that they might go straight. He also told the people how to cook and eat salmon and other food, and showed them tobacco and pipe-stone, and how to smoke. He also taught the people the relationship of the sexes, how to have sexual intercourse, and how to give birth to children. When he had finished teaching them, he bade them good-by, saying, "I now leave you; but if you forget any of the arts I have taught you, or if you are in distress and require my aid, I will come again to you. The sun is as your father, and the earth as your mother. When you die, you will return to your mother's body. You will be covered with her flesh as a blanket, under which your bones will rest in peace." (Teit 1912, 321–3, citing one of several versions; see also 324–7)

The theme of the "world's parents" (Sky-Father and Earth-Mother) – including the creation of the earth from a human, with her bones becoming rock, her hair becoming the vegetation, and her blood the water – is one of the central elements of many origin stories from the northwestern part of North America according to Archambault (2006).

"Nōak.aua and Masmasalā'niq" [Noak'awa and Masmasalániq, or Wise One and Son of Raven] (Oowekeeno)[25]

After Raven had freed the sun, *Nōak.aua* and *Masmasalā'niq* descended from the sky in order to make everything beautiful and well.

Nōak.aua thought, "Oh, if only *Masmasalā'niq* would separate the land from the water." And *Masmasalā'niq* separated the land from the water. And *Nōak.aua* thought, "Oh, if only *Masmasalā'niq* would create the eulachon." And *Masmasalā'niq* created the fish rich in oil. Then *Nōak.aua* thought, "Oh, if only *Masmasalā'niq* would make a path leading up yonder mountain." And *Masmasalā'niq* made it. And again *Nōak.aua* thought, "Oh, if only *Masmasalā'niq* would make a cave in this mountain and would create many berries on its summit. Oh, if only he would carve people from cedar wood, men and women, and would make canoes and paddles for them." And *Masmasalā'niq* executed all the thoughts of *Nōak.aua.*[26]

And *Nōak.aua* thought again, "... oh, if only he would go into the forest to seek a cedar tree; oh, if only he would make the stone axe to cut down the cedar; if only he would make the box to store dried salmon. Oh, if only he would make the firedrill, so that the people would have fire, and if only he would look for the yellow cedar wood which has lain in the water for a long time, in order to set it alight by friction; oh, if only he would make the inner bark beater and would beat the inner cedar bark until it is soft enough to use for tinder." And *Masmasalā'niq* executed all the thoughts of

> *Nōak.aua*. And *Nōak.aua* thought again, "Oh, if only *Masmasalā'niq* would make a net so that the people might catch fish."
>
> *Masmasalā'niq* tried, but without success. So he went to ask Spider to make a net for him ... Spider also made the neck-ring of red inner cedar bark for the winter dance, and the basket, and taught *Masmasalā'niq* how to peel cedars ... *Masmasalā'niq* sharpened a bone and Spider used it to split the cedar bark. Then, at *Nōak.aua*'s wishes, *Masmasalā'niq* made edible roots, houses, paintings, carvings and masks. When everything was done, *Nōak.aua* and *Masmasalā'niq* brought people to life, told them to marry, and told them "When you find no more berries at the foot of the mountain, use the path made by us. Ascend it and you will find many berries up there."

They taught people to make traps in order to catch raccoons and bears. Then the story continues with a narrative of how Deer successfully captured fire for the people by having dry fir wood stuck into his tail: "And the fire dropped to the ground everywhere and the people kept it carefully. Deer called to the wood by the wayside, 'Hide the fire.' It accepted the fire and since that time is combustible." Nōak.aua and Masmasalā'niq then conspired to kill Thunderbird by fashioning a wooden whale coated with pitch to entrap Thunderbird and all of his family. The two heroes eventually went back to the sky but then returned to earth as droplets of blood that impregnated all of the women, "and consequently all gave birth to children" (paraphrased and abridged from Boas 2002, 451–4).

The creation or formation of the world through the efforts and sacrifices of supernatural beings who provide for and teach humans how to survive, ensuring that they have access to sunlight, water, fire, food, and essential materials, is one of the overarching narrative themes throughout the region. Many people recognize particular supernatural figures, especially those who come from above, as ancestors of their own clans, houses, lineages, or families, and it is by way of descent from these beings, traced through hereditary chiefs and recounted in oral histories, that settlements and tribal territories are validated (K.T. Carlson 2001).

PLANTS AS SUPERNATURAL ACTORS AND MAGICAL SUBSTANCES

Many ancient narratives feature either humans with supernatural powers or "animal-people" – animals in human form who are endowed with supernatural qualities – as the main actors. Plants are also often cited in stories but usually in the context of their natural roles as sources of food, materials, or medicines used or created by the supernatural beings and heroes around whom the stories revolve. The applications of plants in fire making, as fuel, as food, or in ceremonial contexts have already been mentioned. Many other and more specific roles appear in various stories: western redcedar limbs twisted to make rope, yew wood for wedges, highbush cranberries and crabapples served at feasts, or some kind of medicinal plant used to gain strength or cure sickness (see table 12-1 for more examples). Such references are highly informative because they are often indicators of some longstanding applications of particular species within a particular culture. Sometimes, too, they can give clues about the origins and relative importance of certain practices related to plants.

Perhaps even more interesting, however, are the stories from many different language groups that give clues about the underlying spiritual or supernatural qualities of plants and that define a different relationship between humans and the plant world – one of gratitude, appreciation, and even mystery – that is the underpinning of peoples' attitudes about plants and even their traditional management practices. A good example of such a story is the Haida narrative "The Artisans,"[27] which features Skunk-Cabbage as a persona. In this story, a young man has been abandoned by his entire village, except for his grandmother. Returning home one day, he comes across a big skunk-cabbage plant in a swampy area. He digs a trench around the plant to allow water to accumulate, in case he should be thirsty another time when he passes. The next day, when he returns to the spot, a sockeye salmon is swimming in the trench he dug. He makes the ditch a little bigger, and then he takes the salmon home to his grandmother, and they eat part of it. The next day when he returns to the ditch, there are two salmon. Again he enlarges the trench. The next time, there are three, then five, ten, twenty, and finally, "a great quantity" of sockeye. Each time he digs the trench a little larger. Then one day he notices that someone has come and is eating all of the salmon in the ditch – a large person with a big stomach. The boy shoots arrows at him. Later, he discovers that this was a supernatural being, Skunk-Cabbage, and that it was he who had been helping the boy by giving him the salmon.

This is only one segment of the story, yet not only does it convey to the audience the essential features and habitat of skunk-cabbage, but it also tells of a survival technique known to the Haida that could be used to obtain drinking water. Most important, it features Skunk-Cabbage in a respected and appreciated role as a helper and provider for humans. The story has an interesting counterpoint in a Kathlamet (Chinookan) story from the Oregon coast recorded originally in the Kathlamet language by Boas (1901, 50), as told by Charles Cultee, one of the last speakers of this language:

> The people of mythical times were dying of hunger. They had only sagittaria-roots [wapato, *Sagittaria* spp.] to eat. They had only small sagittaria-roots and skunk-cabbage [rhizomes] and ... rush roots [possibly cattail, *Typha latifolia*] to eat. In the spring of the year the Salmon went up the river. They had first arrived with many companions ... Then

the Skunk-Cabbage said: "At last my brother's son has arrived. If it had not been for me, your people would have been dead long ago." Then the Salmon said, "Who is that who is talking there?" "Oh, that is the Skunk-Cabbage who is talking." "Let us go ashore." They gave him five elk skins and put war clubs under his blanket, one on each side ... Then they carried him inland and placed him among willows.

Table 12-2 presents examples of the range of plant species featured in "actor" roles in Indigenous narratives. Based on a survey of traditional narratives from across the region, those in which plants are persona or key actors are relatively few. However, these stories are significant reminders that plants, too, are relatives, or kin, of humans – living beings to be treated with respect – with powers to affect humans in various ways. Several of these stories involve trees as humans in a transformed state, and it is often Raven or Coyote who is involved as a co-actor and who enacts their transformation. The Hesquiaht (Nuu-chah-nulth) story of the Yellow-Cedar sisters, told by Alice Paul (cited in Turner and Efrat 1982, 33), is a good example:

> Three young women were down on the beach drying salmon. Raven came along and wanted their salmon, so he kept asking them if they were afraid to be there by themselves – if they were afraid of bears, or wolves or other such animals. They kept saying, "No" to everything he asked them about until he said, "What about owls?" At this they said, "Oh, don't even talk about owls to us. We are afraid of owls."
>
> Raven went away, but hid in the bushes nearby and began to imitate owl sounds. When they heard this sound, the women were so frightened they ran away into the woods. They kept running until they came part way up the side of a hill. They were so tired, they decided to stop and rest. They said to themselves, "We'd better stand here now on the side of the mountain; they will call us *ʕałmapt*." And they turned into yellow-cedar trees. Raven snuck out and ate all their dried salmon. That is why yellow-cedars are always found on the mountainsides, and why they are such nice looking trees, with smooth trunks and few branches, because they used to be attractive young women with long, shining hair.

This story not only tells about the physical and ecological characteristics of yellow-cedar in Hesquiaht territory – not to mention the cunning, deceitful, and gluttonous side of Raven – but also conveys both a fear of owls, which many people hold (because of owls' association with death and witchcraft), and the value of the gift that was bequeathed to people by the Yellow-Cedar sisters. It is no wonder that yellow-cedar bark is so highly appreciated and that basket weavers, even today, are careful when they harvest the bark not too take more

Table 12-2 | Examples of plants or plant products featured in "actor" roles, as well as in an array of supernatural or magical roles, in Indigenous narratives of northwestern North America

Example	*Details of reference*	*Group (reference)*
Pitch	Mink wanted to marry Pitch, but Pitch melted in the sun and stuck to Mink; Pitchman is a fisherman in a number of stories, who has to come home before the sun comes out so he won't melt; Pitchman killed by Raven	Comox (Boas 2002, 180, 189) Nawhitti (Boas 2002, 389) Quwuts'un (Arvid Charlie [Luschiim], pers. comm., 1999) Tahltan (Teit 1919, 210) Tlingit, Ts'msyen, and others (Boas 2002)
Trees	All the trees were formerly people, transformed into trees by X̲éls, the Creator, who said, "you shall all be trees and Fir shall be your boss."	Saanich (Jenness n.d., 63)
ALGAE, FUNGI, AND FERNS		
Bull kelp (*Nereocystis luetkeana*)	Mink wanted to marry Kelp, but in the Great Flood, when he hung onto his wife, he almost drowned	Comox (Boas 2002, 189) Kwakwaka'wakw (Boas 2002, 335)
Tree fungus (*Ganoderma applanatum* or other similar species)	Steersman for Yaał (Massett dialect) (Raven) in Haida "origin of women" story	Haida (John Sky, cited in Enrico 1995, 62–3; and in Swanton 1905, 126)
Horsetail, or scouring rush (*Equisetum hyemale*)	Horsetail helped Coyote by stopping the wind and was paid with dentalia, which he stuck on his body to make the white joints	Nlaka'pamux (Teit 1912, 307)
Giant horsetail (*Equisetum telmateia*)	Formerly a woman who was always drinking water; changed into a plant by the Transformer	Ditidaht (Lena Johnson, cited in Kennedy and Bouchard 1994, 25–7)
Spiny wood fern (*Dryopteris expansa*)	An old blind woman named Skaiya'm was transformed into a wood fern by her angry husband because she had allowed her grandson to be kidnapped; fern root transformed from a girl's body caught in a landslide, with her head becoming the fern root and her body a marmot	Nlaka'pamux (Teit 1912, 283–4) Stl'atl'imx (Hill-Tout 1905, 177) Tlingit (Swanton 1909, 180)
Sword fern (*Polystichum munitum*)	Fern Woman is a prominent supernatural being, often associated with shrews and mice, which are in several stories; Sword Fern was a	Haida (Turner 2004a) Kwakwaka'wakw (Turner and Bell 1973)

/continued

Table 12-2 | continued

Example	*Details of reference*	*Group (reference)*
	supernatural person who controlled the weather and had a hairy face, earrings of dentalia, and cheeks smeared with red ochre	
CONIFERS		
Yellow-cedar (*Chamaecyparis nootkatensis*)	Three young women changed themselves into yellow-cedars	Hesquiaht and Nuu-chah-nulth (Alice Paul, cited in Turner and Efrat 1982)
Western larch, or "tamarack" (*Larix occidentalis*)	A people-eating monster that was vanquished and turned into a medicinal tree by Coyote	Okanagan (Pierre John, pers. comm. recorded and transcribed by Randy Bouchard and translated by Larry Pierre, October 1969)
Sitka spruce (*Picea sitchensis*)	Spruce Tree fathered a son with a young woman	Tlingit (Swanton 1909, 443)
Ponderosa pine (*Pinus ponderosa*)	Pine Cone (presumably this species) is an old man hunting muskrats in one story	Ktunaxa (Boas 1918, 261–5)
Western redcedar (*Thuja plicata*)	A tree who turned himself into a man; cedar bark and cedar wood transformed into Salmon	Nlaka'pamux (Teit 1912, 285; Boas 2002, several stories)
FLOWERING PLANTS		
Red alder (*Alnus rubra*)	Supernatural woman; wife of Coyote (short, dark-skinned)	Haida (Turner 2004a) Nlaka'pamux (Teit 1912, 223; Turner, Thompson, et al. 1990, 57–8)
Cow-parsnip, or "wild celery" (*Heracleum maximum*)	Formerly a person who offended Raven by speaking angrily and was transformed into a food by Raven	Tlingit (Swanton 1909, 20)
Desert parsley, or hog-fennel (*Lomatium macrocarpum*)	This root was said to be the father of one of the Salishan Transformers, Child-of-Hog-Fennel	Nlaka'pamux (Teit 1912, 224, 351; Maud 1978c, 76, 106) Stl'atl'imx (Charlie Mack, pers. comm. to Randy Bouchard and Dorothy Kennedy, translated by B. Ritchie, 1972)
Skunk-cabbage (*Lysichiton americanus*)	Supernatural plant; a fat-bellied person who gave gift of salmon to hero; food for early people and rewarded by the salmon	Haida (Swanton 1908) Kathlamet (Boas 1901)

Table 12-2 | continued

Example	*Details of reference*	*Group (reference)*
Pacific crabapple (*Malus fusca*)	Kusganmelks (Little Crabapple Tree) was one of four Mucus children, who eventually married; Crabapple married West Wind's daughter	Ts'msyen (Cove and MacDonald 1987, 50–5)
Tobacco (*Nicotiana attenuata*)	Transformed by Coyote from a monster into a useful plant	Nlaka'pamux (Teit 1912, 281)
Devil's-club (*Oplopanax horridus*)	Devil's-Club Man is a supernatural being who brings luck in hunting, gambling, etc.; Devil's-Club is a beautiful young woman who helps a young hunter who isn't able to get game by teaching him how to use devil's-club for purification	Haida (Turner 2004a; Boas 2002) Ts'msyen (Boas 2002; Cove and MacDonald 1987, 79–80, 82–3)
Cottonwood (*Populus balsamifera* ssp. *trichocarpa*)	One of Coyote's wives (tall)	Nlaka'pamux (Teit 1912, 223)
Trembling aspen (*Populus tremuloides*)	One of Coyote's wives (light-skinned)	Nlaka'pamux (Teit 1912; Turner, Thompson, et al. 1990, 57–8)
"Raspberry bush" (*Rubus parviflorus* or *Rubus* sp.)	Raven shook a raspberry bush and transformed it into a man, who became his speaker	Tlingit (Boas 2002, 671)
Red elderberry (*Sambucus racemosa*)	In the story "Ghosts of the Volcano," one of the ghosts was Elderberries, a chief with half an elderberry bush protruding through his face; Elderberry Bush was a woman who gave birth before Stone	Ts'msyen (Cove and MacDonald 1987, 149; Boas 1916, 62)
Stinging nettle (*Urtica dioica*)	Nettle was a supernatural woman who stung Tsamsem's penis and made him ill when he tricked her into having intercourse; also widely associated with spiders and spider webs	Ts'msyen (Cove and MacDonald 1987)
Blueberries (*Vaccinium alaskaense*, *V. ovalifolium*)	Berries were "goggle-eyed little boys" who taught a woman proper berry-picking behaviour	Nuxalk (McIlwraith 1948, vol. 1, 691)
Edible valerian (*Valeriana edulis*)	The root, called *mesái*, was a woman with two daughters. She wanted them to marry rattlesnakes, but the rattlesnakes would not accept them because they smelled so strongly	Nlaka'pamux (Teit 1912, 339)

Note: The Indigenous groups are listed more or less in the order of the episodes cited, and their names are those given in the sources. Nahwitti, Nimpkish, and Kwiksootainuk are Kwakwaka'wakw subgroups cited by Boas (2002).

than a strap of bark from any one tree so that the tree will continue to live (see Turner, Ari, et al. 2009; and chapter 13).

Another primary actor in a story about human origins is Tree Fungus Man, featured in the Haida story of the origin of women.[28] In this story, Yaał (Raven) wanted to go on a journey to a certain reef to fetch "certain short objects," which were female genitalia, which he wanted to place on his sister, Siwaas, in order to create the first women:

> Then they say he left there. While he was going along, they say he managed to pick up his sister. He let her off at this wife's place. Then they say he left. He asked Junco to be crew for him and took him with him. He also took a spear. The female genitalia were climbing on each other on the reef there [like chitons]. When the canoe got near it, Junco became deranged. Then Raven went back with him, they say. He asked Steller's Jay to be crew for him and set out with him, too. When they got near it, Steller's Jay too just flapped his wings around. Raven failed with everything [that he asked to accompany him] and he drew a design on a bracket fungus and sat it in the stern. Then, "Keep your eyes open and backpaddle every once in a while," he said to him. Then he left with him. When the canoe got near the reef, the fungus just shook its head.
>
> He [Yaał] speared a big and a little one [female genitalia] and put them aboard and went back with them. He landed and summoned his wife and stuck one on her. Then he put one on his sister too.

In this story, Tree Fungus was the only one strong enough to withstand the power of the female genitalia, and without him, Raven would not have been able to create women.

Another story involving a plant actor – this time one that was harmful to humans and had to be transformed by Coyote – is the Nlaka'pamux narrative "Coyote and the Tobacco-Tree" (Teit 1912, 304). This Tobacco-Tree killed men: "Whoever went under its shade, or plucked its leaves, died. Having made a stone pipe, [Coyote] went up into the shade of the tree, and, plucking some of the leaves, put them in his pipe, and smoked them. He said, 'Henceforth you will be ordinary tobacco, and people will pluck and smoke your leaves without harm.'"[29]

A similar narrative, from the Okanagan, "Tamarack, the People Eater," is about how tamarack (western larch, *Larix occidentalis*) came to be used for medicine:

> Fox traveled along, but he hadn't gone far when he met again with Coyote, and told him about a long, tall tree, growing high up in the mountains, which was a people-eater. This tree was Tamarack. Tamarack killed people by grabbing them when they got too close – around the base of each of these tall trees were the bones of many victims.

Coyote went over to the place where Tamarack could be found. He took with him a large club, so that he could smack the tree when it bent down to grab him. He got close to Tamarack. The tree began to bend down, but before he could hit it, Tamarack smashed him and threw him a long ways. Coyote died.

As usual, Fox knew that Coyote had been killed – he came over to the spot where Coyote lay, and stepped over him, bringing him back to life. Coyote scolded his partner, "Why did you bother me? I was just lying down to rest." "No," said Fox, "You were killed by Tamarack, that people-eater. I'll help you – I'll give you a weapon, tipped with black flint. When Tamarack tries to grab you, hit him with this, and he'll bother you no more."

So Coyote went back to Tamarack, with his new weapon. Tamarack bent down to grab him, but Coyote yelled out, "Don't kill me yet! I have something to say to you!" When Tamarack hesitated, Coyote whacked the top of the tree. Coyote then passed judgment: "Henceforth, when the transformed people come to this land, you will not be a people-eater. You will be used as medicine. Today, the tops of the tallest tamarack, high up in the mountains, are all bent; this is because Coyote made it that way. (Told by Pierre John, Upper Similkameen, cited in Bouchard and Kennedy 1969–72)

Another Coyote narrative theme is of certain trees being wives of Coyote or of Coyote's son: one is dark-skinned (Alder), and one is light-skinned (Trembling Aspen) or tall (Cottonwood) (Teit 1912, 223). There are other stories of men taking wives of wood, such as the Ts'msyen story "The Kitkiata and His Wooden Wife" (Cove and MacDonald 1987, 217–19). In ancient times, according to Haida belief, crabapple trees (and other types of trees) were said to have been human beings as well (Dawson 1880). Western redcedar is also sometimes represented in stories as originally having a human form or, conversely, as being transformed into a generous human, who in one story gives a woman a large bundle of cedar bark (Stewart 1984; Teit 1912, 285).

Even wild blueberries (figure 12-3) are endowed with a human persona, as in the Nuxalk story "The Woman Who Befriended a Wolf" (McIlwraith 1948, vol. 1, 691):

Not many hundred years ago ... a woman [named Ksninsnimdimut] was picking blueberries. As she started to climb a steep bank to a shelf where she saw that the fruit was plentiful, she heard one of the berries speaking to its companions.

"Let's hide," it said, "that foul-mouthed woman is coming." The berry was referring to her ... as "foul-mouthed" because of her habit of

12-3 | Oval-leaved blueberry (*Vaccinium ovalifolium*).

munching berries as she picked instead of putting all in her basket for future consumption.

She hurried up the bank so swiftly that many of the berries were unable to hide, and she saw them in their human forms, a host of goggle-eyed little boys sitting on the berry shoots. On that occasion, she obtained more than any of her companions. Thanks to her sight of the berries in human forms, she was thenceforth able to see them in their hiding places and was accordingly always fortunate. She respected the wishes of the fruit, never eating as she picked, but chewing dried salmon instead.[30]

It is, indeed, considered inappropriate to eat while one is picking berries or harvesting food. Generally, the harvester eats sparingly and only when finished his or her task. Children recall being reprimanded by their grandmother for eating berries before they had filled their baskets (e.g., Mary Thomas, pers. comm., 2001; and Turner and Thompson 2006).

Lena Johnson's story "The Transformer's Travels in Nitinat Lake" (cited in Kennedy and Bouchard 1994, 25–7) provides a final example of a plant-person. The Transformer, *kʷiikʷistupsaap*, was going along the beach, changing things. "And then *baʔax* spoke and said, 'I'll be *baʔax* a horsetail plant. I'll be near the water. I'll be in those puddles because I'm always drinking. I'll be a *baʔax*,' so that's what she became. Then Horsetail put on her big cedar bark cape, *tsiisak̲aadib*, a big cape. That's why *baʔax* has a reddish brown section on top, because of the cedar bark cape" (figure 12-4).

PLANTS WITH OTHER MAGICAL OR SUPERNATURAL PROPERTIES AS DESCRIBED IN NARRATIVES

Aside from the personification of trees and other plants as described in the previous section, there are many more instances in stories where plants and products made from them have magical or supernatural properties attributed to them or are transformed magically from one state to another. For example, in the Haida narrative "He-Who-Got-Supernatural-Power-From-His-Little-Finger," told by artist Charlie Edenshaw to Franz Boas, "an old rotten tree with long roots" belonging to Master Canoe-Builder is transformed into a large canoe full of crying geese as paddlers, simply with the words, "Go into the water, canoe of my father!" (cited in Swanton 1905, 250). Often such supernatural traits are in connection with activities of one of the culture heroes, Tricksters, or Transformers, like Raven, Coyote, or Mink. Table 12-3 provides some examples of magical or supernatural plants and plant products.

There is another story of a magical tobacco tree from which Haida tobacco originated.[31] The story of the Haida tobacco tree, which was apparently not harmful like its Interior Plateau relative in the previous tobacco story, was recorded by Dawson (1880, 152):

> Long ago the Indians[32] had no tobacco and one plant only existed, growing somewhere far inland in the interior of the Stickeen [? Tahltan] country. This plant was caused to grow by the deity, and was like a tree, very large and tall. With a bow and arrows a man shot at its summit, where the seed was, and at last brought down one or two seeds [probably the globular fruits of this type of tobacco], which he carried away, carefully preserved, and sowed the following spring. From the plants thus procured all the tobacco afterward cultivated sprung.

12-4 | Giant horsetail (*Equisetum telmateia*).

Table 12-3 | Examples of plants or plant products having magical or supernatural qualities as reflected in Indigenous narratives of northwestern North America

Example	*Details of reference*	*Group (reference)*
Berries	Berries mixed with crabapples served at Wolves' feast; berry cakes food of supernatural beings at a feast	Ts'msyen (Cove and MacDonald 1987, 275) Haida (Swanton 1905, 300)
Pitch	Sometimes used to impregnate women who chew it; people long ago said to make canoes from pitch	Ts'msyen, Comox, Nuu-chah-nulth, Kwakwaka'wakw (e.g., Boas 2002, 227, 287, 381)
ALGAE, LICHEN, FERNS, AND FERN ALLIES		
Giant kelp (*Macrocystis pyrifera*)	Piled elk hides from undersea shamans turn into kelp fronds when brought to the surface	Haida (Turner 2004a)
Bull kelp (*Nereocystis luetkeana*)	Supernatural kelp stalk leads to house in undersea world; double-headed kelp known as marker of entrance to undersea home of a supernatural being in stories; boxes of grease from undersea shamans turn into kelp floats filled with water	Comox (Boas 2002, 231) Haida (Swanton 1905, 52, 245; Turner 2004a)
Edible seaweed (*Pyropia abbottiae*; syn. *Porphyra abbottiae*) (Sutherland et al. 2011)	Supernatural seaweed spirit cured a man's sick wife by introducing seaweed as a medicine	Haisla/Hanaksiala (A. Davis, Wilson, and Compton 1995)
Black tree lichen (*Bryoria fremontii*)	Created by Coyote from his hair braid	Okanagan (Mourning Dove 1933, 121–5)
Spiny wood fern (*Dryopteris expansa*)	Raven made ferns green and layered, as food for the people, because before, "In Raven's time the butts of ferns ... were already cooked"; hunter taught to watch out for "green fern roots" that grow wherever there are grizzly bears	Tlingit (Swanton 1909, 18–20, 358)
Horsetail, or scouring rush (*Equisetum hyemale*)	Black bears crossed the river in a canoe made of horsetail	Nlaka'pamux (Teit 1912, 221)
Sword fern (*Polystichum munitum*)	Associated with shrews and mice, which are in several stories; fern clump entrance to home of supernatural Mouse Woman	Haida (Turner 2004a)

Table 12-3 | continued

Example	Details of reference	Group (reference)
CONIFERS		
Rocky Mountain juniper (*Juniperus scopulorum*)	Culture hero chopped down this tree ("cedar") as bow wood and scattered it so that the "bow wood of the people" would grow everywhere	Ktunaxa (Boas 1918, 109)
Sitka spruce (*Picea sitchensis*)	Magical spruce tree splits itself into fragments of just the right length for a fire	Ts'msyen (Cove and MacDonald 1987, 315)
Western redcedar (*Thuja plicata*)	Salmon created from cedar wood or bark; bough tips transformed into herring; cedar bark used as a sponge by Raven to bring water for creating creeks and springs	Nuxalk, Heiltsuk, Kwakwaka'wakw, Nimpkish (Boas 2002, 310) Oowekeeno (Boas 2002, 446)
Pacific yew (*Taxus brevifolia*)	Old-One magically created a bow from yew wood	Nlaka'pamux (Teit 1912, 324)
Western hemlock (*Tsuga heterophylla*)	Inner bark highly important food of supernatural beings; needles feature in some stories; Raven changes into a hemlock needle (or other conifer); top of tree magically bent by Wiget, the Transformer	Haida (Turner 2004a) Oowekeeno, Heiltsuk, Nuxalk, Tlingit, and many others (Boas 2002) Gitxsan (Johnson 1997, 140)
FLOWERING PLANTS		
Saskatoon berry, or serviceberry (*Amelanchier alnifola*)	Berries produced magically by the Bear in the story "Sès Ts'íqí Ghàghìndá" (The Bear Who Married a Woman) by slapping his rump; Old-One created a fish spear from a saskatoon berry bush by motioning at it with his knife; stolen from Grizzly Bear by a culture hero and spread around the country to make arrow wood	Tsilhqot'in (L.R. Smith 2008b, 42) Nlaka'pamux (Teit 1912, 324) Ktunaxa (Boas 1918, 93)
Indian-hemp (*Apocynum cannabinum*)	Magical Indian-hemp plants were created in the sky country by Coyote's son	Nlaka'pamux (Maud 1978a, 90–1)
Arbutus (*Arbutus menziesii*)	Served the Saanich people as an anchor during the Great Flood	Saanich (Philip Paul, pers. comm., 1993)
Kinnikinnick (*Arctostaphylos uva-ursi*)	Berries used for eyes by Coyote	Secwepemc (M.B. Ignace, Turner, and Ignace forthcoming)
Silverweed (*Argentina anserina*)	Food of Bear in story; Bear will eat only raw roots	Tsilhqot'in (L.R. Smith 2008b, 42)

Table 12-3 | continued

Example	*Details of reference*	*Group (reference)*
Pacific silverweed (*Argentina egedii. pacifica*)	Food of Goose Woman in story	Haida (Turner 2004a)
Paper birch (*Betula papyrifera*)	Old-One magically created a birch tree (the first tree) and a basket from a roll of birch bark	Nlaka'pamux (Teit 1912, 324)
Spring beauty, or mountain potato (*Claytonia lanceolata*)	Said to have originated from cones scattered in the mountains by Little Dog; planted on Potato Mountain by the wife of Mount Ts'ilʔos; produced magically by the Bear by slapping his rump in the story "Sès Ts'íqí Ghàghìndá" (The Bear Who Married a Woman)	Tsilhqot'in (e.g., L.R. Smith 2008b, 42)
Queenscup (*Clintonia uniflora*)	Leaf was a magic spoon for Wiget, the Transformer	Gitxsan (Johnson 1997, 140)
Bunchberry (*Cornus canadensis*)	Created as food from her own blood by a woman stranded up in a cedar tree by her jealous husband	Hesquiaht (Nuu-chah-nulth) (Turner and Efrat 1982, 63)
Fireweed (*Epilobium angustifolium*)	A fireweed miraculously appears through the snow during a time of famine; fireweed planted by a house grows up to the sky by the next day	Ts'msyen and Gitxsan (Johnson 1997, 140)
Yellow glacier lily (*Erythronium grandiflorum*)	Said to have originated from sticks scattered in the mountains by Little Dog; produced magically by the Bear by slapping his rump in the story "Sès Ts'íqí Ghàghìndá" (The Bear Who Married a Woman)	Tsilhqot'in (e.g., L.R. Smith 2008b, 42)
Northern riceroot (*Fritillaria camschatcensis*)	Used in story to "vanquish" Salmon people	Haida (Turner 2004a)
"Grass"	"Oldest One on Earth" transformed into a big stone; his hair was transformed into grass and earth	Nawhitti (Boas 2002, 421)
Beach pea (*Lathyrus japonicus* or *Vicia gigantea*)	"[Raven's] canoe" in story of X̱uyaa (Raven) travelling	Haida (Turner 2004a)
Giant wildrye grass (*Leymus cinereus*)	Coyote transformed the stalks into dentalium for a necklace	Nlaka'pamux (Teit 1912, 307)

Table 12-3 | continued

Example	*Details of reference*	*Group (reference)*
Dunegrass (*Leymus mollis*)	Featured in several stories, including "Master of Stories" narrative	Haida (Turner 2004a)
Indian celery, or barestem lomatium (*Lomatium nudicaule*)	Smoke from burning seeds is food of Salmon people	Saanich (Jenness n.d.)
Black twinberry (*Lonicera involucrata*)	Gambling stick material in story	Haida (Turner 2004a)
Beach lupine (*Lupinus littoralis*)	Food of ancient supernatural beings	Haida (Swanton 1905, 300; Turner 2004a)
Wild lily-of-the-valley (*Maianthemum dilatatum*)	Food of supernatural beings in story	Haida (Turner 2004a)
Pacific crabapples (*Malus fusca*)	Served as food in the land of the dead (really human eyes); magic comb transformed into a crabapple thicket; crabapples as food of supernatural beings; ceremonial use for women at puberty and in mourning	Oowekeeno (Boas 2002, 460) Heiltsuk (Boas 2002, 494) Ts'msyen (Boas 2002, 586) Haida (Swanton 1905, 300)
Single delight (*Moneses uniflora*)	Plant eaten by people for luck and power by supernatural characters in stories; Cannibal Woman transformed into a "woodland flower," probably this species	Haida (Swanton, 1905, 212–13; 1908, 745) Oowekeeno (Boas 2002, 464)
Haida tobacco (*Nicotiana quadrivalvis*)	Mentioned many times in narratives; derived originally from a magic tree; said to be first planted on Haida Gwaii by X̲uyaa's (Raven's) sister; given out at feast of supernatural beings	Haida (Turner 2004a; Dawson 1880; Swanton 1905, 300)
Water-parsley (*Oenanthe sarmentosa*)	A charm in a narrative on stealing daylight	Ditidaht (Turner, Thomas, et al. 1983, 93–4)
Devil's-club (*Oplopanax horridus*)	Often mentioned in stories as used by shamans for protection against evil and to gain supernatural help	Haida (Swanton 1908, 689–700, 742) Ts'msyen (Cove and MacDonald 1987, 23, 25, 167, 280, 311; Boas 1916, 175)
Fragile prickly pear cactus (*Opuntia fragilis*)	Salmon threw cactus onto the prairie as food for the coming people	Southern Okanagan (Spier 1938, 225)

Table 12-3 | continued

Example	*Details of reference*	*Group (reference)*
Mock-orange (*Philadelphus lewisii*)	Coyote transformed the leaves into shells to make a necklace	Nlaka'pamux (Teit 1912, 307)
Bitter cherry (*Prunus emarginata*)	A Stl'atl'imx (Lil'wat) man had cherry bark as his special power to give him supernatural strength and jumping ability	Lower Stl'atl'imx (Charlie Mack, cited in Bouchard and Kennedy 1977b, 35)
Wild rose (*Rosa* sp.)	Coyote transformed rose hips into beads	Nlaka'pamux (Teit 1912, 307)
Salmonberry (*Rubus spectabilis*)	Swainson's Thrush has magical ability to ripen berries even in the middle of winter; salmonberry blossom transformed into a red snapper fish; salmonberries are the berries of the Ghost people	Comox (Boas 2002, 205) Nahwitti (Boas 2002, 382) Haida (Turner 2004a) Ts'msyen (Cove and McDonald 1987, 152)
Trailing black-berry (*Rubus ursinus*) and/or blackcap (*R. leucodermis*)	Created as food from the blood of a woman stranded up a cedar tree by her jealous husband	Nlaka'pamux (Teit 1912, 286; see also under *Cornus*)
Red elderberry (*Sambucus racemosa*)	"Roots" is the name of a special gambling stick in one story	Haida (Turner 2004a)
"Cut grass" (*Scirpus microcarpus*)	Apparently "Raven's knife" in a Haida story	Haida (Turner 2004a)
Soapberry (*Shepherdia canadensis*)	Common food of supernatural beings; introduced to Bella Coola Valley by Raven	Haida (Turner 2004a) Nuxalk (McIlwraith 1948, vol. 1, 88)
Waxberry (*Symphoricarpos albus*)	Berries are the saskatoon berries of the people in the land of the dead	Stl'atl'imx (Teit 1906)
Springbank clover (*Trifolium wormskioldii*)	Rhizomes eaten by Goose Woman in story; elder brother in the sky country came to a number of blind women sitting around a cooking box cooking these roots	Haida (Turner 2004a) Nuxalk (Boas 2002, 530)
False hellebore (*Veratrum viride*)	Often mentioned in Haida stories as used by shamans for protection against evil and to gain supernatural help; mixed with "fern roots" to make powerful supernatural medicine of Wolverine	Haida (Turner 2004a) Ts'msyen (Cove and MacDonald 1987, 80–1)

Table 12-3 | continued

Example	Details of reference	Group (reference)
Highbush cranberry (*Viburnum edule*)	Cranberries ripened by Swainson's Thrush for sky chief in the middle of winter; cranberry bushes sprouted from his naval and he died; berries highly important food of supernatural beings; most frequently mentioned plant food in Haida stories (e.g., through magical transformation, a single cranberry becomes an entire meal for a supernatural hero); introduced to Haida Gwaii by Raven	Comox (Boas 2002, 185) Haida (Swanton 1905, 300; Turner 2004a)
Eelgrass (*Zostera marina*)	Name of village of T'anu on Moresby Island; food of Goose Woman in story	Haida (Turner 2004a)

Note: The Indigenous groups are listed more or less in the order of the episodes cited, and their names are those given in the sources. Nahwitti, Nimpkish, and Kwiksootainuk are Kwakwaka'wakw subgroups cited by Boas (2002).

This narrative – similar to a version from the story "He-Who-Was-Born-From-His-Mother's-Side" (told by John Sky, cited in Swanton 1905, 233) – is particularly important for those interested in the plant management practices of northwestern North American Indigenous peoples because of its explicit reference to sowing seeds and cultivation. Tobacco has long been considered the only example of plant domestication and cultivation through seed on the Northwest Coast, although we now realize that people had many sophisticated practices for enhancing and promoting the growth of perennial species (see chapter 11).

Raven Stories

Raven, the Trickster and Transformer who evidently came from the north, is credited with many heroic deeds, as well as being ridiculed for his greed and gluttony and his constant efforts to imitate the talents of others. Many of his adventures involve plants in one way or another. Raven has already been introduced as the one who conscripted Tree Fungus Man to help him create women in a Haida story and as the one who scared away the three Yellow-Cedar sisters in a Hesquiaht story so that he could eat all of their food.

Raven performed many important tasks in making the world a good place for humans. In his famous feat of bringing daylight to the people, he travelled to a land where a powerful chief was holding the sun (or, in some cases, the moon) encased in a box. Raven magically turned himself into a small conifer needle (hemlock, pine, or some other kind) and floated on the surface of the water where the chief's daughter came to drink. She inadvertently swallowed the needle as she drank and then gave birth almost immediately to a baby, who was, of course, Raven in disguise. The baby, Raven, cried and cried until his grandfather, the chief who owned the sun, allowed the box holding the sun to be opened so that the baby could play with it and would stop crying. Raven promptly grabbed the sun and, turning into his bird form, flew out of the house with it. He then placed the stolen sun up in the sky so that, at last, there was daylight and people no longer had to go about their lives in the dark. This story, occurring in some version among many different peoples in the study region,[33] was made famous by Haida artist Bill Reid's masterful illustration "Raven Steals the Light" (see Bringhurst and Reid 1996). Boas (2002) concluded that the story originated in the north, with Tlingit, Ts'msyen, and Haida, along with the other Raven "creation" episodes.

For the Ts'msyen, Raven, known as Txamsem (Clever One), was accompanied by a brother, Lagabula (Lazy One). They were two babies found in a kelp patch and adopted by a chief's wife and a supernatural being from the region of Prince Rupert Harbour. Born at a time when there was no daylight, only dusk, the brothers disguised themselves in mallard skins and flew to the head of Nass River country, where they went inside a mountain to get sunlight. Here, Txamsem, who could change freely from animal to bird to human, transformed himself into a "pine" needle, to be swallowed by the daughter of the chief who guarded daylight, who then gave birth to him as a baby. The baby cried for the ball of daylight, and when it was finally given to him, he took on his Raven form and flew away with it. He came up the Nass River, where people were harvesting oulachens by torchlight. Txamsem later burst the ball of daylight, and immediately it became very bright on the Nass River, and the light spread over all the world (Cove and MacDonald 1987, 1).

An Oowekeeno version of the story (Boas 2002, 444–6) contains some interesting variations related to coastal technology. Raven decides to steal the sun, changes himself into a "pine" needle, and drops into the well where the eldest daughter of the chief who owns the sun comes daily for water. She scoops up the water, and Raven slips into her container as a needle, but she blows the needle aside when she drinks. Then, the wily Raven transforms himself into shiny berries. The girl, who sees their reflections in the water, desires to eat them, which she does. Four days later she gives birth to a son, Raven, who grows up rapidly and is able to speak from his first day. He screams and cries to his grandfather to give him various things: a salmon weir so he can catch salmon, then a bow and some arrows, then a paddle, and then a canoe from the skin of sea lions. He plays with all of these things, but then he demands the small box hanging from one of the roof beams, in which his grandfather keeps daylight. He won't calm down until he is allowed to play with the box. Finally, the box is opened, and the sun escapes and lights up the earth.

Raven brought highbush cranberries to Haida Gwaii, as recounted in chapter 3. He also introduced soapberries to the Bella Coola Valley, once more using his magic (McIlwraith 1948, vol. 1, 88):

> Long, long ago, *Sləxləkwailx*, a mountain in the Carrier country above Burnt Ridge, was a chief, possessing human characteristics. Buffalo berries [soapberries] flourished on his slopes, and he wanted to keep these for food for his guests. On one occasion he invited all the animals and birds, including Raven, to a feast and dance. His house was the interior of the mountain, and when all had assembled he carefully closed every opening so that none of the berries could escape. Raven determined to obtain some of this food for the Bella Coola, and accordingly, used his power to force one of the guests to go outside. As soon as a door was opened to let him out, Raven seized some of the whip and flew away, scattering drops of it in his flight. Berries grew wherever the drops fell, and since that time everybody has been able to make this luxury. *Sləxləkwailx* was very angry, but could do nothing.

Raven has many other adventures, using plants and parts of plants in ways that are, in some cases, still reflected in their names. The elongated pods of giant vetch (*Vicia nigra* ssp. *gigantea*) and beach pea (*Lathyrus japonica*), for example, are known in Haida as "Raven's canoe" because of their canoe-like shape and because X̱uyaa (Raven) used them to rescue a supernatural woman from the ocean:

> *X̱uyaa* [Raven] saw a "woman" sticking out of the water at the mouth of Bentinck Arm. He made three attempts to reach her, with three different kinds of canoes, but every time he approached, she sank into the water. Then he opened a wild pea with a stick and used the pod as his canoe. This time when he drew near, she did not sink. He took her into his canoe and set her ashore at Haida Gwaii on the west arm of Cumshewa Inlet. The woman was known as *jilaa kuns* "Creek Woman." (Told by John Sky, cited in Swanton 1905, 115; see also Turner 2004a)

X̱uyaa (Skidegate dialect) or Yáał (Massett and Alaska dialects) (Raven) often used plant materials in magical ways. For example, he took stems of dunegrass (*Leymus mollis*), covered them with moss, and stretched them across a chasm to trick Eagle into thinking they were fallen logs so that he would walk across them and fall to his death. He used spruce needles to trick Low Tide Woman into thinking she was being pricked by sea urchins so that she would make the tide

12-5 | Swainson's thrush, known as "salmonberry bird," whose singing is said to ripen the salmonberries and other berries all along the coast.

go down. Another time, he used spruce cones with grass stuck on rotten logs to simulate canoe loads of enemies (Swanton 1905, 126).

Yet another widely told narrative involving Raven and plants is one of the "bungling host" series, in which Raven tries to imitate Swainson's Thrush, known as "Salmonberry Bird" (figure 12-5), who has the ability to produce and ripen salmonberries (*Rubus spectabilis*) (figure 12-6), which are the first berries of the season to ripen. A good example is the Saanich version, told by Violet Williams and Elsie Claxton (pers. comm., 1991, cited in Turner and Hebda 2012, 103). Swainson's Thrush, called Xʷexʷálesh in the Saanich language, sings "*xʷexʷelexʷelexʷelexʷesh*" in rising notes, and as she sings, the berries ripen:

> One time, Swainson's Thrush invited Raven to her house for a meal. She told her kids to take their baskets out to pick berries. She started singing her song [referring to the four colour forms of salmonberry: *nenel'q'x̱elíqʷ* (darkest), *nenel'pq'íqʷ* (very light), *nenel'kʷemíqʷ* (ruby), and *nenel'pxʷíqʷ* (golden)], then, "*xʷexʷelexʷelexʷelexʷesh*" ('ripen, ripen, ripen, ripen!'). As she sang, her children's baskets filled up. Afterwards, Raven said, "You come to my house." Swainson's Thrush did. Raven told his children to go

12-6 | Salmonberries (*Rubus spectabilis*) of dark, red, and golden colour forms.

> out with their baskets. They did that for their dad. Raven sang and sang and the baskets never got full.

Versions of this story are told from Haida Gwaii to Bella Coola and Puget Sound, sometimes referring to other kinds of berries but mostly to salmonberries. Another example is from the Nahwitti (Kwakwaka'wakw). In this version, Swainson's Thrush holds a feast. When the guests are assembled, he beats his behind four times and then sits on a big kettle, which he fills up with berries for the people to eat. Raven tries to fill up the kettle with berries in this way, but for him, the kettle fills only with his excrement (Boas 2002, 387).[34]

In these narratives, Raven loves berries. He is always trying to find ways to eat the berries of his sisters and other women. Almost every group that features Raven stories has some account of his greediness related to berries.[35] One example is the following Nuu-chah-nulth story:

> Once two women wanted to bring a box of berries to a friend in a distant village and asked Raven to escort them. They started off, with the box of berries in the middle of the canoe, with the women paddling and Raven

steering. On the way Raven became greedy for the berries, but the women wouldn't give him any. He then thought of a ruse. After awhile, he called "There's a canoe coming. I see that they are our enemies who want to kill us!" He told the women to paddle to the shore and run and hide. As soon as they had gone, Raven gobbled up the berries. He pretended he was fighting off the attack, calling out "Oh, save me, save me!" Then he covered himself with juice and smashed the box. When the women returned and wanted to tend his wounds, he wouldn't let them, saying it would only make the pain worse. Raven was sick for several days on account of the many berries he had eaten. (Boas 2002, 259)

Coyote Stories

Coyote in the Interior Plateau is in many ways similar to Raven on the coast. Like Raven, he was both an important Transformer and, at times, a comic figure who was always making mistakes and bungling things. Coyote did not create the world himself; rather, he was sent by "Old-One," the Creator or "Chief of the ancient world," who "sent him to travel over the world and put things to rights" (Teit 1909, 595; see also R.E. Ignace 2008). As already illustrated by the stories "Coyote and the Tobacco-Tree" and "Tamarack, the People-Eater," Coyote is intimately connected with plants of the interior. Like Raven, Coyote had the ability to transform plants and pieces of plants into other objects. In one Nlaka'pamux story, he transformed giant wildrye grass (*Leymus cinereus*) stalks into dentalium, rose hips (*Rosa acicularis*) into beads, and mock-orange leaves into shells to make a necklace (Teit 1912, 307). Coyote and his son are credited with having introduced Indian-hemp fishing nets in some areas, and he also provided food for the people directly, illustrated in an Okanagan story by Mourning Dove (1933), "How Coyote Made the Black Moss Food" (cited in Turner, Bouchard, and Kennedy 1980, 14–15).

The "black moss" in this story is really the tree lichen *Bryoria fremontii*, called *sq^{w}el'íp* in Okanagan, an important food and famine food for Interior Plateau peoples (S. Crawford 2007) (figure 12-7). In the story Coyote tries to catch some swans for himself and his son to eat. The swans fool him by playing dead, and he catches them easily. He then ties his son to the "dead" swans (so that he won't stray) and climbs up a pine tree to get the pitchy top to use as a firedrill so that he can cook the swans. But the swans suddenly come back to life and fly away with Coyote's son still tied on to them. Witnessing this from high up in the tree, Coyote tries to climb down in order to save his son, but his long hair braid becomes entangled in the tree branches. In frustration, Coyote cuts off his hair braid and drops to the ground. He then transforms his hair into *sq^{w}el'íp*, saying, "You shall not be wasted, my valuable hair. After this you shall be gathered by the people. The old women will make you into food."[36]

12-7 | Black tree lichen (*Bryoria fremontii*), *sqʷel'íp* in Okanagan, hanging from western larch (*Larix occidentalis*).

Coyote, like Raven, was always trying to imitate the talents of other animals. Another example is a Secwepemc story of how Coyote tried to imitate Grizzly Bear in harvesting edible cambium and inner bark from pine trees, but not knowing which tree to harvest from and confusing the outer bark (*p'əlen*) with the edible inner tissues (*stiʔqʷ'əl'q*), Coyote was unsuccessful (Teit 1909; see also Bouchard and Kennedy 1979).

A widely distributed series of tales, full of ecological and ornithological lessons, is the cycle that can be called "Coyote Juggles His Eyes." Ida William told a detailed version to Marianne Ignace (M.B. Ignace, Turner, and Ignace forthcoming). In this story, Coyote was gambling and kept winning, so the Bird people envied him. Raven persuaded Coyote to take out his eyes and throw them up in the air. Then the birds stole his eyes from him. Blinded and lost, he travelled a long way and felt around until he found kinnikinnick berries, which he used to replace his eyes, but he still could not see well. To find his way back home, he asked the trees, one after another, to identify themselves so that he would know where he was. Starting in the mountains, he encountered White Pine (*seléwłp*), Engelmann Spruce (*tł'sełp*), and Balsam, or Subalpine Fir (*melénłp*). Then, as he travelled to lower elevations, he came in turn to Cottonwood (*mulx*), Willow (*qʷ'lsélp*), and Trembling Aspen (*meltéłp*). Finally, when he came to Saskatoon Berry Bush (*speqpeqełp*), he knew he had arrived back where he'd started. At this point the Bird people gave his eyes back to him, and he could see once more. Other versions recount a different geographic and ecological trajectory for blinded Coyote,[37] with different tree species, reflecting drier environments than that of Ida William's home area.

Inversion of Characteristics and Roles

Often, in narratives, plants and other products take on traits in other worlds – such as the land of the dead, the land of the Salmon people, or the land of the Wolf people – that are the antithesis of their traits in the "natural" world. For example, waxberries (*Symphoricarpos albus*), which are normally considered inedible, even poisonous, become the saskatoon berries and salmonberries of the people in the land of the dead in some Stl'atl'imx/St'at'imc narratives (e.g., Teit 1906a). Crabapples in the Wolves' world of the Ts'msyen are actually the eyes of dead people, and a man who was taken by the Wolves was warned by Mouse Woman not to eat crabapples when they were served. When food was brought to him by the Wolves – deer meat and salmon and then berries mixed with crabapples – he only pretended to eat.[38] Another example of materials and products transforming as they moved from one world to another is in a Haida story where piles of elk skins obtained by a chief's son from undersea shamans turned into piles of seaweed, apparently giant kelp (*Macrocystis*) fronds, when taken onto dry land. In the same story, boxes of grease were converted into ordinary bull kelp heads filled with water, and the canoe carrying these articles turned into a rotten log (Turner 2004a). In an Oowekeeno story, "In the Land of the Dead," western redcedar trees were very rare, and consequently, cedar wood was so precious and expensive that children were not allowed to play with it. Because there were no cedars for canoes, the supernatural chief, Ménis, gave the heroes his skin canoe to return to the land of the living. The dogs there were seals. The hero crossed back to his own land and brought back cedar and cedar bark "and a canoe filled with all kinds of things unknown to the chief" (Boas 2002, 459–60). In another type of inversion, from an Oowekeeno story, pieces of a Cannibal Woman who was killed by poison were cut up and transformed into the woodland flowers called *k'ak'atsumalás* (possibly *Moneses uniflora*) (ibid., 464); in other versions, Cannibal Woman's remains became mosquitoes.

One of the most ancient episodes of transformation is known as "magic flight" (Boas 2002) – when a hero who is fleeing from a supernatural monster uses various objects, usually gifts provided by a supernatural helper, to thwart the pursuer. This theme has counterparts on the Northwest Coast from southern Alaska to the mouth of the Columbia River, as well as in eastern Asia, such as among the Kamchatka peoples (ibid.). The details vary but often include four objects or substances, such as a wooden comb or a piece of hair, being cast behind to transform magically into some kind of helpful barrier. One example is in the Heiltsuk story of brothers fleeing from the Cannibal Monster, Baxʷbaxʷalanuksiwe. The brothers were first able to slow down the monster's pursuit by throwing a stone behind them, which transformed into a mountain. Then they broke their arrows and dropped them behind, and the pieces grew into a dense forest. After this, they threw back a magic comb, which transformed into a crabapple thicket. Finally, they poured out some water, which transformed into a lake, allowing them to make it home before they were overtaken.[39]

Narrative Networks

Some of the narrative themes from the study region, like the previous "magic flight" episode, are very old, being widely spread geographically, culturally, and linguistically. Boas (2002) carefully analyzed the stories that he and others had recorded from the area and noted many commonalities and similar versions. Some of the prevalent themes and episodes involving plants and plant knowledge have already been mentioned here: Raven stealing the sun, the blind Duck women cooking roots, Deer or some other animal capturing fire for people, Spider teaching people to make fishnets, the berry-picking woman who marries a Bear, the kind grandmother who hides a "slow match" for an abandoned grandchild, Salmonberry Bird ripening the berries, and the woman stranded in a cedar tree who transforms her blood into edible berries. According to Mary Beck (1989), the story of the woman who married a Bear, or the "Bear Mother" narrative (e.g., L.R. Smith 2008b), is very ancient and is actually a reference to an Earth-Mother figure associated with the earliest peoples who occupied southeastern Alaska (i.e., in the Early Holocene). In an act of supreme generosity, the Bear husband in the end usually sacrifices himself to his wife's people, instructing them on the protocols of hunting, and with this knowledge, his own sons grow up to become the greatest hunters.

The "Dog Children" story is likewise very old. In this narrative, a girl is impregnated by a dog and gives birth to several puppies. Her father and the rest of her family are shamed by this and abandon her to her fate, but her grandmother leaves her some live coals hidden in a mussel shell (the "slow match" alluded to earlier), and she is able to survive and provide for her children by digging clams (coastal version) or edible roots (e.g., Tsilhqot'in version). At one point, as she is coming home from harvesting, she hears her children talking and laughing. She sneaks up on them and discovers them in their human forms, their dog skins cast off. When they hear her, they immediately put their skins back on and resume their dog forms. She then tricks her children by leaving her digging stick, with her cape and hat draped over it, at the clam beds or root-digging grounds. The dog children assume that this form is their mother still at work. Meanwhile, she steals back to the house, seizes the children's dogskins before they know what is happening, and throws them into the fire before they can put them back on. Sometimes one, a sister, remains as a dog. All of the children, now as humans, help to hunt and provide for their mother, such that the family soon becomes wealthy in food, and the children grow up as very powerful ancestral figures.

Linda Smith (2008b, 24) considers the "Dog Children" narrative to be "likely the oldest *Tsìnlhqút'ín* story," and she notes that it is shared by most Dene (Athabaskans) as well as by other language groups on the Northwest Coast.[40] It is said in the Tsilhqot'in traditions that Łin Desch'osh (Little Dog) from this narrative was the one who created the two most important Tsilhqot'in root foods: spring beauty, or mountain potato (*sunt'iny*), and yellow glacier lily, or beartooth (*ʔesghunsh*). He made these roots from cones and small sticks, respectively, which he scattered among the snow in the mountains (Quilt 1982, cited in L.R. Smith 2008b). Clan Chief Adam Dick (Kwaxsistalla) (pers. comm., 2007, cited in Turner, Recalma-Clutesi, and Deur 2013) tells a Kwakwaka'wakw version of the "Dog Children" story in association with the *lux̱ʷxiywey* (clam gardens). One of the dog children wants to help maintain the *lux̱ʷxiywey* for his mother, as mentioned in a song that Adam sings in Kwak'wala as part of this narrative.

Another famous and widespread series of narratives involving plants is the "Star Husband" tales, about young women who travel to the sky country and marry stars. In its simplest form, the story is as follows: Two young women are sleeping in the open at night (often while out digging edible roots of one type or another) and see two stars. They make wishes that they may be married to these stars. In the morning they find themselves in the upper world, each married to a star – one of the stars is a young man and the other is old. The women are sent out to dig roots in the sky country and are usually warned against digging too deeply, but eventually they disobey and make a hole in the sky through which they see their old home below. They are seized with longing to return and secure help in making a long rope. On this, they eventually succeed in reaching home (Dundes 1965).

An example of this narrative is from Violet Williams (pers. comm., 1990, cited in Turner and Hebda 2012, 122–3), told to her by her mother, who was a Hul'qumi'num woman from the Westholme area of southern Vancouver Island:

"The Two Girls up in the Sky Country"

There were three girls – two sisters and their cousin or friend – who were out camping with their families, digging camas bulbs. At night, the girls were lying awake. They were looking up at the stars and talking to each other: "I wish I had that Bright Eyes up there for my husband." "I wish I were married to that Red Eyes." They were talking like that, and then they fell asleep. The next morning, the two sisters woke up in a different country. It was very strange and beautiful. "Where are we?" they asked each other. Then, two young men appeared. "Who are you?" the girls asked them. "We're your husbands," was the answer. "But we are not even married!" they said. "Last night, didn't you wish for stars as your husbands? We are the ones you wished for. We are star men."

The girls learned that they were up in the sky country. They were told when they went out to dig camas bulbs not to dig too deeply. If a bulb broke, they were not supposed to dig down to get it. One girl was curious and she did dig deep, and they saw a little hole. Then they saw why they were told that, for through the hole, they saw another whole world far down below. They realized that they were looking down at their own world. They were homesick and wanted to go back.

They thought about how they could get back to their own land, and decided to try to make a rope. From then on, instead of digging camas bulbs every day like they were supposed to, they started gathering cedar bark. They kept making cedar bark rope, rolling it between their hands and their legs [i.e., rolling it along their thigh, with their hand] to make string. They did this day after day, until they eventually figured they had a long enough one to reach bottom. They lowered it down through the small hole they had made, and when they brought it up, a bit of moss was on the end of it, so they knew that it was long enough and had touched earth.

One of the girls volunteered to go down first. She was lowered down and down. Her sister watched her going further and further. She looked like a little spider on the end of its web. Then she reached the bottom, and pulled the rope as a signal to the other. Then her sister came down, and they pulled the rope down after them. It is still lying in the mountains somewhere. From this day on, the girls never wished for star men with bright eyes or red eyes any more. Now, all girls are carefully looked after when they go out, and are told not to wish for stars or anything like that.

Variants of this story, from Saanich ("The Sisters Who Married Stars") (Jenness n.d., 141) and Lekwungen ("The Wives of Stars") are very similar, but the latter mentions "onions" rather than camas as the roots being dug, notes that the sisters were allowed to harvest only the *stalks* of these plants in the sky country (not the bulbs), and identifies the place where the young women came down from the sky as Mount Ñgā'k.un (Nʸgák'un) (Knockan Hill) near Victoria (Boas 2002, 171). In all, there are over 100 versions of this story from all across North America – on the Pacific Coast from southern Alaska to central California,[41] across the Interior Plateau and Plains from the Arizona–New Mexico border, south to Oklahoma, Texas, and Louisiana,[42] north to the Canadian boreal forest,[43] across to the Anishinaabe (Ojibway) of the Great Lakes region, and east to the Mi'kmaq (Micmac) of Nova Scotia (Boas 2002).

Of eighty-six versions cited by Dundes (1965, 449), 95 per cent include a warning given to the women not to dig too deeply in the sky country while they are digging their roots, and 88 per cent have them descending to earth on a rope. In most cases, the repatriating rope is of sinew or skin, but in several versions it is spider web, and some (including Violet Williams's version) feature cedar-bark or cedar-withe rope. In other regions, ropes of *Yucca* (soapweed), grass, roots, or vines were used, in accordance with the predominant local fibre of an area. The edible roots dug by the women vary with the region: "fern roots" (possibly bracken, *Pteridium aquilinum*) in a Lushootseed version, northern riceroot in a Ts'msyen version, "parsnip" (unknown species) in Sarsi, prairie turnip (*Pediomelum esculentum*) in Blackfoot (probably also Cree and Arikara), and "carrots" in Dakota (ibid.). Thus, as noted by Dundes (ibid.), the "Star Husband" tale adapted itself on many occasions to new conditions and took on new forms, but in spite of time and distance, it has maintained its basic pattern.

The "Star Husband" cycle could be distantly related to the "Jack and the Beanstalk" theme from European folklore (Dundes 1965, 104; see also R. Brown 1873), but "Star Husband" seems to hold its own integrity as a North American theme. It does seem congruent, however, with the sky country adventures of Coyote's son, mentioned previously, in which the son of Coyote encounters roots (spring beauty) that are really Stars, meets a Spider couple, creates a fine patch of Indian-hemp for them from his own hair, and is lowered back down to earth in a basket, tied onto an Indian-hemp rope (Maude 1978a, 90–1; see also Boas 2002). The narrative theme is also reminiscent of the Ts'msyen story "The Boy Taken by the Star," in which a boy kidnapped by the Chief-of-the-Skies was rescued by his father and lowered down onto a mountain peak on the cedar-bark rope.[44]

Another "matching" story is the Tlingit narrative on the origin of basketry:

> Back in Raven's time, Sun's beautiful wife, who was human, fashioned the first basket from some roots up in Sky-land at a time when she was longing for the earth-life for herself and her children. Sun, understanding her homesickness, increased the size of her basket until it was big enough for his wife and their eight children and gently lowered them down to the earth. The basket settled at a place near Yakutat on the Alsek River, and that is the reason that the first baskets [of Sitka spruce root] in southeastern Alaska were made by Yakutat women. (Paul 1991, 9)

Perhaps it is no coincidence that the spruce-root hat found with the body of the young man in the Kwäday Dän Ts'ìnchį discovery (see chapter 2, figure 2-9) was so finely woven; it may have originated in this very region.

The theme of baskets being used to lower visitors to the sky country back to earth occurs in stories as diverse as those of the Secwepemc and Chipewyan (a Dene [Athabaskan] language group of the boreal forest) (Boas 2002).

CONNECTING THE PATTERNS IN STORIES

Narrative plots and themes are often far-reaching, many not being confined to single groups, languages, or language families but crossing all boundaries – geographic, ecological, social, and linguistic – while transforming themselves and adapting to new cultural and ecological situations as required. A number of the narratives resonate widely across North America, and in some cases even to eastern Asia – to Japan, Kamchatka, and beyond (Archambault 2006; Boas 2002; Dundes 1965).

The presence of common narrative types and motifs in different societies indicates that the earliest peoples, although spread out in small groups on the landscape, would have maintained at least sporadic contact with each other. They would have exchanged not only oral traditions (Archambault 2006) but also numerous other innovations and technologies, as well as trade goods and even group members as mates. Just as in recent times, different groups of early peoples probably also shared some of their resource-harvesting grounds, meeting seasonally or occasionally at these common territories, and these encounters would have resulted in the acquisition of new goods, information, and ideas, including names, technologies, and beliefs related to plants.[45]

Narratives regarding the origins of certain plant species, such as the Haida story of the origin of highbush cranberry (see chapter 3) or the Nuxalk story of how Raven brought soapberries to the Bella Coola Valley, recounted in this chapter, may actually reflect expansions of the ranges of some species into previously glaciated lands during the Early Holocene. Conversely, such stories could indicate the presence of species that have since disappeared, such as in the reference to soapberries at "Peninsula Point" (Spit Point, at the southeastern entrance to Skidegate Inlet) on Haida Gwaii (Swanton 1905, 301).[46]

Boas (2002), after a detailed examination of the large corpus of stories he had recorded from a wide range of language groups in the study region, concluded that the legend of Raven as shaper of the world originates with the Tlingit, Ts'msyen, and Haida.[47] Mink is another creator, akin in many ways to Raven. Boas suggested that the Mink legends essentially belong to the Kwakwaka'wakw and their immediate neighbours, with the roles of Mink and Raven being interchanged to some extent. Whereas Raven is characterized by his greediness and gluttony, Mink is always trying to seduce and marry different characters, including potential botanical partners: Kelp (*Nereocystis luetkeana*), Horsetail (*Equisetum hyemale*), and Pitch.

The Wanderer, or Transformer, legends, in general, come in infinite variation (Boas 2002, 645). Transformer figures are common in Indigenous cultures across North America but developed most strongly in the southern region of the study area, yielding to the Raven cycle from the Nahwitti territory northward. These cultural deities (sometimes alone and sometimes in pairs or small groups) are creators of animals and plants and often transform individuals into strangely formed rocks that can still be seen around the landscape, such as in the Interior Plateau. The other Transformers do not perpetrate tricks like Raven or Coyote do; other animals like Mink take on this role (Boas 2002). One of the most widely told Transformer stories is the narrative relating how the Transformer encounters Deer and changes him into his animal form, despite Deer's declaration that he has no intention of being changed by this Transformer, whom he has heard is coming. The knives or shells that Deer is holding at the time are put onto his head and become his antlers (see ibid.; and E.R. Atleo 2004).

Dr Ron Ignace (2008) suggests that the Stspetékwll (Stspetékʷł), or Transformers, in Secwepemc oral traditions may have actually been ancient travellers coming from other regions and entering the territory, individually or in groups, and meeting – and possibly settling among – people already living there. They would have brought with them new languages and technologies, new ways of looking at the world, and new stories. Ignace proposes that the Salishan peoples moving eastward from their lower Fraser Valley homeland between about 5,000 and 3,500 years ago merged with people already living in the Interior Plateau to form the Proto-Interior Salish and that together they developed the relatively stable Interior Plateau lifestyle known from the Late Holocene.

In his immense compendium of narratives from across the study area, Boas (2002, 633–74) undertook a comprehensive comparison of the various shared and distinctive elements of the narratives across the groups represented. From this, he drew conclusions about the narratives' historical affinities and evolution. His analysis is relevant here because one can assume that, in cases where story themes and episodes are shared and adapted across different groups, so too is knowledge about plants and other environmental and cultural elements.

Boas (2002, 649) noted, for example, that the Nuxalk (Bella Coola) narratives he recorded still possessed elements from Proto-Salish times, when they were united with the other Salish tribes. The Nuxalk, he observed, share a larger number of episodes in their stories with the Secwepemc (Interior Salish) and with the Comox, Halkomelem, and other Coast Salish groups than with their Wakashan and Dene (Athabaskan) neighbours.[48] Still, although the Nuxalk brought several Salish legends north with them,[49] their whole system shows definite influence from Kwakwaka'wakw narratives; they have woven in the Mink stories from the south and the Raven cycle from the north. Stories of the Nuxalk also show influence from their Dene (Athabaskan) neighbours, which has extended through them to the Heiltsuk on the outer coast. However, the Nuxalk creation story – of the divine brothers who descended to earth – is their own (Boas 2002).

Narratives of the Comox people, Boas (2002) found, show more influences from their northern neighbours than do the narratives of other Salishan groups. In fact, of all the groups that Boas included in his survey, the Comox had the highest level of mixing with other groups in their narratives.[50] In turn, the influence of Salishan narrative themes generally extends northward only to the Nahwitti (except with the Nuxalk, who are Salish). Boas (ibid., 653) suggested that the profusion of languages and dialects in this region, and the close relations of the Comox with Nuu-chah-nulth and Chinookan narrative traditions, cannot be understood without assuming strong, relatively recent displacements among the Coast Salish in the area, probably leading the tribes living south of the Comox into their present sites, as well as possibly bringing about the separation of the Nuxalk from other Coast Salish peoples and their move northward.

In contrast with the Comox, the neighbouring Kwakwaka'wakw share only a very small number of narrative elements with other groups and have many that are exclusive to themselves (Boas 2002, 654). Over time, by consolidation, the clans and lineages of the Kwakwaka'wakw developed, and since each lineage traced its crests and privileges back to an ancient supernatural ancestor, the number of such founding narratives grew very large. This might be why, Boas suggested, the Kwakwaka'wakw narratives are relatively self-contained. They include the Mink cycle and the Wanderer cycle, but most ancestor legends and clan descent stories contain relatively few animals; their ancestors are mostly supernatural beings who come from the sky, the earth, or the ocean.

The Chinookan, Secwepemc, and Ts'msyen also show notable fidelity in their narratives. Boas (2002) found that the Chinookans share considerable material with the Nuu-chah-nulth and Comox alone, and from this he assumed that they were formerly in close contact. The Secwepemc and other Interior Salishan peoples show hardly any evidence of contact with the coastal tribes in their narratives, except with their Halkomelem neighbours on the lower Fraser River. The Fraser River narratives, in turn, show encroachment of narrative cycles belonging to the Interior Plateau and, indirectly, from the Dene (Athabaskans) through the Secwepemc. Both the Secwepemc and Chinookan, Boas suggested, likely share a close connection with the Coyote cycle of the Shoshones.

The Ts'msyen, Haida, and Tlingit show only slight points of contact with the southern groups, with little influence on the development of other mythologies on the coast (Boas 2002). Based on various factors, including the numerous shared elements of Ts'msyen from Haida and Tlingit, Boas judged the Ts'msyen to be relatively recent arrivals on the coast. A number of Ts'msyen narratives support this conclusion.[51]

Boas (2002) noted that the Raven cycle, even in its northern area of origin, has evidently adopted foreign elements, including those originating from southern British Columbia. One of these is the "bungling host" episode, where Raven tries to provide hospitality to his guests using the means he has observed from other animals and fails in this imitation (as in the Saanich Salmonberry Bird story). This theme occurs as a separate legend type all over North America, from the Chinookan to the Mi'kmaq, including some Dene (Athabaskan) versions as well. On the central and northern coast, the personality traits of the Transformer and Trickster are blended, whereas for the Chinookan, they occur in distinct individuals. Boas (ibid., 640) suggested, therefore, that the "bungling host" theme was transmitted from the Chinookan to the Coast Salish "and thus gradually reached the Raven legend, into which it fitted quite well."

Boas (2002) identified a number of other stories that appear to have been transmitted from the Columbia River north into Salishan and Wakashan territories. One story in particular, recounted by Nehwitti storytellers of northern Vancouver Island (Kwakwaka'wakw), embodies a theme that is much more widespread elsewhere in North America, namely the idea of a time when there was only water on the earth and the Creator persuaded various animals to come down from the sky and dive down to bring up soil and plants or leaves in order to create earth as it is known today (ibid., 173). In the Nehwitti version, after many vain attempts, at Raven's bidding, Duck at last succeeds in reaching the bottom of the water covering the earth and brings up a branch, from which Raven forms the mountains, earth, and trees. This episode must have diffused from the Great Lakes and the Mississippi Basin – where it is prevalent and known to peoples as widely dispersed as the Anishinaabe, Sioux, and Cherokee – across the Rocky Mountains, to be woven into the Northwest Coast Raven cycle (Boas 2002). On the Pacific Coast, Boas identified this episode from only three other locations: a group of tribes of the San Joaquin Valley of southern California, the Molalla of the foothills of the Cascade Mountains of Oregon, and the Chinookans at the mouth of the Columbia River (ibid., 642–3).[52]

The various relationships and affinities between groups reflected in oral histories are, as noted previously, significant to ethnobotanical knowledge because – along with similarities in the names of plants across languages and language families – they may show pathways by which botanical knowledge and technology were spread from group to group, just as narratives and themes in oral history have spread and then been incorporated and adapted into already existing oral traditions.

THE ROLE AND IMPORTANCE OF ORAL TRADITIONS IN RELATIONSHIPS WITH THE NATURAL WORLD

The ancient and time-honoured stories also serve to connect people to their communities, their ancestors, and their environments, engendering respect, appreciation, and right ways of behaviour. Linda Smith (2008b, iii) explains the important role of stories in relation to the Tsìnlhqút'ín (Tsilhqot'in) narrative "The Bear Who Married a Woman":

> [T]he story is full of the richness of ancient words, terms from the bear's language, and vivid illustrations of ancient ways. This period, set out originally by mammals and fish to ensure that people continue to prosper and maintain respect for all life forms, is preserved in the term *súwh-tŝ'éghèdúdính*. This documentation sets out to partially shed light on the *Tsìnlhqút'ín* concept of an energy called *nímính* which manifests within individuals at the onset of a life transition (namely at birth, puberty, and death) lingering for varying durations from one week to an entire lifetime, and influencing subsistence items, places, and vegetation. Maintaining balance amidst a web of other lifeforms is an ancient lifeway which now seems a complex undertaking.

Kennedy and Bouchard (1994, 28), discussing the Transformer's travels in Ditidaht, explain,

> Ditidaht myths reveal the models of behaviour that the Ditidaht people regard as good. Implicit in these stories are all their essential beliefs and values. The animal-people's adventures demonstrated the benefits to be derived from being hospitable, unselfish, modest, hard-working, and respectful of one's elders and Chiefs. From earliest childhood on, Ditidaht kids learned that this was how a proper person should behave.
>
> In a culture that maintains order through community approval and disapproval, mythology shows people how to live in accordance with the rules. At the same time, it provides a socially acceptable way to find humour in unacceptable behaviour, such as Raven's greed for food. The

> telling of these stories, then, both instructs and amuses. And it also unites the Ditidaht people with another time, a time when a dangerous world was being set in order for their coming.

Cove and MacDonald (1987, ix–xii), discussing the Ts'msyen narratives, note that these are "a collection of vivid oral histories depicting the rich culture of the west-coast Indians. Much of what is contained in the narratives reflects the relationship these people had with their environment: from their understanding of the spiritual universe to their interpretation of the physical world, the stories help chronicle and trace the unique spirit of these people." Dr E. Richard Atleo (2011, 139) summarizes the relevance of narratives in his Nuu-chah-nulth culture simply by quoting his grandmother's words: "Our stories are true."

Many story themes embody lessons of what can happen if people show disrespect to the environment and the resources on which they rely. In the Nass Valley lava flowing from a volcano filled the valley floor some 300 years ago, wiping out an entire village. Nisg̲a'a storytellers still recount how that deadly eruption was triggered by the cruel antics of boys in the village who captured salmon as they came up the creek to spawn, cut their backs and filled the wound with pitch, and then ignited the pitch and laughed to see the burning salmon swimming away up the river (Chester Morris, pers. comm., 1993). The origin story of the Cumshewa Eagle clan of Haida Gwaii, and of other peoples as well, tells of a mythical woman, variously known as Creek Woman or Volcano Woman – Jilaa K̲uns in Haida – who inflicted her volcanic wrath on villagers after boys had caught and burned her children, the frogs, in a fire (see Turner 2004a, 2005). Such immense and traumatic events have turned into the source of lessons and teachings for people, especially about respecting other life.

According to Mary Beck (1989, 43–4) the figure of Volcano Woman (Creek Woman or Frog Woman, also associated with Fog Woman) is probably the oldest and most revered of the northern coast's mythological figures: "She is believed to have reached Alaska from Asia, with late migrants who came by sea. Her counterparts figure in some Asian mythology, as well as Egyptian, Babylonian, Greek, Phoenician, Roman and Norse. She is the Earth Mother, and as Volcano Woman she is historian, collective memory of tribal tradition. She destroys an entire village for the failure of its people to observe the proper rituals and show respect for living creatures and cherished objects." The Jilaa K̲uns story is echoed by a number of others, including the Ts'msyen story "The Headdress of the Mountain Goat," in which people who were abusing, maiming, and killing mountain goats for fun were punished with death, except for one young man who was respectful to the goats.[53]

Another Ts'msyen story embedding deep messages is that of the "perpetual winter." There are many versions recorded (as cited in Cove and MacDonald 1987), and people today still tell this story (Tina Robinson, pers. comm., 2004).

One example, told by Heber Clifton to William Beynon in the late 1940s, relates to the Ts'msyen migration down the Skeena River:

> There was a thriving village called *Temlaxam* on the upper Skeena River, where many people lived in the olden days. It was "a great place," rich in foods of all sorts. But then, people became reckless. They disregarded the elders and lost respect for themselves. Thoughtless young people killed food animals for no need, just for the sport of killing. One time, in early spring when people were preparing their fish traps and roasting a few fish over the fire, it began to snow, which was unusual, since "The spring salmon was always a sign of the beginning of fine weather ..." A thoughtless young man, *Zamxlao*, stood up with some roasted spring salmon in his hand and rebuked the skies, saying: "What is the matter, Chief-of-the-Skies? Do you not see this salmon that I am eating? Are you trying to make another winter, now that the spring salmon is here?" This disrespect brought disaster, with heavy snowfall, as high as the rooftops, leaving people on the verge of famine. "Day after day, the people were held in their houses and the food was about all gone, and wood was hard to get owing to the deep snow." Then one morning a Steller's Jay appeared and sat on the smokehole of the house of one of the chiefs, *Tsibasa*, holding a bunch of elderberries [*Sambucus racemosa*] in its beak.[54] Then *Tsibasa* realized that it was summer nearby, while the village was still in the throes of winter. He said, "... Let us leave this place and go elsewhere, before we all perish." *Tsibasa* and his brothers all left and travelled down the Skeena River until they reached *Gitseguekla* [? Kitsumkalum]. "Here they found that everything was as in summer." Some of the people stayed here, and others spread throughout the Tsimshian [Ts'msyen] territory, down to the mouth of the Skeena, and north along the coast. (Cited in Cove and MacDonald 1987, 257–8)

This story teaches at once the codes of behaviour by which people should live and explains the origins of the Ts'msyen from the upper Skeena.

Plants also mediate transitions from one world to another in narratives. For example, for the Haida, a two-headed bull kelp leads to the top of the house post of a supernatural chief of the ocean world (Swanton 1905, 52, 245). The sky country has been reached by magically growing trees, as in the story of Coyote's son (see Maud 1978a), with similar episodes occurring in stories from the Pacific Coast to the Ponca and Dene (Athabaskan) territories (Boas 2002). In a Gitxsan narrative, a magical fireweed, *giłast* (*k'ilhaast*, "single fireweed"), was planted in front of a house and grew overnight until by the next day it had pierced the sky. This was the beginning of a new clan, the Gisgahaast (Gisq̱'haast), or People of the Fireweed, one of the four clans of the Gitxsan.[55] In the Mink stories, where

Mink goes to visit his father, the Sun, Mink shoots arrows one after another until they form a chain, which he then ascends. This theme of ascending to the sky world on a chain of arrows is exclusive to northwestern North America, according to Boas (2002), and occurs in stories from the Nuxalk, Heiltsuk, Kwakwaka'wkw, and Nahwitti. Simply thinking and dreaming is another means of travelling to the sky country, sometimes unintentionally, as in the "Star Husband" tales, but then the return to earth is by means of woven ropes, which are sometimes made of plant fibres and are often associated with Spider people (Boas 2002). Mouse Woman, a supernatural guide to those who travel from the Haida mythical world to the land of the living people (Archambault 2006), is often identified with a clump of sword fern (*Polystichum munitum*), which marks the entrance to her home.

The stories are replete with lessons for life – about food, cultural and moral values, the importance of sweat bathing and purification rituals for hunters, proper treatment of hunted and killed animals, and general respect for the environment, including plants and vegetation. For example, Haida and Ts'msyen stories recount how devil's-club should be used for purification by hunters and others needing supernatural strength and power.[56] Many stories also reflect times of environmental and cultural change, with accounts of great floods (see Ludwin, Dennis, et al. 2005; and Ludwin, Thrush, et al. 2005), devastating volcanic eruptions, landslides, endless winters, and massive glaciers, as well as times of war, famine, and separation and intermixing of peoples, with accounts of learning new technologies and lessons and encountering new species (Boas 2002; Kennedy and Bouchard 1994; Teit 1912, 210). They embed peoples' complex knowledge of geography and ecology within their own history and their own resource use and are generally indicative of long-term residency within, and familiarity with, a given location.

Stories also warn people against breaking cultural taboos or about other dangers, such as harmful plants. For example, a number of Central Coast Salish stories (Straits Salish, Nanaimo, and Comox) caution against eating bracken fern "roots" either in the summer or if one is in training because it will cause a person to become "snaked," a fearsome condition where snakes enter one's body, which can be remedied only through extraordinary measures such as supernatural help (see Boas 2002, 145–6).

The stories also recount how knowledge is exchanged and diffused, both at a personal and individual level and between groups and across boundaries. For example, a Nlaka'pamux story, "The Old Woman and the Boy," recounts the teaching process between a boy and his grandmother. When he caught and killed a deer, the old woman

> showed him how the people used to skin the deer, and how they cut it up. She also told him what uses the different parts of the deer were put to ...

> The old woman showed him how to skin and cut up the bear, and told him what the people used to do with the skin, bones, flesh, and fat ... Thus he caught and killed goats and all the other animals and birds; and the old woman told him their names, and what they were formerly used for by the people. [Then they went in search of people.] "... So he put his grandmother in a hollow log and dragged her behind him. She knew the way; and as they went along *she told him the names of all the plants and trees, and what uses the people made of them*. (Teit 1912, 264, emphasis added)

Another example is the story of the origin of the Gitxon Eagle (La̱xsgiik) family of the Ts'msyen, which reflects connections throughout the Northwest Coast region between the Ts'msyen, Nisg̱a'a, Haida, and Haisla. Like the previous story, it also features a hollow log. This account goes back to a great flood, after which the young Eagle chief Gitxon and his band members – the ancestors of this family – drifted around and eventually found refuge with the Killer Whales (Gisbutwada) of Kitsumkalum. They lived there peaceably for some time, sharing the fish trap sites of their Killer Whale relatives. However, Gitxon's traps were more successful and caught huge quantities of salmon, which made the Killer Whale hosts resentful. Eventually, they attacked Gitxon and his band, killing Gitxon and all of his group, except for a young chief and Gitxon's niece, both of whom fled to the Nass River. They were welcomed here and fished for oulachens with their Nisg̱a'a relatives on the Lower Nass. The young woman, Gitxon's niece, looked after a number of the children while the fishing was going on, playing with them in a giant hollow log.

One day, as they slept in the log, a very high tide floated them away, and they drifted for many days and eventually landed on a long sandy beach. They found themselves in Haida territory on Prince of Wales Island. Gitxon's niece married the young chief she met there, and she and the children settled in. She had a daughter and four sons with the Haida chief, but as they grew, her children were teased by the Haida children of the village. Eventually, she decided to send her children, now young adults, back to the Nass River village, after she had given them Ts'msyen names, including the name of Gitxon to the eldest brother. The young people travelled for a long time and came first to Dundas Island and then to the mouth of the Nass, where they were received with great joy by their relatives. They were assisted in taking revenge on the Killer Whales of the Skeena area, with the eldest brother, Gitxon, becoming re-established among the Kitsumkalum.

The story goes on: The daughter of the family married a Kitselas chief but was captured by the Kitamaat during a berry-picking trip and was married to a Raven chief of the Haisla. She already had several daughters with the Kitselas chief and had more children at Kitamaat, thus establishing the Gitxon house among the Haisla.[57]

Children are taught important lessons through stories – for example, to stay close to home and not stray into the forest by themselves because of a fierce Cannibal Woman who will glue their eyes shut with pitch, put them into a large basket, and carry them off to her den. This theme is common and widespread, from the Chinookan, Heiltsuk, and Nuxalk along the Northwest Coast to the Ponca, Mi'kmaq, and other peoples from far to the east (Boas 2002).

Other stories incorporate actual instructions, almost like recounting a recipe. An example is the mainland Halkomelem narrative "Origin of the Siyak, or Salmon Weir," recorded by ethnographer Charles Hill-Tout (in Maud 1978a, 56). The Chilliwack people, who are renowned for their use of salmon weirs, tell the story of how Tamia (Wren) first taught them how to construct one:

> He bade the limbs of the young cedars twist themselves into withes, and short branches to sharpen one of their ends to a point and place themselves firmly in the bed of the stream in the form of a tripod, fastened at the top by the withes, two feet being down stream and one up. He then called upon other boughs to wattle themselves in the lower legs of these tripods, till the weir or dam thus formed spanned the whole stream, at the foot of which the salmon soon congregated in great numbers. He bade the people make their salmon-weirs thereafter in like manner.

Similarly, the instructions for making and keeping fire are conveyed in the Nlaka'pamux story of how Beaver taught the people to make fire, also recorded by Hill-Tout (in Maud 1978c, 101–2):

> As they returned down the river, Beaver threw fire on all the trees they passed, but mostly on the cottonwood trees, and thus it was that the wood from these trees was the best for making fire with from that time onward. He continued to do this till he had reached the coast again and all his fire was gone.
>
> After this he assumed a human form and taught the Indians how to make fire by means of the drill worked between the hands. He also taught them how to preserve the fire when once secured in the following manner. He procured a quantity of the inner bark of the cedar tree and made it into a long rope. This he then covered with the bark of some other trees which burnt less readily. When one end of this rope was lighted it would continue to smolder for several days, according to the length of the rope. When the Indians were traveling and likely to be away from camp several days they always carried one of these fire-ropes, called ... *patlakan*, coiled around their shoulders. After this great gift to them the Indians thought very highly of Beaver, and he was usually called by them "our head brother" because of his wisdom and goodness.

Dr Ron Ignace (M.B. Ignace, Turner, and Ignace forthcoming) provides an elegant explanation of the complex and comprehensive role of narratives in relation to Secwepemc environmental knowledge, but he does so in terms that are relevant to the stories of all peoples:

> It is important to realize that *stspetekʷll* [*stspetekʷɬ*, "traditional stories"] which explicitly or implicitly contain messages about the environment and species in it derive from a time when Secwepemc men, women and children spent the vast part of their annual seasonal round traveling, camping, hunting, plant-gathering and fishing in all parts of the plateaus of the Interior, migrating between highland areas and valley bottoms. In the context of this way of life, the knowledge of the behaviour of species, whether they were directly harvested or were ecological indicators of other species' life cycle events, was crucial to survival and success in pursuing and managing resources. It was also critically important for younger people to become trained in recognizing species in the landscape and understanding the ecological interconnections between species. In Secwepemc society, as in other indigenous societies – but each with its own culture-specific and linguistic ways – the explicit or encoded messages about environment contained in stories represent a way of disseminating and transmitting ecological knowledge needed for the successful harvesting of species, and for survival within the environment. Implicit messages about the connection between ecological indicator species and food sources not yet known to younger people, or individuals who have no extensive knowledge of fauna, flora and landscape, further invite the audience to re-examine the living landscape after hearing the story in the new light of what they have learned.

CONCLUSIONS

The themes and stories cited and recounted here reflect the complexity of history and environment, mediated by cultural ties and adapted to specific groups and localities. They show, for example, that people living in particular territories today have, in some cases, storied accounts of arrival from other places and, in others, experiences of permanent occupation, "since the beginning of time," as expressed by Clan Chief Adam Dick (Kwaxsistalla) (pers. comm., 2001). Some of the episodes are very ancient, connecting peoples from distant places, such as the Chinookan at the mouth of the Columbia River, with the Nahwitti (Kwakwaka'wakw) of the central coast of British Columbia or the Straits Salish of Vancouver Island with the Anishinaabe of the Great Lakes and even the Mi'kmaq of Nova Scotia. Along with story themes, it seems reasonable to believe that other types of knowledge, including ethnobotanical and ethnoecological knowledge, followed similar pathways.

The ancient stories, like the names of plants, have unfortunately been lost to many of the present generations of Indigenous peoples; the very lands and waters in which they take place have in many cases been damaged by industrial encroachment and urbanization and thus alienated from their original inhabitants (see L.R. Smith 2008b). As the fundamental underpinning of peoples' lifeways within the landscapes of their ancestors, these narratives are profoundly connected not only to the places themselves but also to the species and resources of those places and to the activities of food production, technology and use of materials, healing, ceremony, celebration, and social interplay and interconnection with other peoples and other species that comprise the totality of life's meanings. In the following chapter, these various aspects of peoples' relationships with their lands and other life forms are discussed in more detail.

13

Worldview and Belief Systems in Ethnobotanical Knowledge Systems

When we have to honour all the spirits, we acknowledge the spirits of water, all the elements, the fire, light, the earth, plants, animals. We all believe that they all have spirits, the same spirit that we have, all humans. Spirits are no different. (Gilbert Solomon, Tsilhqot'in, cited in Supreme Court of British Columbia 2007, 132)

INTRODUCTION

"*The same spirit that we have.*" Gilbert Solomon of the Tsilhqot'in Nation, recounting his teachings from his mother, Mabel Solomon, and other elders, expressed as part of his court testimony a concept and a perspective that was held by any number of women and men of his mother's generation of the mid-1900s from Indigenous communities right across North America and beyond. Secwepemc elder Mary Thomas (pers. comm., 2001) recalled observing her own grandmother, from an even earlier generation:

> My grandmother, I watched her when we were little. If she went to gather medicine or some kind of edible that was to be preserved so they could survive, she didn't just go tramping in the woods and "chop, chop" or "dig, dig" ... She'd make us walk behind her, and ... she would be chanting a tune as we're going along ... And when my grandmother went up to a tree or a shrub, or whatever she was going to gather, she acknowledged that particular tree ... like it was a human being. She'd put her little hands on the tree or the brush ... and she'd acknowledge it. And as a child we were made to respect our grandmother; we'd be quiet in the back ... That's part of our spirituality.

Many Indigenous people still hold this perspective – that plants and animals are respected persons, like humans. This idea has evidently been a part of Indigenous peoples' understandings for hundreds, probably thousands, of years: "since the beginning of time."

Another reflection of this respectful approach to resource species is in the series of "words of praise"[1] recorded by Franz Boas from Kwakwaka'wakw harvesters, addressed to the beings (plants, fish, and animals) that provided people with the food, materials, and medicines they needed for survival. An example is in the words a medicine gatherer addressed to a salmonberry bush (*q'émdzuxwm'es*) before taking some of its bark: "Don't be startled, Supernatural One, by my coming and sitting down to make a request of you, Supernatural One ... This is the reason why I come to you ... to pray you, please, to [let me] take some of your blanket, Sore-Healer, that it may heal the burn of my child, that, please, may heal up his burn, Supernatural One" (Boas 1930, 218). Similarly, from the Interior Plateau, John Ross (2011, 352) notes that the Spokan practised a strict ritual of praying before gathering any plant foods: "a woman prayed as she moved from one berry bush to another. Praying was done by the person always facing east."

Such perspectives represent part of a people's overarching *worldview*, "a distinctive vision of reality," which is shaped by experiences and events in the collective lives of a people, rooted in ancestral teachings and guided by commonly held values and perspectives. It transcends the lifespan of individuals and provides continuity, identity, and a sense of belonging to a particular cultural group (Ortiz 1973, 91). Examples of a worldview that embraces other species with humans in a collective relationship with all of creation can be identified from virtually every Indigenous group across North America and beyond.[2] The immense congruity of perspectives of peoples who have lived close to their lands and waters for generations, and the integration of values with their practical knowledge and technologies, has been noted by many scholars.[3] According to Nuu-chah-nulth scholar and hereditary chief Dr E. Richard Atleo (Umeek) (2004), of Ahousaht, the eternal and unbreakable interconnections among humans, plants, animals, and the other entities of the environment are drawn directly from their common creation. As Umeek explains from a Nuu-chah-nulth perspective (cited in Scientific Panel for Sustainable Forest Practices in Clayoquot Sound 1995, 15): "The Creator made all things one. All things are related and interconnected. All things are sacred. All things are therefore to be respected." Every entity that was a part of the Creator's overall conception is, therefore, inextricably tied to humans. From the southwestern United States, Tewa (Tesuque Pueblo) elder Vickie Downey of New Mexico echoes this same philosophy: "You give thanks before you even receive your gifts from the Creator. When you ask, you give thanks. Your prayers are thanksgivings for everything – the sun, the moon, the snow, the water, the fire, the rocks. You see them as being alive, having a life of their own. A tree has its own life" (cited in Wall 1993, 6).

For First Peoples in northwestern North America, the Creator, or "Great Spirit" (J.E. Brown 1953), is named and characterized in different ways in the narratives and stories of the ancient myth-times, when the distinctions between humans, animals, plants, and other parts of the environment blurred and when gifted and powerful individuals often transformed themselves and others between these states of being (see chapter 12). For some Interior Plateau peoples, the Creator was referred to as Old-One or Old-Man, and the narratives of how Old-One created the sun, the stars, and the earth were recorded by Teit (1912, 1917) and others in various versions. An Okanagan account, for example, was explicit about the origin of plants:

> An Old-One or chief [whose name also means "sweat lodge"] made the earth out of a woman, and said she would be the mother of all the people. Thus the earth was once a human being and she is alive yet; but she has been transformed and we cannot see her in the same way we can see a person. Nevertheless, she has legs, arms, head, heart, flesh, bones and blood. The soil is her flesh; the trees and vegetation are her hair; the rocks her bones and the wind is her breath. (Teit 1917, 80)

A parallel story was told by one of the Kwakwaka'wakw tribes (Nawhitti), in which the "Oldest One on Earth" was himself transformed into a big stone. In this story, "His hair was transformed into grass and earth" (Boas 2002, 421). This idea of plants as the "hair," or covering, of the earth is reflected in the words for vegetation: in Nlaka'pamux and other Interior Salish languages, grass and low-growing plants are known collectively as "ground-growth" or "ground-hair" (Turner and Brown 2004). Another description of plants, uncovered from James Teit's (1896–1918) field notes on the Nlaka'pamux, dating to around 1900, refers to plants as the "the covering or blanket of the earth."[4]

Mary Thomas (pers. comm., 2001) told another story from her childhood, also featuring grass as the hair of the "Mother Earth." On one occasion, all of Mary's women relatives were sitting around on the ground, working on the *sp'ets'i* (Indian-hemp), extracting the fibres to be spun into twine for making fishing nets and twining mats. The women were all talking among themselves, with the children sitting around and listening. Then at one point the conversation turned to certain sad events and people who had passed away. Before long, the women were all moaning and crying, even wailing loudly, and pulling at the grass at their feet. The children succumbed to the general mood and started to weep themselves. Afterward, Mary asked, "Auntie, why was everyone crying?" Her aunt explained to her that sometimes it really helped people feel better if they cried. She said they had no "doctors" (psychiatrists) around to help, so crying was one way that people healed themselves of emotional distress and tension. Mary's aunt compared this healing process to that of a child who was tired and upset, who would snuggle into its mother's lap, pull at her hair and cry, even for no apparent reason or particular cause. The women, her aunt said, were doing the same thing: sitting on the lap of their "Mother Earth" (using the Secwepemc term), pulling at her hair – the grasses – and crying until they felt better. Thus, from far back, almost a century or so, the elders regarded the grass and other plants in this special way, as the hair of the living earth.

In the Nlaka'pamux "Old-One" narratives, Old-One also instructed the people he had made about how they should live and how to hunt and fish, make baskets, dig roots, and so forth (chapters 5, 6, 7, and 12). He also taught them the proper protocols for harvesting: how to harvest plants like saskatoon berry to make a fish spear or yew wood to make a bow. Before taking anything from a plant, he taught them, one should "pray or talk nicely to it." In other words, right at the inception of humanity, as people perceived this time, they learned that they must treat plants and other living beings with deep respect and relate to them as they would to other human beings.

In this chapter, I discuss the implications of Indigenous peoples' worldviews and visions of reality for the ways that they relate to and use plants. I first consider the perception of plants as close relatives of humans – powerful, but usually generous, beings who can influence humans in different ways, positively or negatively, according to how they are treated. The entire collective system of Traditional Ecological Knowledge of Indigenous peoples across the continent is firmly grounded in the belief that there is a fundamental relationship and interconnectedness between humans, plants, animals, fish, and birds, as well as mountains, water, the moon, and stars. This perspective of plants and animals as our relatives has been termed "kincentricity," or "kincentric ecology," as mentioned in chapter 12. The power of plants as spiritual entities is described in the next section, particularly how certain species – especially prickly plants and aromatic plants – provide special protection against many types of harmful influences. Following this is a discussion of the role of plants in peoples' rituals and ceremonies, and then the reciprocal relationships, taboos, and protocols around the use of plants are described. The chapter concludes with a consideration of the human creative role in transforming plants from one state to another. Table 13-1 provides examples of plant species featured in various ways in belief systems of Indigenous peoples in northwestern North America.

PLANTS AS GENEROUS RELATIVES: KINCENTRIC ECOLOGY

A kincentric belief system, in which all life forms and other environmental entities are seen to be relatives of humans and of each other, is widespread in Indigenous peoples' cosmologies (Martinez 2008; Salmón 2000; Senos et al.

Table 13-1 | Examples of plants related to traditional beliefs of Indigenous peoples of northwestern North America

Species	*Notes (reference)*
ALGAE, LICHEN, FUNGI, AND FERNS	
Bull kelp (*Nereocystis luetkeana*)	Two-headed kelp believed to mark entrance to the supernatural world under the ocean (Haida) (Swanton 1905; Turner 2004a)
Black tree lichen (*Bryoria fremontii*)	Created by Coyote from his hair braid (Okanagan, Secwepemc) (Bouchard and Kennedy 1979; Turner, Thompson, et al. 1990)
Tree fungi (*Fomitopsis pinicola*, *Ganoderma applanatum*, and other spp.)	Tree Fungus was a person in a Haida narrative (see table 12-2); spiritual protection for a household or individuals against evil thoughts (Saanich and other Coast Salish peoples, Ditidaht and others); symbolic crest (Nuxalk and others); shamanic grave markers (Tlingit) (Blanchette et al. 1992)
Puffballs (*Lycoperdon* spp., *Bovista* spp., *Calvatia* spp.)	Said to be associated with stars, thunder, and weather events (Sechelt, Kwakwaka'wakw, and others); associated with corpses and ghosts (widespread) (Turner, Timmers, and Bouchard 1972; Turner and Bell 1973)
Mushrooms (*Tricholoma populinum* and other spp.)	Washing in mushroom juice thought to make babies strong and independent when they grew up (Nlaka'pamux); some people had strong taboos against eating mushrooms (A. Jones 1983; Turner, Kuhnlein, and Egger 1985)
Running clubmoss (*Lycopodium clavatum*)	Considered to be a "bad luck" plant; if handled, will cause one to become confused and lost (Ditidaht, Nlaka'pamux) (Turner, Thomas, et al. 1983, 1990)
Sword fern (*Polystichum munitum*)	Fern Woman was a supernatural being (Haida); Sword Fern was a supernatural person who controlled the weather and had a hairy face, earrings of dentalia, and cheeks smeared with red ochre; used in weather rituals (Kwakwaka'wakw) (Boas 1966; Swanton 1905; Turner 2004a; Turner and Bell 1973)
Bracken fern (*Pteridium aquilinum*)	Associated with snakes; many taboos about digging the rhizomes (Nuxalk, Saanich, Halkomelem, Klallam, and others) (Elsie Claxton, pers. comm., 1989; McIlwraith 1948; Turner 1973; Turner and Bell 1971)
CONIFERS	
Amabilis fir (*Abies amabilis*) and Grand fir (*A. grandis*)	Needles for hair perfume (Nuxalk, Hesquiaht); boughs for purification (Kwakwaka'wakw); branches burned as incense, air freshener, protection against illness (Ditidaht); boughs rubbed on skin of girls for protection against predators, used as a protective wash (Nlaka'pamux) (Boas 1921, 1966; H.I. Smith 1928; Turner 1973; Turner and Bell 1973; Turner and Efrat 1982; Turner, Thomas, et al. 1983, 1990)
Subalpine fir, or "balsam fir" (*Abies lasiocarpa*)	Boughs rubbed on skin of girls for protection against predators; babies washed in bough solution for strength (Nlaka'pamux, Okanagan); powdered bark rubbed on skin as cleansing and for

/continued

Table 13-1 | continued

Species	Notes (reference)
	scent, boughs as bedding, in sweathouse (Okanagan); boughs in sweathouse for purification (Sahaptin); protective smudge from boughs (Ktunaxa, Secwepemc, Okanagan, and others) (Hart, Turner, and Morgan 1981; G. Palmer 1975b; Teit 1906b; Turner, Bouchard, and Kennedy 1980; Turner, Thompson, et al. 1990)
Yellow-cedar (*Chamaecyparis nootkatensis*)	Boughs used in sweat bath, as scrubbers, for rheumatism, arthritis, and protection (Kwakwaka'wakw) (Boas 1930; Turner and Bell 1973)
Common juniper (*Juniperus communis*)	Used as medicine; all had to be drunk, or it would not work (Haida); decoction drunk in sweathouse and used as purifying wash for hunters and others (with *Artemisia* and *Shepherdia canadensis*) (Nlaka'pamux); decoction used as fumigant against death and illness (Okanagan); wash for babies against witchcraft or fever (Sahaptin) (Hunn, Selam, and family 1990; G. Palmer 1975b; Turner 2004a; Turner and Thompson 2006; Turner, Bouchard, and Kennedy 1980; Turner, Thompson, et al. 1990)
Rocky Mountain juniper (*Juniperus scopulorum*) and Pacific coastal juniper (*J. maritima*)	Branches hung in house or burned on stove as incense to keep away germs during epidemics (Saanich, western Washington); decoction as fumigant, incense, and air freshener during sickness, death (Nlaka'pamux); steam inhaled to prevent illness (Secwepemc); wash for hunting gear (Okanagan); boughs burned on stove or in fire as protective incense (Okanagan, Nlaka'pamux, Stl'atl'imx, and others); solution of berries and/or boughs, bark (also *J. occidentalis*) as room disinfectant (Sahaptin) (Gunther 1973; Hunn, Selam, and family 1990; Turner and Bell 1971; Turner, Bouchard, and Kennedy 1980; Turner, Thompson, et al. 1990)
Engelmann spruce (*Picea engelmannii*) and White spruce (*P. glauca*)	Wash of branches for purification by hunters and others (Nlaka'pamux); baby wash made from boughs to give strength (Nlaka'pamux); pitch connected to death and bereavement (general, Nlaka'pamux, Okanagan, Interior Plateau); chewing gum from twin-topped spruce tree will cause a woman to have twins (Tsilhqot'in) (Linda Smith, pers. comm., 1987; Turner, Bouchard, and Kennedy 1980; Turner, Thompson, et al. 1990)
Sitka spruce (*Picea sitchensis*)	Said to bring supernatural protection (Haida, Kwakwaka'wakw, Klallam, Ditidaht); bark mixture in sweat bath, decoction drunk for tuberculosis (Nuxalk); branches placed around sick people's houses with the idea that the sharp spruce needles would keep anything unclean from entering the house; spruce tips rubbed on canoes, digging sticks, and implements to purify them and used to cleanse a person contaminated (*sic*) with menstrual blood (Kwakwaka'wakw); beating with branches said to drive away the sickness and ghosts of the dead (Klallam); said to bring vivid dreams and good luck if you sleep under spruce (Nlaka'pamux) (Boas 1930, 1966; Turner 2004a; Turner, Thomas, et al. 1983)

Table 13-1 | continued

Species	Notes (reference)
Ponderosa pine (*Pinus ponderosa*)	Deodorant, used in sweat bathing (Secwepemc); needles spread on sweathouse floor to help ward off witchcraft, salve to make babies sleep well (Okanagan) (G. Palmer 1975b; Turner, Bouchard, and Kennedy 1980; Turner, Thompson, et al. 1990)
Douglas-fir (*Pseudotsuga menziesii*)	Mixture with other plants for hair wash and skin wash during sweat bathing, needles as hair tonic (Nlaka'pamux, Okanagan); boughs used on floors, in sweathouse, for bereaved persons, used as tea during sweat bathing, cleansing agent (Nlaka'pamux, Okanagan, other groups); boughs used in ritual washing of twins, to construct puberty houses for girls, and in other ceremonial roles (Secwepemc, Okanagan, other Interior Plateau peoples) (Turner Bouchard, and Kennedy 1980; Turner, Thompson, et al. 1990)
Pacific yew (*Taxus brevifolia*)	Tree "talked to" before a piece was cut for bows or other tools (Kwakwaka'wakw); branches used for ritual scrubbing and purification (Ditidaht, Nlaka'pamux, and others) (Turner, Thomas, et al. 1983, 1990)
Western redcedar (*Thuja plicata*)	Strong taboo against harvesting cedar bark when seaweed is being picked (Ts'msyen); bark used in sacred cedar-bark ceremony of the winter dances and for ceremonial cleansing (Kwakwaka'wakw); tree "talked to" before planks or bark removed, or tree cut for canoe, etc. (Kwakwaka'wakw, Haida, Nuu-chah-nulth, and others); mixture with other plants for hair wash and skin wash during sweat bathing, for scrubbing (Kwakwaka'wakw, Okanagan); boughs as ceremonial scrubbers (Saanich); boughs used to sweep house after death and smudge from boughs to scare ghosts (western Washington, Saanich); inner bark used as headband for girl at puberty (Secwepemc) (Sewid-Smith and Dick 1998; Turner 1998, 2003b, 2004a; Turner, Bouchard, and Kennedy 1980; Turner and Clifton 2006)
Western hemlock (*Tsuga heterophylla*)	Boughs used as ritual scrubbers to protect girls at puberty, hunters (Kwakwaka'wakw, Hesquiaht/Nuu-chah-nulth, Ditidaht, Makah, Nlaka'pamux); decoction of branches drunk for ritual purification and used as disinfecting fumigant for houses (Nlaka'pamux) (Boas 1930; Gill 1983; Turner and Bell 1973; Turner and Efrat 1982; Turner, Thompson, et al. 1990)
FLOWERING PLANTS	
Red alder (*Alnus rubra*)	Formerly a woman; you should hug an alder tree before you cut it down (Haida) (Turner 2004a)
Saskatoon berry (*Amelanchier alnifolia*)	Ripening of berries celebrated with First Fruits ceremony (northern Interior Salish, Nez Perce, Sahaptin); Old-One created fish spear from branch (Nlaka'pamux) (Hunn, Selam, and family

/continued

Table 13-1 | continued

Species	*Notes (reference)*
	1990; A.G. Marshall 1977; Teit 1912; Turner, Thompson, et al. 1990)
Red columbine (*Aquilegia formosa*)	Believed to cause rain if picked (Haida); a good luck plant, for love, wealth, etc. (Saanich, Kwakwaka'wakw) (Turner 2004a; Turner and Bell 1971, 1973)
Northern wormwood (*Artemisia frigida*)	Leaves used for incense as disinfectant, smudge (Nlaka'pamux, Secwepemc, Okanagan) (G. Palmer 1975b; Turner, Bouchard, and Kennedy 1980; Turner, Thompson, et al. 1990)
Western mugwort (*Artemisia ludoviciana*)	Leaves used for incense as disinfectant, smudge (Nlaka'pamux, Okanagan) (Turner, Bouchard, and Kennedy 1980; Turner, Thompson, et al. 1990)
Big sagebrush (*Artemisia tridentata*)	Leaves, branches as protective smudge, incense, and house fumigant against illness, death; decoction of branches used to revive unconscious people (Nlaka'pamux, Secwepemc, Okanagan, Sahaptin) (Hunn, Selam, and family 1990; G. Palmer 1975b; Turner, Bouchard, and Kennedy 1980;.Turner, Thompson, et al. 1990)
Wild ginger (*Asarum caudatum*)	A good luck plant, for love, wealth, etc. (Saanich); leaves used as protective wash against illness or evil thoughts (Sechelt); powdered leaves used in baby's cradle (Quileute, Nlaka'pamux); used as love charm and charm to keep families together (Secwepemc) (Mary Thomas, pers. comm., 2001; Gunther 1973; Turner, Thompson, et al. 1990; Turner, Timmers, and Bouchard 1972)
Balsamroot, or spring sunflower (*Balsamorhiza sagittata*)	Many taboos and observances relate to pit-cooking these roots (Nlaka'pamux, Secwepemc, Okanagan) (Teit 1909; Peacock 1998)
Paper birch (*Betula papyrifera*)	Tree ceremonially thanked before bark gathered; special basket used by girl at puberty (Secwepemc); tree said to have been created by Old-One (the first tree), and a basket from a roll of birch bark (Nlaka'pamux) (Mary Thomas, pers. comm., 1994; Teit 1909; Turner, Thompson, et al. 1990)
Poque (*Boschniakia hookeri*)	Said to be a good luck charm (northern Wakashan) (Compton 1993b)
Calypso, or false ladyslipper (*Calypso bulbosa*)	Flowers used as a good luck charm (Nlaka'pamux) (Turner, Thompson, et al. 1990)
Edible camas (*Camassia quamash*)	First Roots ceremony celebrated for these roots in some places (Okanagan, Sahaptin) (Hunn, Selam, and family 1990; Turner, Bouchard, and Kennedy 1980)
Blue harebell (*Campanula rotundifolia*)	Believed to cause rain if picked (Haida) (Turner 2004a)

Table 13-1 | continued

Species	Notes (reference)
Paintbrush (*Castilleja miniata* and other spp.)	Flowers are considered to be sacred; if picked, would cause thunder (Dakelh, Tsilhqot'in) (Kay 1993)
Creeping snowberry (*Chiogenes hispidula*)	Plants used as protective and therapeutic wash for pregnant women (Secwepemc) (Mary Thomas, pers. comm., 1994)
Bull thistle (*Cirsium brevistylum*) and other thistles, including introduced spp. (*Cirsium* spp.)	Root of thistle considered to bring good luck and protection (Saanich) (Elsie Claxton, pers. comm., 1991)
Spring beauty, or "mountain potato" (*Claytonia lanceolata*)	Mature stems (gone to seed) pulled up by bereaved women and thrown onto places where it wasn't growing to help spread the plant (Tsilhqot'in); included in First Roots ceremonies of some Interior Plateau peoples (Mabel Solomon and Gilbert Solomon, pers. comm., 2003; Mellott 2010; L.R. Smith 2008b)
Red-osier dogwood (*Cornus sericea*)	Bush "talked to" before bark, branches removed for medicine (Mary Thomas, pers. comm., 2001)
Black hawthorn (*Crataegus douglasii*)	Charcoal used as face paint in winter dances (Halkomelem) (Abel Joe, pers. comm., 1974)
Shooting star (*Dodecatheon jeffreyi*)	Believed to cause rain if picked (Haida) (Turner 2004a)
Sundew (*Drosera rotundifolia*)	Said to be a good luck charm, especially for fishing gear (Haida) (Turner 2004a)
Pink fawn lily (*Erythronium revolutum*)	Called "sad flowers" and said to cause fog if picked (Ditidaht) (Turner, Thomas, et al. 1983)
Northern riceroot (*Fritillaria camschatcensis*)	Featured in a special ceremonial Flower dance, with other flowers (Hanaksiala) (Compton 1993b)
Cascara (*Frangula purshiana*)	Tree "talked to" and asked permission before bark harvested (many other trees and shrubs were approached similarly) (widespread, coast and interior peoples) (Turner and Hebda 1990)
Sweet-scented bedstraw (*Galium triflorum*)	Plant used for love charm (Ahousaht Nuu-chah-nulth, western Washington) (Densmore 1939; Gunther 1973; Turner and Efrat 1982)
Sweetgrass (*Hierochloe odorata, H. hirta*)	Used as protective smudge, body wash, and incense (Nlaka'pamux, Secwepemc, Okanagan, Ktunaxa, Nuxalk) (Hart, Turner, and Morgan 1981; G. Palmer 1975b; Turner 1973; Turner, Bouchard, and Kennedy 1980; Turner, Thompson, et al. 1990)
Junegrass (*Koeleria macrantha*)	Waving this grass is said to cause rain (Dakelh) (Poser 2008a)

Table 13-1 | continued

Species	*Notes (reference)*
Bitterroot (*Lewisia rediviva*)	Said to have been human beings; many beliefs around picking them; First Roots ceremony held for them (Okanagan, Nlaka'pamux, Sahaptin, Nez Perce, and others) (Bandringa 1999; Hunn, Selam, and family 1990; A.G. Marshall 1977; Turner, Bouchard, and Kennedy 1980; Turner, Thompson, et al. 1990)
Canby's lovage (*Ligusticum canbyi*)	Root used as scent, for smoking and smudging, as good luck charm; used to prevent babies from catching cold and to revive unconscious people or people who have gone into a trance (Spokan, Okanagan, Secwepemc, Nlaka'pamux, Ktunaxa) (Hart, Turner, and Morgan 1981; G. Palmer 1975b; J.A. Ross 2011; Turner, Bouchard, and Kennedy 1980; Turner, Thompson, et al. 1990)
Tiger lily (*Lilium columbianum*)	Flowers used as charm for love and wealth (Nlaka'pamux) (Turner, Thompson, et al. 1990)
Stoneseed (*Lithospermum ruderale*)	Used as charm to call the rain (Stl'atl'imx, Okanagan) (Sam Mitchell, pers. comm., 1974; Turner, Bouchard, and Kennedy 1980)
Geyer's biscuitroot (*Lomatium geyeri*)	First Roots ceremony celebrated for these roots in some places (Okanagan, Sahaptin) (A.G. Marshall 1977; Turner, Bouchard, and Kennedy 1980)
Desert parsley (*Lomatium macrocarpum*)	Associated with Transformer; this root, *qʷ'əqʷ'íle*, was said to have fathered a son, who grew up to be one of the Transformers (Nlaka'pamux, Stl'atl'imx, other Interior Salish) (Turner, Thompson, et al. 1990)
"Wild celery," or barestem lomatium (*Lomatium nudicaule*)	Seeds burned as a smudge in First Salmon ceremony, to "drive away ghosts," to protect against illness and death (Saanich); used widely for spiritual protection (Klallam, Saanich, Ditidaht, Kwakwaka'wakw, Nlaka'pamux, Stl'atl'imx) (Boas 1921; Turner and Bell 1971, 1973; Turner, Thomas, et al. 1983, 1990)
Skunk-cabbage (*Lysichiton americanus*)	Smoke from burning roots inhaled for bad dreams, flu, rheumatism, used in sweat bath (Nuxalk); leaves placed under pillow, or head washed with solution, for "power dreams" (Nlaka'pamux) (Turner, Thompson, et al. 1990)
Tall Oregon-grape (*Mahonia aquifolium*)	Used with wild rose to give protection in illness and death (Okanagan) (Turner, Bouchard, and Kennedy 1980)
Pacific crabapple (*Malus fusca*)	Ceremonial use for women at puberty and in mourning (Haida) (Turner 2004a)
Pineappleweed (*Matricaria matricarioides*)	Used as love charm (Okanagan) (Turner, Bouchard, and Kennedy 1980)
False azalea (*Menziesia ferruginea*)	Placed in coffins and used around illness and death (Haida); said to steal berries if used to cover them in a basket (Ditidaht) (Turner 2004a; Turner, Thomas, et al. 1983)

Table 13-1 | continued

Species	*Notes (reference)*
Monardella (*Monardella odoratissima*)	Infusion of stems, leaves used in a wash to purify weapons (Okanagan) (Turner, Bouchard, and Kennedy 1980)
Single delight (*Moneses uniflora*)	Plant eaten for "luck" and power; used, along with devil's-club, as good luck charm and for supernatural help (Haida) (Swanton 1905, 1908; Turner 2004a)
Indian-pipe (*Monotropa uniflora*)	Said to grow where the wolf urinates; called "wolf's urine" (Saanich, Nlaka'pamux) (Elsie Claxton, pers. comm., 1998; Turner, Thompson, et al. 1990)
Yellow pond-lily (*Nuphar lutea* ssp. *polysepala*)	Used as charm for calling the west wind (Hesquiaht, Nuu-chah-nulth); associated with dreams; can heal a person who is having dreams of deceased relatives (Nuxalk) (Turner 1973; Turner and Efrat 1982)
Water-parsley (*Oenanthe sarmentosa*)	A charm in a narrative on stealing daylight (Ditidaht) (Turner, Thomas, et al. 1983)
Devil's-club (*Oplopanax horridus*)	Supernatural being in many traditions and narratives for sweat bathing and ceremonial washing; bark eaten to bring supernatural protection and luck in gambling, hunting, etc. (Haida); widely used for purification and for washing hunting and fishing gear, as recounted in many narratives (Ts'msyen, Kwakwaka'wakw, Nuxalk; western Washington); charcoal as ceremonial and protective face paint (Saanich, Ditidaht); taboo against shadow of a person touching devil's-club medicine (Okanagan) (Elsie Claxton, pers. comm., 1994; Boas 1916, 1921; Gunther 1973; Lantz, Swerhun, and Turner 2004; Turner 1973, 2004a; Turner and Bell 1973; Turner, Bouchard, and Kennedy 1980; Turner, Thomas, et al. 1983)
Western sweet-cicely (*Osmorhiza occidentalis*)	Root carried by hunters to mask human scent (Okanagan) (Turner, Bouchard, and Kennedy 1980)
Wild poppy (identity not positive) (*Papaver alaskanum*, *P. alboresum*, *P. nudicaule*)	A "good luck" plant that grows in the mountains and opens and closes with the sun is known in the inland area; people with power can talk to it and cause the weather to change (apparently it is poppy) (Dena'ina/Tanaina) (Russell Kari 1987)
White bog orchid (*Platanthera dilatata*)	Widely used in Interior Salish as good luck charm, love charm (Nlaka'pamux, Secwepemc, Okanagan); purifying sweat bath; used to wash guns and as a body wash for hunters (Nlaka'pamux) (Turner, Bouchard, and Kennedy 1980; Turner, Thompson, et al. 1990)
Black cottonwood (*Populus balsamifera* ssp. *trichocarpa*)	Burning the bark causes a strong wind (Squamish) (Bouchard and Turner 1976)

Table 13-1 | continued

Species	Notes (reference)
Smooth sumac (*Rhus glabra*)	Not used to cover berries in baskets, or it will steal some of the berries (Stl'atl'imx) (Sam Mitchell, pers. comm., 1974)
Gray currant (*Ribes bracteosum*) and trailing currant (*R. laxiflorum*)	Rubbed on fishing gear to mask human scent (Haida) (Turner 2004a)
Wild black currant (*Ribes hudsonianum*)	Berries said to turn into trout if they drop into the water (Nlaka'pamux) (Turner, Thompson, et al. 1990)
Swamp gooseberry (*Ribes lacustre*)	Branches burned as protection against snakes and other harmful spirits (Nuxalk, Haida, and others); used as charm to make rain (Secwepemc); rubbed on fishing gear to mask human scent (Okanagan) (McIlwraith 1948; G. Palmer 1975b; Turner 2004a; Turner, Bouchard, and Kennedy 1980)
Red-flowering currant (*Ribes sanguineum*)	Said to cause rain if flowers are picked (Halkomelem) (Mary Thomas, pers. comm., 1989; Turner and Hebda 2012)
Prickly wild rose (*Rosa acicularis*) and Wood's rose (*Rosa woodsii*)	Solution of branches used by hunters as body wash to remove human scent and used as protection (Okanagan); used to "sweep out" graves before burials (Okanagan); branches used to ward off ghosts and used in bereavement ceremonies (Sahaptin, Secwepemc, Okanagan, Spokan); protective face cream for infants, and baby wash, to give strength (Nlaka'pamux) (Hunn, Selam, and family 1990; G. Palmer 1975b; J.A. Ross 2011; Turner, Bouchard, and Kennedy 1980; Turner, Thompson, et al. 1990)
Dwarf wild rose, or baldhip rose (*Rosa gymnocarpa*)	Leafy twigs used to make a protective wash and smudge (Saanich, Nlaka'pamux, Secwepemc) (Mary Thomas, pers. comm., 1994; Turner and Bell 1971; Turner, Thompson, et al. 1990)
Nootka rose (*Rosa nutkana*)	Featured in a special ceremonial Flower dance, with other flowers (Hanaksiala); mixture of leafy twigs with other plants for hair wash and skin wash during sweat bathing (Okanagan, Secwepemc) (Compton 1993b; G. Palmer 1975b; Turner, Bouchard, and Kennedy 1980)
Thimbleberry (*Rubus parviflorus*)	First Shoots and First Fruits ceremonies held for this species (Chehalis, Halkomelem) (Hill-Tout, in Maud 1978a, 1978c)
Salmonberries (*Rubus spectabilis*)	Singing of Swainson's thrush makes the salmonberries ripen (belief widespread along Northwest Coast); dark coloured salmonberries ripen at night, light coloured ones in daytime (Ts'msyen) (Turner and Thompson 2006); featured in a special ceremonial Flower dance, with other flowers (Hanaksiala) (Compton 1993b)
Trailing wild blackberry (*Rubus ursinus*)	Used as medicine for its protective properties (Saanich, Halkomelem) (Turner and Bell 1971)

Table 13-1 | continued

Species	Notes (reference)
Willows (*Salix* spp.)	Inner bark used as headband for girl at puberty (Secwepemc) (Teit 1909)
Gray ball sage (*Salvia dorrii*)	Used in baby's cradleboard (apparently for protection) (Okanagan) (Turner, Bouchard, and Kennedy 1980)
Red elderberry (*Sambucus racemosa*)	Commonly associated with death (Haida, Ts'msyen, and others) (Boas 1902; Cove and MacDonald 1987; Turner 2004a)
Tule (*Schoenoplectus acutus*)	Special ceremony observed by Spokan before harvesting the stems (J.A. Ross 2011)
Snowberry, or waxberry (*Symphoricarpos albus*)	Berries associated with snakes and with corpses and ghosts (widespread); said to be the saskatoon berries of the people in the land of the dead (Stl'atl'imx) (Teit 1906a)
Common tansy (*Tanacetum vulgare*, introduced sp.)	A charm for obtaining luck and wealth in gambling (Straits Salish) (Elsie Claxton, pers. comm., 1994)
Western trillium (*Trillium ovatum*)	Called "sad flowers" and said to cause fog if picked (Ditidaht) (Turner, Thomas, et al. 1983)
Cattail (*Typha latifolia*)	Special ceremony observed by Spokan before harvesting the leaves (J.A. Ross 2011)
Stinging nettle (*Urtica dioica*)	Mixture with other plants for hair wash and skin wash during sweat bathing; used to combat witchcraft (Okanagan) (Turner, Bouchard, and Kennedy 1980)
Blueberries (*Vaccinium alaskaense, V. ovalifolium*)	Featured in a special ceremonial Flower dance, with other flowers (Hanaksiala) (Compton 1993b)
Black huckleberry (*Vaccinium membranceum*)	Ripening of berries celebrated with First Fruits ceremony (Okanagan, Spokan, Sahaptin, Nez Perce, and others) (Hunn, Selam, and family 1990; A.G. Marshall 1977; J.A. Ross 2011; Turner, Bouchard, and Kennedy 1980)
Mountain valerian (*Valeriana sitchensis*)	Whole plant as protective wash for hunters to mask human smell, also a hunting charm (Nlaka'pamux, Secwepemc) (Mary Thomas, pers. comm., 2001; G. Palmer 1975b; Turner, Thompson, et al. 1990)
Indian hellebore (*Veratrum viride*)	Used by shamans for protection against evil and to gain supernatural help (Haida, Ts'msyen); widely used as smudge or wash for spiritual protection and to vanquish monsters like *sisiutł* (double-headed sea serpent) (Kwakwaka'wakw, Okanagan, and many others) (Mary Thomas, pers. comm., 2001; Boas 1921, 1930; G. Palmer 1975b; Turner 2004a; Turner, Bouchard, and Kennedy 1980; Turner and Thompson 2006; Turner, Thompson, et al. 1990)
Wild violets (*Viola* spp.)	Believed to cause rain if picked (Haida) (Turner 2004a)

Note: See also tables 12-1, 12-2, and 12-3.

2006; Turner 2005).[5] The same obligations that almost all humans feel toward their own family members are, in the case of the kincentric view of Indigenous peoples, extended to all life. It is the duty and responsibility of humans to acknowledge, and to look after, *all* of their relatives and to consider the well-being of these relatives as inextricably bound to the health and well-being of humans. It is seen as a sacred trust to care for the earth and all of its inhabitants, as they all care for us. The Mohawk pronouncement of the Haudenosaunee thanksgiving address at the beginning of every major gathering expresses similar relationships and responsibilities to and from all of the entities of the Haudenosaunee world (Turner 2005).

Biologists and geneticists studying our nucleic acids will tell us that it is largely true, at least for living things, that we are all descended from the same common ancestral forms and that much of our genetic makeup is, indeed, virtually identical to that of plants. However, most people in western urbanized, industrialized society consider humans to be apart from, and above, other life forms, even those like chimpanzees and other primates who are genetically closest to us. In this predominant perspective, even where humans may feel some connections to other mammals such as dogs, cats, horses, or dolphins, plants are conceptually far distant. For the most part, they are seen as "objects," strictly resources, that are available for our use and capacity to gain wealth, not as sentient beings with a sensory capacity equivalent or even superior to that of humans.

Even rocks, mountains, rivers, and lakes are recognized at some level in the kincentric worldview as powerful and sentient beings. Some of the mountains of Tsilhqot'in and Ulkatcho country, for example, were said to have been human beings at an earlier time; two individuals near Chilco Lake, T̲s̲'il?os̲ and ?Eniyud, were a chief and his wife, who didn't get along and ended up separating. The couple had six children; two of them are behind T̲s̲'il?os̲ and one is on his lap. The three other children are right beside ?Eniyud. In a land claims court testimony in Victoria in 2003, Xeni Gwet'in chief Roger William (2003) recounted the story, explaining the present locations of the mountains:

> *?Eniyud* was heading towards ... Tatlayoko. She planted them wild potatoes [spring beauty, *Claytonia lanceolata*, at Potato Mountain] on her way over. And *T̲s̲'il?os̲* just stayed there [figure 13-1]. Some of the stories I hear was that the beartooth, *?esghunsh* [yellow glacier lily, *Erythronium grandiflorum*], that he planted those around there. And in our stories in our history, when ... you're married, they're not supposed to separate, and what happened there, there was a separation. So when they turned to stone, they – you know, they're going to look after us, the Xeni Gwet'in Tsilhqot'in people, all the time.

13-1 | T̲s̲'il?os̲ (Mount Tatlow), in Xeni Gwet'in Tsilhqot'in territory, who used to be a bad-tempered man according to Tsilhqot'in traditions. It is considered disrespectful to point at this or other mountains.

The stars themselves are included in this immense family of creation. For example, the stars of the Pleiades constellation were said by the Secwepemc formerly to have been a group of women roasting roots in an earth oven who were transformed into these stars (Dawson 1892, 39). The well-known "Star Husband" narrative (see chapter 12) recounts how two sisters married Stars, identified in one version (R. Brown 1873) as Aldebaran and Sirius and in another (Boas 2002) as the planets Mars and Jupiter. Interestingly, the Navajo name for "star," *sǫ́si̱*, means "my ancient ancestor" (Larry Emerson, pers. comm., 2009). The sun, too, is seen as a powerful person who was once a great chief. Of course, animals, birds, and fish of all kinds are all major players in this familial relationship, as reflected in their portrayal in narratives, dances, songs, and ceremonies. For example, the Saanich ceremonial names for the different kinds of salmon, used during the First Salmon ceremony, are a reflection of their perceived kinship: Sockeye is addressed as "Honoured One"; Coho as "Parent of your daughter-in-law or son-in-law"; Humpback, or Pink Salmon, as "Adopted niece or nephew"; and Spring, or Chinook, as "Sister-in-law or Brother-in-law" (E. Claxton and Elliott 1994).

Kincentricity for plants, too, is reflected in the diverse stories, many recounted in chapter 12, in which plants – in the ancient times before the transformation –

were originally people, or in other cases, they were transformed from their previous botanical form into humans. Other plants were created by transformations of certain objects or substances, or they were converted magically into other entities. In other words, the lines that divide people, animals, trees, ferns, flowers, rocks, mountains, rivers, and stars into different and distinct categories were not fixed in myth-times as they are mostly today. Plants in those ancient days were actually very similar to people, living in communities with their own social organization as humans still have. For example, in the Saanich narrative of the Tree people, all of the trees were formerly human-like and were transformed into their present form by Qäls (X̱e'ls), the Creator. Douglas-fir (*ch'sey'*) was said to be the "boss" of the Trees (Jenness n.d., 63).

In one Okanagan cosmology, all of the different creatures, plants, and rocks were described as having their own societies, each with its own chief, overseen by Old-One. Three Yellow-Cedar sisters of the Hesquiaht, who were scared away from the food they were preparing by Raven, voluntarily placed themselves into the service of humans. Others were compelled into their new roles as being useful to humans. Such is the case with Cow-parsnip in a Tlingit story; she was a woman who offended Raven by speaking angrily and was therefore transformed by Raven into a food for people. Similarly, Western Larch, or Tamarack, and Tobacco (*Nicotiana attenuata*) were said to have been people-eating monsters, transformed by Coyote to be of service to humans. Other plants were created by individuals from parts of themselves: black tree lichen by Coyote from his own hair braid, bunchberries and trailing blackberries by a dying woman from her own blood, and Indian-hemp plants by Coyote's son from his own hairs for the elderly Spider couple of the sky country. Others, like the old blind woman who was transformed into an edible wood fern and the young girl caught in a landslide whose head became a wood fern root and body a marmot, were simply apparent victims of transformations (see chapter 12, table 12-2). Plant actors in the ancient stories, although not as common as animal actors, can be identified here and there. The Kwakwakw'akw and Central Coast Salish stories about Mink are a case in point. In different episodes of these stories, Mink tries to marry Horsetail, Resinous Bark, and Bull Kelp (Boas 2002, 130, 131, 190). Another is the role of *qʷ'əqʷ'íle* (desert parsley, or hog-fennel, *Lomatium macrocarpum*), a root that was said to be the father of one of the Salishan Transformers, Child-of-Hog-Fennel (see Teit 1912, 224, 351) (figure 13-2). To this day, the *qʷ'əqʷ'íle* root is associated with fertility and fathering children (Turner, Thompson, et al. 1990, 155).

Viewing plants as sentient, and even spiritual, beings – sharing the same ancestors as humans and with lives as worthy as human lives – requires a different, more respectful way of approaching, recognizing, and especially harvesting and utilizing them. Their role as resources must be reconciled with their role as persona. They must be addressed (as Mary's grandmother did) with appreciation,

13-2 | *Qʷ'əqʷ'íle* (desert parsley, or hog-fennel, *Lomatium macrocarpum*), an edible root that was said to be the father of one of the Salishan Transformers, Child-of-Hog-Fennel.

as generous relatives. Their relationship must be acknowledged and their help requested with supplication and humility, not just assumed.

The Kwakwaka'wakw refer to a standing cedar tree from which a board has been cut as "begged from" and treat it as a sentient being (see chapter 6). For the Nuu-chah-nulth, a similar approach is made to trees – for example, when sought for making a canoe or for some other purpose. "Talk to it like a person. Explain to the tree the purpose, why you want to use it – for the people at home and so on. It may seem like you're praying to the tree, but you're praying to the Creator" (Roy Haiyupis, pers. comm., 1995; Scientific Panel for Sustainable Forest Practices in Clayoquot Sound 1995). Dr E. Richard Atleo (pers. comm. and lecture at University of Victoria, 28 September 2006) added that a canoe builder would come to visit a potential canoe cedar, and the plants living around it, four different times before even touching it. He said, "So much respect is given to all the plants around the tree, even the little plants, before focusing on the tree itself. It is recognized that the tree will participate in the dynamic nature of life; in reciprocity, giving, taking and transformation."

Another example reflecting kincentricity is from Klallam (Clallam), a Central Coast Salish group. In addressing the berries they like to pick, people name them with the first person singular possessive prefix (*nə-*, "my"), a term of affection and thankfulness used when addressing a younger person: *nət'éy'əqʷ* ("my strawberry"), *nəts'qʷ'úʔmə* ("my raspberry"), *nəyíʔx̱əm* ("my blue huckleberry"), *nəpíxʷ* ("my red huckleberry"), and so on (L.C. Thompson and Thompson [1975–80]). Mary Thomas (2001), on many occasions, emphasized the respect people gave to the plants they wished to use – for example, to the birch tree: "You never just went up and cut off the bark. You always talked to the tree, asked its permission to use its bark, and gave an offering of tobacco or similar material as a token of acknowledgement." Once, when she harvested some *tseqʷtseqʷélqʷ* (red-osier dogwood, *Cornus sericea*) bark to demonstrate how it is prepared as a medicinal poultice for sores, she asked us to wait, and then she approached and talked to the bush and offered it some tobacco, just as her grandmother had taught her to do (pers. obs., 1995).[6] Yet another testimony to the close affection people had for their plants is in the Sm'algya̱x (Ts'msyen) name for black mountain huckleberry, *maay*, which means simultaneously "berry" and "great grandmother" (J.A. Dunn 1978).

Other specific ceremonies that involve plants and relate to their spirituality are described in later sections. Perhaps the most important aspect of kincentricity is that it fosters an ethic of stewardship between humans and their environment where humans are responsible for protecting the land and resources that sustain them, as they would their own kin. Humans and nature are part of an extended ecological family that shares ancestry and origins (Martinez 2008; Salmón 2000; Teit 1909; Turner 2005).

PLANTS AS SPIRITUAL ENTITIES

Not only are plants seen in many Indigenous worldviews as kindred to humans, with common ancestries and origins, but they are also seen has having a spiritual or metaphysical dimension, as do all their other relations, human and otherwise. There is no separation of the spiritual dimension from the physical in the worldviews of some (E.R. Atleo 2004, 2011). Nuu-chah-nulth hereditary chief Dr E. Richard Atleo (Umeek) (pers. comm., 2008) explained: "The coincidence of these two dimensions is so complete that everything, including all matter, is alive. There is an absence of inert matter. What is also apparent is that the spiritual dimension is considered the source of the physical dimension. That is why deliberate interaction with the spiritual realm is thought necessary to ensure the maintenance and continuation of necessary provisions. This is the reason 'You give thanks.'"

As entities of the spirit world, plants have innate powers to influence human destiny and fortune. This influence can be negative or positive, depending upon the actions and attitudes of the people who encounter them. Thus plants can bring good fortune and success in hunting, fishing, root digging, berry picking, basketry, or canoe making. Plants can also protect people against evil forces, illness, or even death. On the other hand, if people neglect to show appropriate respect and appreciation in their actions, and if they disregard certain taboos or constraints, plants can cause serious harm to them. The same is true of animals, rivers, and mountains.[7]

TABOOS RELATED TO WEATHER AND PLANTS

Disrespect toward plants, too, can result in bad weather or misfortune, as explained in the quotation cited above from James Teit's (1896–1918) field notes on the Nlaka'pamux about flowers, grass, and other plants being the blanket of the earth, which could cause "rain, fog or bad weather." There are many other beliefs related to plant use that pertain to the weather. For example, the Gitga'at (Ts'msyen) maintain that peeling cedar bark will cause rain, and therefore, people should never undertake this activity until the seaweed harvest has been completed and the seaweed has been dried in the sun and stored away.[8] Picking seaweed in the rain is itself a taboo and would cause continued rain. Helen Clifton (pers. comm., 2001; see also Turner and Clifton 2009) explained further:

> A lot of people like to eat those giant mussels. For some reason Nature always works together with all other forms of life, and so, with the mussels ... you're not supposed to [collect them] until after your seaweed is finished, after your halibut is dried. Because it's going to rain. Same way, you will feel, this is the time of year when the older women go to get their cedar bark, and prepare it for their winter weaving. And, so, they're not allowed to do that until all the food that requires the sun and needs drying is done. Especially the halibut and the seaweed. Because, Nature works with the trees and as soon as you take that bark off ... all the rain will come, a mist will come ... because the tree has a burn on it. So, to heal the tree, to keep the tree alive, a mist will come and the rain will come to soothe the burns on the tree.

In other traditions, certain flowers were known as "rain flowers," and people would not pick them because doing so would cause bad weather. For example, the Massett Haida names for red columbine and blue harebell are *dall(-xil)-sg̲id*

13-3 | Trillium (*Trillium ovatum*), called "sad flowers" (*chaachaawaʕs*, "sad ones on the ground") in Ditidaht. Picking these flowers is said to cause fog and bad weather.

("red rain-flowers") and *dall(-xíl)-gu̲łał* ("blue rain-flowers") respectively; Haida children were warned that picking these flowers would bring rain and make it impossible for people to dry their seaweed (Turner 2004a). For the Ditidaht on the west coast of Vancouver Island, two kinds of flowers, pink fawn lily (*Erythronium revolutum*) and trillium (*Trillium ovatum*) (figure 13-3), are called "sad flowers" (*chaachaawaʕs*, "sad ones on the ground"); picking them was said to result in fog and bad weather (Turner, Thomas, et al. 1983). Interestingly, Nlaka'pamux elder Annie York said that trillium, which was a medicinal plant used for eye troubles, should never be picked, except by a serious healer who truly needed it as a medicine; otherwise, bad luck would befall the one who picked it (Turner, Thompson, et al. 1990).

A Hul'qumi'num version of "rain flowers" is red-flowering currant. Violet Williams and her sister Mary Thomas (not the Secwepemc elder Mary Thomas) (pers. comm., 1991), who grew up around Westholme on Vancouver Island, were warned by their parents that picking these flowers would cause rain. For the Ulkatcho and other Dakelh (Carrier) people, paintbrush (*Castilleja miniata* and other spp.) was known as "thunder flower." One elder explained, "Long ago, when we were children, our parents forbade us to pick Indian paintbrush [called *beuk'enułtsih* or *ditnikʷun*, also 'lightning, electricity'] because it was considered to be sacred" (cited in Poser 2008a, 98). One Ulkatcho woman (pers. comm., 1992) recounted that people had once picked a bouquet of paintbrushes for a funeral, and a big storm ensued. Some Dakelh people just considered these flowers to cause bad luck if picked, without specifying bad weather particularly. Interestingly, paintbrush (*Castilleja* spp.) is also called "thunder flower" in Selish (Flathead), Interior Salish, and Sahaptin.[9]

In the Interior Plateau, people used other plants to "call" or invoke rain at times of drought. During long bouts of hot, dry weather, Okanagan people invoked rain by beating on a wooden log or, for even more effect, a wooden cross in a cemetery with the flowering stalks of stoneseed (*Lithospermum ruderale*), called *kłts'áleqʷtn* ("instrument for beating on wood") (Turner, Bouchard, and Kennedy 1980, 91) (figure 13-4). Ray (1932, 214) stated that the leaves of this plant could cause rain by being mashed and immersed in a basketful of water and that the flowers of the same plant would bring wind if they were pounded between rocks, dampened, and spread out.

Secwepemc elder Mary Thomas (pers. comm., 1998) recalled her own mother breaking off the prickly branches of swamp gooseberry (*Ribes lacustre*) during a summer drought and then dipping these in the water of a mountain creek and calling on the rain to come. Russell Kari (1987, 157) describes a "good luck" plant that grows in the interior mountains of Dena'ina (Tanaina) territory and opens and closes with the sun, suggested to be a poppy (*Papaver* sp.). It was said that people with power could talk to this flower and cause the weather to change. On the Northwest Coast, usually it is wind – or lack of it – rather than drought, that causes problems. The Kwakwaka'wakw had a ritual using sword fern to call the northwest wind, needed for canoe travel from Fort Rupert to Alert Bay,[10] and some Nuu-chah-nulth people used yellow pond-lily leaves – called *hachłsmapt* ("west wind plant") in Hesquiaht – to summon the west wind.[11]

Not only mistreatment of plants but also any disrespectful acts – overhunting, unwarranted harm of animals, killing a fish with a stone, playing with one's food or with parts of a dead animal – are seen as likely to cause bad weather events and bad luck in general. For the Tlingit, "Any abuse of a living being, especially needless killing or harming a helpless creature, or insulting an animal or bird, was likely to bring bad luck" (Emmons 1991, 103; see also Emmons 1991, 427–8; and Turner, Ignace, and Ignace 2000). The relationship between lack of respect

13-4 | Stoneseed (*Lithospermum ruderale*), used by Okanagan and others to invoke rain during droughts.

and bad luck for the perpetrators is reflected over and over again in narratives like the "perpetual winter" story of the Gitga'at and other Ts'msyen peoples and the Ts'msyen story "The Origin of the Name of Chief Mountain" (Cove and MacDonald 1987, 340; see also chapter 12). The Jilaa K̲uns (Creek Woman or Fog Woman) story of the Haida, Tlingit, and Ts'msyen and the Nisg̲a'a story of the Nass Valley volcano are other examples (Chester Morris, pers. comm., 1993; see also chapter 12).[12]

Today, some people see climate change itself as an outcome of disrespectful practices or breaking of taboos: "It's hard to say [what is the cause of differences in vegetation from the past] because the weather has changed *so* much, it's hard to say what's happening to the natural growth ... The old people knew about the weather. I just wonder if they were alive what they'd say about this weather that we're having now ... They would say somebody did something. Somebody *did* something, why the weather is the way it is" (Helen Clifton, pers. comm., 2001; see also Turner and Clifton 2006).

Plants That May Be Harmful

Many other taboos are, or were, observed that relate to the proper, respectful use of plants and other resources and ways of behaviour toward the environment. Another plant that is at times seen as potentially harmful is bracken fern. Although its black, branching rhizomes were a well-known food, especially for Northwest Coast peoples (Norton 1979b; Turner 1995), there are some negative elements associated with their use. If the rhizomes are dug at the wrong time of year (i.e., summer) or by those who are ritually vulnerable (e.g., men who are in seclusion for training), there is a danger – seen among Halkomelem, Straits, Clallam, and other peoples – of being "snaked," or having one's body contaminated with snakes, a condition that can be ameliorated only through extraordinary measures. If you eat too much bracken root, you might be "snaked" (Jenness n.d., 142), or you will become too heavy. In a Klallam story, one girl who ate nothing but bracken fern gave birth to little snakes and was put onto an island (Laurence C. Thompson, pers. comm., 1975). Snake Island near Nanaimo is said to have been named because of the snakes that were stranded there when a man infested with snakes because of digging bracken roots at the wrong time of year was desnaked by his friends (Elsie Claxton, pers. comm., 1994).[13] The Stl'atl'imx believed that eating the central fibres of bracken rhizomes would cause paralysis. The Kwakwaka'wakw believed that only old women should dig these rhizomes; if young women dug them, they would become sick. Bracken rhizomes are generally roasted or pit-cooked, but they contain a number of harmful compounds (Turner and von Aderkas 2009), and the various taboos associated with their use may relate, at least indirectly, to their potential toxicity.

Bracken fern fronds and a number of other plants, including false azalea (*Menziesia ferruginea*) on the coast and smooth sumac (*Rhus glabra*) in the Interior Plateau, are known as materials one should not use to cover berries in a basket because they will "steal the berries," or "cheat you," and you will end up with fewer berries if you cover them with these particular plants (Laurence C. Thompson, pers. comm., 1975; Turner 2004a; Turner, Thompson, et al. 1990). Another plant that is seen as potentially harmful is running clubmoss (*Lycopodium clavatum*) (figure 13-5). Ditidaht children were warned not to pick or handle this plant, which is called *haayaapadatssiʔ* – literally "something (medicine) that gets you confused in the woods" – and is said to bring bad luck (Turner, Thomas, et al. 1983, 60–1, 104). The Nlaka'pamux also considered this plant a "bad luck" plant (Turner, Thompson, et al. 1990). For others, it apparently featured in stories that related its use by Deer and other animals as a belt (see chapter 3).

13-5 | Running clubmoss (*Lycopodium clavatum*), said by some to bring bad luck or confusion if handled.

In the Interior Plateau, one species that was considered particularly sensitive to inappropriate preparation techniques, and therefore subject to a number of different rituals and taboos, is arrow-leaved balsamroot, or spring sunflower. This plant was highly esteemed, as indicated by a "prayer" addressed to "the Sunflower-Root" by young people partaking in their first plant foods of the season: "I inform thee that I intend to eat thee. May thou always help me to ascend, so that I may always be able to reach the tops of mountains, and may I never be clumsy! I ask this from thee, Sunflower-Root. Thou art the greatest of all in mystery." Omission of this prayer was said to "make the person partaking of the food lazy" (Teit 1900, 349; see also Peacock 1998).

While cooking and digging balsamroots, women were to abstain from sexual intercourse. When going out to dig the roots, the women often painted their whole faces red or black or painted a large black or red spot on each cheek. When the roots were being pit-cooked, no men were allowed to come near the cooking pit. Sometimes the women cooking the roots spread out the small ends of four long, thin fir branches in different directions near the bottom of the oven where the roots were and then tied the thick ends together and raised them above the centre of the oven so that they protruded a little from the filled-in oven. When the cooking was finished, these branches were pulled out; according to their colour, the women could determine whether the roots had been successfully cooked. If the fir boughs were black or dark-coloured, the roots would be well cooked, but if the boughs were spotted or light-coloured, the cooking would not be successful. It was sometimes said, when sunflower roots had been cooked successfully, that Coyote had caused success by urinating on them (Steedman 1930; Teit 1900). There were various other rules or taboos against certain behaviours related to pit-cooking, seaweed harvesting, and berry picking. For example, young Nlaka'pamux people, as a rule, did not eat berries until more than half of the crop was ripe (Teit 1900, 349; see also Steedman 1930, 509).

Good Luck Plants

Many plants were also regarded – by virtue of their spiritual powers – as having highly positive qualities and powers to assist people. For example, some were used as special charms to ensure long life, obtain friendship, or bring high status, love, wealth, or success in hunting, gambling, and other endeavours. Brightly coloured flowers such as red columbine, tiger lily, and calypso orchid were widely valued as charms, as were scented plants like wild ginger and white bog orchid, which is said to release special vapours in wet montane meadows (Hunn, Turner, and French 1998; Turner, Thompson, et al. 1990; see also table 13-1). Even bracken fern, identified above for the taboos and potential harm related to its use, brought good fortune in at least one instance. There is a Nuxalk narrative about a woman who dug up a particularly large bracken fern rhizome, which, when it was in her basket, transformed into a writhing snake. This turned out to be the "Mother" of the Bracken Fern, and it gave her extraordinary powers (McIlwraith 1948, vol. 1, 92). Other plants and parts of plants served as guardian spirits for young people at puberty and for others who undertook spirit quests, although animals were more usually identified as guardian spirits. Young women in particular, especially at puberty, often sought the spiritual help and guidance of particular plants.

"Sharp Plants"

One class of plants that has been attributed particularly potent powers for protection and assistance are those with spines, thorns, or sharp parts. Of these, devil's-club (figure 13-6) was, and still is, probably the best known and most widely used plant for spiritual protection, for luck in gambling, fishing, and so forth, and for its overarching health-giving properties (see chapter 7). Devil's-club was a well-known persona in the narratives of Ts'msyen and other

13-6 | Devils'-club (*Oplopanax horridus*), highly respected and widely used in healing and for spiritual protection.

northern coast peoples. In "The Origin of the Wolf House of Hawaao,"[14] Devil's-club appears as a beautiful young woman to a young man named Gwaxskkik, who had lost all of his possessions in gambling. After he had purified himself through fasting and bathing and eating devil's-club stalks for many years, the young woman appeared. He took her for his wife, and she showed him how to use devil's-club to make gambling sticks and rubbed her hands over his eyes, enabling him to become a great gambler. Eventually, he gambled with the man who had vanquished him before, and he regained all of his property.[15] The Haida name for devil's-club is evidently derived from the name for gambling sticks or vice versa.[16] Sometimes, for luck and to obtain supernatural powers, Haida shamans ate devil's-club inner bark together with the bitter-tasting leaves of another plant, single delight (Swanton 1905, 1908).

Bathing in a solution of devil's-club is said to make one strong, and those wishing to succeed in contests of strength or hunting might cleanse themselves by bathing, fasting, and eating devil's-club roots or inner bark (Cove and MacDonald 1987; Johnson 1997; Turner 2004a). Many precise rituals were developed for the proper use of devil's-club, both in general training and especially for hunters. In fact, in Ts'msyen and other traditions, it is Devil's-club as a supernatural being who instructs humans about how to make snares and traps, as well as how to purify themselves before hunting using devil's-club (Cove and MacDonald 1987, 80). Even today, people of the central and northern coast use devil's-club to wash their hunting and fishing gear, as well as for ritual bathing and for a range of different medicinal purposes (see chapter 7).

Although it may not be mentioned explicitly, it seems that the thorny nature of devil's-club is one of the features – along with its potent scent and medicinal qualities – that makes it so desirable as a spiritual protector. Devil's-club charcoal, for example, is used to make face paint, worn by dancers and others at winter dance ceremonials and said to protect one from any bad influences or evil thoughts directed toward them by anyone present. Ditidaht elder John Thomas explained that a person wearing devil's-club face paint is so powerful that one cannot look them in the eyes (Turner, Thomas, et al. 1983). Many people still place devil's-club stalks in the corners of rooms of their house, over the doorway, or under their mattress to protect the household from any kind of negative influence. Powerful animals such as bears are commonly associated with devil's-club and other spiny plants. For example, the black bear is said to use devil's-club as bedding for protection in the night (Boas 1898, 111).[17]

Swamp gooseberry (*Ribes lacustre*), another very prickly shrub, is attributed protective powers similar to those of devil's-club. Its prickly stems can be used in the same way – placed over a doorway or under a person's pillow to give protection against evil thoughts or "witchcraft" being directed toward members of a household. As noted earlier, smoke from burning swamp gooseberry branches is said to deter and protect one against snakes. One Nuxalk man who was infested with snakes was fumigated with smoke of burning devil's-club and gooseberry together and was cleansed of this affliction (McIlwraith 1948, vol. 1, 75). Other prickly species that are used variously for their spiritual protective qualities, whether in winter dances or in times of illness or death, include thistles (*Cirsium* spp.) (Straits Salish), trailing blackberry (Straits Salish, Hul'qumi'num), black hawthorn (Hul'qumi'num), wild rose (*Rosa* spp.) (Straits Salish, Interior Plateau), Oregon-grape (*Mahonia aquifolium*) (Okanagan), Sitka spruce (Central Coast Salish, Ditidaht, Kwakwaka'wakw, and others), and Rocky Mountain juniper (*Juniperus scopulorum*) (Interior Plateau and others) (see table 13-1).

Scented Plants and Other Protective Species

Another class of plants that have particularly strong spiritual properties, and that were therefore widely used for protection and for bringing luck and good fortune, are the scented, or aromatic, plants. These include a number of coniferous trees, like western redcedar, subalpine fir, and Douglas-fir, as well as junipers (*Juniperus* spp.), all of which are also recognized for their aromatic qualities. In all, thirty-five or more different species have been identified as being used in some way in ceremonial protection or for bringing success or good luck in hunting or other pursuits (see table 13-1).

Tree fungi – especially artist's fungus (*Ganoderma applanatum*) and pine conk (*Fomitopsis pinicola*) – were, and are, known along the coast as powerful spiritual entities. They are called "echo-maker" in some languages because of their reflective powers.[18] Not only could they reflect sound, such as echoing someone yelling, but they were also said to reflect any evil thoughts that were directed toward someone back to the originator. People who had tree fungus as their spirit protector could thus use it to maintain their personal safely (Turner, Thomas, et al. 1983). The spiritual and protective qualities of tree fungi may be why the Tlingit used them to carve grave effigies (Blanchette et al. 1992).

Smoke from burning tobacco and some other plants, especially aromatic ones, is also considered sacred because it has the capacity to mediate between the earth and the spirit world (Dr Ron Ignace, pers. comm., 2008). For example, the smoke from burning the seeds of *q'əx̱mín* ("wild celery," *Lomatium nudicaule*) was said to be the food of the Salmon people and was therefore burned ceremonially when the Salmon were first escorted to the Saanich homelands, as described in the "Origin of Salmon" narrative in chapter 12 (Jenness n.d.; Turner and Hebda 2012). For some coastal and interior Salishan peoples, the downy seed fluff of cattails is also considered sacred and was used in burials (Ostapkowicz et al. 2002). It may be that, like smoke and eagle down, cattail down is considered to be a linking agent between the earth and the spirit world.

CEREMONIAL RECOGNITION OF PLANTS

For traditionally trained Indigenous people, there is no real separation of the secular and the spiritual. Plants, animals, people, natural objects, and supernatural entities are not distinct but are all linked to each other and to their surroundings through cultural traditions and interactive, reciprocal relationships (E.R. Atleo 2004, 2011). Every secular pursuit, whether it is hunting, making a basket, or picking berries, has a spiritual side to it. Any given activity thus involves not only practical preparation – assembling hunting gear, harvesting and processing basket materials, preparing berry baskets, and making berry-drying racks – but also spiritual preparation, such as training, ritual purification, sweat bathing, or cleansing. Any harvesting of a plant or animal also requires proper and respectful behaviour. For example, women should not eat berries while they are picking, as was revealed in the Nuxalk story "The Woman Who Befriended a Wolf" (McIlwraith 1948, vol. 1, 691; see also chapter 12). Klallam children were taught, "When you're picking blackberries, don't eat the berries or you'll turn into a bear" (L.C. Thompson and Thompson [1975–80]).

Sometimes, too, relationships between people and the other entities of their environment must be mediated through ritual or ceremonial means to restore and maintain the balance and harmony of people's lives (E.R. Atleo 2004). There were many different ritual practices and ceremonies that related to peoples' relationships with plants. Some – such as formally or informally addressing a plant from which medicine was sought or from which bark, wood, or some other material was to be harvested – have already been mentioned (see Boas 1930). These small-scale and private rites, recognizing the spirit of the plant and the Creator, still take place on an ongoing basis.

The ceremonial complex that came to be called the "potlatch" is one widely known traditional cultural institution in northwestern North America that reflects the general recognition of the sacredness of environmental entities and their ties to human health and well-being.[19] The term is derived from the Nuu-chah-nulth verb *pachitle* ("to give") (Turner and Atleo 1998). The potlatch not only embraces social, economic, and political aspects of people's lives but also reflects the kincentric worldview and other spiritual aspects of life. From an economic perspective, the potlatch and associated feasts incorporate investment systems, as well as helping to redistribute food resources and other wealth. They function to promote the well-being of the community without doing damage to the environment, without violating the principles of balanced stewardship over the land, and in such a way as to render respect and honour to chiefs, nobility, people, plants, animals, spirit powers, and through all of these, the Creator (E.R. Atleo 2011).

Ancient stories, teachings, laws, songs, dances, names, and lineages are emphasized during the potlatch to validate sovereignty over territories, as well as highlighting leaders' and communities' accountability to lands, habitats, and other species. Potlatches have been occasions for the joyful sharing of resources and the renewing of friendships and goodwill, all in the attitude of respect and honour. Also reflected in the potlatch is the institution called *hahuuli* in Nuu-chah-nulth, with comparable terms in other languages, pertaining to hereditary leaders' rights and stewardship over their territories and resources, as well as the responsibilities they incur for the environment, for the resources, and for the people of their communities to sustain all of them wisely and fairly (E.R. Atleo 2011). Ownership and proprietorship of lands and resources often have a strong spiritual basis (Turner, Smith, and Jones 2005). In his address to the Supreme Court of British Columbia in 1987, for example, Gitxsan hereditary chief Delgamuuxw began by outlining the *ayaawx* (ancient oral histories): "For us, the ownership of territory is a marriage of the Chief and the land. Each Chief has an ancestor who encountered and acknowledged the life of the land. From such encounters come power. The land, the plants, the animals and the people all have spirit – they all must be shown respect. That is the basis of our law" (Gisday Wa and Delgam Uukw 1989, 7).

Although each group has its own particular approaches and traditions, the process of acknowledging and thanking the Creator and the plants and animals that provide for people is very widespread. Linda Smith (2008b, 89) explains, from a Tsilhqot'in perspective,

> There is no *Tsìnlhqút'ín* [Tsilhqot'in] term for "sacred" or "spiritual" likely because everything is considered to be sacred. The natural environment stimulates an energy of gratitude, and for people who have lived for centuries in one area, nature provides close connections to one's ancestors. Stones, creatures, and vegetation are sacred, and water from the river and mountains are ancient ancestors. The yearly journey of salmon, our relationship with everything, and all our activities on the land are spiritual. *Tsìnlhqút'ín* life is full of rituals and ceremonies. A hunter begins the day with a purification ceremony, and makes a request for a successful hunt, and calls upon the ancestors. A successful hunt ends with a thanksgiving ceremony where food is shared with the ancestors.

Ceremonial recognition can be as straightforward as simple words of thanks. Mary Thomas's (pers. comm., 2001) mother, Christine Allen, whenever she was given wild strawberries or other kinds of berries, never neglected to hold them up in her hands and say, "*Kʷukstsámxʷ, Kʷukstámxʷ, Kʷukstsámxʷ!*" (Thank you, thank you, thank you!). She did this over her entire life, even when she was in her late nineties. Leaving a small gift of appreciation was likewise a very common practice. If people took hazelnuts, whitebark pine seeds, spring beauty corms, or mountain potatoes, or *masru* ("Indian potato," *Hedysarum alpinum*) from the cache of a mouse, squirrel, or vole, they would also leave a gift, usually of grain or some other edible material, and most said that you should never "clean out" an animal cache but only partially empty it so as not to seriously deprive the animal of its food (see A. Jones 1983). In recent years, a small amount of tobacco from a cigarette is offered as a gift of respect to a birch tree for providing its bark or to a medicine plant for giving up its roots, leaves, or bark for medicine (Mary Thomas, pers. comm., 1995). People will also leave coins or other items of value in the hole where a medicine root is removed or at the base of a tree.[20]

Collecting and administering medicinal plants was, and is, widely considered a sacred act, and usually a very private one, with the ceremonial or ritual aspects shared only with close relatives or apprentices. Usually medicine harvesters went to a distant location, away from human habitats, to collect medicine, even if the same kind of plant was growing closer at hand. The harvesters would often go early in the morning, before having eaten, taking great care to cleanse themselves and remaining focused on their mission to gather medicine and on the patient who would use it.

Not being "greedy" in harvesting medicine was also important. It is still a common practice, especially on the Northwest Coast, to harvest only four pieces of a medicine: four devil's-club stalks, four strips of bark from four individual trees, four roots of medicinal plants, and so forth (Turner and Hebda 1990; Turner and Thompson 2006).[21] When the medicine is prepared, again, special words of thanks are addressed to the spirit of the medicine itself, requesting its help to heal the patient. It is considered bad luck to waste medicine; all of what is prepared should be used, even if it is over a period of days (see chapter 7).

Knowledge of specific medicinal preparations, even involving very widely used plants, may be restricted to a single community, or even to a single family. Many people recall their elders or relatives being very careful not to reveal the ingredients of a recipe, although they were always willing to provide it for anyone who needed it. It is often said that the medicine will lose its power if too many people know about it. Since around the 1960s, however, many plant specialists, like Ida Jones (cited in Turner, Thomas, et al. 1983) and Annie York (Turner, Thompson, et al. 1990), have been concerned that their traditional remedies will be lost and have asked to have the details of their preparation and use recorded. Others continue to practise and pass on these medicines in traditional ways (Sewid-Smith and Dick 1998; Turner 2004a; Turner and Hebda 1990; Turner and Thompson 2006). Always, those collecting and administering medicine stress the strongly spiritual nature of medicinal plants and how important it is to approach them with respect and to ask their permission before harvesting them (Boas 1930; Mary Thomas 2001). Those harvesting basket materials, berries, or roots undertake similar, usually private, rituals to recognize these resources.

Ceremonies Involving Cedar

Western redcedar is a tree that has had particularly strong spiritual associations. For the Kwakw̱aka'wakw, before cutting cedar wood or gathering cedar bark, the traditional harvester would purify himself by washing and would say a "prayer" to the spirit of the tree. One such ritual was undertaken by a man when he wished to have the tree he was cutting down fall in the right direction. He would take four chips of wood, throw them behind the base of the tree, and then say to the tree, "O supernatural one! Now follow your supernatural power" (Boas 1930). Another example is the well-known Kwakwaka'wakw "words of praise,"[22] spoken to a young cedar tree by a woman who wished to take some of its bark for a basket:

> Look at me, friend! I come to ask for your dress, for you have come to take pity on us; for there is nothing for which you can not be used, because it is your way that there is nothing for which we can not use you, for you are really willing to give us your dress. I come to beg you for this, long-life

maker, for I am going to make a basket for lily roots out of you. I pray, friend, not to feel angry with me on account of what I am going to do to you; and I beg you, friend, to tell our friends about what I ask of you. Take care, friend! Keep sickness away from me, so that I may not be killed by sickness or in war, O friend! (Cited in Boas 1921, 619; see also Stewart 1984; and Turner 1998)

Harvesters of cedar bark or of cedar-wood planks were similarly respectful in approaching the trees they wished to use (Turner 1998, 2004a).

The sacred Cedar Bark ceremony, enacted during the Kwakwaka'wakw winter dances, was explained in Dr Daisy Sewid-Smith's narrative about *ts'eqamey* (see chapter 12). At the time of the Great Flood, *ts'eqamey* was told by the Creator to hollow out and enter a giant cedar tree, where he and his family survived, and from which various Kwakwaka'wakw lineages descended (Sewid-Smith and Dick 1998). It was while he was inside the cedar tree that he was instructed about the Cedar Bark ceremony:

> While [*ts'eqamey*] was inside the cedar tree, he was told to weave the first cedar headdress, the first cedar neck ring, wrist rings, and ankle rings. And he was told to put the faces of men all around his cedar head ring and his cedar neck ring, and this was to signify two things, and that was that he was saved, from out of all his tribe, and also that from that day on, he was to honour and respect the cedar that saved his life, and that he was going to be the first one to, ever to perform what we now call the Cedar Bark ceremony. And he was told how it was going to be performed ... And so, when he came out, this is what happened, that through him, the Cedar Bark ceremony spread among all the Kwakwaka'wakw people. (Dr Daisy Sewid-Smith, pers. comm., 1994)

The Cedar Bark ceremony takes place at the beginning of the Kwakwaka'wakw winter dance season each year. A broad ring of specially harvested inner cedar bark, often dyed with red alder bark, is carried into the big house in a solemn procession, encircling a ritually trained woman of high standing who has entitlement to this role, who walks in with it. Everyone in the big house is expected to maintain complete and respectful silence as the cedar bark is taken four times around the fire. It is then ceremonially cut into narrow strips, which are distributed as headbands to each person attending, as a symbolic recognition of this sacred material and the tree from which it came. This cedar bark is so sacred that a mere touch of it has the power to "tame" and purify a *hamatsa* ("wild man") who has been possessed by the man-eating spirit Bax^{w}bax^{w}alanuksiwe (Cannibal Who Lives at the North End of the World). The Hamatsa dance, which can be performed only by initiates of the Hamatsa Society, is a re-enactment of a young man's possession by this spirit, in which he is eventually humanized by being repeatedly approached and then touched with the sacred cedar bark.[23]

Cedar had many other ritual associations and applications (Boas 2002). It was often connected with salmon, and in fact several narratives describe how salmon were actually created from cedar bark (see chapter 12). In the First Salmon ceremony of the Kwakwaka'wakw and others, the first salmon to be caught was laid upon a piece of western redcedar bark during the welcoming ritual (Boas 1935). The Straits Salish laid their first salmon on cedar boughs, as well as sword fern fronds, which are also considered to be a sacred plant for the Saanich (Turner and Hebda 2012).

Spiritual Aspects of Food Harvesting and Food Preparation

Food sharing and giving recognition to the plants and animals that provided food were generally practised within communities and extended families as a matter of course.[24] For some peoples, sharing food and paying tribute to the spirits of the different food species was formalized in annual ceremonies known collectively as the First Foods ceremonies. Most commonly and widely practised was the First Salmon ceremony, mentioned previously in connection with cedar and sword fern, which was held at the time of the first sockeye or spring salmon runs (E. Claxton and Elliott 1994; de Laguna 1972, 105; Maud 1978c). However, there were – and are – also a number of ceremonies that recognize key plant foods. Most Interior Plateau peoples celebrated First Fruits and First Roots ceremonies, along with their First Salmon and First Game ceremonies, to acknowledge and thank the spirits of important fruits like saskatoon berry and black huckleberry and of important roots like bitterroot, spring beauty, biscuitroot (*Lomatium* spp.), and camas at the beginning of their harvest season (see table 13-1).[25] Roy Carlson (1976, 27) suggests that the First Salmon ceremony is "a widespread and very old practice related to first fruit ceremonies all over the world." If he is right, it is likely that the First Fruits and First Roots ceremonies are also very old and reflect attributions of spirits to plants and animals that extend far back into the depths of time.

Alan Marshall (1977, 138) notes that the Nez Perce held a First Fruits ceremony, *qéʔuyit*, for "canyon serviceberries" (a variety of saskatoon berry), which was presided over by a respected old man, who gave thanks for life and food, and in which four highly respected elder ladies directed young women in food gathering, preparation, and serving. Young men supplied the deer and other animal foods for the feast. Marshall (ibid.) describes the social importance for the Nez Perce of these gatherings to celebrate special foods and other resources, noting that they tended to mark the change-over from the resource-poor early spring period to the time of year when there was a profusion of resources. These favoured foods were gathered and enjoyed in plenty, before the work of gathering and storing large quantities of food for winter began. The First Fruits ceremony for the Nez Perce also marked a change in area of resource utilization from lowlands and canyons to upland areas (ibid.).

A Nlaka'pamux First Foods ceremony was described by ethnographer Charles Hill-Tout:

> As far as I could learn, the hunting, fishing and berry grounds of the Thompson were common property. But no one under penalty of a severe punishment could take a fish, pick a berry, or dig a root until after the Feasts of First Fruits had been held ... Messengers are then sent to the neighbouring villages, calling a meeting of the people on a certain day, at which all must attend at the appointed place. When the day has arrived ... the head chief, attended by the other lesser ones and the elders, opens the ceremony at daybreak by a long prayer. (In Maud 1978a, 46)

Hill-Tout also described a rather unusual First Shoots ceremony as observed by the Chehalis (Chilliwack), a Stó:lō group of the Fraser Valley (Harrison River) region:

> Another of these ceremonies [aside from the First (sockeye) Salmon ceremony] was kept in connection with the satske [*sthe?thqəy*], or young succulent suckers of the ... [thimbleberry][26] which the Indians of this region eat in large quantities, both cooked and raw. When cooked, I am told they eat like asparagus. The time for gathering these was left to the judgment and determination of the chief. When ready to gather, he would direct his wife or daughter to pick a bunch and bring them to him; and then, the people all being assembled, a ceremony similar to that connected with the salmon ceremony would take place. After the ceremony anyone might pick as much as he liked. A similar ceremony took place later in the summer, when the berries of this plant were ripe. (In Maud 1978c, 116)

Charles Hill-Tout obviously thought carefully about the meaning of the ceremonies he observed, recognizing the importance of the spiritual aspects of the food species, as well as tying the ceremonies to the origin stories for these foods:

> Concerning the meaning and object of these ceremonies I have been led to the opinion by my studies of the Salish and other tribes that they were always propitiatory in intent. They were intended to *placate the spirits of the fish, or the plant, or the fruit*, as the case may be, in order that a plentiful supply of the same might be vouchsafed to them ... if these ceremonies were not properly and reverently carried out there was danger of giving offense to the spirits of the objects and being deprived of them. The

> myths which speak of the first coming of the salmon ... make this quite clear. In the compact made by *Qäls* [X̱éls, the Creator] on behalf of the Indians and the salmon chief, the Indians are pledged to treat the salmon according to the instructions of this chief ... For it must be remembered that, in the mind of the [Indigenous person], the salmon, or the deer, or the berry, or the root, was not merely a fish or an animal or a fruit, in our sense of these things, but something more. (In Maud 1978c, 116, emphasis added)

The Yakima people, who were said to use eighteen different kinds of berries, of which huckleberries (*Vaccinium membranaceum, V. deliciosum*) were the most important, celebrated a Yakima girl's first basket of berries, along with a counterpart ceremony for a boy's first salmon or deer (Curtis 1911, cited in Kirk and Daugherty 2007, 69). The Secwepemc and others also celebrated the first fruits with a special dance, in which women carrying baskets and branches of saskatoon berry imitated the gathering of berries. The Dakelh (Carrier) and Tsilhqot'in also held a Saskatoon Berry dance, as well as a Salmon dance (Solomon 2006, 20; Teit 1909, 578). In 1847 artist Paul Kane described a special dance performed by the Songhees Coast Salish of Victoria that included ceremonial recognition of camas gathering, an activity he portrayed in one of his paintings, "The Canoes Returning from Gathering Camas to the Esquimalt": "These Indians have a great dance, which is called 'The Medicine Mask Dance'; this is performed both before and after any important action of the tribe, such as fishing, gathering camas, or going on a war party, either for the purpose of gaining the goodwill of the Great Spirit in their undertaking, or else in honour of him for the success which has attended them" (cited in Harper 1971, 101–2).

On the central and northern coast, harvesting and food use were also ritually enacted, and the availability of particular foods was recognized and celebrated in feasts.[27] The Hanaksiala of the Kitlope Valley celebrated the New Year (*h'isdlàm hs h'snx*) around the end of March when the northern riceroot flowered. They performed a ceremonial Flower dance in which "costumes were covered with flowers of the Nootka rose, salmonberry [figure 13-7], blueberries, riceroot, and any other plants that were blooming then" (Compton 1993b, 197). This time of year was known as *q^{w}'àxilaqus* ("growing month") and *zàx^{w}ilaqus* ("oulachen month") and was associated both with the Flower dance and with the time of the oulachen harvest (Lantz and Turner 2003).

As well as the First Foods ceremonies, there were various rituals and observances that related specifically to food harvesting and preparation. For example, it was customary for Nuu-chah-nulth root diggers to paint their faces with specific designs before harvesting particular kinds of roots, such as springbank clover, Pacific silverweed, and camas bulbs, as well as before whaling, seal hunting, and medicinal plant gathering, as a part of their spiritual and physical

13-7 | Salmonberry (*Rubus spectabilis*) flower, one of the blossoms featured in the Hanaksiala Flower dance (Compton 1993b).

preparation, called *oosumich* (E.R. Atleo 2004, 2011; Pukonen 2008; D. Thomas 2000). Without this careful preparation, a woman might encounter only small and poor-tasting roots.

Even everyday experiences with food often took on ceremonial aspects. Chief Earl Maquinna George (2003) recalled from his childhood at Ahousaht how, when a couple of women went out to harvest thimbleberry shoots, as they approached the village, they called out, and all of the women in the village would come down and join in a "feast" of these fresh springtime foods.[28] The same type of spontaneous feast occurred for the Kwakwaka'wakw when people brought a canoe-load of fresh eelgrass back to a camp or village (Clan Chief Adam Dick and Dr Daisy Sewid-Smith, pers. comm., 2009). Soapberries were also celebrated with feasts and special songs that recognized this valued and important food.

Some of the ceremonies surrounding food harvesting and use go back to ancient times, with underlying narratives, as noted by Charles Hill-Tout (see Maud 1978c), describing how the foods were first provided or given as gifts. For example, a Sinixt (Lakes Okanagan) story describes how the Mountain Goat brothers first gave the Eagles a gift of black huckleberries, which have been a treasured food ever since (Turner, Bouchard, and Kennedy 1980, 103). Rituals for food harvesting and use also sometimes represented engrained systems of resource management (see chapter 11).[29] For example, some First Fruits ceremonies, such as the First Huckleberry ceremony of the Klamath, included tossing berries to the ground as a ceremonial way of "giving back" to Mother Earth, which probably helped to perpetuate the berry populations (Douglas Deur, pers. comm., 2008).

Not only foods but also important material plants were celebrated and recognized ceremonially. For example, John Ross (2011, 248) describes a special ceremony preceding the actual cutting and gathering of cattail and tule by Spokan women: "This ritual was led by an elderly woman who was recognized as being the most knowledgeable of such harvesting, and, prior to cutting and gathering, she would give an oration of gratitude for the resources as well as recognize the presence of all the women in attendance ... votive offerings were once given for each person's harvest, but prayers varied with each individual." Ross also notes that while gathering tules the women often sang songs, directed by this recognized elderly woman, who also instructed young girls and helped them with their first harvesting efforts. These songs were about the enjoyment the cattail and tule mats would bring to others. The older women who helped to direct the ceremonies and songs were recognized later with gifts of finished mats (ibid.).

Protocols and Ceremonies for Ritual Protection

A number of rituals relate to protection of individuals against illness, death, and other negative influences, using in particular the groups of spiritually powerful plants mentioned previously: the "sharp plants" and the strongly aromatic plants. For example, Central Coast Salish and Wakashan peoples have commonly used Sitka spruce branches to protect against illness. Following a death, the Klallam ritually "beat" the bed and walls and the arms of surviving relatives with *sqolp* (spruce branches) to drive away any residual illness (Laurence C. Thompson, pers. comm., 1975). In the Interior Plateau, Okanagan "Indian doctors" used branches of wild rose and tall Oregon-grape during funerals to "sweep out" the grave before the coffin was lowered in and to prevent others from being drawn into the underworld with the one being buried (Turner, Bouchard, and Kennedy 1980, 131, 87). The Okanagan have also used these plants and stinging nettle to combat witchcraft (ibid., 142).

Juniper boughs were, and still are, widely used for protection, both by the Coast Salish (using the maritime species, *Juniperus maritima*) and in the Interior

Plateau (using Rocky Mountain juniper, *J. scopulorum*). Farther south, the Sahaptin, Great Basin, Southwestern, and Californian peoples used western juniper (*J. occidentalis*) and other species in similar ways. In Alaska and elsewhere, common juniper (*J. communis*) was used: "Some people toast the leaves on the back of the stove in a room where there is a sick person so the rest of the family won't get sick" (A. Jones 1983, 110). Solutions of juniper were used to wash the clothing, bedding, walls, floors, and possessions of a recently deceased person in order to protect bereaved friends and relatives from their spirit. People placed juniper boughs in the fire or on a hot stove as a household fumigant for the same purpose, and juniper solution was poured across doorways and pathways to the house and outbuildings, and along window openings, to prevent the spirit of the deceased from re-entering the house (Hunn, Turner, and French 1998; Turner and Bell 1971; Turner, Bouchard, and Kennedy 1980; Turner, Thompson, et al. 1990).

Mountain valerian is another strongly scented plant used ceremonially. For example, a Nlaka'pamux hunter might prepare a decoction from the whole plant of mountain valerian to wash his gun and other hunting equipment. He would also drink the solution as a tea and wash his skin with it, often accompanying bathing or sweat bathing. Through this practice, the hunters believed that the deer would become "tame" and easier to approach (Steedman 1930). Another example of the ceremonial importance of aromatic plants was the widespread use of Douglas-fir boughs in puberty rites, hunting rituals, and other ceremonies (Teit 1900, 1906a, 1909; Turner, Thompson, et al. 1990, 109–10).

Smudging and Decontamination

Smudging, mentioned above in connection with devil's-club and juniper, is evidently a practice of some antiquity both on the coast and in the interior. Although it may be better known in association with the plains and eastern First Nations' pipe ceremonies, and their burning of sweetgrass (*Hierochloe odorata*) braids and sage (*Artemisia tridentata, A. ludoviciana*) bundles, northwestern North American peoples also ceremonially burned different plant materials and used them as an incense for spiritual protection. The Selish (Flathead), Ktunaxa, Sahaptin, and some other Interior Plateau peoples sometimes used sweetgrass, and even the Nuxalk on the Pacific Coast used sweetgrass (*H. hirta*) ceremonially, going back at least to the 1800s, if not earlier (Dave Moody, pers. comm., 1972; Sam Moody, pers. comm., 2005).

Kwakwaka'wakw hunters routinely smudged any new implements with the smoke of burning spruce boughs to cleanse them (Clan Chief Adam Dick, pers. comm., 2009). The seeds of "wild celery" (*Lomatium nudicaule*) are mentioned in a Klahoose narrative as being burned as a disinfectant (Boas 2002, 223) and are still well known for this use among the Central Coast Salish.[30] False hellebore root was, and still is, used to smudge and fumigate a house in order to ward off suspected evil influences, as well as for ceremonial cleansing of implements such as fishing gear. This root was also used as a "charm" – a piece of root kept in a small pouch hung around the neck or placed in the gear box of sea hunters and fishers – to protect a child or a hunter from illness, accident, or any kind of negativity (see Turner and Bell 1973).

As well as being used in smoking rituals, tobacco (*Nicotiana attenuata*) smoke – and smoke from burning other powerful and fragrant plants like yarrow – was, and is, used for ritual smudging and protection against illness, for purification, and for bringing good luck and fortune. Smudges of yarrow, sagebrush (*Artemisia* spp.), and other plants are also used for repelling mosquitoes and other biting insects (see chapters 6 and 7). Canby's lovage root was smoked ceremonially, at least by the Secwepemc and Spokan (Aimee August, pers. comm., 1992; Mary Thomas, pers. comm., 1998; J.A. Ross 2011; Turner 2009), and the toasted leaves of kinnikinnick were widely smoked, both recreationally and ceremonially (although apparently people on the Northwest Coast did not start smoking them until after Europeans arrived; see chapter 4). Sumac (*Rhus glabra*) leaves were smoked with kinnikinnick and native tobacco in some Secwepemc ceremonies (Lilly Harry, pers. comm., 1990). Within the historic period, commercial tobacco gradually supplanted the ceremonial role of native tobacco and kinnikinnick (Hunn, Turner, and French 1998).

Ritual Scrubbing, Sweat Bathing, and Cleansing

Use of tree boughs and other plant materials to scrub the skin during ritual bathing was very common. For example, western hemlock is called "scrubber plant" in the Nlaka'pamux language (Turner, Thompson, et al. 1990) since its boughs were commonly used for this purpose. Young men at puberty generally went through a period of "training" that involved seclusion, fasting, and bathing. Sometimes the scrubbers they used were graduated, ranging from very soft willow roots for young boys, for example, to western hemlock or Douglas-fir boughs and finally to western redcedar or yew boughs, which are very tough and could be used only on skin that was well conditioned (Boas 2002, 239; Turner, Thomas, et al. 1983). Hunters and fishers also bathed and scrubbed themselves with these and other plant materials for strength and purification and to bring them success in their endeavours.

Use of plants and fungi in their supernatural or spiritual roles is generally considered very private knowledge and is not widely shared. In fact, as with some medicinal remedies, many people believe that if others know about such protective rituals, they will lose their effectiveness. Each individual and each family has had its own particular practices involving plants and other charms that they might use for gaining luck and protection, and these go together with their own rituals, songs, and ways of invoking spiritual protection. This privacy itself is very widespread and represents a general consensus across many different languages and cultures.

Many plants (some of which could be considered toxic if taken in excess) were drunk in the form of infusions or decoctions for their emetic and/or laxative qualities (see chapter 7) to purify and cleanse the body internally. This was often undertaken in conjunction with spiritual training and external cleansing through bathing and scrubbing. Those seeking special protection or strong powers for hunting, fishing, plant gathering, or healing would often undertake such treatment. For example, an infusion of tobacco was drunk as an emetic after the death of a close relative (Hunn, Selam, and family 1990, 230).

These rituals involving bathing, internal cleansing, or smudging and washing implements with aromatic plants or their solutions to bring luck and protection may well be related at some level to the capacity for masking the human scent on hunting and fishing gear or on the skin of hunters and fishers handling the gear. In this case, the role of aromatic plants in increasing the success of hunting and fishing would be profound in terms of survival. The spiritual protection provided by aromatic plants such as juniper and wild rose was – and is – no less important for peoples' well-being and overall health than other, more physical protection. Ritual and ceremonial practices often embrace ancient knowledge metaphorically and in ways that are deeply meaningful within a culture, leading to overall respect and conservation of critically important species and habitats (Turner and Berkes 2006).

Guardian Spirit Power

The concept of individuals seeking a special guardian spirit to assist them in all of their endeavours is part of an overall belief system, especially expressed in Salishan cultures but also recognized more widely in various practices. In these traditions, the guardian spirit is said to be responsible for all of one's achievements and successes in extraordinary ventures or in everyday subsistence activities, whether gambling, hunting, root digging, or basketry. Any individual – especially young men and women at puberty but also anyone else who aspires to excellence in a particular feat or vocation, such as being a powerful shaman, a successful hunter, or a skilled basket weaver, root digger, or berry picker, as well as anyone who simply aspires to gain luck and good fortune – can undergo purification and associated training and seclusion. This generally involves ritual bathing, scrubbing with tree boughs or other plants, fasting, prayer, rigorous exercise, and sometimes consuming medicines such as devil's-club bark or other plants. If all goes well, individuals will have a vision of their guardian spirit, which they will hold and carry with them, usually for their entire lives (H. Robinson and Wickwire 1994; J.A. Ross 2011).

Charles Hill-Tout (in Maud 1978c, 49) described how a Halkomelem shaman would seek his *sulia* ("guardian spirit powers") through training: "To acquire these [shaman's] powers, he usually underwent a long and secluded training in some lonely spot in the forest or on the margin of some lake. This training consisted in prolonged fasts, trances, body washings and exercises, accompanied by invocations of the mysteries. This medicine, or power, it is believed, was bestowed upon him by his guiding spirit, or spirits." The guardian spirit may have been a particular plant or botanical object, such as a type of flower or root, an animal, or some other feature, and could be represented with a spirit drawing, with a special symbol or design on the face, on one's digging stick or drinking tube, or on wood or rock, or with a song, a dance, a crest, or a totem pole figure. Hill-Tout (ibid.) emphasized that any animal or object could become one's guardian spirit, noting: "This is a fact of special significance, clearly showing that every object in Nature, animate or inanimate, possessed for them active and sentient powers and qualities."

For the Haida, a similar situation was noted by Swanton (1905, 29): "Besides bird and beast powers the existence of supernatural beings in the trees ... was recognized ... It would appear that not only animals and trees, but bushes, sticks and stones, were alive with spirits; for when one of the great heroes is about to break a taboo or commit some similar error 'everything in the forest cries out to stop him.'" As explained by Vickers (2008, 67) for the Nisga'a and other Northwest Coast peoples,

> All – humans, plants, the land, supernatural beings, are alive and intimately related to each other – impacting each other. The continued use of carved and painted crest designs in Northwest coast art in the form of totem poles, chief screens, ceremonial pieces such as masks and button robes that give account of supernatural encounters or an individual's origins, practical implements such as soapberry spoons, ladles, feast bowls, vests, capes, jackets and shawls remind us of the continued connection between humans, the land and the supernatural world.

Some keep the identity of their guardian spirit secret, whereas others might reveal it publicly on a ceremonial occasion. In any case, guardian spirit power becomes part of one's social standing and status and is often linked to special skills and success in particular tasks.

Puberty and Other Ritual States and Plant Use

At certain times of one's life, or in certain conditions, a person may be ritually more powerful or more vulnerable, and at these times, special precautions must be taken. Puberty, especially for girls, is one of these times. A young woman at puberty or during her menstrual period was considered highly potent and capable of damaging hunting, fishing, or harvesting gear or causing misfortune for others. Therefore, she was usually kept in seclusion and observed a range of taboos and protocols, avoiding particular foods (Laforet, Turner, and York 1993; L.R. Smith 2008b; Teit 1909; Mary Thomas 2001) and practising skills she might use in her adult life. Interior Plateau women were usually housed in lodges of Douglas-fir boughs, constructed at some distance from other habitations, for a major period of time during their puberty, as well as for a few days each month during menstruation. The only people they encountered during this time were older women relatives, who would visit and instruct them on what they needed to know in order to participate fully in the community as food harvesters and processors, as weavers of mats and baskets, and eventually as mothers. Teit (1909, 587) described the puberty rituals for a young Secwepemc woman:

> When a girl had her first menses, she was considered a "mystery," and had to live in a small lodge apart from the people. The lodge was conical in shape, and made of fir-branches placed very thickly together, or of mats ... She wandered abroad at dusk, and spent every night in the mountains roaming about exercising herself and praying. About daybreak she bathed in running water, washing herself with fir-branches, and praying to the Day Dawn. She had to be back in her lodge before it was properly daylight. If she had wandered far and daylight overtook her, so that she might possibly meet people on her way home, she made a veil of fir-branches to hide herself; for no one, excepting her instructor or nearest relatives, was supposed to see her face during her training ... While out at night, girls practised running, climbing, carrying burdens, digging trenches, the last so that in after years they might be expert at root-digging. Each morning they brought home a small load of fresh fir-brush or fire-wood to their lodge.

The girl also carried a small birch-bark basket, decorated with dentalium shells. "Inside her lodge she suspended grass, bark, rushes, and other materials that she had gathered for the making of miniatures; strings for threading seeds on, that she might become an adept at stringing roots; fir-branches, from which she plucked the needles in her spare time, that she might learn to move her fingers nimbly; and pieces of skin, on which she practised tanning and dressing" (ibid., 588). In general, girls of the Interior Plateau during their puberty seclusion kept themselves busy weaving baskets or digging trenches as preparation for serious root digging, and the parallel lines found on many digging stick handles may signify these trenches (M.B. Ignace, Ignace, and Nord 2009).

Connected to a girl's puberty rituals were associated rites of passage such as weaving her first basket or completing her first harvest of roots. Strong (1959,

180) reports that "a digging stick handle was made for a young girl [of the Columbia Plateau] when she was first able to help with the food gathering." John Ross (2011, 333) reports for the Spokan, "a rite of passage was always recognized when an older girl received her first digging stick. After digging her first root, usually balsamroot, it was presented to an elderly person, who offered a prayer for her continued success." For the Sahaptin, Hunn, Selam, and family (1990, 209; see also D.E. Walker 1998, 273) noted that a young woman presented with a digging stick and with bags and baskets when she started harvesting was expected to give these objects away to the elder who helped her during her puberty rites but that they would soon be replaced by return gifts from the elder. Nlaka'pamux elder Mabel Joe (pers. comm., 1985) remembered making her first basket as a girl and being told by her mother to fill it with black huckleberries when they went berry picking near Coldwater and then to give the berry-filled basket away as a gift to her uncle. She remembered feeling proud to present this gift and was well praised for her generosity.

Root digging and berry picking also signified important rites of passage for Nez Perce girls, and a girl's first basket of roots or fruit included a large tribal celebration and feast recognizing her transition into womanhood. Root harvester Gwendolyn Carter recalled that a girl gave her first-dug camas bulbs away to someone important in her life, such as a grandmother, aunt, or tribal elder (cited in Gilman and Ronda 2003; see also Driscoll 2003). On such an occasion, at the end of her puberty training, a young woman might have her face painted or have tattoos placed on her arms and wrists, signifying her new status as a harvester (Teit 1930, 404; see also Maud 1978c, 55). Pictograph images might be painted on boulders and cliffs to mark the occasion as well, some of which represented fir branches and related objects (Teit 1930, 404–5). Some of the images found on digging stick handles are the same as those found in rock paintings and in basketry, and it is said that young women were given their digging sticks after puberty (M.B. Ignace, Ignace, and Nord 2009; D.E. Walker 1998, 273).[31]

Boys at puberty – when their voices changed or "when they commenced to dream of women, arrows, and canoes" – also underwent training and instruction but as warriors, hunters, and runners (Teit 1909, 589). From two to three days to many weeks at a time, they moved away from the community to seek a guardian spirit or other supernatural power and trained for their chosen profession. These periods of training could occur over eight or ten years. They fasted and undertook a regime of purification and sweat bathing. They wore special clothing and spent long hours running, jumping, climbing up mountains, shooting arrows, and other physical training, along with ritual bathing, internal cleansing (using emetics and laxatives such as decoctions of soapberry or Oregon-grape twigs), cutting their bodies to help them withstand pain, and praying. They also undertook contests, like walking in shoes made of balsamroot leaves, to test their ability to tread softly and carefully. Or "They tied long fir-branches to their feet, or made rough snowshoes of fir-branches and ran in shallow water with them on, praying to the Day Dawn that in after years their limbs might be strong and tireless when they walked on snowshoes, and that their shoes might not sink deep into the snow" (ibid.). Both boys and girls of the Interior Plateau at puberty often wore bark headbands of willow (*Salix exigua* or other spp.), silverberry, or western redcedar.

For Interior Plateau peoples, including Secwepemc and Tsilhqot'in, twins are ritually powerful, and both twins and the parents of twins, at least in the past, underwent a number of ceremonial observances, some related to plant use, to protect them and to protect others from their powers. Twins were not allowed to be around other people, except their parents, for a period of four years after their birth. The entire family lived in a conical lodge of Douglas-fir boughs, and their father or mother washed the babies daily with fir boughs (Teit 1909, 586–7). Twins also were said to have particular powers over the weather. For example, it was thought that if a twin bathed in a lake or stream, this would cause rain; thus twins were washed by dipping fir branches in a basketful of water and rubbing these on the skin or by using a dipper to pour water over themselves.[32]

For the Tsilhqot'in, it was believed that twins should never harvest anything in the normal course of events because they could ruin a harvesting site, like a good berry patch, due to their supernatural energy. However, because of their powers, twins could also lend themselves to certain ceremonies. Their hair could be used ritually to enhance the growth of berries and increase the harvest of an area. To increase the growth of spring beauty corms, or mountain potatoes, people would have twins plant one of the corms, or they would take the hair from a twin and tie it around the stem of the plant. The powers attributed to twins also extended to their parents and families; everyone had to be careful around twins (Mabel Solomon and Gilbert Solomon, pers. comm., 2003; L.R. Smith 2008b).

For the Spokan and others, a "two-spirited" person (a male with female qualities) was also considered to have special powers and might become a shaman and perform important rituals around plant harvesting. For example, a two-spirited shaman who held bear power would have special knowledge of locating and harvesting black huckleberries, and one with weasel power would be asked to help women to collect tules and cattail leaves. In the case of tule harvesting, a two-spirited person with water power might direct a ceremony following the tule harvest, in which unharvested tules were burned over to ensure an abundant supply the following season. He would first sing his power song, joined by all of the women attending, and then loudly recite a prayer of thanksgiving (J.A. Ross 2011, 249).[33]

Certain other circumstances, such as encounters with animals like bears or wolves, can make one more vulnerable or subject to harm. An example is from the Okanagan narrative "The Bear Woman" (Teit 1912, 90–1), in which a woman was kidnapped by a bear while she was out digging roots. Her husband came to rescue her from the cave where she was living with the bear. "Her face was painted and she carried her root digger." Her husband killed the bear and restored her, but she would have died if he had not "rubbed her with fir-tops and veratrum [false hellebore, *Veratrum viride*]." Even after this treatment, she needed to be isolated from the other people of her community, and she instructed her husband to build "a lodge of fir brush for me some distance away from the people."

Resource use taboos, such as the avoidance of certain foods at certain times of the year or under certain conditions, occur widely across the cultures of northwestern North America but seldom pertain to the use of plant foods, except perhaps when fresh berries are avoided by girls at puberty in some circumstances (Laforet, Turner, and York 1993) or when certain foods are harvested by widows or others during a period of bereavement. Colding and Folke (2001, 584) note that informal institutions such as social taboos (e.g., food taboos pertaining to menstruation and puberty) are seen in many traditional societies throughout the world and frequently guide human conduct toward the natural environment. Other taboos, such as those noted previously surrounding the use of bracken fern or the avoidance of eating berries while picking, may be imposed because of seasonal inedibility or potential harm, in the case of bracken rhizomes, or for increased harvesting efficiency, in the case of berry picking. Still others – like never picking seaweed in the rain or when it is under the water at mid or high tide – may relate both to the taste of the seaweed and to concerns about the safety of the pickers (Helen Clifton, pers. comm., 2003).

Bereavement

The loss of a child or spouse placed one in a ritually vulnerable state and generally required special purification rituals to be undertaken. Depending on the area, these might involve bathing and cleansing with aromatic plants such as juniper (*Juniperus scopulorum*), Douglas-fir boughs, or sagebrush (*Artemisia tridentata*) or with prickly plants such as wild rose or spruce boughs, as described previously, to protect one from being haunted by the ghost of the departed one. A bereaved parent or spouse might also be banned from eating certain foods and might take only "medicine" made from herbs (Maud 1978c, 54).

Linda Smith (2008b, 142) describes the Tsìnlhqút'ín (Tsilhqot'in) concept of *nímính*, which includes at least ten ritually powerful and vulnerable categories, including birth, puberty, motherhood, widowhood, and death. During any of the *nímính* states, one was considered "different" and might be set apart or emotionally, if not physically, isolated, as though having entered a different world.

These states and the different routine they invoke also allow people to contemplate and renew and reorient themselves. "From this perspective," Smith surmises, "abiding by these ancient laws is a privilege." Individuals carrying the energy of *nímính* must take care of themselves, and protect others, in particular ways, such as respectful observances in relation to game animals and fish, including the gear used to obtain these resources. Animals and fish were always treated with particular respect, but handling fresh meat and fish was believed to be actually "dangerous" if done by one who was *nímính* (ibid., 70–1). The energy of *nímính* also affects wild plants, berries, and even garden vegetables, and any negative impacts must be ritually ameliorated. For example, *nímính* can cause saskatoon berry and soapberry bushes to wither and the berries to fall off of the bushes. Wild roots, corms, and garden vegetables do not grow well under the influence of *nímính*. During their menstruation, for example, some women will wear gloves when they pick berries so as not to harm those who will eat the berries. As a protective measure, they will also tie pinegrass around four branches of any berry bushes from which they are gathering, saying "become plentiful" to the bushes in the Tsilhqot'in language: "*Lhàn tsínsh nághùndláx*" (ibid., 133).

A grieving widow (in the state of *nímính*) was instructed to go to Potato Mountain or other places, pull out the stems of spring beauty, or mountain potato, after they had finished flowering, and throw them around in places where they were not growing. This was said to keep these plants plentiful (Mabel Solomon and Gilbert Solomon, pers. comm., 2003). There have been occasions when women did not heed the taboos and failed to follow the required rituals, with resulting long-term damage to the berry bushes (L.R. Smith 2008b). Other plants, including common juniper and false hellebore, can also be used to ritually treat berry bushes, fishing gear, and people who have been negatively affected by the *nímính* energy (L.R. Smith 2008b, 135).[34]

TRANSFORMATION: PLANTS AS GIFTS AND SYMBOLS OF SPIRITUAL LIFE

The spiritual nature of plants – and their interconnectedness with humans and other animals, as well as with rocks, water, and objects that are, in a Western worldview, thought of as inanimate – meant that they could readily transform into other states and other dimensions. The Nuu-chah-nulth phrase *h̲ishukish tsawaak* ("everything is one") effectively encompasses the concept of the interrelatedness of all things (E.R. Atleo 2004, 2011; Scientific Panel for Sustainable Forest Practices in Clayoquot Sound 1995). The profound implications of this concept are reflected in the notion of kincentricity and in the innumerable transformations and conversions that are recounted and reinacted in the narratives and ceremonies, as described in this chapter and in chapter 12. In fact, the entire concept of grass and flowers being the "earth's blanket" is a transformative one.

Not only are plants seen as related to humans, but – at least in the past – they could also undergo transformation from a plant state into a human state or from a human state into the physical form of a tree or other plant. The dramatic dance performances of the Kwakwaka'wakw and other Northwest Coast peoples, such as the Hamatsa dance and the Atlágimma (Spirits of the Forest) dance, with their masks, rattles, and spiritual regalia, represent and reinact transformations and interconnections between species and ecosystems through the medium of the sacred (Dubin 1999). Plants and animals can be either direct players in such transformations or intermediaries that assist in the processes.

Narratives of Transformation

The stories and beliefs around the association between the song and actions of Swainson's Thrush and the ripening of salmonberries and other berries, as described in chapter 12 (see Boas 2002, 117, 177), are one example of the transformational attributes of plants, animals, and people. The case of Swainson's Thrush involved a temporal transformation as well, with Thrush being able to expedite the ripening of salmonberries and highbush cranberries, even in the middle of winter. Other examples from stories are the origin of salmon from cedar bark through the actions of individuals like Beaver and Woodpecker (e.g., Boas 2002, 129), trees turning themselves into humans, humans being transformed into trees, herring originating from yellow-cedar bark or redcedar boughs (Teit 1912, 223, 285; Boas 2002, 147, 310), Pitch impregnating women (Boas 2002, 227, 287), Raven transforming himself into a hemlock needle (Boas 2002; Turner 2004a), Raven converting the pod of giant vetch or beach pea into a canoe (Turner 2004a), and Coyote transforming the stalks of giant wildrye grass into dentalium for a necklace (Teit 1912, 307).

There are instances, too, of important plants being spread or introduced by spiritual beings, such as in the Nuxalk story of soapberry whip, scattered by Raven, transforming into soapberry bushes (McIlwraith 1948, vol. 1, 88) and the Ktunaxa story of Rocky Mountain juniper growing everywhere from pieces of "bow wood" being scattered around (Boas 1918, 109). The gift of her own blood, transformed into edible berries (bunchberry or blackberry), by the woman stranded up a western redcedar tree by her jealous husband in "The Revenge of the Brothers" narrative is another example, as is the transformation of an old woman into a wood fern (*Dryopteris expansa*) plant, to be used as food for people (Boas 2002). The gift of abundant root vegetables in Botanie Valley in Nlaka'pamux territory, as described in a narrative recorded by Teit (1930, 477), is another case in point: "Their abundance [edible roots] is accounted for by the

Thompson River tribes as due to a powerful woman who lived at Lytton. She was taken away by a great chief, some say the sun. She wanted to leave provisions for her people so she dropped edible roots at Botanie [Valley] saying that, 'roots will grow in abundance in this place and all my children shall repair here to dig them.'"

The well-known and ancient theme of culture heroes throwing various objects behind them while fleeing from a monster, and these objects turning into barriers to thwart the pursuer, provides another example of transformations: a magic comb growing magically into a crabapple thicket and a small cedar stick transforming into a giant log (Boas 1921, 2002). Another theme involving transformation is in the Haida story where a single highbush cranberry expands by magic into an entire meal for a supernatural hero (Swanton 1905), and yet another, in a Nlaka'pamux tradition, is when black currants (*Ribes hudsonianum*) dropping into the water become trout (Turner, Thompson, et al. 1990). Perhaps the ultimate in transformations is the desert parsley (*Lomatium macrocarpum*) root's fathering of a boy, named *q*^{w}*'əq*^{w}*'íle* after his father, who eventually became one of the Transformers in Interior Salish traditions, travelling around the country with other supernatural beings and changing many different objects, plants, and animals into their present form (Teit 1912; see also chapter 12).

The ability of plants to grow magically in other worlds – like the camas, spring beauty, western redcedar, and Indian-hemp in the sky country of the "Star Husband" tales and related themes and like various species in the underworld – is worthy of note. Plants were involved in transformations in these places as well. For example, in some traditions, as noted in chapter 12, waxberries are seen as the saskatoon berries of the spirit people in the land of the dead (Kennedy and Bouchard 2010), and in other narratives, the eyes of dead people are crabapples. Perhaps even more compelling, however, is the mediating role that many plants take in linking between worlds. Some of the heroes of ancient stories travelled to the undersea world by following the stalk of a two-headed bull kelp (Turner 2004a). Heroes like Mink travelled to the sky country by climbing up a chain of wooden arrows shot into the sky. Roots like camas bulbs, pulled up by those in the sky country, revealed the earth country below through the openings they left. And being lowered – with or without a basket – from the sky country back to the earth on a fibrous rope of cedar bark, nettle fibre, or Indian-hemp is a common means of re-entry (see chapter 12 for specific examples). In a Tlingit version of the story about the boy who ascended to the sky on a ladder of arrows, "The Boy Who Shot the Star," the amazing transformational qualities of berry bushes are featured:

> [The boy who shot arrows and ascended to the sky] took various kinds of bushes and stuck them into the knot of hair he wore on his head. He

> climbed up his ladder all day and camped at nightfall upon it ... When he awoke early on the second morning his head felt very heavy. Then he seized the salmon berry bush that was in his hair, pulled it out, and found it was loaded with berries. After he had eaten the berries off, he stuck the branch back into his hair very much strengthened. About noon of the same day he again felt very hungry, and again his head was heavy, so he pulled out a bush from the other side of his head and it was loaded with blue huckleberries [blueberries, *Vaccinium alaskaense* or *V. ovalifolium*]. It was already summer there in the sky. That was why he was getting berries. When he resumed his journey next morning his head did not feel heavy until noon. At that time he pulled out the bush at the back of his head and found it loaded with red huckleberries. (Swanton 1909, 210; see also Thornton 1999)

In all of these instances, plants serve as mediators across spiritual barriers or boundaries. As noted previously, smoke from burning tobacco, "wild celery" seed, false hellebore, or other types of incense and cattail seed fluff can also assist in communications with the spirit world or sometimes help to ward away negative or evil spirits from a house or a vulnerable individual, which is why they are considered to be sacred.

From Living Plant to Art for Living

One of the most compelling types of transformations, also involving careful training and assistance from the spiritual realm, is the conversion by skilled artisans of raw plant materials like cedar bark, spruce roots, and cedar wood into baskets, canoes, and other objects, at once beautiful and utilitarian. In a predominant worldview from the Northwest Coast and Interior Plateau, baskets, canoes, digging sticks, and hunting gear were regarded as having their own personalities, their own consciousness, their own spirits, and their own ability to bring "luck." Like the plants from which they were derived, they were said to have powers and capabilities to influence human affairs. People could talk to them and ask for their help, as they would from a relative. Canoes, paddles, and baskets were often given names. They were seen to have a life of their own, and their first use was celebrated ceremonially. Saanich elder Dave Elliott (E. Claxton and Elliott 1994) described how a paddle was treated at the end of the harvesting season: "When November comes around, people would put away their paddles. They would ceremonially hold their paddle up in the air and speak to it: 'We'll travel again together.'" Having respect for and recognizing the agency of these items are part of the predominant worldview of relationships and interconnections. Similarly, as noted by John Ross (2011, 250, 333), "Even when working with other women, a [Spokan] woman might talk to the plant (*ch'iʔtm'iłt*) as she made a tule or cattail mat," and "years ago it was not unusual to see a solitary elderly woman quietly speaking to her digging stick as she worked."

The decorative motifs that adorn implements, boxes, baskets, houses, digging sticks, spindle whorls, rock paintings, and many other objects and settings also serve as symbolic representations of the dynamic nature of environments and the spiritual interchanges across species and planes of existence. The designs have definite symbolism, often linking the users of decorated objects with their intended endeavours – such as the previously noted parallel lines on some digging stick handles that evidently represented trenches a girl at puberty would dig during her training and presumably connected her spiritually with her task and her vocation (M.B. Ignace, Ignace, and Nord 2009; MacDonald 1976).

Not only food but also the vessels that contain it have long been considered sacred. The Tlingit story of the origin of spruce-root baskets from a woman who was married to the Sun up in the sky country is a testimony to the spiritual nature of baskets for these people (Paul 1991). Baskets also have an important place in the First Fruits and First Roots ceremonies of the Interior Plateau peoples (Schlick 1994; Teit 1909).[35]

CONCLUSIONS

The concepts of spirituality, kincentricity, and interconnectedness of all things must have ancient origins, extending back in some form to the "beginning of time" since they are firmly embedded in many origin stories of families and clans and in peoples' ceremonies and other traditions. The manifestations of spiritual power, in the form of symbolic designs and representations of birds, animals, humans, or supernatural figures on tools and rock faces, extend back at least to the Mid Holocene. As noted in chapter 2, ornamental objects and designs proliferated on the Northwest Coast starting about 3,500 years ago and are especially evident in many sites from about 2,000 years ago to the present. Ritual power and strong connections with supernatural entities in the form of guardian spirits and totems were also linked – through artistic representations and good success in hunting, fishing, and other endeavours – with increasing social complexity, growth of wealth and status, and expansion of hereditary prerogatives, as well as with increasing occupational specialization and possibly an amplification in shamanism, spirit quests, and ceremonial enactments of groups and individuals. It stands to reason that an individual who could gain strength and success from spiritual sources through advanced training, and thereby provide more food for his or her community, was more likely to be given high status and to be looked upon as a role model and leader. A person who could claim close relationships with particular important plants or animals would likewise be accorded high esteem and probably be recognized as a gifted person and leader.

The three-way spiritual connection seen to exist among humans, plants, and animals is also quite common and widespread. Bears in particular are strongly associated with plants, both practically and spiritually, as seen in the narrative theme of "The Woman Who Married a Bear." The parallels in bear and human foods – especially berries and roots – is remarkable, and it is not surprising, therefore, that human women who, in narratives, are taken as wives by Grizzlies or Black Bears spend their time picking berries and digging roots for their husbands (L.R. Smith 2008b; Teit 1912). Geese and ducks (who relish springbank clover, Pacific silverweed, and eelgrass roots), swans (who eat camas bulbs), and muskrats (who eat wapato tubers) may well be the original spiritual and practical mediators between these roots and their human consumers. Similarly, beavers and woodpeckers may have been the original inspirations behind the sophisticated woodworking technologies that developed in the area, and both are associated in narratives with the creation of dugout canoes and other carpentry. Beaver is also associated with causing floods and, conversely, with providing fire to humans (see chapter 12).

The attribution of sacred or spiritual qualities to trees (like western redcedar) and other plants, as well as to special places, is not confined to northwestern North America. All over the world, people have identified sacred trees, sacred forests, sacred mountains, and sacred waterways (Chandran and Gadgil 1993; W. Davis and Henley 1990; Mauzé 1998; Pandey 1998; Turner, Ari, et al. 2009). All around the circumpolar region, for example, trees are accorded special status.[36] This reverent attitude people have toward trees and, by extension, forests has significant implications for biodiversity conservation. In many parts of the world, the only places where certain rare and threatened trees and other species can be found are in sacred groves and temple forests where cutting is prohibited.[37] Some of these ideas, including a kincentric approach to ecology, have made their way into more recent Western thought as well, such as in Leopold's (1990) notion of a "land ethic."

Places where spiritual work and special events have taken place are also an important part of overall management and conservation. The valley where Dr Daisy Sewid-Smith's ancestor *ts'eqamey* was said to have resided within the giant cedar tree during the Great Flood and the pools in the Nuu-chah-nulth territory where whale hunters and others training to accomplish great tasks undertake their spiritual training (*oosumich*) are elements of ecological and historical intersection with human thought and spirituality that are fully integrated into community and culture and into the maintenance of environments (E.R. Atleo 2011; Scientific Panel for Sustainable Forest Practices in Clayoquot Sound 1995, 25). In short, these are places from which human participants enact their responsibilities to others, human and otherwise, and to past, present, and future generations.

The various taboos against picking certain flowers, and against any disrespectful treatment of fish, animals, and trees, also have implications for conserving these species and may in fact serve as deeply rooted prescriptions for sustainability, so profoundly embedded in cultures that people may not even have been aware of the broader effects of their prohibitions. The prohibitions against wasting any food or medicine – supporting the tenets of taking only what one needs and no more and of sharing any excess with others rather than "hoarding" – are part of the overall suite of practices and cultural mores that underlie resource use and promote conserving practices (Turner and Berkes 2006; Turner, Ignace, and Ignace 2000). Together, beliefs and associated codes of conduct guide the practices for sustaining and enhancing resources and habitats that might be embraced under the title of "management." However, such practices are mostly much less prescriptive than conventional resource management and much more based on broad-scale complex customary approaches. Beliefs and worldviews tend to motivate a general sense of responsibility toward resources, which seamlessly merges with the inherited roles of chiefs and designates in overseeing, caretaking, decision making, and intergenerational accountability that go with leadership and associated social organization (see chapter 9).

These beliefs and cultural norms that relate to proper use of and care for the environment and to appropriate social behaviour are routinely self-enforced, with monitoring of compliance occurring across generations and within and among family and clan members. They are embedded in strong social and religious teachings that are enacted and re-enacted many times over an individual's lifetime, being reinforced through stories, ceremonies, and songs, as well as by example (M.R. Atleo 2009). Sanctions against violations of taboos might also be determined and meted out by chiefs and leaders, with serious offences being punishable by exclusion from the community, abandonment, or even death for violators in cases such as polluting a salmon river.[38]

A good example of the link between beliefs and sustainability is with the harvesting of cedar bark. Weavers all up and down the coast are taught that one should remove only a portion of the bark from a cedar tree: only one or two "straps," or around one-third of the circumference of a young cedar tree. This convention was drawn from the knowledge that the tree would die if the bark was peeled off all around the tree: "Even when the young cedar-tree is quite smooth, they do not take all of the cedar-bark, for the people of the olden times said that if they should peel off all the cedar-bark ... the young cedar would die, and then another cedar-tree near by would curse the bark-peeler so that he would also die. Therefore, the bark-peelers never take all of the bark off a young tree" (Boas 1921, 616–17; see also ibid., 619; and Schlick 1994).

Another manifestation of the intersection of beliefs, rituals, and management is with the previously mentioned Tsilhqot'in tradition of instructing a woman who had recently lost her husband, and who was having a difficult time adjusting to his death, to ascend Potato Mountain, pull out the mature stems of spring beauty, or mountain potato, and strew them around in places where they did not grow, a practice that at once helped the woman in her grief and effectively extended the range of this important edible root, increasing its overall productivity. As explained by Mabel Solomon and Gilbert Solomon (pers. comm., 2003), "The woman [widow] had a special energy. That's why she was asked to work to make more *sunt'iny* (mountain potato) grow; the other women would keep asking her, 'Have you gone up yet? Have you gone up yet?'" Examination of these mature stems, which are quite succulent, revealed that they ripened within a short time and produced quantities of viable seed that would soon have exploded from their capsules, germinated, and started to grow in bare patches in the high country. Thus spiritual renewal in this case goes hand in hand with ecological renewal. The Xeni Gwet'in and other Tsilhqot'in maintain that it was their own practices, such as this tradition, that created the dense and productive mountain potato meadows within their territories at places like Chunoz̲ Ch'ed (Potato Mountain) (Mabel Solomon and Eileen William, pers. comm., 2003). The Tsilhqot'in philosophy is summarized in the idea that "If you take care of and respect the land you live on, the land will take care of you" (Glavin and People of the Nemiah Valley 1992, 150). The statement of Nuu-chah-nulth elder Roy Haiyupis that "Respect is the very core of our traditions, culture and existence" (cited in Scientific Panel for Sustainable Forest Practices in Clayoquot Sound 1995, 6–7) reflects the understanding and expression of the reciprocity that exists broadly, still today, in many Indigenous people's relationships to their environment and the resources that they depend on, such as plants for food and shelter. This type of "coming to understanding" and the many similar lessons instilled by elders and leaders about developing relationships of respect and reciprocity over the generations may have originated from some long-ago catastrophic depletion of an essential resource resulting from an individual's or community's overexploitation or improper behaviour (Turner and Berkes 2006).

Ancient stories and ceremonies as conservation tools may well be rejected out of hand by ecologists and land managers since they do not fit the model for direct interventions implied by some definitions of the term "conservation." Yet they may be far more effective in communicating complex concepts and key principles, in promoting effective learning and understanding, and in motivating people to participate in implementing more sustainable strategies for long-term resource use than approaches based on scientific knowledge alone. Collectively, they gently instil and effectively reinforce the codes of proper behaviour, consideration, and restraint required to live well in a kincentric world.

Plants are important elements of an overall belief system, and their spiritual dimension is reflected in virtually every aspect of Indigenous cultures, from the "words of praise" spoken to the plants and other resource species before people harvest them to narratives, songs, and ceremonies that show appreciation for them. Ultimately, the careful and respectful approach to plants and other living beings, in a "sentient ecology" perspective, guides and directs people's behaviour toward them and the care that they receive when they are harvested and used. They truly are "re-sources" in this context, to be used and renewed over and over again, through time and generations.

14

Ancient Pathways and New Pathways for Retaining and Renewing Botanical and Environmental Knowledge Systems for the Future

Well, I heard a lot of my Elders ... talk about the spirituality of the environment. And they would say that we as human beings are not superior to Mother Nature's creation. *We are only a strand in it.* And what we do to the environment, we do to ourselves. (Mary Thomas 2001, 48)

INTRODUCTION

In a single sentence Mary Thomas, Secwepemc elder and cultural specialist, portrayed the essence of an Indigenous perspective that is widespread within the cultural groups of the study area: *We humans are only strands in the immense fabric of the universe.* This is a critically important idea – one that has been evidently little recognized in our predominantly urbanized, industrialized, and increasingly globalized societies. Most of us seem to behave as though we humans are all that counts – as though the environment no longer matters. We are increasingly appropriating more and more of the earth's production. Yet it is at our own peril that we ignore our fundamental connections to the environment and to all of the other species of the planet. No one understands this better than the Indigenous peoples of long-resident communities around the world who have always relied upon their close and direct ties to their territories and local ecosystems for their very survival. In many cases, they see the consequences of their own harvesting and resource use practices directly.

Generation after generation – in some cases, for thousands of years – Indigenous peoples, dwelling in their own home localities, have developed, learned, and adapted the knowledge, skills, and values that have enabled them to subsist as active players in their own habitats, not just taking what they can through random opportunism but also finding ways and developing technologies and belief systems that have allowed a mutual dependency to grow between themselves and the habitats and species that surround them – and with which they have surrounded themselves.[1] Of course, a key aspect of this interdependency is peoples' relationships with the plant world.

This book has examined these relationships for Indigenous peoples of northwestern North America, as reflected in their history, their botanical vocabulary, their adaptations to the immense changes of the period since European contact, their uses of plants for food, materials, and medicines, their spatial, temporal, and social relationships, their interactions with one another as mediated by their available resources, their systems for sustaining and enhancing the plant species and habitats on which they have depended, their oral histories involving plants and botanical knowledge, and their belief systems as they relate to their various associations with plants and environments. As outlined in each chapter, all of these relationships have been dynamic – ever changing and adjusting to fluctuating and sometimes unpredictable conditions. They are also diverse, with each aspect linked to the other but no two situations exactly the same. Nevertheless, patterns exist and can give important signals about the historical development of human-environment relations and, in particular, about effective interactions that can sustain human societies and environments together over long periods of time.

An overarching problem to be addressed, based on findings in this and other studies that link biological and cultural studies, is: *Why are both cultural diversity and biodiversity failing*? This is a daunting question because it is so complex, but I believe that the answer is to be found in examining the inextricable linkages, such as those examined here, that underpin human cultural systems and ecological systems. For hundreds of thousands of years, humans have been major participants in the evolutionary dynamics of species and ecosystems throughout the world. In northwestern North America, people have been living within particular regions and relying almost completely on local resources over centuries and millennia. Yet local and Indigenous peoples, who are closely tied to their home places through generations of interaction with their environments, are often excluded from mainstream society's decision making and policy formulation, including efforts to conserve biodiversity, even though the methods used by many resource managers – narrowly based on scientific principles of conservation biology – are all too often ineffective.[2]

Furthermore, efforts to maintain the integrity of Indigenous cultures through conventional education and revival of language instruction have also often been unsuccessful, at least in part because they fail to take into account those same inherent linkages between humans and their environments (Battiste and Henderson 2000; Cajete 1994; Kawagley 1995, 2003; J.C. Thompson 2004). Finding and identifying the multiple ways that people and their habitats are united –

through their languages, belief systems, foods, technologies, practices, and relationships with other species or ecosystems – and then directing our attention to these specific intersections will be, I believe, the most effective means of ensuring and renewing both ecological and cultural integrity.

In this chapter, I examine insights and processes identified in the earlier chapters – the legacy of knowledge, practices, and beliefs that relate to the people-plant and human-environment interactions of Indigenous peoples in northwestern North America – to help identify effective ways of reconnecting the human strand with the rest of "Mother Nature's creation" (in Mary Thomas's words),[3] in which we are embedded and on which we are completely reliant, whether we realize it or not. Chapter 4 in particular shows the effects of sudden, relentless, and externally imposed changes on Indigenous peoples' abilities to retain and maintain these relationships with their home places. The effects of eroding connections, of displacement, and of loss of control over their lives and territories have resulted in increased vulnerability and reduced resilience both for the people and for native species and habitats. Can peoples' capacity for resilience and adaptation be strengthened in this dramatically changing world, based on past wisdom and on a sense of place and identity, as expressed in many of these chapters (e.g., chapters 12 and 13)?

To answer this question, I build on themes and lessons exemplified in the previous parts of the book. I first reintroduce and summarize some of the key elements related to patterns of people-plant relationships from the various facets of peoples' history and knowledge systems. This section is followed by a discussion of the diverse ways that knowledge and skills are learned and transmitted. The next section identifies a series of ideas and strategies, drawn from the teachings and examples described in the book, that could contribute to our collective efforts to maintain and renew biocultural diversity; a brief concluding section completes this analysis.

Finally, I reflect on the changing perspectives on ethnobotany and ethnoecology that have developed over the course of recent decades as an expanded appreciation has been formed for the complexity, role, and importance of plants and their ecosystems for the First Peoples of this vast and beautiful part of North America.

PATTERNS IN ETHNOBOTANY AND ETHNOECOLOGY IN NORTHWESTERN NORTH AMERICA

Historical Relationships

Peoples' deep connections with plants in the study area extend back to the first entry of humans into the New World, some 14,000 years ago, and even before. Those small groups of hunters, fishers, and plant gatherers making their way across Beringia or around the coast along the "kelp highway" would already have had strong and deep associations with a wide range of plants: berries, roots, greens, and the inner bark of trees as food; grasses, sedges, cattails, willows, and other woody materials for fuel, construction materials, matting, and basketry; stinging nettle and willow for cordage; and likely sagebrush, marine algae, and numerous other plants as medicines. Although plants are sparse, to say the least, from the earliest archaeological records, inferences about the use and cultural importance of plants can be made from their identification in palaeoecological contexts, their occurrence in later archaeological situations, and their widespread salience among contemporary Indigenous peoples, including those of northeastern Asia, indicating a probable long time frame for ethnobotanical and ethnoecological knowledge. The difficulty of plant preservation in many archaeological contexts has meant that they have often been overlooked by archaeologists in considerations of early peoples' lifeways. Yet, based on findings by Croes (2012; Croes et al. 2008) from relatively recent wet sites, like Ozette and Hoko River on the Olympic Peninsula and Sauvie Island on the Columbia River (sites where plant remains *were* well preserved), plants and plant materials constitute as much as 90 to 95 per cent of the total inventory of species and materials used by people at Northwest Coast sites. This situation is probably not limited to recent centuries; more likely, plants have always played a dominant role in peoples' lives. Other findings from archaeology and ethnographic research in the study region also refute any assumption of minor roles played by plants in Indigenous lifeways (see chapters 2, 5, and 6 in particular).

One inference that can be made is the consistency of the types of species used over the millennia of human occupation of the study region. Over 14,000 years ago, the people of the Monte Verde site in Chile, who had evidently travelled down the Pacific Coast from northern North America, were using digging sticks and other wooden tools, making twine, living in wooden houses, eating fruits and seaweeds, and using a range of medicinal plants (Dillehay et al. 2008). On Haida Gwaii wooden stakes, wedges, twine, and implement hafts have been found dating back 10,000 years or more. In fact, Ames and Maschner (1999) suggest that virtually all of peoples' resources from 10,000 years ago were also resources of 200 years ago. The beginnings of plant-based technologies, too – nets, wedges for splitting wood, spear throwers, barbed points for fishing and hunting, and even basic watercraft – were also likely underway in the earliest of times, as were basic food storage processes of smoking, wind and sun drying, and food caching. People probably learned some of their early knowledge by observing animals. In any case, the basic resources and techniques have remained. What has changed over that time is the increased specialization and sophistication of exploitation, manipulation, and preparation of the resources. Rather than discovering entirely new resources, people refined and built upon the original suite of species and developed novel and improved methods and processes of harvesting and preparation. Some resources were intensified through application of more refined harvesting, production, and/or processing and storage methods; yet peoples of the area have always maintained a wide diversity of food and material supplies to allow alternatives if and when needed.

Similarly, tools – including those made of wood and other plant materials, as well as implements for harvesting plant resources – started out in the Pleistocene and Early to Mid Holocene as relatively basic and general-purpose. In composition, they were made from whatever was readily available. Over the millennia and especially in the Late Holocene, they became more sophisticated, precise, and focused in their use. As the particular properties of different kinds of woods and fibres came to be appreciated through discovery, experimentation, and sharing of knowledge, the materials used became increasingly specialized according to the required properties of the specific tools. Even different parts of a single species of tree had specific applications: knot wood for fishhooks, compression wood for tool handles, roots for fibrous stitching and binding materials, sapwood for splints, and saplings for poles and long implement handles.

Some plants have a particularly long and consistent history of high cultural salience, judging from their widespread occurrence in palaeoecological sites, as well as other evidence from their names, stories, and more recent and extensive use. Some of these species are known from the glacial refugia along the coast or possibly existed in nunataks (unglaciated peaks) in the mountains, and they were evidently present in the region even in the Late Pleistocene.[4] Others may have first been encountered south of the ice sheet along the ice-free coastlines or interiors of Oregon and California.

Another insight from the archaeological records is that even long ago, people were highly mobile, and individuals and communities were interconnected on a range of scales. Well-developed trade networks going back nearly 10,000 years are indicated, for example, by the presence of obsidian at On Your Knees Cave, Alaska, likely from Mount Edziza around 200 kilometres away.[5] Farther south, affiliations between the early Interior Plateau peoples and those of the Great Basin to the south and east are suggested by the presence of Great Basin–style twined sagebrush-bark sandals of a unique weave in caves in south-central Oregon, most notably Fort Rock Cave, dating from over 10,000 years ago to just over 9,000 years ago (Connolly and Barker 2004). These connections are particularly significant because they reflect peoples' abilities to learn from each other – to share not only materials but also innovations in technology, as well as ideas, stories, and beliefs that have, together, resulted in some of the patterns of similarity and congruence reflected in the ethnobotanical and ethnoecological knowledge systems of today. Other evidence of people converging and exchanging goods and knowledge exists in the prime fishing locales of the Columbia and Fraser Rivers, for example. The Dalles was such a place of ancient gathering and exchange, dating back at least 9,000 years. Camas meadows on the Columbia

Plateau similarly became sites for gathering, first apparently in the Willamette Valley and vicinity and a little later, just under 7,000 years ago, in the Calispell Valley; the technologies for pit-cooking camas and other foods, possibly originating farther south, evidently developed further and spread out from such sites. Many other innovations – new styles of dwellings, improved methods for canoe construction, new basketry materials, and different weaving techniques – would have been learned and adopted at such places.

The palaeoecological and archaeological records tell us, too, that plants and other resources were hardly ever constant; their ranges were continually shifting with environmental change over decades, centuries, and millennia: sea level rises, floods, volcanic activity, climatic warming or cooling, and accompanying changes in the extent of wetlands or the frequency of fires. People had to respond to the changes in their environments and in available resources, learning to rely on new species or on new ways of harvesting and preparing old species. This required observing, testing, and learning new skills.

The integration of different materials from different sources and locales is another major aspect of ethnobotanical patterns reflected in the archaeological record. Fishing and marine-based resource exploitation is a good example, linking plants with animal resource procurement. The fish and sea mammals could not be taken without implements and gear, mainly originating from plants and mostly from terrestrial environments; the much-celebrated salmon and other food from the ocean and rivers must be considered in the context of wooden fishing weirs, wooden spear shafts, and plant fibre netting and twining. As well, animal resources required various plant materials to enable their transport, preparation, cooking, and storage: knives, spreaders, wooden racks, baskets, fuel for cooking and smoking, and so forth. The components of peoples' food systems – including both the foods and the materials and knowledge required to obtain and process them – cannot be separated, and plants have been integral to these systems from the earliest times in northwestern North America.[6]

Since people's arrival in the study area, human populations and settlements have more or less continually expanded, especially starting in the Mid Holocene, between about 7,000 and 3,500 years ago, with increasing diversification in languages, in cultural traits, and – of relevance here – in plant resources. People were settling down more into larger, more integrated communities, but they continued – and perhaps even expanded, with increasing mobility – their seasonal rounds for food harvesting. By the Late Holocene, there is no question that peoples were not only more successful in providing themselves with the food resources they needed – whether wapato or salmon, camas or halibut – but were also becoming stronger social entities, with greater occupational specialization, more interchange, and larger extended family dwellings and permanent villages. Pithouses may have been the most ancient form of permanent shelter, but their construction became increasingly sophisticated in the Late Holocene, culminating in the Plateau Pithouse Tradition beginning at least 3,500 years ago, possibly as early as 5,000 years ago, with large pithouse villages being established along the major rivers of the Interior Plateau in the latter part of the Late Holocene. On the Northwest Coast, the iconic western redcedar plank and post big houses came into their own around this same time period, as the large, monumental cedars became more common.

Another lesson of history is that, despite their adaptability and ingenuity in times of immense environmental change, people seemed to thrive more in the times when climatic conditions – and sea levels – were generally more stable. Within the past 3,000 to 2,000 years in particular, the coastline in the region has been fairly consistent, and this has allowed the rich coastal ecosystems – clam beds, estuarine tidal marshes, and rocky shorelines – to develop without excessive shifts in tidelines. These conditions, in turn, may have enabled people to access their fish, shellfish, estuarine root gardens, and other coastal resources more predictably. Extensive, metres-thick shell midden deposits all along the Northwest Coast at this time are a testament to the abundance of intertidal resources and the intensity of their use.[7] With the increased populations and larger settlements allowed by the relatively steady resource base, more complex social organization and occupational specialization ensued, as reflected in the elaboration of art and ceremonial practices, for example. People were able to travel more, and this enhanced trading and exchange.

Also in the Late Holocene, people evidently started to identify more with particular places, affiliating themselves more with home territories, and increasingly developing more predictable resource bases and following more predictable seasonal rounds, while still maintaining ties and trading relations with other groups. There was a relative prosperity, a cultural and linguistic blossoming, in the latter half of the Late Holocene, and one could argue that the relative stability of climate and sea levels was a major contributor to this situation. Significantly, too, the more stable coastline and climatic regimes of the past few thousand years also probably enabled people to develop methods of habitat enhancement and resource management (Caldwell et al. 2012; Deur 2000; Deur and Turner 2005; Recalma-Clutesi 2005).[8]

Expansion in knowledge and use of plant resources was a major element of the general elaboration of subsistence strategies characterizing the entire Late Holocene period, including harvesting and storage technologies, housing, social structures, art, ceremonial practices, and transportation and trade connections, right up to and including the time of European contact. To say that many people were flourishing in a rich and developing environment would not be an understatement – and the contributions of plants to this relative prosperity were numerous and diverse.

Language and Vocabulary

The high significance of plants for the Indigenous peoples of the study region is reflected in their botanical vocabularies, particularly in the profusion of names applied to plants in the various languages. People give names to things they converse about – things that are important to them – including the most common and salient entities in their environments, of which plants are a prime example. Almost without exception, the species of plants that provide people with their food, fuel, materials, medicines, decorations, and so forth are named in any language, as are plants that are particularly notable by virtue of a characteristic shape, colour, texture, or importance to a particular animal. Names thus reflect a plant's abundance, visibility, distinctiveness, seasonality, potential utility, and history of cultural association. Seldom do people give specific names to small, undistinguished species that are not used, even though some people, at least specialists, may recognize them at some level. The plant names of the Indigenous peoples of the study region, as described in chapter 3 (see appendix 2), show these general trends very well.

The survey also reflects the dynamics of plant naming – how names develop and disseminate over time according to a group's historical roots and relationships with other groups, as well as with the floristic history of the region. The body of plant names in any particular language represents a complex of words: original terms that were retained more or less unchanged with their primary referent species; terms that were expanded, modified, or adapted in meaning and reference from earlier forms; names that were borrowed from neighbouring or more distant languages (through trade, intermarriage, or other means); and names that were themselves products of shifting or expansion from some other term. The acquisition of new names, loss of old ones, and semantic changes of existing terms in a language reflect the function of the language itself: to allow communication about things that matter to people in a dynamic, constantly changing world.

In orally transmitted languages, without written records, particular changes in names and other terms over time are not necessarily evident; they must be inferred from known patterns of sound shifts and other factors, such as the occurrence of a plant in a given area or trade patterns between groups. For clusters of related, cognate terms – terms developing from a common ancestral root word within a language family – linguists can sometimes track an existing term back to the original form in the language that must have been used by the ancestral peoples of the language groups, before the peoples and languages diverged: the proto-language (see Kari and Potter 2010a; Kuipers 2002). Knowing how long ago groups of people sharing a common ancestral language separated from each other – and therefore the approximate time when their languages started to develop as separate entities from the original language – can give some indication about the antiquity of the names and about the assumed continuity of cultural importance and uses of a particular plant in a given language.

Some of the plant names used by Indigenous peoples in northwestern North America are judged to be extremely old, as reflected by the existence of proto-language forms and by the widespread dissemination of a term over many languages, both within and across language families. For example, names for soapberry (*Shepherdia canadensis*) show ancient origins, occurring both in Proto-Salish and early Dene (Athabaskan) languages in association with a term meaning "to foam" or "foaming," and possibly extend back in some form to Early Holocene times (Turner and Burton 2010). If this is the case, it implies, too, that the popular use of soapberry for making a whipped confection may be similarly ancient.

The assumption is that the broader the distribution of a term and the more widely it is shared, the more important is the plant resource named by the term. The situation can be extremely complex, however, since the original meanings and referents of a name can shift and expand.[9] Other plant names are assumed to be very old because they no longer reflect any particular meaning; their original meaning has been obscured over time, such that the words are no longer analyzable, except perhaps by deep and careful linguistic analysis. This is often the case with plants that are highly salient and culturally important; their names might include a suffix pertaining to "plant," but the root of the word has no other meaning than as a plant name.

However, another reason why a name may not have any identifiable underlying meaning is that it has been borrowed from some other language. A significant number of plant names, for example, were apparently borrowed from Proto-Salish into early Wakashan languages. All nineteen of the Wakashan terms identified as originating from Salish are culturally important, and thirteen of them (including highbush cranberry and soapberry) apply to food plants. The borrowing of terms can usually be detected by surveying names in neighbouring languages, especially when the languages are in different language families, but (as with the Haida and Ts'msyen names for highbush cranberry discussed in chapter 3) the direction of borrowing may not be obvious, unless there are known proto-antecedents in one of the languages but not in the other.

As pointed out by Leslie Main Johnson (1997), borrowing plant names and other terms bespeaks what must have been extended periods of intimate contact between groups of very different origins, especially in the more peaceful contacts of trade, intermarriage, and feasting exchanges. The plant names exemplified in appendix 2 (and in the database from which it is drawn) reveal just how much communication has occurred across geographic and linguistic boundaries (such as between Salish and Wakashan peoples) in the study area over many millennia. The "top 25" (most frequently named) plant species and groups of related taxa (see chapter 3, table 3-8) are all widespread across the language territories of the study region, and all have a high level of cultural importance as foods, materials, and/or medicines. All of these species also possess names that reflect a substantial degree of retention from proto-languages and that have common uses and cultural features that indicate considerable exchange between cultural groups. Notably, too, eighteen of the twenty-five are woody plants (trees or shrubs), reflecting both a high level of visibility and significant utility.

Much of the linguistic diversification within the major language families evidently occurred around the beginning of the Late Holocene, when human settlements were expanding and when people were becoming increasingly mobile. This was evidently the time, for example, when the early Salishan peoples started leaving their original homeland in the lower Fraser Valley and moving eastward and southward into Interior Plateau regions and Puget Sound, as well as westward and northward to Vancouver Island, the Sunshine Coast, and beyond. The Salishan migrants may well have amalgamated with peoples already residing in some of these places and blended their vocabulary and cultural features with these peoples to some extent. This type of mixing and merging of distinct groups of people apparently occurred in many different parts of the study area, creating the remarkable convergences and distinctive characteristics in languages, vocabularies, and cultural practices existing today, or at least in recent times. Plant names, along with other vocabulary, reflect this dispersal and integration but with the added complication of uneven distributions of plants, and botanical referents sometimes changed as people encountered new and different species and left others behind. For example, camas (*Camassia* spp.),[10] whose bulbs are a "root vegetable" and food staple of the First Peoples of the southern part of the study region, does not occur in the lower Fraser Valley or the northern part of the Interior Plateau at present and likely moved northward onto Vancouver Island and the Gulf Islands from south of the Pleistocene ice sheet along the coast (Tomimatsu, Kephart, and Vellend 2009). Therefore, the original Proto-Salish term *qʷlawl*, whose descendant terms apply to both camas and nodding onion (*Allium cernuum*) in different Salish languages today, likely referred first to nodding onion and then shifted to camas as Salishan peoples moved westward to Vancouver Island and vicinity, where camas would have been encountered (see discussion in the preface and chapter 3).

Change with the Newcomers

When the Europeans arrived in the area with all of their new goods and technologies, numerous semantic expansions and shifts occurred, almost simultaneously, as people adopted many new species and tools and integrated them into their languages (see chapter 4). In some languages, for example, the original name for "arrow" was transferred to "bullet," and in others the name for "firedrill" shifted to "match." New types of fruits and vegetables were often designated with the same names as their most similar indigenous counterparts: domesticated onions were called by the same name as wild nodding onions, domesticated rhubarb by the name used for western dock, and so on (see the example of the potato in appendix 3 and other examples of names for introduced species in the database from which the appendix is drawn).

The entire system of knowledge, practice, and belief that was fundamental to Indigenous peoples' lifeways was put to an extreme test with the opposing influences and practices of many of the European newcomers. At first, the goals and needs of the European explorers and traders fitted within the existing social, cultural, and economic norms of First Peoples, but soon the newcomers' impacts, demands, and cash-based economy outstripped the capacities of the Indigenous occupants to maintain many aspects of their traditional lifeways. The cumulative consequences of disease epidemics, multitudes of new goods and introduced species, newcomers' appetite for lands and resources, landscape transformations, and the imposition of new languages and religions, a wage economy, and labour practices – enacted through a compulsory and often cruel and coercive education system – were profound.[11] Knowledge and practices around plants and environments were suppressed along with other cultural knowledge as people struggled simply to survive in a drastically changing world.

Fortunately, however, there were knowledgeable and determined individuals, families, and communities who were able to acquire and retain key elements of their cultural knowledge and languages, including botanical and environmental knowledge, over decades and generations despite the many changes and pressures surrounding them. These people, who were themselves taught about their ancestral wisdom and traditions by their elders, recognized the value of these teachings and understood the necessity of preserving and maintaining this knowledge for the benefit of future generations. Many of these individuals have been quoted and cited in this book, and countless others have contributed to the ethnographic and linguistic work referenced here. Their contributions, recognized or not, have been critical elements of peoples' resilience and have allowed a general renewal and revitalization of ethnobotanical, ethnoecological, and other cultural knowledge and practice that is still ongoing (see Turner, Robinson, et al. 2012).

Food

Although they sometimes go unrecognized, plants have played a long-standing and critically important role in the food systems of Indigenous peoples over all of northwestern North America: the Interior Plateau, the Northwest Coast, and even the northern, sub-boreal regions. Plant foods provided – and continue to provide – essential vitamins and minerals, as well as carbohydrates, some protein, and in the case of seeds and nuts, nutritious fats. Plant foods also provide dietary fibre, which is essential for proper digestion (Kuhnlein and Turner

1991). The diverse types of plant foods – berries, root vegetables, greens, tree cambium, and seaweed – were undoubtedly known to the earliest peoples entering the region, and it is likely that many of the actual species that were used in these early times have continued as food resources up to the present. Obviously, new foods were added to the original selection available – species that are restricted in their distribution to the New World and that people arriving in the area would have encountered either in ice-free refugia around the end of the Pleistocene and Early Holocene or in the unglaciated area south of the ice sheets.

Camas, mentioned previously, also presents an intriguing example of how ancient people may have acquired access to new foods through innovative ways of processing. In recent times and for the past several millennia, Indigenous peoples of the southern Interior Plateau, across Washington to Montana, and on southeastern Vancouver Island (as well as in California, parts of the Great Basin, and Texas) have harvested and processed camas bulbs by cooking them for prolonged periods in underground pits, or earth ovens. As described in chapter 5, this special processing technique using heated rocks was required to break down the plant's main storage carbohydrate, inulin – a complex sugar that is largely indigestible – into sweet-tasting and digestible fructose. As noted previously, camas itself appears to have largely migrated northward after the Pleistocene by natural means from south of the Cordilleran Ice Sheet.[12] However, without the proper cooking to free up the fructose in the bulbs, camas bulbs would not have contributed substantially to peoples' diets in terms of food energy.

Pioneering research by Thoms (1989, 2008a, 2008b, 2009) provides clues to the origins of the earth oven technology that allowed camas – and some other inulin-containing root vegetables – to become a staple food throughout most of the Holocene. The earliest known earth ovens with rock-heating elements in North America are located along the ecotone between the southern Plains and southeast Woodlands, dating to around 9,500 to 8,000 years ago (Thoms 2008a). Earth ovens occur – with remains of camas bulbs – in the Willamette Valley region of Oregon from around 8,000 years ago and in the Calispell Valley of Washington from between 6,000 and 5,000 years ago. The revolutionary earth oven technology apparently spread rapidly from these early sites throughout the Interior Plateau and onto the entire Northwest Coast, where their use is well documented in relation to cooking camas as well as many other foods (see Turner, Bouchard, and Kennedy 1980; Turner and Kuhnlein 1983; and Turner, Thomas, et al. 1983). Not only did pit-cooking produce more palatable and digestible bulbs, but it also produced a sweet coating over the bulbs that helped to preserve them as edible products, when dehydrated, for exceptionally long periods of time – in some cases, for many years.

Thoms (2008a, 2008b, 2009) suggests that earth ovens represent a relatively sudden dietary transition toward use of more carbohydrate-rich root vegetables during a time of relatively dry climate, when availability of game was decreasing and people had to learn to exploit alternative foods, resulting in a land use intensification that he terms the "carbohydrate revolution." Pit-cooking also adapted readily to the large-scale processing of many other foods, including balsamroot, nodding onions, yellow glacier lily bulbs, black tree lichen, and even various types of berries, as well as clams, venison,[13] and other animal foods, thus demonstrating how one technology can be modified for many purposes.[14]

There was a general trend of sophistication in methods of food harvesting and processing over the Holocene, culminating in the Late Holocene, with a substantial, increasing emphasis on food production and storage throughout the study region. High-quality basketry and, on the coast, bentwood cedar boxes brought new possibilities for cooking and storing food, as well as for trade – by allowing standardized quantities and containers for transport and consistency in exchange measures. Other innovations that related to food included improved digging stick design (in some cases, with elaborate handles), presumably enabling more efficient root harvesting, and the development of other specialized equipment and improved processes for dehydrating berries (such as the trenches of the Indian Heaven huckleberry sites near Mount Adams). Acorn-processing technologies, such as those at the Sunken Village site on Sauvie Island,[15] are also part of the suite of innovations that swept the region in the Late Holocene, but in this case there are intriguing linkages both to the Great Basin and to the Jomon culture of ancient Japan (Croes, Fagan, and Zehendner 2009a, 2009b). All of these improvements occurred, not coincidentally, in concert with growing populations.

Also by the Late Holocene, peoples were developing ways of enhancing the growth and productivity of plant foods, using fire, selective harvesting, and probably other forms of management such as clearing and pruning berry bushes. Intensively used and managed estuarine root gardens, camas meadows, and subalpine root-harvesting and berry patches also would have become common during the Late Holocene. Ownership by chiefs and particular families of prime productive patches of plant foods within designated territories, especially along the Northwest Coast, probably also began at this time, with increasing populations and more permanent settlements. It is possible, too, that people were moving plants (e.g., stinging nettle, cattail, berry bushes, and young cedar trees) and plant propagules (e.g., hazelnuts, camas bulbs, wapato, and acorns) from place to place, as well as developing arrangements with neighbouring groups to harvest some resources in their territories. More detailed genetic research of plant populations might reveal the extent to which humans were involved in plant dispersal in the study region.[16]

Significantly, the majority of the plant food resources in the region were probably harvested and processed predominantly by women and children in the past, as has been the case in recent times. The times spent together in food-related activities would have provided significant opportunities for women to socialize, pass on information, recount stories, and teach the children, not only about the particular plants being harvested or worked on but also about important cultural practices, songs, games, and other traditions (Turner 2003b). Many of the elders of recent times throughout the study area reminisced about childhood times of picking berries or digging roots with parents and grandparents and recounted how they had learned the details of how to select prime bulbs, where to find the biggest berries, and how to care for and respect the plants and the places where they grew.

Although most of the plant foods identified in this study were used by people throughout their range, wherever they were found, a few were used by only one group or in one restricted area. The reasons for this are not always clear but may be related to historical factors. Some might be regarded only as famine foods: skunk-cabbage rhizomes, for example, were eaten occasionally in early spring by the Chinookans and Lil'wat (Stl'atl'imx/St'at'imc) of the Mount Currie area and by the Secwepemc but evidently not by others. These same peoples also ate cattail rhizomes, although throughout most of the study area, cattail was used only in technology – its leaves for making mats and bags and its fluffy fruiting heads for diapers, tinder, and ceremonial materials. *Q'əx̱mín* ("wild celery," *Lomatium nudicaule*) seeds were used by Coast Salish and Kwakwaka'wakw as a flavouring and ceremonial incense, but the Lil'wat and Interior Plateau peoples also ate the young leaves, raw or cooked, as a springtime green, rich in vitamin C. The young vegetative leaf bases of arrow-grass (*Triglochin maritima*) were relished as greens by the Squamish, Comox, and Sechelt Coast Salish (who have recently called them "sea onions") but apparently not by other coastal peoples, even though the plant is common all along the Northwest Coast. Indigenous peoples from Alaska to California gathered and ate, and still do, red laver seaweed (*Pyropia abbottiae* and related species), but evidently the Nuu-chah-nulth and Ditidaht on the west coast of Vancouver Island did not eat it.[17] Grains and "seeds" in general were widely gathered, processed, and eaten by peoples of the Great Basin and California but much less so in the study area.[18]

Edible mushrooms present a particular puzzle. Over much of the study area, they were generally avoided as food, as in Siberia and across the Canadian north (Kuhnlein and Turner 1991; Yamin-Pasternak 2011). Yet the Interior Plateau peoples ate, and still do, several different types of mushrooms, including pine mushroom (*Tricholoma magnivelare*) and cottonwood mushroom (*T. populinum*) (Turner, Kuhnlein, and Egger 1985). California Indigenous peoples also gather and eat a number of different kinds of mushrooms (M.K. Anderson 2005). The use of mushrooms as food in the Interior Plateau region may therefore have spread – like pit-cooking and camas use – from peoples of the south.

There are many commonalities and correspondences of food species across cultural, linguistic, and geographic boundaries, with phytogeographic patterns of plant distribution largely – but not entirely – determining which species are used. In all, the plant foods of peoples in the study area represent a complicated mix: some already known to people entering the area from northeastern Asia, some existing in situ since the late Pleistocene and Early Holocene, and others encountered and adopted later. Some of the food plants, such as hazelnut and springbank clover, were evidently moved to different locations or expanded in range by the people themselves. As well, various innovations associated with intensified food production have arisen and spread relatively quickly across wide areas, although there are still some notable regional differences.[19]

Materials and Technology

From their earliest arrival in the New World, people would have relied heavily on plant materials as sources of fuel, for weaving and matting, in construction, and for other technological purposes. The trees encountered by the new immigrants – either travelling overland across Beringia or moving along a cold and glacier-lined coastline (much lower than the present one) – were probably mostly stunted and few in number. People likely used whatever wood they could find – willow, aspen, birch, and possibly lodgepole pine, spruce, and hemlock, as well as driftwood – for fuel and construction, including frames for houses, racks for drying food, and even watercraft, spear shafts and atlatls (spear throwers), and other implements for hunting and fishing. Woodworking would thus have been a central occupation from the earliest times, and it continued to develop and flourish right up to the present, especially on the Northwest Coast, where opportunities for working with large cedars created the pinnacle of woodworking in the study area.

As noted previously, the massive western redcedar trees that provided planking, beams, and posts for the large Northwest Coast dwellings of recent centuries (see Ames 1996, 2005) evidently did not appear widely on the coast until around 4,000 years ago. By 3,000 to 2,000 years ago, the Northwest Coast peoples had become masterful architects in cedar wood, constructing immense multifamily dwellings with removable planks that could be taken to provide shelter at summer resource-harvesting sites. These accomplished woodworkers and engineers – most of them men – also crafted dugout redcedar canoes of many sizes and styles that allowed greater possibilities for travel, just at a time when the coastline had stabilized, populations were growing, and marine resources were intensifying. The Haida and Heiltsuk became famous for their canoes, which they used to travel as far along the coast as California and which they traded to their respective neighbours. It was said to be only a three-day journey from Haida Gwaii to Victoria – a testament both to the fine quality of the craft and to the strength and prowess of the paddlers. In the interior and in the north, Dene (Athabaskan) peoples were famous for their wooden-framed birchbark canoes, similar to those made by the Cree and Anishenaabe to the east. Salish and Ktunaxa peoples on the Interior Plateau also made birchbark canoes, as well as vessels from other types of bark, such as white pine. Interior peoples also made dugouts canoes but usually from cottonwood since cedar was not always available.[20]

Woodworking skills expanded in the Late Holocene to the crafting of bentwood boxes on the Northwest Coast, as well as of fine and elaborate tools: canoe paddles, digging sticks, spindle whorls, bows and arrows, long-handled dip nets of Douglas-fir saplings, fish traps, and complex fishing spears.[21] Carpentry and carving, seemingly a universal men's craft and occupying a substantial proportion of time for most males, became more specialized by the Late Holocene, with certain individuals becoming renowned for their skills in canoe making, crafting bentwood boxes, constructing houses, fashioning various tools such as spears and wooden fishhooks, or carving the immense representational figures and totem poles so well known from the post-European contact era.

Whereas men were, for the most part, the hewers and crafters of wood, women were the supreme weavers, producing diverse sizes and styles of basketry from numerous different fibrous materials, including grasses and sedges, roots and withes of spruce and western redcedar, inner bark of red and yellow cedars, Rocky Mountain maple, and willow, and bark of bitter cherry or pin cherry. They also cut sheets of birch bark, spruce bark, cottonwood bark, and some other tree barks, which they incised, folded, and stitched into buckets and other types of containers. They also used birch bark and coiled basketry techniques to make baby cradles.[22] The women also sewed and twined mats – some of them enormous – of cattail leaves, tule stems, and inner bark of cedar. Cattail and tule mats, in particular, were used as coverings for summer lodges and even for large multifamily dwellings in the Interior Plateau, as well as for drying berries on and for sleeping mattresses, room dividers, and other purposes. Other fibrous materials used by Interior Plateau women in weaving and for cordage include Indian-hemp, inner bark of silverberry, bark of big sagebrush, and even black tree lichen.[23] Baskets, in particular, often reflect not only regional differences in twining, decorative overlay, patterns, and materials used but also styles of individuals and female lineages, such that, in some cases, the baskets of an individual weaver and her descendants can be recognized from her own signature patterns and stitching.[24] Basketry styles were remarkably constant, retaining regional fidelity in weaving and stitching types in some cases for thousands of years, while accommodating variation in detail (Croes, Fagan, and Zehendner 2009a, 2009b).

The ways that the materials for baskets and other weaving projects were harvested reflect a detailed knowledge of plant growth and propagation. From an early age, girls were taught to avoid overharvesting bark or roots from a particular tree, and they also learned that cattails, tules, and basket sedge (*Carex obnupta*), as well as stinging nettle and Indian-hemp, actually grow better when the patches are cleaned and maintained by regular harvesting. Keeping and maintaining basketry and fibre plants were as important as maintaining the edible roots and berry bushes since food could not be transported or stored without these containers.

Every stage of the basket-making process requires an intimate knowledge of the materials and of the plants and environments from which they are derived. For example, if western redcedar roots are harvested from a place that is too rocky, the roots will be too crooked to use. If the soil contains too much clay, the roots will not pull out easily. If the tree has too many branches, it is said that the roots themselves will be short and branching. The best sites for cedar roots are moist, sandy riversides, where the trees are tall, straight, well spaced, and not too bushy. Old logs on the ground also yield long, straight cedar roots growing through the rotten wood. If the roots are harvested at the wrong time of the year, the bark will not peel off well, and the roots will not split properly. Wooden splints for basketry must be cut from young sapwood; if it is too old, the wood will not be flexible enough to split or mould to the desired shape. Birch bark and cherry bark can be easily removed from the trees only within a short window of time. Furthermore, as described in chapter 6, the cut must be made carefully, and the bark taken off gently, so as not to damage the growing layer (cambium tissue) of the tree.

The amount of time women devoted to harvesting and processing their materials and to weaving baskets, mats, blankets, clothing, and hats was substantial, especially if extensive travel to particular harvesting sites was required. Many of these items – such as the decorated coiled cedar-root baskets of the Nlaka'pamux and Stl'atl'imx/St'at'imc and the wrapped twined trinket baskets of the Nuu-chah-nulth and Ditidaht – are constructed of two, three, or more different materials, each requiring its own harvesting expeditions and processing methods. Although sometimes baskets could be fashioned roughly and quickly as needed from folded sheets of cedar or other bark, most took many hours, or sometimes days or weeks, to complete.[25] Not surprisingly, such objects were treasured and used with care over many years; some were repaired to make them last even longer.

Depending on the region, women in a single extended household might possess as many as twenty or thirty medium- or large-sized baskets for transport, storage of roots and berries, cooking, and other purposes, as well as many woven mats and bags (Schlick 1994; Turner 1996; Turner, Thompson, et al. 1990).[26] Thus basket making, including gathering materials, processing, and weaving, may well have consumed from one-quarter to one-half of a woman's working hours each year in some regions. Some women were specialists in basketry arts and may have spent even more of their time as weavers. Weaving can also be a very sociable activity, a time for teaching girls the required skills and for stories and conversation among friends and across generations.

Some items might be made by both men and women working together. For snowshoes, men might harvest and shape the wood for the frames – of yew sapling, birch, or maple – and women might prepare and lace the rawhide webbing. Similarly, both men and women might make fibre twine from nettle or Indian-hemp and work together to prepare fishnets. Both men's and women's technologies were as essential for people's survival as the food they procured, providing together the prime means of transport, shelters of all kinds, implements for obtaining food, and vessels not only for carrying goods but also for cooking and storing food. Children assisted in all of the harvesting, processing, and manufacturing activities, the tendency being for boys to help their fathers, grandfathers, and uncles and for girls to help their mothers, grandmothers, and aunts.

Technological pursuits of all types became increasingly specialized over the millennia following the Pleistocene, with increasingly enhanced opportunities for trade and exchange of canoes, baskets, implements, and other items, as well as the techniques and designs used in their construction. Gathering the materials and constructing these various items were not simply practical activities; they also provided opportunities to learn about other species and reflected people's spiritual connections with the plants and their environments. Harvesting, carving, and weaving all required deep understandings of the protocols that would minimize damage to the plant resources and pay appropriate tribute to individual plants, such as cedar trees and birch trees, for giving their lives or parts of themselves and for allowing themselves to be transformed into canoes, houses, or other items of beauty and utility. Consider, for example, the relationship between a woman and a birch tree from which a sheet of bark is removed. The point at which harvesters reached an understanding that they could remove the outer bark without killing the tree would have been pivotal, both in terms of developing practices that were sustainable and in terms of fostering an awareness of nonhumans as living entities. When, and how often, this realization was achieved is not known, but the concept of trees and plants as sentient living beings is undeniably universal within the study area, indicating an idea carried forward and disseminated from deep time, possibly right from the time of people's entry into the region.

Medicine and Healing

Treating diseases, ailments, and injuries and providing required assistance in childbirth or for the aged are necessary activities for any group of people. Although it is difficult to prove the antiquity of medicinal plant use (Leonti, Sticher, and Heinrich 2003), it is highly likely that the earliest peoples in northwestern North America were already using many different plants – as well as some animals – for various medicinal treatments. Use of medicines – in contrast with food or materials – is arguably more contained and exclusive within the knowledge and practice of individuals or families.[27] Yet a host of different medicinal plants have widespread and similar usage across the cultural, linguistic, and geographic boundaries of the study region, indicating considerable common knowledge and exchange of medicinal practice over millennia. Some medicinal species, like devil's-club and false hellebore, are particularly distinctive and potentially drastic in their actions, and it is conceivable that their medicinal qualities may have been discovered and put to use independently by different groups. More likely, however, the knowledge of their potency would have been spread widely and rapidly with its initial discovery. Numerous aromatic plants (e.g., sagebrush and wormwood, Labrador tea, yarrow, spruce, and juniper) are widely used medicinally around the northern hemisphere. They may have been discovered and developed as medicines on many occasions in different places, but, alternatively, they may well have been known to ancestral Eurasians even before people entered the New World, their use simply evolving and spreading from an ancient tradition.[28]

Considering that many of the medicinal plants used in northwestern North America also have names that are related across languages and even language families, the likelihood is that – despite the degree of privacy around medicines in recent times – knowledge of medicines was fairly widely shared through one means or another and that the sharing extended back at least four or five millennia to the time of Proto-Salish and other ancestral language groups. Another indication of active exchange of medicinal plant knowledge is the congruence in the ways that these medicines are prepared and administered. Many different peoples, for example, make salves from tree pitch – most notably pitch from coniferous trees like spruces and true firs and resin from cottonwood buds – heated with animal fat. Drinking infusions or decoctions of plant materials, especially of tree barks and leafy twigs of trees and shrubs, is a predominant way of administering medicine internally. Plant materials are sometimes mixed together to make particular medicines, and it seems more than just coincidence that both the Saanich (Coast Salish) and the Saik'uz Dakelh, or Carrier (Dene, or Athabaskan), are known to make a medicinal mixture containing ten different ingredients.[29] At some point, through some venue, it seems that there was a connection, direct or indirect, between these two distant and unrelated peoples.

Bathing in plant infusions or decoctions is widely used as a treatment for arthritis and muscular aches and pains. Sweat bathing – in a framed sweat lodge covered with skins or matting (commonly used by Interior Plateau peoples) or on a bed of aromatic plant materials laid over hot rocks (a usual coastal method) – is another well known treatment for arthritis and rheumatism. Additionally, aromatic plant materials are widely applied in the form of smudges or incense, inhalants, or washes for disinfecting a household, furnishings, and clothing during times of death and disease. Scrubbing and ritual purification during bathing or sweat bathing, and using plant materials as washes for hunting and fishing gear, are also widely practised.

There tend to be two different types of healing practised throughout the study area: *herbal healing*, mainly using plants and other natural products, which is often practised by specialists who are particularly knowledgeable about the medicinal properties of plants; and *spiritual healing*, which may involve the use of plants, too, but also has a strong supernatural component and is administered by a shaman or "medicine man" (or woman), usually someone who has undergone prolonged training and may undertake spiritual journeys to ascertain the cause of a physical or psychological problem.[30] In almost every family, there are herbalists who are called on for advice or to administer medicine for what would be considered mainly physical ailments, injuries, or conditions (e.g., assisting women in childbirth and with care of infants and children). Many of these herbal specialists are women. Practising herbal medicine also reflects spiritual values; herbal practitioners throughout northwestern North America and beyond uphold that, for a medicine to be effective, it must be gathered with care, focus, and respect, as well as with words of thanks given to the plants providing the medicine, both when the medicine is harvested and when it is being prepared and administered (Boas 1930; Turner and Hebda 1990, 2012).[31] As with harvesting plant materials, the harvesting and use of plant medicines, intertwined with spiritual beliefs, would ultimately lead to sound resource management practices.[32]

Few of the hundreds of different medicines used by Indigenous peoples in northwestern North America have been clinically tested, but many are known to contain chemically active ingredients, most notably alkaloids or glycosides but also other compounds such as tannins, resins, and volatile oils. Many of the same or related species are used medicinally in other parts of North America and even in Europe and Asia (Blumenthal, Goldberg, and Brinckmann 2000; Moerman 2003), and it is possible that there are extremely ancient roots to some of these medicines.[33]

In summary, particular medicinal recipes and precise applications of individual medicine species tend to be closely held as private knowledge and not as widely shared as food and material plants, at least in the recent past. However, in terms of general processes of harvesting, preparation, and application, as well as of the belief systems underlying harvesting and use, herbal medicine practice seems to follow patterns similar to those of food and technology plant use. Overall approaches seem to be widespread across the entire study region, with many of the most intensively used medicinal plants being known and used in analogous, if not identical, ways, indicating both longstanding use of medicinal plants and extensive sharing of knowledge about which species are effective as well as methods and concepts of application.

Resource Use: Places and Seasons

Variation in resource availability over the seasons and across space, as well as within different habitats and ecosystems, has always been an overriding determinant of peoples' lifeways throughout northwestern North America. Starting in earliest times, people had to accommodate short windows of opportunity when many of the products that they needed to survive were available at a particular place and had to travel to many different locales to obtain the total required complement of resources. People probably learned in the earliest times, too, that they could move upward in elevation with the progression of the harvest season, enabling them to obtain different resources – or sometimes the same types as at lower elevations – later in the year.

Within certain margins of variability, the times and places to obtain the different resources would have been predictable to those who knew and understood seasonal and ecological patterns, making it possible to plan and to some extent anticipate the places within a given landscape and within a given time frame that yielded the greatest benefits in terms of resources. To allow enough time for harvesting and processing a sufficient quantity of a particular resource, whether salmon, roots, or berries, it was often necessary for people to remain in a harvesting location for many days or weeks. During this interval, not only did they have to provide themselves with food and shelter, but they also had to harvest and store enough excess resources to meet their winter requirements.

Since there are normally limits to the productivity of any resource at a given place, having small groups of people – families or clan members, for example – spreading out over the territory is usually an excellent strategy for harvesting foods and materials without depleting the available supply. In a few cases, as formerly with the immense wapato patches of the lower Columbia Valley, the camas prairies of western and central Washington, and the salmon fishing sites along the Columbia, Fraser, Skeena, and other great northwestern rivers, the resources were abundant enough to meet the requirements for hundreds – or sometimes thousands – of people at a time. In some cases, too, resource sites can be rich in multiple resources – fish, edible roots, berries, and basket materials, for example – that are all available for harvesting at around the same time. Whether known for a single highly sought after resource or for a whole constellation of different resources, these particularly productive places, such as certain estuarine wetlands and subalpine parklands, became renowned centres of convergence for people (Turner, Deur, and Mellott 2011). These "cultural keystone places" not only provided families and communities with important resources but also gave them opportunities to socialize, trade goods, and exchange stories and cultural knowledge such as basketry or food preparation techniques. Over time, these places became nodes, or "interaction spheres" (Ames and Maschner 1999), of immense social networks that extended across the entire study region and beyond.

Peoples' regular movements over the lands and waters of their territories for resource harvesting, known as "seasonal rounds," are described in detail in chapter 8. People made their seasonal rounds both on the coast and in the interior, in the north and in the south – across the entire study area; such harvesting and travelling patterns still continue to some extent for some peoples today. To help determine the best times for their travels, people relied on seasonal indicators, including the relative positions of the sun, moon, and stars, the changing shape of particular snowbanks, changes in wind patterns, and tidal cycles, as well as on biological indicators (also known as phenological indicators), such as the growth stage of certain plants (e.g., the blooming of oceanspray or wild rose), the ripening of particular berries (e.g., salmonberries or saskatoon berries), and the appearance of certain migratory birds (e.g., geese or robins). Often, the knowledge of the timing of key resources is encoded in Indigenous languages as names of the moons or months of the year. These important indicators, however, have been shifting with global climate change, such that resource-harvesting times at particular locales are not always as predictable as they once were (Turner and Clifton 2009).

The patterns of seasonal rounds followed by the First Peoples at the time of the arrival of the first Europeans were, in many cases, established hundreds, if not thousands, of years ago. Around the beginning of the Late Holocene, when populations started growing and permanent winter settlements and houses became larger both along the Northwest Coast and in the interior, it became increasingly necessary for people to disperse across a wide landscape (or seascape) to obtain their resources.[34] These predictable rounds would have been undertaken in concert with more intensive management of habitats and resources, as well as with greater assumption of control over resource-harvesting locales by individuals and lineages. Higher levels of investment of time and energy in resource production made it necessary for people to make claims on the results of their focused work. This also probably resulted in greater specialization of individual families and lineages in terms of the particular resources under their control and further encouraged and strengthened relationships of trade and exchange between families, including ceremonial, political, and formal economic activities such as with the potlatch and intermarriage arrangements. These relationships, in turn, would have resulted in even greater rates of knowledge exchange – including the sharing of ethnobotanical and ethnoecological knowledge – both across generations and between different groups of people.

Social Organization and Plant Use

Division of labour between males and females has evidently been an ancient and widespread practice not only in the study area but also virtually worldwide, with men undertaking most of the hunting and fishing, carpentry, and implement manufacture and women most of the plant food and shellfish gathering and the processing and preparing of food for winter storage, as well as most of the harvesting of plant fibres and the weaving and manufacture of baskets, mats, and clothing (Barber 1995; Howard 2003). Associated knowledge and skills were thus passed down from mother to daughter, grandmother to granddaughter, aunt to niece, father to son, and so forth, in some cases over hundreds of generations.

Across the study region, as elsewhere in the world, there have been trends toward greater complexity and sophistication in resource use strategies and technologies since Pleistocene times. Social organization became increasingly complex in terms of societal stratification, intensified leadership positions, greater specialization of roles and activities, and social inequity, particularly over the last half of the Late Holocene. Larger populations and more focused settlements throughout the area meant a need for more resources both locally and regionally, which in turn required greater oversight and a higher level of economic and institutional control and regulation. Ceremonial and ritual ways of instilling societal understandings of cultural and ecological interactions also became more elaborate, as did artistic enterprise. Increased specialization in leadership, shamanism, woodworking, weaving and basket making, and various other occupations was a logical outcome.

Along the Northwest Coast these developments occurred in concert with more stabilized sea levels and, evidently, greater predictability of coastal resources. Over the entire region, fluctuating but generally more stable environments – relatively similar to those of the modern era – were in place starting about 3,000 years ago. Intensification of resource use through focused management on various scales and correspondingly stronger systems of land tenure also occurred. Alliances and ties between communities and even across different language groups helped to spread the availability of particular resources through enhanced trade and exchange, including intermarriage. Inevitably, too, resource use conflicts resulted in wars and other forms of hostility. Wealth, prestige, and authority accrued for some families and communities, with sufficient resource accumulation to allow leisure and the development of elaborate artistic imagery, even on utilitarian objects. At the other end of the social scale, slavery was a common institution (Donald 2003; Martindale 2003), with slaves being treated as more or less low-class members of a family in some cases or as disposable property, or "goods," in others. The labour of slaves and the knowledge and skills they contributed to overall societal development cannot be dismissed, nor can the contributions of women, children, and elders. Each and every individual had a role to play – varying somewhat with relative prestige, power, wealth, age, gender, life stage, inherited place in the nuclear family and community, intelligence, political and social skills, ability, and creativity – in building and maintaining the overall well-being of the family, clan, and community as a whole – including the nonhuman components: the plants, animals, and environments on which humans rely.

All of these developments – from increasing social and economic complexity to intensification of resources – influenced the ways that plants were used for various purposes, how the uses were organized, and how knowledge about plants and environments was communicated and disseminated. Ultimately and collectively, they led to the patterns of plant use seen and documented by the European newcomers entering northwestern North America.

Trade and Exchange

The archaeological record in the study region is rife with examples of how goods and information were exchanged across long distances over thousands of years. Obsidian – a volcanic glass that can, with skill, be worked into remarkable and effective tools and implements with edges sharper than a razor – is traceable with chemical analysis to its source volcanic mountain or formation and therefore provides a superb means of discovering the connections and travel routes of people. Other minerals, too, can be traced to source deposits, as can shells to some extent. The presence in archaeological contexts of obsidian and other rocks hundreds or even thousands of kilometres from their sources of origin, as well as of marine shells like dentalium, pecten, and abalone far from the coast in deposits dating back thousands of years, serves as a proxy for human interactions across landscapes and generations.

Plant remains and plants as trade goods are not as readily discernable from the archaeological record, and when found, their origins are not necessarily easily determined, but it can be assumed that plants, too, were traded, in the form of both raw materials and processed goods – and possibly also as living propagules – back into the distant past, probably along the same routes and through the same venues as the more durable minerals and shells. Along with these botanical goods, associated information, names, and stories about the species and their applications would have been acquired and exchanged, resulting in the patterns of naming of plants reflected in recent and existing languages.

We can obtain hints of how transmission of such knowledge occurred through accounts from more recent times (see chapter 4) of women who married into different communities and introduced new basketry techniques and new food-processing methods to their husband's people; of Saanich families who travelled to Interior Plateau locations like Yakima for agricultural work and traded strings of smoked clams from the coast for goods and clothing from the interior peoples; of Nuu-chah-nulth and Cowichan families who transplanted cattails and wapato from the Fraser Valley to wetlands in their respective territories on Vancouver Island; and of Stl'atl'imx/St'at'imc families who travelled coastward from the Lillooet area on the mid Fraser River to the Pemberton Valley in order to pick huckleberries with their Mount Currie Lil'wat relatives and then hosted these same relatives for the soapberry harvest up in the Fountain Valley. Each one of these occasions represents the face-to-face contact of individuals at a particular place and time. These multiple, focused, small-scale interactions, repeated year after year and generation after generation, accumulate to reflect the intricate web of similarities (and differences) in ethnobotanical and ethnoecological knowledge, practice, and belief that exists today.

Resource Management Systems

A key attribute of plants – especially perennial species (which comprise the overwhelming majority of species used by First Peoples of northwestern North America) – is their versatility in regrowing and reproducing themselves, especially through regeneration by meristematic tissues and through vegetative reproduction.[35] Combinations of strategies for encouraging plant regeneration, applied in concert with practices that mimicked natural processes, led to the maintenance and enhancement of individual plant populations and their habitats. Humans learned these strategies increasingly over time and by at least the Late Holocene had developed a whole suite of practices that can be termed "management" of plant resources: creating clearings and removing competing species through burning or weeding; harvesting at the optimal season for a plant's regrowth and renewal; selectively removing roots or other parts of a certain size or growth stage to allow continued growth of the remaining parts; partially harvesting bark, wood, or pitch and then allowing the tree or shrub to heal itself; pruning back branches of berry bushes or fruit trees or cutting back stems to ground level to activate growth and renewal; finding ways to channel moisture or nutrients into the plants' habitats to stimulate their growth and productivity; and scattering, replanting, and otherwise dispersing seeds, fruits, or propagules. All of these methods were employed in one way or another by Indigenous peoples in the study region, as well as elsewhere in North America and beyond.[36]

Collectively and on different scales, these actions resulted in sustained and enhanced plant resource production, as embodied in a Kwak'wala word shared by Clan Chief Adam Dick (Kwaxsistalla): *q*^{w}*'aq*^{w}*'alaʔowk*^{w} ("keeping it living") (cited in Deur and Turner 2005). The "keeping it living" approaches can be considered – along with innovations in plant processing and storage – to be strategies for resource intensification that must have built up gradually over the entire Holocene but predominated in the recent millennia. Ultimately, people applied these diverse modes of resource management and intensification over a wide spectrum of ecological zones and habitats, from intertidal and estuarine environments to subalpine parklands. How they learned to manage resources and environments in these ways is not really known, but astute observation of the effects of natural fire, wind, floods, and other environmental processes – including the disturbances caused by the digging and browsing of birds and animals, as well as those caused by people themselves – was undoubtedly a major source of learning and inspiration. Disseminating and exchanging observations, ideas, and results through social means and incorporating references and descriptions of activities and outcomes of management into narratives, ceremonies, and songs would have helped people to retain and enact key information and practices that supported a community's survival and well-being (Turner and Berkes 2006).

As noted previously, investments of energy in intentional resource management seem to have strengthened synchronously with increased populations, more complex social organization, occupational specialization, and formalized land tenure systems within the Late Holocene. One of the keys to peoples' success may well have been their reverent attitude toward the other species that served as their resources, embodied in their spiritual and kincentric belief systems, in which the lives of trees, berry bushes, and all other plants and animals truly mattered. The life of another living being was not taken heedlessly or wastefully or else bad luck would ensue. The species and habitats involved varied from place to place, but the underlying ideas were held in common throughout the entire area, and they pertained to animals as well as plants. Some plants – for example, western redcedar and yellow-cedar, western larch, or "tamarack," spiny wood fern, skunk-cabbage, and devil's-club – had direct associations with original supernatural ancestral figures. Others – like trailing blackberry and bunchberry in the "jealous husband" narratives – were specifically known for the legacy of their creation and transformation, sometimes through profound sacrifice on the part of an ancestor. Thus using the bark and wood of trees without killing them, grooming berry bushes to make them produce more and larger berries, and maintaining the camas prairies and estuarine and subalpine root gardens without depleting them were not just practical measures but also had profound and compelling spiritual foundations.

Narratives as Reflections of Botanical Knowledge

Knowing which plants are named, what the names are, and how they compare and relate to plant names of other languages and cultural groups can give many clues about peoples' past history and their interrelationships with the environment and with one another. Oral histories or traditional narratives add yet another dimension to the picture. Not only do they indicate – through common story themes and cycles of ancient stories – how people have connected with each other in the past, but they also provide insights into aspects of peoples' history, as well as into their perspectives and worldviews – the belief systems that shape their actions and decisions even to the present day (Cruikshank 1998; R.E. Ignace 2008). The origin stories – those "from the beginning of time," to use the words of Dr Daisy Sewid-Smith (Mayanilth) and Clan Chief Adam Dick (Kwaxsistalla) – are recounted over and over and serve as teachings for children and adults alike. Important lessons are embedded within them: proper conduct toward animals and plants, as well as respect for the powers of the environment.

The stories inspire generosity by recounting the consequences of greedy behaviour, and they instil a sense of gratitude for all of the gifts bequeathed to humans, often at great cost, by those who have gone before. The kincentric view of the world is reflected particularly in the transformation stories, in which the boundaries between humans and other life forms are eliminated and all beings – humans, animals, plants, and even rivers, mountains, and stars – are seen to be in one big society, each with potential spiritual powers to help or impact the lives of others, and in which humans are by no means the most powerful force. The idea of Skunk-Cabbage, Tree Fungus, Yellow-Cedar, or Blueberries as people who variously walked around, talked, paddled canoes, or dried salmon may seem strange, but the stories in which these individuals show their human forms are reminders of their close relationships to us and of their generous contributions to the lives of humans today. The ancient and time-honoured stories thus serve to connect people to their communities of kin, their ancestors, and their environments, engendering respect, appreciation, and right ways of behaviour. The consequences of breaching these protocols, as warned about in the narratives, are severe and can result in suffering and even death, not only for individuals but also for entire communities.[37]

Sometimes stories – such as those about the antics of Raven trying to imitate other animals or attempting through trickery to get more berries or food than he deserves – are amusing and entertaining, but they also carry serious lessons about socially acceptable behaviour. Other stories inform the audience about techniques and ways of doing things: how to make a "slow match" with which to carry fire, how to make a fish weir, how to use pitch as a protective coating for implements, how to treat a killed animal or fish, how to show respect for berries or other resources, how to burn the seeds of *q'əx̱mín* ("wild celery," *Lomatium nudicaule*) in order to feed the spirits of the salmon, or how to follow appropriate training, such as sweat bathing and purification with devil's-club, in order to gain luck, supernatural strength, and power. Many stories embody local ecological and geographical referents, so they teach people lessons about their own home places, including key historical and sacred landmarks, expected distributions of plants, which plants are useful (such as stinging nettle or Indian-hemp as sources of fibre), and species on which one can rely in times of food scarcity. They also remind people about how their ancestors acquired knowledge from other animals, like bears, mountain goats, and mallard ducks, who taught them about edible roots and other food they could eat. Origin stories like the "Old-One" narratives of Interior Salish traditions also explain how deeply engrained peoples' knowledge and customs are, demonstrating the division of labour between men and women, the kindling of fire, the making of spears and birch-bark baskets, the construction of mat lodges, the search for food during famine, and the digging of roots and drying of berries. As the narratives recount, all of these practices are based on instructions from the Creator and are therefore embraced as cultural norms.

Although many of the stories are programmed to particular localities and environments, they also carry common themes that resonate widely across cultures and language groups, sometimes in far-distant places. For example, versions of the story of the origin of sunlight – in which Raven transforms himself magically into the tiny needle of a hemlock or other conifer – occur up and down the coast, from the Tlingit and Haida to the Coast Salish and Nuu-chah-nulth. Similarly, the stories of Salmonberry Bird and his magical song that ripens salmonberries and other berries, of Spider teaching people to make fishnets, of the blind Duck women cooking roots in the sky country, of the "dog children," and of the berry-picking woman of high status who married a Bear extend throughout the Northwest Coast and beyond.

Some of the stories from the study area also align with narratives from other parts of North America and even from Asia and Europe. There are over 100 versions of the "Star Husband" tale, in which young women root diggers are transported into the sky country as the wives of stars and manage to lower themselves back to earth by making a long rope of fibre. This theme extends all along the Pacific Coast from Alaska to California, across the Interior Plateau and the Great Plains, east to the Great Lakes and the Maritimes, and north to the boreal forest (Boas 2002). The diffusion and integration of themes and stories across such a wide geographic and cultural space are a testament to the remarkable network of communication and exchange that has existed for millennia in northwestern North America and right across the continent. It is quite likely that other cultural traits, including knowledge, practices, and beliefs that relate to numerous types of plants, were similarly spread, transformed, and integrated into diverse languages and cultures; the patterns of influence across various linguistic and cultural boundaries are probably as relevant for botanical and ecological knowledge as they are for stories and other forms of cultural expression.

Philosophy and Worldviews

Recognition of the close relationship between humans and other entities of the environment – kincentricity, or kincentric ecology – is fundamental to the belief systems of Indigenous peoples in northwestern North America. Another, related aspect is the deeply spiritual relationship people have built with their environments. Every activity – from gathering foods and basket materials to harvesting cedars for canoes, preparations for hunting, and administering medicines – has an essential religious and spiritual side to it, reflecting the perceived powers and influences seen in the plants, animals, mountains, and environments that must be acknowledged and respected. If these established protocols are not followed and if proper expressions of appreciation are not directed to the spirits of the salmon, the fruits, and the other life forms that give themselves to people, these beings will withdraw themselves as supernatural benefactors and may, instead, actively harm those individuals and communities that commit transgressions. Peoples' stories, dances, First Food ceremonies, and symbolic representations carved in wood or painted on rock also serve as reminders of the spirit within and underlying every part of the universe.

Social constraints against wasteful, greedy, or thoughtless actions toward the plants, animals, and powers that provide for humans – communicated over generations through parables, rituals, and art – may be even more powerful and more enduring than formal laws or regulations. Community-mediated controls against improper resource use, embedded in cultures and institutions of local and Indigenous peoples worldwide, exemplify alternative models for conservation and management (L. King 2004). As argued by political scientist Elinor Ostrom[38] (1990; see also Dolšak and Ostrom 2003), such community-based cultural controls feature in the management and regulation of so-called common property resources and are potentially effective and formidable forces in our current global efforts to conserve biodiversity, manage natural resources and protected areas, and even reverse trends in climate change (see E.N. Anderson 1996; and Berkes 2012).

ACCUMULATING BOTANICAL KNOWLEDGE: A SYNTHESIS

The immense diversity of knowledge, practice, and belief around plants and environments among the many Indigenous groups in the study area had its roots among small numbers of people entering the region by land or, more probably, along the coast toward the end of the Pleistocene glaciation over 14,000 years ago. Indications are that the people arriving here were already well acquainted with many plants – species that are widespread and still important to Indigenous peoples of the past few hundred years. These include food plants like northern riceroot, lingonberry, soapberry, fireweed, cow-parsnip, spiny wood fern rootstocks, bracken rhizomes, and pine cambium; material plants like birches, spruces, lodgepole pine, poplars, willows, bull kelp, sedges, and stinging nettle; and medicinal plants like yarrow, sagebrush, and tree pitch and barks.

The Monte Verde archaeological site in Chile, dating to this early time, gives us some understanding of the lifeways and technologies of the contemporary people of northwestern North America since it is likely that the Monte Verde people travelled along the Pacific Coast via the "kelp highway." The earliest peoples in the study area, from this and other evidence, used fire and different fuels, knew how to cook their food, used wooden implements like wedges and digging sticks, made cordage and matting from plant fibres, including stinging nettle and willow bark, probably stored at least some of their food in woven or stitched bark containers as well as in caches and caves, and probably had some type of boats. These early peoples would have had names for the plants they used, probably shared stories and songs about them, and may well have held worldviews that included reverence for plants and that imbued them with spiritual powers, as they did animals. They probably practised smudging, bathing, and purification using plant materials, and this would have enhanced their hunting success and helped to protect them from predators. Shamans, as well as herbal healers, likely would have been active then as in more recent times.

Knowledge about new useful plant species encountered and about effective new techniques in weaving, woodworking, harvesting, fishing, or hunting would have spread from family to family and group to group – just as obsidian, shells, and other valued materials did – along trade routes, trade networks, and nodes of gathering that were established very early on, as documented in the archaeological record. Likewise, new names and vocabulary, stories, and ideas about plants and their use would have been continuously developing and spreading across linguistic and cultural boundaries.

Life would have been undeniably harsh in the early millennia, with residual ice, torrents and floods, and changing sea levels requiring flexibility and strong adaptive capacity, although many of the seemingly dramatic changes actually occurred over centuries, rather than days, years, or decades. Judging from some of the origin stories, however, famines and food shortages would have been frequent, and in these narratives, knowledge of plants as more predictable food sources – especially root vegetables like bracken and spiny wood fern and fruits like rosehips, highbush cranberries, and kinnikinnick berries that remain on the bushes over winter – often meant the difference between life and death. Knowledge about firewood, tinder, and ways to keep fires burning was also a common theme of the stories from deep time.

Human populations remained sparse and relatively mobile throughout most of the area, although some places, such as around river estuaries and lakeshores, where people had access to a variety of different ecosystems and resource species, would have supported higher numbers of people in more permanent settlements. Some of these places, such as Namu on the central coast of British Columbia, have remained more or less continuously occupied to the present, probably with some of the very same plant species being used initially as more recently, although the plant diversity would have been significantly lower since some species – like western redcedar – apparently did not extend their ranges that far north until around the Mid Holocene.

Ancestral languages – and plant vocabularies distinctive to those languages – were initiated at least by the Mid Holocene and possibly earlier: Proto-Haida in the area of Haida Gwaii today; Proto-Na-Dené (including what would become the Eyak, Tlingit, and Dene [Athabaskan] languages) in central Yukon, Alaska, and north-central British Columbia; Proto-Salishan in the lower Fraser Valley;

and so forth. These early languages would not have been as complex or nuanced as they later became but would have had terms for firewood, for common trees like willow and spruce, and for edible roots, berries, and other types of plants of cultural salience. Many of these names probably reflected key characteristics of colour, shape, texture, taste, or odour, as well as how they were used or the animals with which they were associated. A good example is the name for soapberry, evidently derived in both Proto-Salish and early Dene (Athabaskan) from a term meaning "to foam" or "foaming" (Turner and Burton 2010). Even languages like Haida that have retained much of their unique character over several millennia show evidence of borrowing of plant names from other people, indicating intergroup contact of one form or another. The similarity between the Haida name for western redcedar, *ts'uu*, and the name for spruce in several Dene languages and the congruence of the Haida and Ts'msyen names for highbush cranberry are two examples, although the time frame and direction of borrowing are uncertain in both of these cases.

Likely, as in more recent times, men and women had generally separate roles in the very early days, with men undertaking most of the hunting, fishing, travelling, woodworking, and heavier construction and women undertaking the harvesting of plants and shellfish, processing of food, and harvesting and weaving of fibrous materials, as well as the care and teaching of infants and children.

It seems, from all accounts, that around the beginning of the Late Holocene – from about 4,000 to 3,000 years ago – was a time when the patterns of plant knowledge and use as known and documented from the past few centuries were firmly established. Sea levels were generally stabilized, and resources became more predictable. The vegetation and forest composition seen today were largely in place, although there were intervals of fluctuation in temperature and precipitation, in the extent of grassland and upland meadow versus forest, and in the relative size of wetlands and water bodies compared to terrestrial habitat. Human populations expanded, houses became bigger (at least in some cases), and permanent settlements grew in size, accompanied by greater social stratification and occupational specialization, especially evident on the Northwest Coast. More people meant the need for more resources of all kinds and perhaps limits to certain key resources – including some berries, roots, cedar bark, or stinging nettles – which then had to be managed and regulated. Diversification of resource use through seasonal rounds – regular movements between different elevations and resource locales – with sites being accessed by smaller family or clan groups, was one strategy to limit stress on resources. Another was developing more sophisticated food-processing and storage techniques.

Of particular interest is the development, over time, of management systems for plant and animal resources, including a suite of practices to maintain and enhance key resource species and habitats: burning over areas to create and maintain openings and mosaics of successional stages; selectively harvesting greens, root vegetables, and tree cambium; pruning berry bushes; replanting, transplanting, and scattering propagules; tilling and fertilizing the soil; weeding undesirable grasses and other species; and in some cases constructing and maintaining terracing for optimal plant growth and better access (Croes 2003; Deur and Turner 2005; Lepofsky 2009a; Peacock and Turner 2000; Turner, Deur, and Lepofsky 2013). Along with these practical techniques for encouraging productivity of resources were associated institutional, social, and cultural developments – including protocols for managing and sharing the resources.

Especially on the Northwest Coast, but also to some extent in the Interior Plateau, increased social complexity – with the creation of a leadership class – was evidently yet another outcome of denser populations and resource limitations. Chiefs and nobility often took control and ownership of key productive areas, assuming not only the rights to particular resource species and locales but also the responsibility for their care and maintenance. They oversaw the harvesting, processing, and storage of the resources, and through various means, including feasting, they shared the resources with their communities, as well as exchanging them with others as gifts and trade goods, thereby bringing prestige and support to themselves, their families, and their communities. Thus, at some point, probably in the Late Holocene, prime crabapple groves, prime stands of cedar, prime patches of salalberries, highbush cranberries, and huckleberries, and prime root garden areas for springbank clover and edible camas became "owned" resources, which were subjected to more intensive management and enhancement under the oversight of chiefs or other recognized leaders or specialists.

Combined with highly productive sites and habitats situated strategically at nodes or crossroads of travel and interaction, systems of social and ecological management and intensification of key resources led to the creation of "cultural keystone places" – places throughout the study area and beyond that are known as resource hubs and as centres for social, cultural, and economic exchange. Some of these, like The Dalles on the Columbia River and the Fountain area on the mid Fraser River, are extremely old; The Dalles shows evidence of being a focused harvesting site going back around 9,000 years at least. Some of these "cultural keystone places" are in montane regions (Turner, Deur, and Mellott 2011); many are situated along the coast, beside waterways, or at river estuaries. These places are usually rich in multiple resources like fish, edible roots, berries, and basket materials, many of which are available for harvesting at around the same time. Not only are these places important as nodes of trading and social interaction, but in terms of transmission and exchange of botanical and environmental knowledge, their role has also probably been critical. Likely, it was at such nodes that new technologies (e.g., new designs for canoes, adzes, fishhooks, or digging sticks; basketry and spinning techniques; and bows and arrows), new ideas (e.g., cooking in earth ovens, creating clam gardens or root gardens, and pruning to enhance berry production), and novel products (e.g., edible roots, berries, basketry materials, and medicines from other places) were acquired, either directly from the original sources or indirectly through a second or third party travelling from some distant location. Along with these new technologies, ideas, and products came related names and vocabulary, as well as stories, songs, and ceremonies. Since spouses were met and slaves bought at such gathering places, they served as important sources of such information, resulting in even greater rates of knowledge exchange.

Some of the travellers drawn to these nodes of interaction evidently came from far-distant places, just as peoples from the study region travelled to gathering places and communities far outside of the region, such as in the Haida and Heiltsuk excursions to California. Interior Salish and Nuu-chah-nulth peoples, too, may have travelled long distances to meet at some place like The Dalles and somehow transfer information, such as the name for skunk-cabbage, from one group to another; and this must have occurred far enough back for such a term to have been widely disseminated among both language subfamilies. Similarly, Nisg̱a'a or Gitxsan people may have travelled far south, long ago, to the Proto-Salish homeland around the lower Fraser Valley area, where they somehow acquired the Salishan name for hazelnuts. There are other intriguing linkages: pit-cooking, basketry technologies, and other traditions that connect people in the study region with those of the southern Plains and Great Basin regions, possibly via the Willamette Valley region of Oregon or the Calispell Valley of eastern Washington; stories from the Northwest Coast that have themes similar to those from eastern Canada and even the southern United States; and practices around harvesting and use of birch bark from the Secwepemc and Dakelh peoples that are similar to those of the Anishenaabe of eastern Canada (Turner, Davidson-Hunt, and O'Flaherty 2003). There are even ties through foods, technologies, story themes, and in the case of the Dene (Athabaskans), apparent language relationships with northeastern Asia (Croes, Fagan, and Zehendner 2009a, 2009b; Kari and Potter 2010a, 2010b).

In terms of the future of humanity, perhaps the most significant element to emerge from this study on the development of knowledge, practices, and beliefs that relate to plants and environments is the widely embraced belief system – common to virtually all of the Indigenous groups in the study area and beyond – that underpins peoples' relationships with and use of the environment. Its fundamental nature is reflected in peoples' food systems, technologies, and medicinal practices throughout the region (see F. Brown et al. 2009; Turner 2005; and Turner and Atleo 1998). Termed "kincentric ecology," or "kincentricity," it is the understanding that all plants and animals – and other elements of the environment – are sentient living beings, closely related to humans and often with the power to influence human lives positively or negatively depending on how we behave toward them. It brings an imperative for respect and reciprocity, including acknowledging and giving thanks to those life forms that help and provide for us, using them with care, and not wasting them. This philosophy, or worldview, is reflected in stories, songs, dances, and ceremonies – such as the First Foods ceremonies – throughout the study region. Although we associate traditional resource management practices with the past few millennia, the philosophical underpinnings of these practices must have ancient roots, as enshrined in narratives and other traditions, likely going right back to the time of people's entry into northwestern North America and earlier. These belief systems have fostered a caring attitude toward plants, animals, and the environment that seems to be lacking in mainstream society. If such a worldview could be more widely shared and adopted, we might have a better chance of maintaining ecological integrity and biodiversity in the world.

PATHWAYS AND VENUES FOR LEARNING AND KNOWLEDGE EXCHANGE

We possess all of the information reported in this book that relates to cultural knowledge, names of plants, species that are edible or have medicinal properties, and technologies and approaches for using plant materials, enhancing plant growth, and maintaining habitats because people passed it along from previous generations, teaching, demonstrating, modifying, and adapting it. Even in the times of great change and loss, some individuals were able, through desire and circumstance, to retain the knowledge of the past and bring it forward to the present (see Turner, Robinson, et al. 2012). Given the crucial role of these remarkable individuals – these keepers of culture – in acquiring, adapting, and disseminating ethnobotanical and ethnoecological knowledge, and given the rich legacy of knowledge about plants and environments revealed from ancient times, what lessons can be learned from examining the modes and circumstances of this learning and sharing? Furthermore, how can these lessons be applied as components of ongoing cultural revitalization and the maintenance of biocultural richness? The remaining sections of this chapter address these questions, starting with considerations of pathways and approaches to knowledge transmission and adaptation.

Acquisition, transmission, and adaptation of cultural knowledge, practices, beliefs, and language that relate to plants and environments are an integral and essential component of Traditional Ecological Knowledge systems (Berkes 2012; R.E. Ignace 2008; Turner, Ignace, and Ignace 2000; Wyndham 2004). The patterns of ethnobotanical and ethnoecological knowledge of northwestern North American First Peoples identified from this research have resulted from intricate combinations of social-ecological processes, some of which had their beginnings even before humans first stepped onto the North American continent with their botanical and environmental knowledge from their ancient

Asian homelands. In terms of ecological processes, the same forces that affect patterns of distribution and abundance in plants and animals can also influence humans. For example, the fortunes and opportunities for survival of all life forms, from ancient times to the present, are directly affected by sea level fluctuations, climatic shifts, volcanic eruptions, droughts, floods, and the abundance and productivity of biotic resources. Additionally, peoples' names, uses, and cultural practices that relate to other life forms depend on their familiarity with them and thus on the immediacy or availability (i.e., abundance, frequency, and distribution) of these organisms as living entities, as raw materials, or as finished goods. In turn, humans are active participants in the system, and their activities can directly influence the abundance, frequency, and distribution – and even the genetic composition and evolutionary trajectories – of other biota.

Ethnobotanical and ethnoecological patterns align well with biogeographical patterns and processes; application of the latter can facilitate our understanding of the tremendous complexities, congruences, and variations in the former. Concepts such as endemism, isolation, association, disturbance regime, microclimate, vector, geographic barrier, and corridor can all be adopted into the realms of ethnobotanical and ethnoecological knowledge transmission, as can terms related to biological change and transformation: centre of origin, parallel and convergent development, selection, adaptation, replication, fragmentation, dissemination, affinity, incursion, and diffusion. Metaphorically, processes like propagation, germination, cross-pollination, hybridization, and symbiosis can also be applied in the context of patterns in ethnobotanical and ethnoecological knowledge.

As with the patterns of biological species, the existence of any particular cultural knowledge – the name or application of a given plant in a particular language or culture, for example – is dependent not on one thing but on a combination of many influences. Some of these predominate, however, and may apply simultaneously to suites or clusters of cultural knowledge. Identifying those factors that, in combination, are the most influential in determining ethnobotanical and ethnoecological patterns in northwestern North America requires a detailed examination of many different phenomena on different spatial and temporal scales. Although this research provides the beginnings of such an examination, it will not provide conclusive answers but rather will suggest ideas to be further considered and built upon.

Contact between two or more individuals at a single place and at a single time is generally required for information to be transmitted orally – although there are exceptions, such as with new knowledge acquired through dreaming or spiritual training, as explained by Dr E. Richard Atleo (Chief Umeek) (2004, 2011; see also Turner and Atleo 1998).[39] Many Indigenous specialists in herbal medicine or art can attest to gaining inspirational teachings through fasting and ceremonial purification in the form of visions and spiritual guidance, which they have then enfolded into their personal accumulating store of knowledge, practice, and belief. Furthermore, considered through the lens of kincentric ecology, individuals participating in knowledge transmission need not necessarily be human since animals and plants – and, some would say, even rocks – can be teachers and conveyors of critically important information. Coming to a deep understanding of humans' place in the ecosystem can thus involve diverse routes and pathways even on the most restricted scale.

There are many different interpersonal relationships involved in the transmission and acquisition of ethnobotanical and ethnoecological knowledge. Intergenerational teaching, as noted previously, has tended to take place within the circle of a family – between spouses, parents and children, or grandparents and grandchildren. Each stage in an individual's life has had its own balance between learning and acquiring knowledge, on the one hand, and sharing, teaching, and enacting knowledge, on the other. For First Peoples of the study region, infants, even while still unborn, began their journey of learning through songs and stories from their mothers, shaped in some cases by intended roles determined through inherited positions or prophetic dreams. Training and instruction continued in many forms once the baby was born. Dr Daisy Sewid-Smith (pers. comm., 2009) described how, in Kwakwaka'wakw traditions, a newborn infant's umbilical cord would be made into a bracelet, dried, and given to a talented person admired by the parents – a canoe builder, basketmaker, or healer, for example – to wear in order to create a bond between the child and the expert and to encourage the talents that would be offered and taken up as the child grew. Being confined to a cradle and carried around or being positioned on a tree branch while its mother dug roots or picked berries nearby taught an infant patience and contemplation (Mary Thomas 2001). Washing a baby with mushroom juice gave it strength and independence, according to Nlaka'pamux traditions (Turner, Thompson, et al. 1990). Ashes from sea palm (*Postelsia palmaeformis*) rubbed on the spine of a Ditidaht infant destined to be a whale hunter prepared him for the difficult training and harsh trials he would endure in his future life (Turner, Thomas, et al. 1983).

Children of nobility, especially in the highly stratified societies of the Northwest Coast, were groomed as chiefs and matriarchs from the time of their infancy and often underwent intensive and strenuous training in oratory and other skills of leadership (E.R. Atleo 2004, 2011; Recalma-Clutesi 2007). Toddlers and children in general were carefully observed by parents, grandparents, aunts, and uncles as they grew up in order to determine their aptitudes and interests so that by the time they were eight or nine years old their future specialization in a profession could be supported and encouraged as much as possible. All children were provided with opportunities to observe and practice the basic skills they would need as adults: hunting, fishing, and carpentry for boys; and plant gathering, food processing, basketry, and so forth for girls. Experiential learning through participation in a family's or clan's essential activities was an overriding method for children to acquire all kinds of cultural knowledge. Even young children participated in food harvesting, for example, and many provided themselves and even their families with a substantial portion of their required daily nutrition by foraging for greens, berries, and clams, by fishing, and by snaring birds and small mammals (Gerald Amos, pers. comm., 2008; Adam Dick, pers. comm., 2008; Turner and Thompson 2006). Times of travelling and camping out with their families were also effective learning opportunities for children and youth. Key landmarks, sacred ancestral sites, important harvesting places, springs of water, and species of plants and animals not normally encountered were pointed out on such travels, often emphasized by associated place names and stories, to be recalled and passed on many decades later. Children and young adults also witnessed and participated in management practices like landscape burning, pruning of berry bushes, replanting of roots, and ceremonial scattering of berries when they were out on the land with parents and elders.

Focused instruction of children and youth during teachable moments, especially around food preparation and eating, was another time-honoured mode of passing on information. As children take food into their bodies, it is often said, they are particularly situated for taking in knowledge. As recounted by Tlingit elder Betty James, "They usually talked to younger people after dinners in the evenings, make them sit down so they can listen, sometimes usually around a fire and make them sit, make sure they listen. And later on, they're through eating, sometimes still around the table, they used to do that to us, my dad. He made us sit so we listen to what he's telling us. And they never used mean words when they're trying to explain anything. It's usually told in a nice way" (cited in Kake 1989, 20–1). Stories, rich with cultural lessons and teachings about plants and environments, were told to children over and over again by grandparents and other elders in a household, with some individual narratives continuing over three or four nights. Children were rewarded with stories for tasks that they undertook, like pulling the bones out of dried fish, helping to pick berries, or bringing in fuel or water. Many of today's elders, like Clan Chief Adam Dick (Kwaxsistalla), remember these occasions and the stories that were told to them, which they are now recounting to their own grandchildren and great grandchildren (Turner, Recalma-Clutesi, and Deur 2013).

Formal oration by knowledgeable elders – historians and experts in narratives and traditions – at feasts, potlatches, and other gatherings was another means of transmitting important knowledge: origin stories of families and communities, accounts of the boundaries of a nation's or clan's territories, stories of war, conflict, and resolution, and instructions on cultural protocols and the dangers of transgressing them. Even young children were made to listen carefully to these speeches, as recalled by Chief Earl Maquinna George (2003).

Children were also brought into the spiritual and ceremonial life of families and communities from an early age. Young children participated, and still do, in dances and were sometimes key participants in First Foods ceremonies – in the Saanich First Salmon ceremony, for example, where young children were the ones to bring the first salmon up from the water to an elder to be ceremonially welcomed. Children were taught through stories and by example to respect and appreciate other life forms and never to torment or needlessly kill animals, never to point at certain mountains, and not to pick certain flowers. As they grew older, they learned to offer formal thanks to medicinal plants and other resource species before harvesting, and they were taught associated protocols such as only removing one piece of cedar bark from each tree, not cutting too deeply into a birch tree when removing the outer bark, or leaving a gift of food behind if they took nuts or roots from a rodent cache.

Children also learned through contests, competitions, and games, such as the Nuxalk game with *tsayamuus* (paintbrush, *Castilleja unalaschcensis*) (see chapter 6) or the Ditidaht *pilaa-pilaa* contest with sword fern fronds (see chapter 9), which helped children learn to control their expressions of emotion, in the first case, and trained boys to hold their breath for long periods of time, in the second. Puberty was a special time in a young person's life that included training focused on the various tasks they would be responsible for as adults.

As already noted, grandparents often had an immense influence on their grandchildren's acquisition of knowledge and skills. Children were often cared for by grandparents while parents were otherwise occupied, and grandparents provided many opportunities for detailed and ongoing instruction, including participatory learning (Turner 2005; Turner, Robinson, et al. 2012). The memories of grandparents held by knowledgeable elders of recent generations – Clan Chief Adam Dick (Kwaxsistalla), Dr Daisy Sewid-Smith (Mayanilth), John Thomas, Joan Morris (Sellemah), Dr Arvid Charlie (Luschiim), Dr Margaret Siwallace, and Dr Mary Thomas – are invariably filled with affection, and the teachings they imparted have remained clear and compelling – in some cases, over six or seven decades.

Marriage presented opportunities both for learning and imparting new knowledge from, and to, the clan or community of one's spouse, as well as for eventually passing on teachings and skills to one's children. For example, Gitga'at matriarch Helen Clifton came to the Gitga'at community of Hartley Bay as a young bride from the Prince Rupert area, and in the early years of her marriage, she was taught many critically important methods of food preparation by her mother-in-law, Lucille Clifton, and by her sisters-in-law, who were older than she was: how to cut and sun dry halibut and salmon, how to smoke fish and seal meat, and how to harvest and process seaweed. Now Helen, herself an elder, is a major instructor of other women in these skills. Many women, like Helen, moved after marriage to their husband's community, which might be some distance away and sometimes even from a different language group, since marriages were often arranged to provide opportunities for resource access and trading alliances in different places.[40] These women would become functionally bilingual or, later, multilingual since, in the time following European contact, they usually also spoke Chinook Jargon and English. They, in turn, often taught their children some of the language and traditions of their original community, as well as of their adopted community, and served as conduits for channelling information and resources – including plant resources like wapato, camas, seaweed, bitterroot, huckleberries, and soapberries – from one group to the other.

Cultural exchanges from place to place could, at times, be fleeting and unpredictable. Sometimes a single traveller or group of travellers, visitors from a distant location – perhaps like the young man in the Kwädąy Dän Ts'ìnchį discovery described in chapter 2 – might encounter others on the way and exchange observations, experiences, and stories. They might also stop and stay at another community for a period of time, such as over a winter, to be hosted by a family and, in turn, contribute new techniques, new stories, and new ideas to their hosts. Such stops might sometimes become routine, resulting in the establishment of "cultural edges" and augmenting opportunities for both learning and introducing novel ideas, terms, and innovations (Turner, Davidson-Hunt, and O'Flaherty 2003).[41]

One can also imagine chains and networks of learning that must have taken place, where innovations, stories, or new vocabulary acquired by one or more individuals in one location would be passed on to others and then retransmitted – sometimes very quickly – over and over until they became widely known and practised. Some of these ideas and practices would be adapted to new situations as they were transmitted, applied to different species and products, or altered in various ways to meet different cultural norms. Thus variations were produced – and reproduced – until, over decades, centuries, and millennia, the original source of a particular procedure, name, or application that related to plant use, for example, was altogether obscured. One can assume that the use of the bow and arrow as a hunting implement represented such a rapidly transmitted innovation since, over a relatively short period of time, bows and arrows replaced the spear and atlatl (spear thrower) almost everywhere. Most people were quick to adopt a new and more effective technology, as they are today. The variations existing in the designs and materials used for bows and arrows in different areas reflect the adaptations to particular environments and cultures.

Learning from others also took place at shared harvesting grounds and nodes of convergence – those "cultural keystone places" where people routinely gathered to harvest resources like salmon, oulachens, camas bulbs, or huckleberries, as well as to trade with each other and share cultural events. Wintertime feasts and potlatches also served as institutionalized networks of social and cultural exchange, providing chances for people to visit each other and to bring novel goods and ideas, as well as vocabulary, songs, dances, stories, techniques, and technologies, thereby increasing the resilience of individual family units and reciprocating communities. Less exotic but equally important opportunities for learning took place in the context of peer groups during harvesting expeditions, food preparation, weaving, carving, or other tasks – times of mentoring – when older people instructed younger ones and participatory learning occurred naturally and consistently, turning novices into experts over the course of the years.

One can assume that most of these traditions and venues for knowledge transmission go back to earliest times. The routes or pathways of dissemination of names and knowledge, and possibly even the plants themselves in some cases, would have followed major transportation routes: along the Pacific coastline; up and down the major riverways, namely the Stikine, Skeena, Bella Coola, Fraser, and Columbia; across peninsulas, headlands, prairies, plateaus, and lakes; and through the mountain passes. The actual vectors of dissemination – travellers, chiefs, slaves, brides, or combinations of people – can only be speculated about or inferred from oral traditions or from more recent experiences. Of course, the types of knowledge transferred would have varied with the individuals involved.

Groups of people travelling, moving, and settling into new territories would also bring their knowledge, practices, beliefs, and language with them, possibly blending these with the cultural knowledge and practices of people already inhabiting the destination area and adapting them to new environments and new vegetation as encountered. Such blended knowledge systems reflect elements from the two – or more – original sources, but often these elements may be difficult to distinguish unless they are obviously based on species not occurring in one or the other locales. Similar patterns may emerge for suites of characters, such as groups of plants that have similar distribution and use patterns across similar areas.[42]

Perhaps the most important lesson to be deciphered from examining modes and pathways of learning about plants and environments is that practical knowledge and cultural teachings can accumulate and convert themselves into mechanisms for social and environmental adaptation and change. As those who study political economy and human ecology can confirm, we humans are linked inextricably to our environments and are capable of causing profound changes to species and ecosystems through deforestation, overharvesting, introducing species, and other forms of disruption. Sometimes human impacts on resources and environments are cumulative and indirect or are manifested across different scales of time and space, making it even more difficult to detect the interplay of causes and effects. Nevertheless, key indicators – including an alarming loss of biocultural diversity – tell us that we need to change our behaviours if we wish to stem the tide of global deterioration that we have, collectively, let loose. Currently, many of us in the so-called "developed" world are using resources and degrading ecosystems beyond the capacity of the earth to recover and restore itself. We are depriving other species and future generations of the habitats they likely will need to survive. Furthermore, because of past colonial practices and current government policies, Indigenous peoples' rich and complex knowledge systems, as discussed and reflected in this book, have been eroded, and their once-strong ties to their lands and territories have been weakened and in some cases lost altogether. Over the past several generations, children have not been learning the details and levels of knowledge about plants and environments that their parents and grandparents knew. A two-pronged approach for mainstream society is therefore needed that will, first, reverse the environmental degradation we humans are causing on the earth and, second, strengthen and renew the planet's rich cultural diversity, including local and Indigenous knowledge systems. Thus finding ways to learn new and better behaviours for living within our means and developing new and better relationships within the "global commons" are crucial (E.R. Atleo 2011; Berkes 2012; K.M.A. Chan et al. 2007; W. Davis 2009; Dolšak and Ostrom 2003).

CULTURAL RENEWAL IN THE TWENTY-FIRST CENTURY

Most of the circumstances of recent change described in this book, particularly in chapter 4, have not been positive for Indigenous peoples in terms of maintaining their cultural identity and overall health and well-being. In some ways, one could argue, their botanical knowledge has expanded, with the influx of many new species and plant products and with the increased opportunities for exchange of knowledge across language and cultural divides, both among Indigenous groups and between them and the newcomers, including Europeans, Euro-Americans, and Euro-Canadians, as well as those of other ethnic groups such as Chinese, Japanese, Hawaiians, and Mexicans. For many Indigenous groups, dozens of new names for introduced vegetables and other plant species were incorporated almost seamlessly into vocabularies, along with many other new terms based on all of the novel devices and species made accessible (Kay 1993; L.C. Thompson and Thompson 1996; Turner, Thompson, et al. 1990). This seeming language enrichment, however, occurred only in the short term, until the languages started to disappear, to be replaced by English.

As well as the wide array of new and diverse plant foods that were introduced to the First Peoples throughout the area, new materials and implements allowed woodworking and other forms of art and sculpture to flourish.[43] In terms of travel, first horses and then cars, trains, and new kinds of watercraft of all sizes, from small motorboats to fishing boats, large steamboats, and later diesel ferries, made transportation of both people and products easier and faster. Imported woven fabrics of wool, cotton, burlap, and canvas, Hudson's Bay blankets, and sewing machines – along with European styles of dress – increased the possibilities for clothing, shelter, household furnishings, and art. New types of tools, containers, and vessels for cooking and storage enhanced food harvesting and production methods. The advancement of electricity – with its lights, electric stoves, freezers, and refrigerators – facilitated life in many ways, not only for the Indigenous peoples but also, equally or more so, for the newcomers, depending on economic circumstances and proximity to power generation sources.

Sadly, all of these innovations and introductions were accompanied by an overt policy by the colonial government to eliminate Indigenous languages and cultures and to assimilate Indigenous peoples as much as possible into the European mould. The newcomers were almost successful in their endeavours, but thanks to the staunch efforts of many people who struggled to maintain their cultural traditions against all odds, these rich cultures and many of the languages still endure. Generations of Indigenous people, who loved and believed in their cultures (the foods, the stories, entire ways of life) and saw the value and beauty in their languages (their place names, their plant names, their songs) have worked hard in many different ways to keep them alive and to pass them on to others. Their efforts can be seen in the publications and notes of ethnographers like Franz Boas, James Teit, John Swanton, Thomas McIlwraith, and Marius Barbeau, none of whom would have been able to accomplish their detailed descriptions and reports of the various languages and cultures with which they worked if not for their Indigenous teachers, colleagues, and friends who spent their time and energy meticulously explaining, translating, correcting, and answering what must have been countless questions and requests for information. The commitment of these people to keeping viable the rich botanical knowledge of their ancestors can be seen in the many books and other publications, theses, and films related to ethnobotany and ethnoecology in Northwestern North America that they have authored or directed or that have cited them as major consultants and collaborators (see Turner and Lepofsky 2013).

Some of these dedicated people have been recognized with awards and honorary degrees at universities and colleges for their work in maintaining and teaching about their language, culture, and knowledge of plants and environments.[44] Other Indigenous scholars have undertaken postsecondary education, achieving graduate degrees in areas related to ethnobotany or ethnoecology.[45] Countless others, equally committed to the maintenance, revitalization, and renewal of their botanical traditions and environmental knowledge, are known only to their own families and communities but must nonetheless be recognized. Without all of these people and their individual and collective work – in many cases alongside non-Indigenous scholars, supporters, and allies – the immense richness and diversity of these cultures and the languages within which they are embedded would have long disappeared, and humanity would be much poorer for this loss.

Today, in the twenty-first century, there are an increasing number of Indigenous people of all ages who are interested in learning or relearning their own languages (J.C. Thompson 2012). There is also continuing and expanding interest in Indigenous arts like basketry – with immensely beautiful spruce-root and cedar-bark baskets and hats being made by Haida and Tlingit weavers, for example, that are valued as priceless gifts or sold at a fair market value as works of art. Weaving traditions in other areas and with other materials are also strong.[46] Of course, the economic value of such items can in no way account for their cultural or intrinsic worth as iconic art works of immense and lasting importance that reflect the innovations, adaptations, and interconnections between peoples and environments over thousands of years.

The use of natural resources in local economic development has come to be known as the "bioeconomy" (R.B. Anderson et al. 2008). Indigenous peoples have always harvested and used their local resources, including trading goods and items to which they may have ready access or for which there is a particular abundance to neighbouring families and communities, sometimes over fairly long distances (see chapter 10). Today, there have been various initiatives within Indigenous and other local communities, sometimes assisted by academic institutions, governments, and nongovernmental organizations, to develop small-scale industries and markets for products from wild-growing plants and other species found in forests, wetlands, and montane and marine environments. Such products have been called "nontimber forest products" (NTFPs), among other terms (Turner 2001).

Despite their potential to provide jobs while still sustaining the growth and productivity of important species and habitats, these initiatives to market NTFPs are regarded by some Indigenous people as a dangerous commercialization of their cultural property and knowledge. There are fears that commercial use of culturally valued plant products like black huckleberries and edible seaweed will ultimately deplete them and reduce their availability as traditional foods, as well as impact the lives of bears and other wildlife that depend upon them. Small-scale harvesting of basketry materials such as cedar bark (both western redcedar and yellow-cedar), cedar and spruce roots, and birch bark, as well as the use of specialty woods such as yellow-cedar and red alder for carving, has generated significant income for Indigenous artisans, but given the destruction of these species through industrial forestry, there is still a fear that the supplies of these products are limited and that they are becoming more and more scarce.

Large, monumental cedars are increasingly rare. This is mostly due to "high grading" – selectively harvesting only the biggest and best trees – and to often wasteful use of old growth cedar and other timber species in the forest industry. As early as 1918 Whitford and Craig (1918, 182) wrote, "The waste in the manufacture of lumber in this province is appalling to one accustomed to more conservative methods. Huge slabs of absolutely clear wood, 3 to 6 inches thick, are sent to the fuel pile [to be burned for fuel at the mill or perhaps sold for heating] ... It is estimated that at least 25 per cent of the tree is left in the woods and another 30 to 35 per cent is wasted in the mill." Recently, an Indigenous canoe maker flew by helicopter with forestry officials over an entire coastal watershed and was unable to find a single redcedar tree large enough to make

14-1 | Haida dugout canoes, on the beach at the Haida Heritage Centre at K̲aay Llnagaay, Skidegate, BC, in 2008.

an oceangoing canoe. The spokesman for the Gitga'at at the McKenna-McBride Commission hearing of 1916 was mistaken in assuming "the day of the canoes is passing if it has not already passed" (McKenna-McBride Commission 1913–16). In fact, the day of the canoes is returning, and people are again crafting large, elegant dugout canoes and seeking the big cedars for these projects (figure 14-1).[47]

The potential marketing of traditional medicinal plant products brings a fear of overharvesting and disappearance of the medicinal plants and a fear of misuse of the medicines by those who do not understand the complexities of how they are to be administered. The spiritual aspects of their harvesting and use must be considered, for example, when publicizing cultural and intellectual knowledge that is normally kept private and that is subject to potential misappropriation. All of these concerns are valid, and some have already been borne out in one way or another. In the first part of the nineteenth century, as mentioned previously, some people earned a living by harvesting cascara bark or other products for the pharmaceutical industry, but there was a widely recognized depletion in cascara populations following this intensive harvest, and eventually the pharmaceutical companies started to grow their own cascara.

The lessons learned in these cases are reflected in the more recent "gold rush mentality" harvest of Pacific yew when a potent anticancer compound, Taxol, was isolated from the bark. To make a profit and fill orders from the pharmaceutical company that began the manufacture of Taxol, many people started to harvest immense quantities of the bark without regard for the Indigenous peoples who had used it medicinally for generations and had relied on the tough, dense wood for a whole range of tools and implements (Turner 1998, 2001). Fortunately, commercial growing of Pacific yew and new technologies to synthesize Taxol from more basic molecules allowed a reprieve for the remaining yew trees. However, the lesson of what can happen with commercialization of a forest-based medicine remains. Another lesson could be that any harvesters of products like yew bark should follow the protocols established long ago by First Peoples: harvest only a strip of bark from the tree trunk or take bark only from some of the branches to allow the tree to heal itself and continue to grow. With this caution in mind, especially for medicines that might potentially treat serious illnesses like diabetes or cancer, some Indigenous people believe that as long as they are used with care and respect, it is acceptable to harvest or grow certain types for commercial use. Others consider such a practice to go against their fundamental belief that "medicines are for healing, not for sale" (Barbara Wilson [K̲ii'iljuus], pers. comm., 2008).

Other bioproducts that are being harvested commercially by both Indigenous and non-Indigenous people include floral greens from species like evergreen huckleberry and salal (Turner and Cocksedge 2001). Here again, however, although the foliage of these species is not "used" in any quantity by Indigenous peoples (except for use of salal branches in pit-cooking), the berries of both species are highly valued, and there is a concern that heavy and prolonged harvesting of these greens will eventually reduce their potential to produce fruit.

Another, somewhat novel type of nontimber forest product, at least novel for the Nisg̲a'a in the Nass Valley, is the pine mushroom, or American matsutake (*Tricholoma magnivelare*). The Nisg̲a'a, like many other peoples of the boreal forest and Northwest Coast, did not eat mushrooms originally.[48] In the late 1980s and early 1990s, however, pine mushrooms from North America gained high value in Japanese markets, where the original matsutake (*T. matsutake*) was becoming scarce. Pine mushrooms were identified in the valley by a non-Nisg̲a'a resident and plans were made to develop a mushroom-picking industry there. This case provides an interesting example of adaptive management and governance of a new resource, in which protocols for sustainable harvesting and jurisdiction of gathering sites were developed based on Nisga'a conventions. Whereas pine mushroom harvesting in other areas was taken over largely by Asian immigrants, in the Nass, the Nisga'a – with their newly acquired treaty rights – gained control of the mushroom harvest. In the end, it provided a new venue for younger Nisga'a to regain knowledge of their lands and historic harvesting sites owned by their *waaps* ("houses"). Contrary to a "tragedy of the commons" situation, which has developed in some other places around mushroom and other NTFP harvesting, the Nisga'a pine mushroom harvest serves as a promising model for sustainable land and resource use (Menzies 2006a).

Perhaps one of the biggest concerns expressed by contemporary elders about the future of their botanical knowledge is how to impart the spiritual values and beliefs that they have always considered to be a requisite part of plant use and relationships with the natural world. They believe that cedar-bark harvesters need to continue to pay their respects to the tree before they remove any bark, that people gathering medicine still need to pay tribute to the medicine plants ceremonially and ask for their help and assistance if they are to be successful in healing, that the spirits of the mountains and rivers should be recognized and respected, or some misfortune will ensue, and that resources should never be wasted and always generously shared among the group.

Some elders are worried that younger people, caught up in the consumer-oriented society of today, are not learning the values of their cultures, including a respect for others and proper treatment of fish, game, and other foods. Some attribute the notable declines in berries and other plant resources to the lack of attention to these important protocols. The beliefs of Tsilhqot'in elder Helena Myers (Ìnkél) are cited in her daughter Linda Smith's (2008b, 135–6) thesis on Tsilhqot'in oral traditions:

> *Ìnkél* recalls that in her youth, the Saskatoon bushes across the lake from *T'èt'ághíntíl* (a place along Konnie Lake, *Xèní*), used to get so heavy with berries that one could mistake the berry patches for black bears, and the flowers from the spring beauty patches appeared as if covered in snow. Now, there are many Saskatoon berry patches and spring beauty grounds in the *Tsìnlhqút'in* [Tsilhqot'in] territory that are less productive and the plants in some areas have completely died out. *Tsìnlhqút'ín* elders often blame this scarcity on people who are *nímính*, saying they are ignoring traditional codes. There are many who are not receiving the essential cultural education and some simply do not believe in these cultural restrictions, therefore few practice it.

Elders also point to wasteful and selfish behaviour by some that transgresses the social order. For example, with the convenience of freezers, some families put away more fish than they need, and then they have to throw it out the following year because it is too old. Another observation that relates to the use of freezers is that since families are now able to freeze an entire side of deer or all of the berries they are able to pick, some do not share these foods with other families as much as formerly.

There are many others, however, who are carefully observing cultural protocols and continuing and strengthening their cultural traditions, including ceremonies and dances that reinforce unique perspectives and relationships with the natural world. In the summer of 2008, for example, the Tsawout community of Saanich hosted a traditional food feast, with pit-cooking of root vegetables, and a traditional First Salmon ceremony to show respect and thanks for the salmon.[49] This ceremony, in various forms, has been enacted throughout the Northwest Coast as well as in some interior areas. For the Saanich, children were selected to bring the salmon from the ocean on a tray of sword fern fronds. The children, wearing sword ferns on their heads and around their waists, solemnly carried the salmon along a pathway of fern fronds and then presented it to the elders to cook and share with all of those present, along with prayers of thanksgiving.

Similarly, the Atlágimma (Spirits of the Forest) dance, or Grouse dance, recently performed under the authority of Clan Chief Adam Dick (Kwaxsistalla), both at the big house at Kingcome Inlet and at the Society of Ethnobiology meetings at the University of Victoria in the First Peoples' House,[50] brought the magic of the deep ecological teachings of reciprocity and interrelationship among all things and all actions to a new generation of Tsawataineuk youth and young adults – and to a broad international audience – who had never witnessed this dance (figure 14-2).

In many different communities, as well as through academic institutions, there are educational programs and publications intended to help both document and pass on traditional knowledge about plants and environments (e.g., Nuxalk Food and Nutrition Program 1984; and Turner and Thompson 2006). Courses in ethnobotany have been offered in Indigenous-led institutions such as Wilp Wilx̱o'oskwhl Nisg̱a'a ("Nisg̱a'a House of Wisdom"), the postsecondary education institution of the Nisg̱a'a Nation, which offers a bachelor of arts degree in First Nations Studies–Nisg̱a'a in cooperation with the University of Northern British Columbia (figure 14-3). Demonstration ethnobotanical gardens have been established in many places, and new buildings and park areas are being landscaped with native species, which function simultaneously as beautiful ornamentals, as living examples for educational purposes, and as potential resources for culinary and artistic endeavours (Mackin 2004; Pukonen 2008; Turner and Wilson 2006). Native plant nurseries are producing quantities of culturally important indigenous species like wapato to aid in landscaping and restoration projects (Senos et al. 2006). On a broader scale, in a program called Recovering America's Food Traditions (RAFT), Nabhan (2006; see also Nabhan

14-2 | Grouse dancer, during the performance of the Atlágimma (Spirits of the Forest) dance under the authority of Clan Chief Adam Dick (Kwaxsistalla) in May 2010.

14-3 | Diana Smith, Nisga'a elder, in a Wilp Wilxo'oskwhl Nisga'a (Nisga'a House of Wisdom) ethnobotany course in 2012, holding a sample of northern riceroot on a fieldtrip to Gingolx at the mouth of the Nass River, BC.

and Rood 2004) and others are working toward documenting the biological diversity of indigenous and other local foods, especially those that are endangered, either through scarcity or through diminishing use.

The revival of the cedar canoe culture through a program known as Tribal Journeys is another manifestation of the many enterprises in cultural renewal. In the words of one participant in the canoe journeys, "We are taking our canoes out of the museums and putting them back in the water – and showing the world our culture is alive" (Celletti 2009). Described as "one of the most significant cultural movements of our time," the Tribal Journeys program serves as an example of healing through tradition, in which coastal Indigenous people are literally following the paths of their ancestors and, in so doing, are regaining their knowledge, skills, identity, and pride, as well as participating in demanding physical exercise and group training that has many health benefits (ibid.; see also Halber Suarez 2012).

Treaty rights and land negotiations continue at a number of levels, especially in British Columbia. In some cases, as with some of the Salish communities who signed Douglas treaties, efforts are underway to better define the extent and nature of rights granted under these treaties and to incorporate more up-to-date understandings of Indigenous cultivation practices and land tenure than were recognized at the time the treaties were established. For the most part, the British Crown (and later the federal government of Canada) did not conclude treaties with First Nations west of the Rocky Mountains, leaving the subject of Aboriginal title to lands and resources largely unresolved in British Columbia for the past 150 years. First Nations never accepted that their rights to lands and resources were extinguished, and many used civil disobedience, court action, and political lobbying to force the public governments to reassess their position on the question of Aboriginal lands and rights (Tennant 1990).

In the past three decades, the federal government, and later the provincial government as well, entered into treaty negotiations. Until these treaties are concluded, the governments are legally obliged to consult with First Nations before allowing any activity on publicly owned lands within their territories (the majority of lands in British Columbia) that may infringe upon Aboriginal rights arising from original use and occupancy. Although an awkward, and often acrimonious, process for protection, the legal acknowledgment of Aboriginal rights and title has opened up interest in and research on traditional knowledge of plants and animals of First Nations in British Columbia. Some First Nations in the province have opted to exchange their general and undefined Aboriginal rights for a clearly defined set of rights within defined geographical areas through modern-day treaties.[51] These modern treaties, such as the Maa-nulth Final Agreement,[52] provide some examples of how First Nations' traditional plant and animal resources are being managed.

Even without treaties, but backed by Aboriginal rights recognized in the Canadian Constitution and other legal venues, many First Nations throughout British Columbia are initiating their own land use plans that take to heart their needs for plant resources and land bases on which to practise and sustain their cultures. Others have negotiated with provincial and federal governments, and with supportive nongovernmental organizations, to collaborate in the sustainable use of traditional lands and resources – for example, in the development of marine protected areas and in the co-management of national park reserves.[53] Much more needs to be done in developing co-management and co-governance systems, in establishing laws that effectively protect Indigenous peoples' lands, waters, and resources from overexploitation, and in giving a real voice to Indigenous peoples' needs and perspectives.

As modern treaties are being slowly negotiated, however, there is a policy and legislative gap in providing protection for medicinal plants and other culturally valued species, as well as for culturally important lands. Currently, for example, the Heritage Conservation Act of British Columbia (HCA), despite its stated purpose "to encourage and facilitate the protection and conservation of heritage property in British Columbia" (British Columbia 1996, part 1, 1), is limited in its application and does not explicitly protect medicinal plant populations or, necessarily, their habitats or culturally valued habitats in general – even when such areas are known from oral history to have cultural significance over many generations – unless these have some verifiable associated archaeological feature. Recently, for example, a large patch of obviously very old *q'ex̱mín* ("wild celery," *Lomatium nudicaule*)[54] was destroyed by construction of a new bridge at the Little River ferry terminal near Comox on the east-central coast of Vancouver Island. As the site did not contain physical evidence (in the form of a culturally modified artifact) of past human use, it was not afforded automatic protection under provincial heritage legislation.

The HCA appears to accommodate the heritage of First Peoples, but the legislation's criteria of what is worthy of protection were selected through a decidedly European lens, which favours and values the built environment (e.g., monumental structures, works of art, and archaeological sites). Significantly, even an archaeological site can be legally demolished once it has been "scientifically" documented and the artifacts and/or human remains have been excavated. Further, although the HCA protects sites where there is physical evidence of past human use or occupation, this legislation was written at a time before archaeologists and anthropologists recognized the physical evidence of Indigenous-managed root gardens, berry gardens, orchards, and clam gardens. As a result, there is no explicit policy or practice associated with the HCA to include these unique and highly significant heritage features within the province's heritage site system.

Although there is other provincial and federal legislation in Canada that addresses indigenous plants, as well as threatened species and ecosystems, none explicitly considers the protection of First Nations' cultural keystone species or managed landscapes; thus protection is by accident rather than by purposeful inclusion.

Frustrated by the lack of legislative avenues to protect their traditional food, medicinal, and material plants and unique sites, First Nations have often asserted their Aboriginal rights and title through the courts. Although this action has provided some real success for First Nations, proving they have Aboriginal rights and title to a given area is an expensive, time-consuming, and acrimonious solution of last resort.

Another venue that might help in this endeavour is the United Nations Declaration on the Rights of Indigenous Peoples (United Nations 2007),[55] which Canada and the United States have both endorsed. Along with the United Nations

Convention on Biological Diversity (United Nations 1992), to which Canada is a signatory, this agreement requires Canada and the United States to conform to international standards in supporting their Indigenous citizens, including their rights to continue their traditions, languages, and use of their territories and resources.

STEPS TOWARD A HEALTHIER WORLD

How can the lessons of ethnobotanical and ethnoecological knowledge and its modes of dissemination, transmission, and adaptation be applied as components of ongoing cultural revitalization and maintenance of biocultural richness? In the previous section, some of the diverse ways that knowledge is acquired and shared in Indigenous societies were identified. The process of knowledge building is not a static one but is cumulative, dynamic, and adaptive, as seen from the means used by First Peoples to fit themselves into the new situations imposed by the arrival of Europeans and other newcomers in the study region and from how they managed to retain key elements of their cultural practices and, in many cases, their languages through a period of immense and profound change. The endurance and strength of their belief systems and spiritual practices is particularly notable because worldviews and philosophies guide behaviours toward other humans and other species (Berkes 2012; L. King 2004; Nigh et al. 2002).

In promoting ecocultural knowledge and practice, engagement and empowerment of individuals and communities of all types are fundamental. Both old and new modes of knowledge generation must be employed. Learning and teaching activities on diverse geographic and temporal scales, and through a range of formal and informal institutions, will assist in the appropriate and respectful dissemination of ecocultural knowledge and in the reconstruction and renewal of culturally and ecologically sound practices and processes. Successful models will provide lessons and tools and help to build ever more flourishing, effective, and resilient systems of transformative learning and practice, whereby we might be able to change attitudes and behaviours of mainstream society to less destructive and more conservation-oriented ways of living.

A series of strategies for achieving long-term biocultural conservation – drawn from the teachings and examples of ethnobotanical and ethnoecological systems of Indigenous peoples of northwestern North America and supplemented from other sources – is presented below. These strategies provide, I hope, some initial steps for reversing the current trend of biocultural erosion and loss and for restoring a more balanced relationship between humans and their home places. Taken together, they comprise a blueprint for more diverse, resilient, mutually supportive, and environmentally aware human communities.

Although embedded in time-honoured perspectives, the guiding principles are presented in a contemporary context, being intended for use by educators, planners, decision makers, and policymakers at all levels. Although they are categorized for convenience, the inextricable connections between them are acknowledged. They should be considered collectively and not in any particular order.

Strategies Supporting Positive Change in Biocultural Systems

Connecting with Communities and Places

It is important for any individual to identify with – and to love – particular places where they feel at home and particular people (and other life forms) whom they view as family and community. These personal connections are vital because they motivate people to care about the future, not just for themselves but also for others, and instil in them a desire to learn about seasonal and biological cycles and how to sustain and support those people and places that they care about (Cajete 2000; Feld and Basso 1996; Helin 2008; McKibben 2007; Thornton 2008). Furthermore, ethnoecological knowledge and learning are "situated" and contextualized in place, right across the spectrum of aesthetic, ceremonial, economic, personal, familial, historical, political, social, spiritual, and practical considerations (Feld and Basso 1996, 27; Nazarea 1999). When an individual's or a community's connections and relationships with a central place have been severed, they need to be renewed or rebuilt.

Celebrating Intergenerational Continuity in Learning and Teaching

Learning about sustainable lifestyles and environments takes place at all stages of life, and each contributes to the knowledge system in unique ways. The knowledge, teachings, and guidance of elders are essential because of their long experience and the wisdom and judgment they provide. They are uniquely placed to pass on knowledge, skills, and perspectives to children, and on many occasions they are major contributors to a child's learning experiences, in terms of both gentle instruction and providing opportunities for practice and participation. Often elders are also the storytellers and the memory keepers of a family and community. Young and middle-aged adults and parents are both teachers and receivers of instruction from peers and elders. Men and women each have their separate perspectives and areas of knowledge and specialization that should be recognized and incorporated into any strategies for biocultural renewal. Children represent the future; they take their acquired teachings, skills, and values with them through their entire lives. As Wyndham (2004, 175) writes, "Children's home educational environments and peer interactions are central to understanding processes of knowledge transmission and acquisition and recognizing changing knowledge structures." In addition, as noted by Ramirez (2007), long-term collaborative educational and research programs developed with young members of communities can lead to many decades of involvement and commitment – decades in which real changes can be realized and enacted. Facilitating all of these aspects of intergenerational teaching and learning – especially between elders and children – is an effective contribution toward greater cultural continuity, sustainability, and resilience (Turner and Thompson 2006).

Recognizing Holistic Relationships and Interacting Processes

All things – whether in human or nonhuman communities, whether biotic or abiotic – are connected through relationships with each other and with the land in a web of interactive entities and processes, as reflected in the Nuu-chah-nulth phrase *hishukish tsawaak* ("everything is one") (E.R. Atleo 2004, 2011; Scientific Panel for Sustainable Forest Practices in Clayoquot Sound 1995). Effective learning must engender this understanding: every action and every decision has consequences and outcomes, both for human endeavours (the "ethnosphere") and for ecosystems at large (the "biosphere"), because they are inextricably connected (W. Davis 2001). The priorities and choices that we make, collectively, must reflect this reality. How humans treat each other and their environments will reflect back on their lives in various, sometimes very complex, ways.

Instilling Gratitude, Accountability, and Responsibility

Everything we humans use – our foods, medicines, building materials, and water – is a gift, but with the gifts come obligations; we need to give back in some way whatever we take. Engendering a sense of appreciation and responsibility, even in young children, is a key to careful and wise use of resources and provides motivation for promoting biocultural sustainability. Teachings of respect are complex and require contextual knowledge and reflection, and they might be reinforced through participation in ritual and ceremony.

Valuing and Supporting Diversity

Diverse species and populations, habitats, cultures, and languages provide greater opportunities and choices, thus increasing our collective resilience in the face of change. In the words of one of Henry Lewis's (1977, 38, cited in A.M. Miller 2010, 34) informants in a study of traditional burning practices in the boreal forest, "it is better to have all kinds of places not all the same." Similarly, recognizing diversity in strengths, skills, and aptitudes of individuals and communities is important in allowing people to reach their full potential and to make the best possible contributions to society, as skilled artists, healers, or wise leaders, and in allowing each community to contribute in its own way.

Some individuals seem to be biologically predisposed toward conservatism and others toward seeking adventure and innovation; having these opposing forces within a population can lead to a healthy balance of cultural stability and adaptation to change. To gain as many potential solutions to problems as possible, it is necessary to ensure that everyone and every view has a voice and an ear and that everyone is empowered to participate. This includes particularly Indigenous peoples and their perspectives because of their unique knowledge and experience. As Ts'msyen author and lawyer Calvin Helin (2008, 168) explains, "In the same manner as we take ownership of our problems, we must also take ownership of our future."

Embracing Different Teaching Approaches and Learning Styles

Cultural and environmental learning and teaching occur through many venues and forms, as described previously. Therefore, employing multiple styles and strategies for imparting knowledge and skills will achieve a greater overall effect than restricting instruction to a single method. Learning is dynamic, cyclical, and cumulative. It involves instilling memory in all of the physical senses as well as emotions. Stories, creative arts, music, dance, ceremony, and ritual are thus important elements in a holistic learning process. Place-based learning, away from routine situations (e.g., at fishing or harvesting camps) can also be effective. Transformative education – taking people to new levels of understanding – occurs through building on learning stages over time and providing a range of experiences, including allowing times for contemplation and reflection, for observation, for practice, and even for learning from mistakes.

Integrating Old and New Materials and Technologies

Although Traditional Ecological Knowledge is grounded in a people's history and traditions, it is by no means static or fixed in the past. One of its key features is its ability to incorporate and integrate new information, ideas, and skills with those already existing. This has happened over and over again, all through human history, as demonstrated in many examples presented here. Some refer to this process as "cultural evolution." It is thus not only possible, but also completely appropriate, to embrace new and different materials and technologies, for example, in supporting learning and teaching. With the availability of a whole range of educational and interpretive tools – board games, books, magazines, photographs, posters and printed material of all kinds, films, digital recordings, computers and other electronic media, geographic information systems, analyses of nutrients, of phytochemicals, and of genes, interactive multimedia, and Internet sources – opportunities for effective learning have, in some sense, expanded significantly.[56] A caveat is that these devices sometimes result in a further separation of children and others from direct human interaction and its associated opportunities for learning, as well as from the natural world on which they depend, and may reduce their familiarity and empathy with other living things.

Recognizing the Importance of Scales of Interaction

Human activities that relate to environments are manifest on many different scales, resulting in complex and nonlinear outcomes of actions and processes. These must be linked and integrated to achieve a fuller understanding of problems and opportunities. For example, on a small spatial and temporal scale, fire that people might use to clear a patch of forest can result in severe damage to trees and other vegetation and to wildlife. Over time, however, it can bring renewal by encouraging the growth of a range of new, early successional species and by creating ecological edges that tend to be highly productive.[57] Across a wide landscape, many burns occurring at different times can enhance the overall biodiversity and productivity of a forest by creating a mosaic of diverse successional stages of vegetation. Examining only the negative short-term outcomes of a single fire will not give the full range of understanding needed to develop effective resource management strategies. In terms of supporting ethnobotanical and ethnoecological knowledge transmission, too, multiple scales of learning and interaction must be developed – ways of linking short-term personal experiences to broader understandings at the community, societal, and intergenerational levels.

Developing and Utilizing Diverse and Interactive Institutions to Support Knowledge Transmission

Schools and universities are major institutional venues for education in Western society, but they are by no means the only type of institutions that support learning. There are many institutions in Indigenous societies that represent norms and conventions that are embedded in cultural belief systems, thus serving as effective means of imparting knowledge and skills related to peoples' cultural and environmental relationships (Berkes and Folke 1998; Natcher and Hickey 2003). For example, the clan system of central and northern Northwest Coast cultures provides an entire alternative kinship network complementary to that of the extended family. Access to resources and important learning opportunities are provided through one's clan and clan chief. Various types of formal and informal apprenticeships and mentorships are another example of institutions that can promote a person's knowledge of cultures and environments. Potlatches and feasts serve as key institutions for environmental and cultural teaching through demonstration and oratory. Any of these could serve on a broad scale as models for developing new venues for environmental and social-cultural learning across society in general and could add power and effectiveness to biocultural conservation programs.

Finding Ways to Revitalize Languages and Retain Linguistic Diversity

Language is a critically important component of cultural transmission and embodies key information about environments and biocultural relationships (T.J.S. Carlson and Maffi 2004; Chipps-Sawyer 2007; J.C. Thompson 2012). When a language disappears – or even when people who speak the language forget its specialized vocabulary, such as names for places, for plants and animals, and for different practices related to resource use, processing, and management – much important knowledge is lost. When language revitalization programs are developed, care must be taken to preserve and restore these special components of languages, along with the grammatical and phonological as well as regular speech elements. Bringing back terms, such as the recently officially recognized name "Haida Gwaii" to replace the European name "Queen Charlotte Islands" in the gazetteer, is an excellent step toward renewing these languages. More could be done to reinstate such original Indigenous place names as well as terms for plants, animals, and environments.[58]

Using Food for Biocultural Renewal

Because it is such a basic human resource and need, and because it is environmentally dependent and culturally mediated, Indigenous food can serve as an important and effective element in supporting biocultural renewal and transmission of biocultural knowledge. For Indigenous peoples of northwestern North America, as well as elsewhere in the world, overwhelming changes in their food systems have had serious consequences for their overall health. Society in general is suffering from a "fast food" addiction that has, often indirectly, placed many environments at risk. As Heron (2008) describes, "We love fast food ... and it's really bad for us. It's bad for our health, our culture, the environment. In short, it's unsustainable. But our once-diverse food lore and skills have been scattered to the four winds. Our taste buds have been jammed on salt, sugar and every conceivable molecular permutation of corn. We literally eat petroleum-derived substances, and ask for more ... How do we kick our fast-food addiction and re-establish a relationship with what's good for us and good for the planet?" Maintaining and restoring key elements of our original food and accompanying lifestyles, wherever possible, are considered to be desirable goals by a number of authorities (Kuhnlein, Erasmus, Creed-Kanashiro, et al. 2006; Kuhnlein, Erasmus, and Spigelski 2009; Kuhnlein, Erasmus, Spigelski, et al. 2013; Parrish et al. 2007). Renewing local and Indigenous food systems can be undertaken at many levels and in diverse ways. One collaborative strategy has been the "Feasting for Change" initiative on Vancouver Island (Devereaux and Kittredge 2008).[59] Reinstating First Foods ceremonies (*CUJ* 2008) and publishing cookbooks featuring Indigenous foods and recipes (Nuxalk Food and Nutrition Program 1984; Uu-a-thluk 2008; Watts and Watts 2007) are other ways of renewing and restoring connections with food systems and local environments.

The confederated Umatilla tribes (*CUJ* 2008) have developed a "First Foods mission statement" to help guide their efforts "To protect, restore, and enhance the First Foods – water, salmon, deer, cous [*Lomatium cous*], and huckleberry [*Vaccinium membranaceum*] – for the perpetual cultural, economic, and sovereign benefit of the Confederated Tribes of the Umatilla Indian Reservation. We will accomplish this utilizing traditional, ecological and cultural knowledge and science to inform population and habitat management goals and actions, and natural resource policies and regulatory mechanisms."

Renewing and Recreating Vibrant Ethnoecosystems

Recognizing that humans need not always be destructive to the environments in which they dwell or to the resources on which they rely can foster new perspectives and approaches to resource management as well as to educational processes, planning, and decision making. Engagement and empowerment at the community level – for example, through school projects and cultural programs related to environmental relationships – are fundamental to reconnecting the crucial cultural ties with our environments and resources and, ultimately, to increasing environmental and cultural sustainability. Activities ranging from developing relevant school curricula to assisting with the dissemination of ecocultural knowledge and creating projects that focus on the reconstruction and renewal of culturally and ecologically sound practices will assist in restoring the processes of cultural-environmental interactions at various levels and on different scales. In turn, these activities can be employed as models for other endeavours, to be tested, adapted, and applied in diverse situations.[60]

As another aspect of this renewal, parks and protected areas should continue to be reassessed and reconfigured since not only do natural features and native species and habitats need to be maintained and preserved, but so, too, do the time-honoured relationships and interactions people have developed with their environments. These relationships and the knowledge systems that support them are, in many cases, as endangered as the places and species themselves, and frequently it is the very human presence and activities that have helped to develop certain habitats and populations (M.K. Anderson and Barbour 2003; Hunn et al. 2003; Senos et al. 2006; K.L. Turner and Bitonti 2011). Conversely, places of high cultural importance, featuring culturally important and culturally maintained species, should be restored and preserved in their own right – along with the dynamic processes that are essential in their maintenance – as significant and recognized legacies of our common human heritage.

In the spirit of adaptive management – and given the complexity and unpredictability of these systems – actions and initiatives aimed at promoting and maintaining biocultural diversity should be monitored and frequently assessed in order to ensure that their effectiveness is maximized. Successful biocultural management and conservation rest on our capacity to understand and adapt to environmental and cultural change over time and across geographic and cultural space. Hence ecological knowledge and understanding among those people most closely affected plays a key role in this context, and the means by which this knowledge and understanding are shared and accepted are critically important (Berkes, Colding, and Folke 2000).

CONCLUSIONS

What we have come to understand in recent decades is that there is a vast complexity in the many and varied interactions between the First Peoples of northwestern North America and the biological diversity of this region. In particular, in the case of this study, the role of plants in supporting and sustaining Indigenous cultures and lifeways for countless generations has been immense and diverse. It was not so long ago that those of the dominant Western society, including many academics, understood little about how plants have been fundamentally important to First Nations. Political leaders, educators, policymakers, and bureaucrats often recognized little beyond the importance of salmon, and perhaps deer, as the key resources for Indigenous peoples of this area. Across their traditional territories, "reserves" and "reservations" were designated to accommodate their "needs," but these were generally restricted to immediate village sites and, in some cases, to places used for fishing or hunting. Metaphorically, communities were left without their gardens, their pharmacies, their hardware centres, and their sacred places.[61] The reserve and reservation systems in Canada and the United States respectively reflect policies often based on greed, but perhaps more on ignorance, under a philosophy that First Peoples in northwestern North America and elsewhere were not really "using" most of the land they claimed as their territory. Similar notions were taught widely in the public school system as recently as a couple of decades ago.

Gradually, this incomplete understanding is changing. Of course, many First Nations individuals, particularly culturally knowledgeable elders, knew that this Western view was deeply flawed, but their voices were seldom heard and usually ignored. At last, however, through their efforts, and through a growing awareness of society in general, this situation is changing. Many recent initiatives at the global, national, and regional levels have focused on the significance of Indigenous peoples' knowledge and on their rights to influence policy formulation and to exert some control over activities occurring within their territories.[62] The imperative for consultation, listening to their perspectives, and respecting their deep knowledge and wisdom has resulted in a change and a growing appreciation of First Peoples' histories, cultures, perspectives, and knowledge.

In these endeavours, knowledge and understandings of plant resources and culturally valued environments have been prominent. However, there is a long way to go. Now as perhaps never before there is pressure for conformity, not only at a community level but also at a global level. Languages are disappearing and cultural distinctions are blurring across the planet as the large multinational companies offer the same products and as media spread the same messages far and wide. People will have to work hard to maintain their cultural distinctiveness in the face of these forces, but diverse knowledge systems and perspectives bring advantages in terms of offering more and potentially better solutions to problems and thus increasing our resilience as humans in these times of great change (W. Davis 2009). By no means do Indigenous peoples have all of the answers for sustaining humans and other life forms over long time periods, but they have had – and still do in some measure – distinctive approaches and different worldviews and related values that might help humanity collectively to find a better and less destructive relationship with our environment.

As a global community, we must come to a better understanding of our collective impacts, both on the environment and on countless local and Indigenous cultural groups in different parts of the world. Not only are these peoples being deprived of their cultural and linguistic heritage, but we are also depriving humanity as a whole of the richness and resilience that biocultural diversity can bring; we are losing a major portion of our combined environmental knowledge, practices, beliefs, and relationships, which are such an integral part of human existence (W. Davis 2009; Hunn 1999). The ability to access, use, care for, and perpetuate the plants, animals, habitats, and landscapes of their own homelands is essential for the continued health and well-being of Indigenous peoples and for the continuation of Indigenous cultures, languages, and identities. These peoples, together with their cultures, languages, stories, and ceremonies, are inextricably linked to their traditional territories and to the other species that dwell there. Consequently, their identity and their ability to maintain and use this environment are critical to their future success and survival as distinct peoples.

Ethnobotany and ethnoecology have finally come into their own in northwestern North America. The recognition that these knowledge systems are vastly more complex and important than was appreciated only a few decades ago is reflected in many ways – not only in publications such as this one (see also Turner and Lepofsky 2013) but also in the very vibrancy of programs being developed by Indigenous communities and their collaborators in schools, in local economic development initiatives, in videos and educational productions shown on television, and in the celebrations of Indigenous foods and feasts that are occurring in many places on Vancouver Island and beyond. The message is now loud and clear: Indigenous peoples in northwestern North America were not just casual wanderers through a bountiful land, hauling salmon from the rivers and perhaps plucking a few clusters of berries as they were encountered. They were active participants in managing their environments and in using their resources sustainably, treating them with care and consideration. This newly recognized paradigm serves as an excellent starting point for all of us to create better, more respectful, more meaningful, and more careful relationships with our local environments, with other species, and with the entire natural world on which we rely.

Helin (2008, 168) writes, "In going forward, Aboriginal people must understand the realities of their ancestors' success and self-reliance over millennia (a foundation of independence, interdependence, self-discipline, ethical leadership, and cultural cooperation)." More than this, *all* of us need to recognize this history of success and self-reliance. Indigenous peoples' knowledge, skills, values, and approaches can contribute substantially to humanity's efforts to achieve the "paradigm shift" that is needed if we are truly to reduce our impacts on the earth and to reverse the tide of global environmental and cultural change and loss. Indigenous perspectives – about the interconnectedness of life, about plants and animals as relations or kin, about the importance of a long-term view of the future, about the linking of human health and well-being to the health of the environment as a whole, and about what is truly important to the human condition – need to be communicated effectively to Western urbanized society so that they can be considered and potentially adopted more widely.

I hope that this synthesis will contribute to the advancement of knowledge and understandings both about cultural adaptations to specific places and environmental situations and about influences of people on these places and ecosystems. Moreover, the inclusion of scales of time and space over which patterns have been identified in this project will hopefully lead to a clearer understanding of humans' capacity for learning, exchanging, retaining, and regenerating knowledge. This understanding is critical to our efforts to reformulate our behaviours in ways that will allow us to live more sustainably in the world we have inherited and to bequeath it to the future in a fully functioning, healthy, vibrant, and diverse state.

NOTES

CHAPTER EIGHT

1. For British Columbia, *E-Flora BC: Electronic Atlas of the Plants of British Columbia* (Klinkenberg 2013) provides comparative information on the frequency of occurrence of different plant species in the various biogeoclimatic zones of the province, with distribution maps based on actual records and collections. This has been an invaluable source of information here.
2. Human strategies for management and enhancement of plant resources – including transplanting some species on occasion – are discussed further in chapter 11.
3. This variance may relate to weather conditions during bud formation and flowering, availability of insect pollinators, or other factors, or it may simply be part of an organism's innate reproductive pattern.
4. For example, sometimes, in years when saskatoon berries are less productive in the Interior Plateau, people will harvest more black hawthorn fruit, which many would normally use mainly as a secondary, or alternative, resource.
5. Edible "root" species include bulbs, corms, tubers, and rhizomes, as well as true roots, that are harvested and eaten.
6. For some reason, apparently no one has reported finding this sugar for some decades, and some elders have suggested that introduced honeybees or ants may be preventing it from accumulating.
7. This name possibly refers to the reserve name at Neskonlith in Salmon Arm, Switzmalp, but another interpretation, provided by Marianne Ignace (pers. comm., 2011), is *sxitseméłp* ("cut or shear plants"), which refers to people gathering grass or hay there (see also Mary Thomas, Turner, and Garibaldi forthcoming).
8. Both ʔeniyud and T̲s̲'ilʔos̲ were later turned into mountains, part of the homeland of the Xeni Gwet'in and other Tsilhqot'in (L.R. Smith 2008b).
9. This is probably due to lower levels of phototoxic furanocoumarins in shade-growing cow-parsnip (Kuhnlein and Turner 1987).

10 Today, however, climate change is affecting the window for harvesting and processing seaweed since, for the past decade or more, early May has been generally rainy, making it impossible to pick the seaweed then (Turner and Clifton 2009).

11 The earliest spring salmon are indicated for the Stl'a'tl'imx/St'at'imc by the blooming of the sagebrush buttercup (*Ranunculus glaberrimus*), called "spring salmon eye," whose bright yellow blossoms are said to resemble the eye colour of these fish.

12 For Tlingit, see Thornton (2008) and Thornton et al. (2004); for Heiltsuk, see Rath (1981); for Nuu-chah-nulth, see Bouchard and Kennedy (1990); for Gitxsan, see Johnson (1997); for Dakelh, see Poser (2008a, 2008b); and for Sahaptin, see Hunn, Selam, and family (1990).

13 Secwepemc named habitats and geographic features include alpine, alpine meadow, on the high plateau, in the mountains, mountain, hill (also mound and domed shape), little hill, knoll, prairie, clearing, meadow, valley, waterfall, well, cave, sidehill, gully, canyon, burned-off area on a mountainside (through landscape burning), rockslide, sunny/south side of a mountain, shady/north side of a mountain, little lake, outflow and drainage creek of a lake, melt-water runoff, mouth of a river, river, island, creek, and shore, as well as a suite of lexical suffixes that occur in place names and terms for geographic features that denote parts of the human or animal body (R.E. Ignace 2008).

14 "Oulachen" is also spelled "eulachon" or "ooligan."

15 The responsibility of Kwaxsistalla, as oulachen guardian, was to carefully monitor the fish coming up the river and to ensure that the river was kept clean and that the fish had a chance to spawn along the banks far up the river, before allowing the people to catch them on their return to the ocean. He would also ensure that people kept to their own harvesting sites and did not encroach on those of others.

16 Some people celebrated the first harvest of springtime greens, like thimbleberry shoots, with a special ceremony, similar to the First Salmon ceremony (see chapter 13).

17 Giant kelp is called *ngaal* by the Haida, *gyoós* by the Ts'msyen, and *q'áxq'əllis* by the Kwakwaka'wakw.

18 Sometimes people stayed upriver until fall, harvesting berries, hunting, and fishing, before returning downriver for the winter to stay in well-protected areas like Baker Inlet, where they could harvest salmon, groundfish, and other winter foods (McDonald 2003).

19 They broke the eelgrass rhizomes into short lengths, peeled off all but the finest central leaf, and wound this around the rhizome to make a bite-sized piece, to be dipped in oulachen or seal oil and eaten fresh (Cullis-Suzuki 2007).

20 Besides western hemlock, the cambium and secondary phloem of Sitka spruce, amabilis fir, black cottonwood, and red alder were eaten by various coastal peoples.

21 Besides western redcedar and yellow-cedar, these species included bitter cherry bark; willow bark (*Salix lucida* ssp. *lasiandra*), used by the Saanich for their reefnets; and the inner bark of Rocky Mountain maple.

22 The Secwepemc also call this time *snexʷixuyt* ("leaving") (Mary Thomas, pers. comm., 2001).

23 These plants reach their peak blooming period in the northern Interior Plateau montane regions around mid-June and are considered best when collected just before or just after blooming.

24 An archaeological complex at Cattle Point on San Juan Island dating back over 2,500 years was likely an ancient reefnet site, judging from the presence of net weights,

perforated anchor stones, and a whole assemblage of tools and evident fish-drying systems (Kirk and Daughtery 2007).

25 An alternate name for saskatoon berry is juneberry.

26 Black huckleberries were the main fruit around which the Indian Heaven archaeological site near Mount Adams and other sites in the northern Cascade Mountains were centred (see Mierendorf 1999). In fact, throughout the interior, black huckleberries and their relatives (e.g., *V. caespitosum*, *V. deliciousum*) were – and are – highly favoured, and the places where they grew were destinations of many, from the Nisg̱a'a and Gitxsan and peoples farther north to the Ulkatcho Dakelh, Tsilhqot'in, Ktunaxa, Sahaptin, and all of the Interior Salish groups.

27 In a diary entry for 19 August 1843, missionary H.K.W. Perkins noted, referring to a Tenino-Wishram group from a village at The Dalles on the Columbia River, "People moving to the mountains for berries. They obtain at this season the large mountain huckleberry [*Vaccinium membranaceum*] ... They are usually absent on these excursions [away from their village at The Dalles], from four to six weeks during which, each family lays in, for winter use, four or five pecks of nice dried berries" (cited in Boyd 1996, 275–6).

28 Women were also known to fish, on their own or in small groups, for trout and other species.

29 Edith O'Donaghey, quoted at the beginning of this chapter, recalled that her family harvested three or four large gunnysacks of this lichen each year.

CHAPTER NINE

1 Moss (2011) and Dana Lepofsky (pers. comm., 2011), however, both caution that the seeming sudden appearance of these large middens may be a result of other factors: the obscuring of earlier middens by changing sea levels, increased preservation of shell once a certain threshold of accumulation is reached, and/or people piling up shells to create house construction sites. Similarly, the high visibility of salmon in the archaeological record may be due, in part, to the relative size of their bones, as compared with, for example, herring, which may have been as important but not as visible. More finely focused research approaches may change these perceptions.

2 The large plank and frame houses of the ancestral Makah at Ozette (see chapter 2) illustrate one model of multifamily living arrangements from the latter part of the Late Holocene. These houses were around 18 to 20 metres long and about 10 to 11 metres wide, constructed of split cedar planks lashed onto a framework of upright cedar posts. Each building probably housed six to ten nuclear families, along with any visiting relatives and slaves, and each family had its own living area, with raised platforms around the living space providing places for beds and for storage. Beneath one excavated platform, a storage basket, a loom, a canoe paddle, two seal clubs, and a bow were found together. On top of the platform was a sleeping mat, still in place, and on top of it was a flattened box (Kirk and Daugherty 2007, 109).

3 As Moss (2011) points out, however, it is important to remember that the respective roles of men and women varied according to cultural norms and circumstances. As a result, the stereotype of women undertaking plant food gathering, cooking, and work with fibres and of men being the hunters, fishers, and woodworkers was not always accurate, and there were many exceptions.

4 Weaving spruce-root baskets and hats has continued unbroken in Isabella Edenshaw's family for five generations, ever since Florence Davidson became a well-

known weaver, as did her daughters, including Primrose Adams, and Primrose's daughters, including Isabella's namesake, Belle, and her sisters Joyce and Alice, and Primrose's granddaughters, including Alice's daughter, Ariane Medley, who recently made an exquisite spruce-root basket for a friend to use as a gift (see Turner 2004a).

5 For example, a Spokan woman berry picker knew that raising her voice and shouting was considered disrespectful to the plants in an area, but group singing was acceptable and reinforced a sense of consciousness of kind. Women had to know about controlled burning and how to care for the traditional huckleberry patches (J.A. Ross 2011).

6 In summary, the following tasks and responsibilities, along with their associated tools, techniques and skills, and spiritual knowledge, have been largely relegated to women in the study area (Turner 2003b, 2006b): harvesting (plant food, shellfish, eggs, basketry materials, firewood, medicinal plants); processing (all types of food, fibrous materials, hides, medicines); storing (all types of food, materials, medicines); manufacturing (basketry, mats, hats, etc.; buckskin work; tools for these projects; temporary shelters); provisioning (food for family meals, feasts, travel; use and care of baskets; medicinal herb administering; general family health); managing land and resources (clearing, selective harvesting, weeding, pruning; gardening of food plants; management, trimming, etc. of basketry materials); leading and participating in ceremonial and spiritual activities related to men's hunting and fishing and to women's harvesting rites (First Fruits ceremonies, medicine gathering rituals); maintaining and monitoring animal and plant populations; keeping fires and hearths; coordinating household activities; bearing and nurturing children (caregiving, storytelling; educating children, especially girls and young women); and contributing to the cultural life of family and community.

7 One of the most skilled basket makers on the northern coast, Ts'msyen artist William White (Tsymiyaanbiin) of Prince Rupert, was taught weaving by his aunt. He spends much of his time teaching others – both men and women – how to make cedar-bark baskets and how to weave wool and other materials as well to make Chilkat blankets.

8 For example, Dr Arvid Charlie (Luschiim) (pers. comm., 2011) noted that in some Hul'qumi'num families, only men pit-cooked the camas bulbs.

9 People peeled and ate these green shoots with oulachen grease and a little sugar sprinkled overtop; they were a much sought after treat in the spring (Turner and Turner 2008).

10 Many of the digging stick handles found in archaeological settings are marked with geometric zigzag designs, and these may be symbolic representations of the practice root trenches dug by girls at puberty as described by Teit (Kirk and Daugherty 2007; Nord 2009; M.B. Ignace, Ignace, and Nord 2009).

11 Alan Marshall (1977, 138) notes that the Nez Perce ceremony held when the canyon serviceberries (saskatoon berries) were ripe was presided over by a respected old man, who gave thanks for life and food, and that four highly respected elder ladies directed the younger women on gathering, preparing, and serving the food. When Simon Fraser descended the river named after him in 1808, he and his men encountered, around the present site of Lytton, a venerable elderly man who was blind and had to be carried but who obviously had a high position of authority. This man was an eloquent orator and spoke a great length, obviously taking on a role as a welcomer to the strangers (Lamb 1960; Turner 2005; see also Teit 1912, 414–15).

12 Lopatin (1945, 90) suggests that for the Kitamaat region, the price paid for a woman slave might be two or three boxes of oulachen grease (approximately or somewhat more than the value of one canoe).

13 The salmon were counted out in twos; if there was an uneven number, the extra salmon was returned to the ocean. Also, there was an opening at the far end of the net that always allowed a certain number of salmon to escape and continue on their way to the spawning grounds of the Fraser River (Dr Earl Claxton Sr, pers. comm., 1999).

14 Ida Jones described how a group of Ditidaht women would cook their silverweed and clover roots together in the same pit, but each woman tied her roots in bundles using her own unique knot so that when the pit was opened, everyone would recognize whose roots were whose (cited in Turner, Thomas, et al. 1983; see also Turner and Kuhnlein 1982).

15 At least, this has been the practice until recently, when rainy weather has caused a change in the timing of seaweed processing. If there is too much rain in June, the women now might have to wait until there is a long enough stretch of sunny weather to continue processing their seaweed: compressing, chopping, and drying it. In 2011, for example, Helen Clifton (pers. comm., 2011) was not able to spread her seaweed to dry until mid-August.

16 This food has not been commonly harvested within the past two or three generations. In Hartley Bay, for example, until the spring of 2004, when Gitga'at elder Archie Dundas (pers. comm., 2004) demonstrated harvesting *ksiiw*, the edible inner bark of amabilis fir and western hemlock, the last time anyone from the community had harvested *ksiiw* was around 1945. The Gitga'at practised a communal activity for harvesting, scraping, and processing *ksiiw* that was similar to the method described for the Kwakwaka'wakw (Turner and Thompson 2006).

17 According to Keith Carlson (2001), the island dialect, Hul'qumi'num, is a mixture of two Fraser Valley Halkomelem dialects, reflecting the exchange networks in which they partook.

18 Until the mid-1900s, camas bulbs were routinely served at potlatches in Duncan and Nanaimo (Ida Jones, pers. comm., 1979).

19 In Kwak'wala, for example, the name for nobility, *ninogad* (plural), essentially translates as "knowledgeable ones" (Dr Daisy Sewid-Smith, pers. comm., 2009).

20 Later on, the people moved to Kitsumkalum and Port Essington at the mouth of the Skeena River. Robintown became overgrown, but it still reflects remnants of this thriving village, with a series of level terraces staged down the embankment, each formerly occupied by different clan groups (Downs 2006; McDonald 2003). The entire slope is still a mosaic of "orchard gardens," with many important edible and medicinal species (discussed further in chapter 11).

21 Unfortunately, as noted by Croes (2003), perishable artifacts like baskets, which can serve as a major source of information about prehistoric exchange, are rare in the archaeological record, especially from the earlier millennia.

22 Note that *ü* is an unrounded *u*.

CHAPTER TEN

1 These berries are called *məxáz'* in the Pemberton (Lil'wat) dialect and *ʔúsaʔ* ("egg/berry/round object") in the Fraser River dialect of the Stl'atl'imx.

2 This railway, running from North Vancouver to Prince George and other parts of central and northern British Columbia, was originally known as the Pacific Great Eastern Railway. It was renamed the British Columbia Railway in 1972 and later became known simply as BC Rail. The railway operated passenger service, which from the mid-1950s on used Dayliners or Budd Rail Diesel Cars, and many people living in communities along the railway used it as a major form of transportation for visiting, going to school, and travelling to larger urban centres. Passenger service ended in 2004.

3 Ames and Maschner (1999, 172) provide maps of Late Holocene (Early and Middle Pacific) obsidian exchange.

4 Contemporary accounts confirm such long-distance trading. In a general discussion about how children learned what they needed to know in their culture, the late Heiltsuk cultural specialist Cyril Carpenter (pers. comm., 2002) said that, for the Heiltsuk of past generations, until about eight years old, a little child was simply steeped in the stories, songs, and culture. Then, from the ages of about eight to twelve, the child started to show his or her own talents and participated in activities according to those talents, whether it was canoe making, art, fishing, or carpentry. Then, "in the old days," if a young person really worked hard for about three years, he was given a "holiday" and was allowed to participate in a canoe trip to trade. They used to go as far south as Catalina, taking Heiltsuk dried berries, herring eggs, salmon, halibut, seaweed, and other goods to trade for abalone and goods from other regions. The large oceangoing canoes used to travel about 30 kilometres offshore, where it's not too rough; they would have two large sails set.

5 Peoples of Ukshi Lake Kamchatka sites as early as 11,000 years ago had a similar pebble tool culture based on use of salmon (R.L. Carlson 1976).

6 In fact, Oregon may well owe its name to the Cree word for "oulachen" or "ooligan": *ourigan* (Byram and Lewis 2001).

7 Another example, from just south of the study area, is Iwamkani, or "Huckleberry Mountain," just west of Crater Lake in the southern Cascade Mountains in Klamath territory (Deur 2002a; Turner, Deur, and Mellott 2011).

8 Oberg (1973) distinguishes between the concepts of "gift exchange" and "trade" or "barter," noting that the former takes place within the context of social relationships and networks, whereas the latter pertains to strictly economic interchange without regard for social or cultural values. In reality, it would be difficult to develop systems of reciprocity in purely economic terms without some type of cultural and social context.

9 The women who came would give the first bucket of berries they picked to Ambrose Robinson, as the owner of the patch, and could then keep the rest for their own use. Later, with the highbush cranberries he received, Ambrose would host a feast for the entire village.

10 More recently, such gifts might be in the form of money, although blankets and food are still used in many circumstances to recognize the witnessing of events.

11 Gitga'at matriarch Helen Clifton (pers. comm., 2005) recalled an occasion when Lucille Clifton, the Eagle clan matriarch and Helen's mother-in-law, requested that the Eagle clan use some of Helen's half-smoked coho salmon for a feast that they were hosting. Helen, who is of the Blackfish clan, willingly contributed her fish. Although she did not receive any immediate compensation, when Christmastime came that year, she received an exceptionally fine gift from her mother-in-law (see also Turner, Robinson, et al. 2012).

12 A similar situation might occur between those people along the coast who spent their time and energy harvesting and drying edible seaweed and those who were mostly fishing and drying halibut (Helen Clifton, pers. comm., 2005; Clan Chief Adam Dick, pers. comm., 2005).

13 Nitinat Lake is a long, narrow tidal lake, connected to the ocean on the west coast of Vancouver Island by a narrow channel, on the south side of which Whyac village (now mostly unoccupied but still a designated reserve) is situated. The West Coast Trail crosses Nitinat Narrows at this point, and hikers are taken across the channel via a small ferry.

14 The village, called Tl'uqtinus, was located on Lulu Island, on the South Arm of the Fraser River, adjacent to a cranberry peat bog, where the Cowichan tribes lived for significant periods of time throughout the year, obtaining berries in the summer and fall, harvesting cattails, tule, and many other plant resources, and fishing for salmon and sturgeon.

15 This scenario might have occurred with breakthroughs in food production – for example, in the intensification of salmon with improved storage techniques that allowed people to thrive, to develop more complex social structures, and to gain prestige as individuals and communities by hosting feasts. Matson and Magne (2007) suggest that such a situation may have allowed Athabaskan peoples moving south from the British Columbia region to occupy areas of northern coastal California at a time when salmon were plentiful and the peoples already in the region didn't mind having an influx of newcomers.

16 The Secwepemc, for their part, were known to have traded dentalium from the coast to both the Stoneys and the Ktunaxa (Teit 1909).

17 Dr Daisy Sewid-Smith's ancestor lived for an extensive period of time with the Nuxalk but eventually returned to her Kwakwaka'wakw relatives. Many years later, a peace treaty was signed and a valued ceremonial "Echo" mask that had been taken by the Nuxalk at that time was replicated so that each community had its own (Reid and Sewid-Smith 2004).

18 Nephrite is a highly localized stone, found in only a few places, such as along the middle Fraser Canyon, along the Rogue River in southern Oregon, on the lower Nooksack River, and in the upper Skagit Valley. The production of finely crafted nephrite celts, therefore, was probably a specialized occupation, with their rarity in most parts of the Interior Plateau and Northwest Coast leading to their particularly high prestige value (Schulting 1994; Lepofsky and Lenert 2005).

19 For example, in May 2008, at the Kitsumkalum House of Simoygets in Terrace, along the Skeena River, there were large zip-topped bags of dried Metlakatla seaweed for sale at fifty dollars apiece. Although this may seem expensive to some, considering the time and effort required to produce this amount of dried seaweed, it is a reasonable price.

20 Compare these to the animal product equivalents: herring grease was valued at fifty cents per gallon, seal grease at seventy cents per gallon, and oulachen grease, the most expensive, at one dollar per gallon. A cake of goat grease cost two dollars, a cake of deer grease was one to two dollars, and half a dried deer was $1.50. One dried salmon was worth ten cents, and a biscuit box of herring eggs cost upward of $2.50 (Emmons 1991).

21 Indian-hemp cordage is said to equate with the strength of modern synthetic fibre, with a test weight of 100 kilograms or more.

22 Prime Indian-hemp locations are open moist swales where the plants can grow very tall and straight.

23 When raphia (fibre of a palm, *Raphia farinifera*) and corn husks were introduced by Europeans, Indigenous weavers also readily incorporated these materials into their work. Similarly, Coast Salish women travelling from Vancouver Island to pick hops in the Fraser Valley or at Yakima in central Washington were happy to use cotton hop-growing strings in their tule mats in lieu of stinging nettle twine (Elsie Claxton, pers. comm., 1994).

24 Later, a new market for their baskets developed with the white settlers of the region. These women traded their baskets for second-hand clothes, potatoes, beef, or "anything we could get" (Margaret Lester, pers. comm., 1984). In fact, within the past century, many Aboriginal women maintained their families almost entirely by selling their baskets to non-Aboriginal travellers and neighbours (Turner 1996; Turner and Kus 2002).

25 The Nisg̱a'a and Gitxsan names for hazelnut bush, *'tsak'a tyaýtk*^{w} and *sg̱an-ts'ek'* or *sg̱an-ts'ak'* respectively, apparently have an ancient derivation from the Proto-Salish word *s-ts'ik'* or *s-ts'ik*, which refers variously to "nut, acorn, or conifer cone or seed" (see Kuipers 2002; see also reference to seeds of whitebark pine, *Pinus albicaulis*, in chapter 3). For example, the Squamish name for hazelnut is *ts'ichn*, and the Halkomelem name is *st*th*'itsm*.

26 Cattail (*Typha latifolia*) possibly followed a similar trajectory of introduction. The Nisg̱a'a name for cattail mat, *t'uuna'awk*^{w} (Burton 2012) or *t'una'ax̱*, also apparently derives from Proto-Salish *t'unx*^{w}*n* or *t'unx*^{w} (Kuipers 2002), and cattail's distribution in the Nass-Skeena region is quite restricted (Klinkenberg 2013).

27 Variants reflected in basketry include: material, weaving style, age, rims, ornamentation, context, size, shape, element widths, weaving gauge, attachment, placement and material of handles, design, decoration, and colour contrast (Bernick 1998c).

28 For example, three spruce-root hats were twined in the Haida-Tlingit style, up-to-the-left, and employed an ornamental skip-stitch typical of Haida and Tlingit work. Three complete coiled cedar-root baskets of distinctive Salish style are presumed to have come from the Strait of Juan de Fuca, and there were also fourteen intentionally cut coiled basket pieces, which Croes (2003) surmises may have been potlatch gifts. From the south came distinctive forms of small Salishan overlaid-twined cylindrical baskets, common to the Quinault, Chehalis, Squaxin, and Skokomish (Croes 2003, cited in Matson, Coupland, and Mackie 2003). Croes (2003) suggests these baskets were traded to Ozette rather than being made there by women marrying into the Ozette community.

29 Of twenty-five species of trees and shrubs whose bark was used medicinally, all except *Oemleria cerasiformis* (bird cherry) reportedly have been used medicinally by other peoples of North America. A perusal of Moerman's (2003) ethnobotanical database for North American First Peoples confirms a widespread commonality of medicinal plant applications.

30 These "seeds" (fruits) of *Lomatium nudicaule* are called by variants of the name *q'ex̱mín* by Salish and Wakashan peoples from Puget Sound to northern Vancouver Island.

31 This was a difficult treatment, but Annie recovered well from this accident.

32 Soapberries (*Shepherdia canadensis*) are called russet buffaloberries by McIlwraith (1948, vol. 1, 88).

33 Soapberries are particularly variable, geographically and seasonally; even where they grow, their populations might be scanty, and they often vary in abundance from year to year, so some trade was necessary to ensure a constant supply. In some years, for example, if the local bushes were not productive, the people around

Lillooet and "The Fountain" (Xaxl'ep) had to go to Bralorne, or up in the mountains, to get their supply of soapberries (Sam Mitchell, pers. comm., 1974). Saanich elders remembered travelling to the Malahat area, to East Sooke, or over to the San Juan Islands to pick soapberries. Notably, the south side of Mitchell Bay on San Juan Island is called Sx̱ʷésm (Soapberry) (E. Claxton and Elliott 1994; Turner and Hebda 2012).

34 Teit (1909, 535) also lists as interior trade goods glacier lily bulbs, fibre and woven bags of Indian-hemp, silverberry-bark bags, whitebark pine seeds, hazelnuts, wolf lichen, "woven baskets of the best type," and smoke-tanned moose hides.

35 Other goods included Labrador tea leaves, Pacific crabapples and highbush cranberries preserved in water or oulachen grease, cedar-wood boxes and other cedar products, and yew wood and vine maple (*Acer circinatum*) wood for bows (Turner and Loewen 1998).

36 Teit (1930) noted that for the Okanagan, the introduction of horses resulted in a shift of trade routes to uplands rather than valley bottoms since canoes and water transportation became less essential with the option of travel by horseback.

37 These items are given in recognition of the guests' service as witnesses to important events for which the potlatch is held, those of highest value going to the chiefs and matriarchs visiting or invited from other villages, houses, or clans. Potlatches are occasions for public and formal recounting of a group's history, the extent of its territories, and the rights of access to and control over resources held by its chiefs and their designates, as well as occasions for re-enacting origin stories through dance and song, presenting and upholding names, holding memorials for those who have passed on, and recognizing special accomplishments or experiences of people.

38 For example, as noted previously, names for edible camas (*Camassia* spp.) among a number of Coast Salish groups and for nodding onion (*Allium cernuum*) among the Interior Salish are derived from a common ancestral term, which shifted semantically at one or more points in its evolution and divergence. A similar story exists for some of the Coast Salish and Interior Salish terms for hazelnuts (*Corylus cornuta*) and whitebark pine (*Pinus albicaulis*) seeds respectively (see chapter 3).

39 Procuring soapberries (called *nexʷeskén*) through trade is well remembered by contemporary Kwakwaka'wakw elders, including Clan Chief Adam Dick (Kwaxsistalla) and Dr Daisy Sewid-Smith (Mayanilth) (pers. comm., 1996, 1998), who described trading for them both to the south (Parksville-Courtenay region) and to the north (Bella Coola region) of the Alert Bay–Campbell River–Kingcome region. An illustration of the importance of soapberries as a contemporary product of gifting and exchange is the fact that, before my work on this book in Dawson City, Yukon, where I would be surrounded by soapberry bushes as a major boreal forest understory species, at least six Indigenous friends and colleagues, from Straits Salish, Kwakwaka'wakw, Haida, and Ts'msyen communities, specifically requested that I bring them soapberries from the north, if at all possible.

40 This time depth is based on findings of the Automated Similarity Judgment Program (n.d.), as of 3 April 2009 (Cecil Brown, pers. comm., 2010).

41 The usual practice for harvesting the berries is to hold a berry-laden branch taught and hit it sharply with a stick, causing the ripest berries to drop en masse onto a mat or into a container.

42 References for this information on soapberry include 'Ksan (1980), Kuhnlein and Turner (1991), McIlwraith (1948, vol. 1, 88), Turner (1995, 1997a, 1998, 2004a), Turner, Bouchard, and Kennedy (1980), Turner and Burton (2010), and Turner, Thompson, et al. (1990).

43 For example, "timbergrass," or pinegrass (*Calamagrostis rubescens*), is called "soapberry grass" in Tsilhqot'in.

44 In one Massett Haida story (told by Walter, a member of the Rear-Town-People of Yan), a chief (lit. *nangʔiitlagadaas*, "wealthy/respected one") of the town of A'nagun hosted a feast in which "They brought different sorts of food down out of the Stikine. They brought down [highbush] cranberries, soapberries, crabapples and dried fruits (or berries)" (cited in Swanton 1908, 537). In another story, "A-Slender-One-Who-Was-Given-Away," the heroine was taken in a magical canoe: "the canoe was filled with good food, with cranberries, berries in cakes, soapberries, and the fat of all kinds of animals, grizzly-bear fat, mountain-goat fat, deer fat, ground-hog fat, beaver fat – the fat parts of all mainland animals" (John Sky, cited in Swanton 1905, 164).

45 There is oral evidence that one Haida group may have originated on the mainland.

CHAPTER ELEVEN

1 Merriam-Webster, online dictionary, http://www.merriam-webster.com/dictionary/manage (accessed 11 July 2013).

2 This is reflected, for example, in the writings of E.N. Anderson (1996), M.K. Anderson (2005, 2009), Beckwith (2004), Boyd (1999b), Deur (2000), Deur and Turner (2005), Hunn, Selam, and family (1990), Johnson (1994, 1997), L. King (2004), Lake (2004), Lepofsky (2004, 2009a, 2009b), Loewen (1998), McDonald (2003), Peacock (1998), Peacock and Turner (2000), Senos et al. (2006), L.R. Smith (2008b), Thoms (1989, 2008a, 2008b), Thornton (2008), Turner (2004a, 2005), Turner and Berkes (2006), Turner, Deur, and Mellott (2011), Turner, Ignace, and Ignace (2000), and Turner, Smith, and Jones (2005).

3 For discussions on this topic, see also Beckwith (2004), Lutz (1995, 2008), Natcher (2004), Trusler and Johnson (2008), Turner (1999), and Weiser and Lepofsky (2009).

4 This aspect of resource management is discussed in chapter 4, on changes in plant use and knowledge since the arrival of Europeans.

5 See M.K. Anderson (2005), Bandringa (1999), Beckwith (2004), Chambers (2001), Chambers et al. (2002), Cullis-Suzuki (2007), Deveau (2010), Joseph (2012), Lloyd (2011), Loewen (1998), Mellott (2010), Peacock (1998), Proctor (2013), Pukonen (2008), and Wyllie de Echeverria (2013).

6 See also chapters 5, 6, and 7 for discussions on practices of harvesting plant foods, materials, and medicines respectively.

7 An example is the Atlágimma (Spirits of the Forest) dance, whose performance is a prerogative of Kwakwaka'wakw clan chief Adam Dick (Kwaxsistalla). He was given the rights to the performance by the family of Heiltsuk and Wuikinuxv elder Evelyn Windsor of Bella Bella. This dance, a visual and auditory lesson in conservation and human relationships with the forest and forest processes, was performed under Kwaxsistalla's direction in May 2010 at the Society of Ethnobiology's annual meetings in Victoria, British Columbia. Members were astonished and moved by the deep lessons of environmental stewardship embodied in this dance.

8 This is indicated, for example, by the quantities of charcoal layered in the soil, the presence of fire scars on trees in greater numbers than expected naturally, and evidence of a high abundance of grasses and forbs in extant populations or in the pollen record (K.J. Brown and Hebda 2002a, 2002b; Lepofsky and Lertzman 2008; Weiser and Lepofsky 2009; White 1980, 1999).

9 Examples are soapberries, huckleberries, saskatoon berries, camas, wapato, bitterroot, stinging nettle, Indian-hemp, and cedar bark (Turner and Loewen 1998).

10 For further examples of possible berry intensification, see Budhwa (2007), Gill (1983, 2005a, 2005b), Lepofsky (2004), Lyons (2000), Norton (1981), and Trusler and Johnson (2008).

11 See, for example, Alcorn et al. (2003), E.N. Anderson (1996), M.K. Anderson (2005), Baker, Davies, and Young (2001), Balée (1994); Berkes (2008), Berkes, Colding, and Folke (2000), Blackburn and Anderson (1993), J. Ford and Martinez (2000), Hunn et al. (2003), Minnis and Elisens (2000), Turner, Ignace, and Ignace (2000), N.M. Williams and Baines (1983), and N.M. Williams and Hunn (1982).

12 See Boas (1921, 186–94), A. Jones (1983), Turner (1997b), Turner and Berkes (2006), and Turner and Kuhnlein (1982).

13 It is even conceivable that, in the ancient past, the long claws of the digging grizzly bears – or other, earlier bears – inspired the development of root-digging sticks as implements for humans to extract roots more readily.

14 In Ontario some Anishenaabe elders believe that bears are capable of starting fires by scraping their claws against rocks to throw a spark and that they do this because they know that berries will grow after a fire and be available as food (Andrew Miller, pers. comm., 2009; A.M. Miller 2010).

15 This narrative, "Deer and Beaver," was told by Walter of the Rear-Town-People of Yan.

16 James Douglas to Sir E.B. Lytton, Secretary of State to the Colonies, 14 March 1859, in British Columbia (1875).

17 See Peacock (1998, figure 4.5); see also Peacock and Turner (2000), and Turner and Peacock (2005).

18 This total is based on oral evidence of elders like Christopher Paul of Tsartlip, just north of the Songhees area (cited in Babcock 1967; see also Turner and Hebda 2012).

19 This estimate is based on a calculation of 100 bulbs per litre and on accounts of families in the early 1900s digging several "50-pound" potato gunnysacks, or bushels, of camas bulbs in a season (see Deur and Turner 2005).

20 They went up to Potato Mountain in groups of four or five families together, the men to hunt and the women to harvest the corms. Once one group had enough corms to meet their needs, they would start to help other families until everyone had enough. As of the 1940s and 1950s, people used large cotton flour sacks to transport their mountain potatoes, and they had coffee cans tied around their waists with a scarf in which to place their immediate diggings.

21 Now, however, if immediate drying is not possible, people can store the fresh seaweed in the freezer, to be worked on later as time or weather permits (Deveau 2010; Turner and Clifton 2006).

22 The five days would include two full days for harvesting and initial drying, one and a half days for cleaning, wetting, and pressing the seaweed into boxes, and one and a half days for chopping and redrying the seaweed and storing it away; these are minimum time requirements under ideal conditions. The seaweed itself was used in many ways, including being toasted and eaten dry as a snack and being used as an ingredient in soups and various other dishes throughout the year (Turner and Clifton 2006; Turner, Robinson, et al. 2012; Turner and Thompson 2006). In addition, a portion of it was set aside to be exchanged for oulachen grease and other goods from trading partners in other communities – for example, the Nisga'a of the Nass Valley and the Haisla and Hanaksiala of Kitamaat, in the case of the Gitga'at.

23 The cedar bentwood boxes people are still using to press their seaweed at Hartley Bay have been in service for 100 years or more; these are passed down from generation to generation and are highly valued and used with great care (Turner and Clifton 2006).

24 As with the seaweed, cedar bark was generally gathered by groups of women going out together and harvesting many strips of bark over a period of hours. They would take time to break off the brittle outer bark and then bring bundles of the raw, uncleaned inner bark back to the village or camp, to be further cleaned and split apart into inner and outer sections.

25 For example, burning over an area may simultaneously increase the reproductive capacity and productivity of berry bushes and root vegetable species, reduce fungal and insect pests, initiate new growth and provide better forage for deer and other game, and create clearings, allowing easier movement and access to these resources. Cutting trees around a village site not only provides wood for house construction but also creates openings for defensive purposes and produces tree stumps within the clearings that are an ideal substrate for the production of berries such as red huckleberries, blueberries, trailing currants (*Ribes laxiflorum*), and salal (Johnson Gottesfeld 1994; Turner 2004a; Turner and Peacock 2005).

26 For example, the use of fire to burn over a berry patch or root-harvesting site requires tremendous knowledge and experience if one is to know when, where, and how to kindle the fire so that it will yield the maximum benefits in the years to come and will not negatively impact the species and habitats being managed.

27 This tobacco, discussed previously, is also sometimes called Tlingit tobacco. It is closely related to *Nicotiana quadrivalvis*, a species cultivated by Indigenous peoples of southern Oregon, California, and some other parts of the southwestern United States. Rather than being smoked, the leaves of this tobacco were usually chewed, sometimes in elaborate ceremonies, together with lime produced by burning mussel, clam, or abalone shells. According to Dawson (1880) and Swan (1972), one elderly woman grew the tobacco at Cumshewa on Haida Gwaii up to the 1880s. However, commercial tobacco (*N. tabacum*) introduced by European traders and sold in plugs soon replaced native tobacco.

28 The leaves of this tobacco (*Nicotiana attenuata*), after harvesting, were finely cut and then dried, thoroughly mixed with mountain ram's grease, and ground between two stones or with a mortar and pestle. It was said to be milder than the commercial "white man's" tobacco (*N. tabacum*), which was generally mixed with kinnikinnick (*Arctostaphylos uva-ursi*) leaves or red-osier dogwood (*Cornus sericea*) bark to cut its strong taste and effect (Dawson 1892; Teit 1906a, 1909). It is unclear whether people actually cultivated it, but Teit (1909, 574) noted, "In some places where tobacco was scarce, it was gathered under the surveillance of the chief, in order to give every person an equal chance. It grew luxuriantly on some village sites, and some Indians think the seeds must have been planted in these places." Samples of *N. attenuata* were collected by early botanists at places like Lytton and Spences Bridge up to the turn of the twentieth century, but the plant has not survived well in modern times in the region and is known recently only at Osoyoos in the Okanagan Valley (Klinkenberg 2013).

29 The most likely *Chenopodium* to be represented, and perhaps propagated at least incidentally, is strawberry blite (*C. capitatum*), whose brilliant red fruits, in which hundreds of small, black seeds are embedded, are used as a face paint and stain (Turner 1998).

30 Even today, distinctive smooth patches on the trunks of Douglas-fir trees around southern Vancouver Island reflect the legacy of long-ago harvesting for fuel (Mathews and Dady 2008).

31 Mary Thomas (pers. comm., 2005) noted that the Secwepemc preferred to use windfall trees when they needed logs for pithouse construction. When they needed trees, her mother, Christine Allen, said, they would "wait for Mother Nature." The people sought out big cedars or firs that had been blown down in the winter storms. They would then cut the limbs off and pull the trunks out onto the river ice in wintertime and allow them to float down to campsites or village sites after the spring thaw.

32 Picking the seaweed by grasping, twisting, and pulling off the lengths of fronds by the handful left residual portions that were said to regrow readily. The second growth of edible seaweed was said to be finer and of better quality and was ready for harvest within about three weeks to a month after the first harvest from the same site (Turner 2003a; Turner and Clifton 2006).

33 Dr Daisy Sewid-Smith (pers. comm., 2004) explained that the Kwak'wala term that refers to burning back a berry bush is the same verb used in reference to singeing the hair of a ten-month-old baby during the naming ceremony: "Our berry bushes were pruned, weeded, and they also singed them a bit. Just enough ... so that it would come back and the following year it would have more berries. Now, when our babies are ten months old, we cut the baby's hair, and we singe it as well ... Because that baby now at ten months [is] ... 'out of danger.' It's going to live now. And we singe the baby's hair so the hair would grow back in abundance. And he will be handsome – or she will be beautiful. So that is similar to what we used to do with the berry bushes."

34 This information was provided by Tsartlip elder and plant specialist Christopher Paul (cited in Babcock 1967), although in this case, size preference may actually have indicated a preference for bulbs of *Camassia quamash* over those of *Camassia leichtlinii*, whose mature bulbs are considerably larger. However, Hul'qumi'num plant specialist Dr Arvid Charlie (Luschiim) (pers. comm., 2011) stated that people preferred the bulbs of *C. leichtlinii* because they were larger.

35 The Kwak'wala name for the process of weeding, *siix̱a*, also means "cleaning" and was further used in reference to practices like separating the brittle outer bark from the inner bark of western redcedar and picking out bits of dirt and sticks from roots and berries after they were harvested (Clan Chief Adam Dick, pers. comm., 2008; see also Deur 2000).

36 Exceptions are cases in which these practices can possibly be proven from DNA analysis of given individuals and populations of plants.

37 The First Huckleberry ceremony of some First Nations south of the study area, such as the Klamath of southern Oregon and northern California, entailed throwing handfuls of huckleberries to the ground, or "giving them back" to the earth, as a means of recognizing and showing respect to the earth. As Douglas Deur (pers. comm., 2007) explained for the Klamath ceremony, "Though conceived of as a cosmological act, it is clear that by fanning out to remote picking areas, regrouping at campsites, and tossing berries to the ground each year, these people effectively reseeded huckleberries around their camps." This, alongside burning practices that cleared the ground each year, tentatively explains the anomalously high densities of huckleberries immediately surrounding traditional campsites (see Turner, Deur, and Mellott 2011).

38 As a child, Mary Thomas (pers. comm., 1994) watched her father and the other men undertake this task in the late spring and early summer in the Vernon area. They used a piece of smouldering birch bark tied onto the end of a pole to smoke out all of the tent caterpillars.

39 I am indebted to Jennifer Carpenter, director of the Heiltsuk Cultural Education Society and culture and heritage manager at the Heiltsuk Integrated Resource Management Department, for permission to include her late husband's account of the berry gardens here.

40 See, for example, accounts in Boas (1921, 1934), Edwards (1979), Turner and Kuhnlein (1982, 1983), Deur (2000), Deur and Turner (2005), Turner (2005), Turner and Efrat (1982), and Turner, Thomas, et al. (1983).

41 For example, Pacific hemlockparsley (*Conioselinum gmelinii*) was also sometimes planted and tended (Compton 1993a). For the Kwakwaka'wakw and Nuxalk, Nootka lupine was produced in their *t'əkkillakʷ* (estuarine root gardens), and for some Nuu-chah-nulth, especially the Huupachesath and other peoples of the Port Alberni area, edible camas was also grown (see chapter 5).

42 The families used to travel downriver from Gwa'yee to the gardens at the mouth of the Kingcome River in large dugout canoes when the tide was going down and then return to the village upriver when the tide came up so that they wouldn't have to fight the current.

43 There is a place called Miłgeelde, the name of a territory of the *g̱anhada laxyuup* ("Raven lands") of the Kitsumkalum band, that can be translated as "burned shrubs" or "burned mountain tops," evidently reflecting this practice. People still remember burning cranberry patches (either *Vaccinium oxycoccos* or *V. vitis-idaea* or possibly *Viburnum edule*) at the Lockerby Creek camp and on the Ecstall River in this area (Mcdonald 2003, 53).

44 See Deur (2000, 2004), Deur and Turner (2005), Johnson (1999), McDonald (2003), Norton (1979a), Thornton (1999), Turner (1999), and Turner and Thompson (2006).

45 As noted previously, this has occurred for many of the prairies that have been documented from oral history and other lines of evidence as having once been much more extensive (M. Kat Anderson, pers. comm., 2009; Lepofsky, Heyerdahl, et al. 2003).

46 Chiefs and designated leaders were also placed in charge of other harvesting and management endeavours, including deer drives (Interior Plateau), reefnet and weir fishing (Straits Salish and Northwest Coast peoples in general), and oulachen fishing (Kwakwaka'wakw, Nuxalk, Haisla, and Nisg̱a'a), as well as overseeing various activities in plant harvesting and management.

47 The *sxʷ'pʕápm* ("burner") also directed the manual clearing of Indian trails clogged by unmanageable debris or trees blown down, as well as the clearing of debris from important springs and water seeps and from naturally occurring game salt licks within the group's territory, along with overseeing the driving of deer into a "deer surround" (J.A. Ross 2011, 271).

48 For example, Yup'ik Inuit would traditionally bury some food in the tundra, usually fish, before going berry picking in order to feed the "little men," or berry spirits, so that they would provide a rich harvest (Thornton 1999).

49 As noted in chapter 9, some sites, such as the estuarine root gardens along the San Juan River, owned by an ancestor of Chief Charlie Jones (Queesto) of the Pacheedaht First Nation, were so valued that they were actually guarded by slaves against intruders (Turner, Thomas, et al. 1983). Tlingit berry gardens were sometimes similarly guarded by slaves or other sentinels (Thornton 1999).

50 For example, the Katzie (Halkomelem) families of the Fraser Valley, who owned large patches of wapato on the west bank of the Pitt River, traded these tubers to the Straits and Halkomelem Salish on Vancouver Island, to the Squamish, and, inland, to the Lower Nlaka'pamux (Thompson) in exchange for camas bulbs and, in the last case, products from the interior. Similar trading of wapato occurred around the lower Columbia River in Chinook territory (Darby 2005; Galloway 1982; Kuhnlein and Turner 1991; Suttles 1955, 1987a, 1987c).

51 See, for example, E.N. Anderson (1996), M.K. Anderson (2005), E.R. Atleo (2004, 2011), Berkes (2012), Johnson Gottesfeld (1994), Minnis and Elisens (2000), J. Ford and Martinez (2000), Fowler (2000), Deur and Turner (2005), Turner (2005), and Turner and Berkes (2006).

52 For example, Yakima women traditionally sweat bathed and refrained from sexual relations for five days before beginning to dig roots, and only certain, ritually clean, women were allowed to prepare the earth ovens for roasting roots. Men, conventionally, did not participate in root roasting or even look in the direction of the earth ovens while women were tending them (Kirk and Daugherty 2007; Peacock 1998).

53 This description is related to the scattering of spring beauty seeds by recently bereaved women as recounted previously for the Tsilhqot'in.

54 This protocol continues to the present day.

CHAPTER TWELVE

1 Mayanilth is from the Mamaliliqela and Qʷiqʷasutinuxʷ of Village Island and the Broughton Archipelago.

2 The cracks in the tree were smeared over with spruce pitch, and bumblebees were assigned to guard any potential entrance so that no one else could break in to the cedar tree.

3 See, for example, E.R. Atleo (2004, 2011), Barbeau (1961), Boas (2002), Bouchard and Kennedy (1979), Cove and MacDonald (1987), Kennedy and Bouchard (1983, 1994), Mackin (2004); Maud (1978a, 1978b, 1978c), Mourning Dove (1933), Wilp Wilx̱o'oskwhl Nisg̱a'a (1995), J.A. Ross (2011), Swanton (1905, 1908), and Teit (1898, 1909, 1912, 1919).

4 Some of the Boas (2002) stories were translated from one or more dialects of Kwak'wala, Nuu-chah-nulth, or Secwepemc-tsín into Chinook Jargon, then into German, and much later into English (translated by Dietrich Bertz).

5 At a conference of the Society of Ethnobiology at Anchorage, Alaska, in May 2005, an apprentice Tlingit storyteller related, in the presence of his teacher and mentor, how he had to work hard over many years, memorizing hours of text, so that he could repeat, word for word, the stories of his ancestors. The ability to do this was considered a gift, an art, and a heavy responsibility. As well, having heard Dr Daisy Sewid-Smith and Clan Chief Adam Dick recounting stories on many occasions, even in English, which is not their first language, I have been struck by the accuracy of replication of the wording and conversations embedded in the stories about Mink and the other powerful characters who made the world into its present form.

6 Deer is well known as one who obtains fire for people, with stories on this theme recorded in Boas (2002) from the Tlingit, Ts'msyen, Heiltsuk, Nahwitti, Nuu-chah-nulth, and Comox; the Heiltsuk name for Deer is "torch bearer" because he stole fire with the aid of wood, which he had tied to his tail (ibid., 495).

7 A "slow match" is comprised of a wad of tinder, sometimes of dried root fibres of bracken fern or other plant material or sometimes of shredded cedar bark, which is ignited and encased as a smoldering coal within bivalve shells or a wrapping of birch bark or cedar bark and then buried in the ground. "Slow matches" were commonly used until quite recently and are easier to use than having to start a fire with a drill. As well as being buried on site, they can be carried by travellers and used to rekindle fire at a new camping spot.

8 Similarly, with supernatural help, the young man meets all of the other challenges set by his father-in-law and at last obtains his acquiescence.

9 The use of fire in cooking, smoking, and dehydrating food, technologies related to fire and fuel wood, and the use of fire in traditional vegetation management are all topics addressed in this book (see chapters 5, 6, and 11).

10 A common element in several of these stories involves a situation in which the brother takes revenge on his sister's husband for her death and is then "discovered" when he tries to imitate his sister pounding bracken fern rhizomes to eat because he does not hit the roots in the same way a woman would. The two women to notice this were Mallard Duck sisters (Boas 2002).

11 *Ngexʷ* is a delicious "pudding" comprised of roots, berries, fat, and other ingredients, including combinations of yellow glacier lily (*Erythronium grandiflorum*) bulbs, spring beauty (*Claytonia lanceolata*) corms, tiger lily (*Lilium columbianum*) bulbs, onions (*Allium cernuum*), bitterroot (*Lewisia rediviva*), saskatoon berries (*Amelanchier alnifolia*), and black tree lichen (*Bryoria fremontii*), along with deer fat and sometimes salmon eggs (Turner, Thompson, et al. 1990).

12 Told by Walter, a member of the Rear-Town-People of Yan (cited in Swanton 1908, 448).

13 Oil and fat are also commonly mentioned foods in many stories, sometimes being served as an accompaniment to berries or other plant foods.

14 Fog Woman is an Earth-Mother figure who appears to have evolved from Volcano, or Frog Woman, also known as Creek Woman, or Jilaa K̲uns in Haida.

15 *Q'əx̲mín* refers to the seeds (fruits) of "Indian celery," or Indian consumption plant (*Lomatium nudicaule*).

16 Possibly this is Chatham Island, right beside Discovery Island, called Tl'ches.

17 The Saanich name for willow, especially Pacific willow (*Salix lucida* ssp. *lasiandra*), is *sx̲ʷələʔ-íłch* ("reefnet-plant/tree").

18 Cured stipes of bull kelp furnished the long lines used for the gear for halibut and other groundfish (chapter 6).

19 This narrative was told by Mrs Harriet Hudson of Kitselas to Ts'msyen ethnographer William Beynon in 1948 (cited in Cove and MacDonald 1987, 84–8; see also Johnson 1997, 138–9).

20 In early spring, the nettles normally would have died back for the winter, but the seasonal timing of the story isn't necessarily in accordance with modern times.

21 For example, McIlwraith (1948, vol. 1, 89) reported the following for the Nuxalk oulachen nets: "In the beginning of time, Raven told two men to make a net according to a pattern which he showed them and gave them some [nettle] fibre. It was insufficient and when the workers asked for more, Raven told them to draw it from their own intestines. This they did and ever since, spiders have had the same ability." The Heiltsuk have a similar tradition (Boas 2002). In these last narratives, there is a direct association between spider webs and fishing nets.

22 It is not surprising that for some of the cultural groups in southern California, Arizona, and New Mexico, Spider is equivalent to the Creator, or first being (Archambault 2006).

23 Although the species of fern in this case is not mentioned, it is probably the rootstock of spiny wood fern (*Dryopteris expansa*), a well-known wintertime food and famine food (Turner, Johnson Gottesfeld, et al. 1992), although possibly the "roots" are bracken fern (*Pteridium aquilinum*) rhizomes, which are also known as a famine food (Norton 1979b).

24 Told by Walter of the Rear-Town-People of Yan (cited in Swanton 1908, 574).

25 I wish to acknowledge Randy Bouchard and Dr Dorothy Kennedy, David Stevenson and his Oowekeeno consultants, and Evelyn Windsor and Clan Chief Adam Dick (pers. comm. to Randy Bouchard and Dorothy Kennedy, 2001) for explanatory annotations related to this narrative.

26 As Nōak.aua wished, Masmasalā'niq also made a box with four dancing flutes that had the voices of various birds, as well as a dancing staff.

27 Told by Walter of the Rear-Town-People of Yan (cited in Swanton 1908, 460–6).

28 Told by John Sky to John Swanton within the series of episodes of "Raven Travelling" (cited in Enrico 1995, 62–3; and in Swanton 1905, 126).

29 A Secwepemc story about Tłéesa, the eldest of four Transformer brothers (who was changed into a dog), also alludes to such a tree: "They now crossed the river. Soon they saw a 'tobacco tree.' As soon as anyone tried to get tobacco, one of the tree's branches began swinging around in circles and killed him. *Tlē'esa* [Tłéesa] took a small stick and walked up to the tree. When the branch began swinging, he lopped it off with the stick and tossed it into the river. Then, using his stick, he dug up the tree and pushed it over. Now the brothers came up and picked off all the tobacco. *Tlē'esa* did not get anything" (Boas 2002, 64).

30 This same woman later encountered a wolf giving birth and assisted her in delivering the cubs. In gratitude, the wolf gave her midwife the power of a shaman (McIlwraith 1948, vol. 1, 691).

31 This is a different species from the tobacco (*Nicotiana attenuata*) created for the Nlaka'pamux people by Coyote. The Haida tobacco is a now-extinct relative of *N. quadrivalvis*, whose closest known relatives today grow in Oregon and California (Turner and Taylor 1972).

32 Here, "Indians" refers to the first people or ancient people (*tlldluu x̱àadag̱aay*) or to their more contemporary equivalent (*tllsdaa x̱àaydag̱aay*) (John Enrico, pers. comm., 2002; see also Turner 2004a).

33 Included are the Tlingit, Ts'msyen, Haida, Nuxalk, Heiltsuk, Oowekeeno, Nahwitti, Nuu-chah-nulth, and various Coast Salish peoples (Boas 2002; see also Barbeau 1961, 78–9).

34 A similar version is told by the Nuxalk (Boas 2002, 507).

35 In regard to Raven's craving for berries, Boas (2002) cites stories from the Tlingit, Ts'msyen, Heiltsuk, Nahwitti, Nuu-chah-nulth, Nuxalk, and Comox (for the Ditidaht, see also Kennedy and Bouchard 1994).

36 Mourning Dove (1933, 125) adds, "That was Coyote's ruling near the Beginning. That is why his hair, the long black timber-hair, hangs from trees in the mountains. It is called *sqʷel'íp*. It is the black moss that the people cook in pit ovens." Secwepemc elder Ike Willard told a similar story to Bouchard and Kennedy (1979, 22) about the origin of black tree lichen, but in this case, Coyote tried to imitate Spider sliding down to the ground from the top of a tree on a strand of his web: "Coyote climbed

up the tree and lay on his back on the top of the branches. He ... used his fur as a web and started to slide down it, but when he was halfway down, he became stuck. Spider unhooked Coyote from the tree and helped him back to the ground. 'Coyote, when the coming people live on this land, your fur that is still hanging from the branches will be black tree moss. The people will gather it and cook it to eat.'"

37 One such version is told by Ike Willard (also Secwepemc) (cited in Bouchard and Kennedy 1979, 15–18; see also M. Dixon, Palmantier, and Kuipers 1982).

38 He stayed many years with the Wolf people, and later when he returned to his village he was helped by the Wolves (told by Nathan Shaw, Gitxaata, to William Beynon in 1952, cited in Cove and MacDonald 1987, 275).

39 Boas (2002, 494). See also Boas (ibid., 538), Swanton (1905, 499, 503, 521), and Cove and MacDonald (1987).

40 These language groups include the Tlingit, Kwakwaka'wakw, Comox, Halkomelem, Quinault, Nuxalk, and Kathlamet (Chinookan). Other Dene (Athabaskan) peoples who carry similar stories include the Apache of the Southwest, who re-enact it in puberty ceremonies, and the Dogrib of the Northwest Territories, who believe themselves to be descendants of the people in this story (Boas 2002; L.R. Smith 2008b). Similar stories are also found among Inuit peoples from Greenland, Hudson Bay, and Alaska, as well as in Cree and Cherokee (Boas 2002).

41 Examples are the Tlingit, Ts'msyen, Heiltsuk, Straits Salish, Klallam, Quinault, Lushootseed, Chehalis, Quileute, Cowlitz, Maidu, and Washo.

42 Examples are the Shoshone, Ktunaxa, Blackfoot (Siksika), Assiniboine, Gros Ventre, Arapaho, Arikara, Hidatsa, Cheyenne, Crow, Dakota, Kiowa, and Ponca.

43 Examples are the Tahltan, Tsetsaut, Dakelh, Tsilhqot'in, Sarsi, and Cree.

44 Told by Emma Wright, Gitlaxdamks, to William Beynon (cited in Cove and MacDonald 1987, 306–7).

45 Exchanges of the past are reflected today in similar plant names across language groups, even between groups that are linguistically unrelated or only distantly related (see chapter 3). Names and specialized terminology tend to be transmitted with goods and information for speakers of different languages.

46 Soapberry is apparently not known to grow anywhere on Haida Gwaii today (see Calder and Taylor 1968).

47 Boas (2002, 639) observed, "The self-contained structure of the Raven Legend, as found in the region's north, and its gradual shrinking towards the south with simultaneous modification of the traits common to the north, prove that a gradual transmission of a legendary cycle from the north southward has taken place here, which essentially found its conclusion in Newettee (Nahwitti) although, even from there, isolated traits, strongly modified, still penetrated southwards ... it can safely be deduced that the Raven cycle was diffused southward along the coast as a self-contained structure and gradually lost cohesion."

48 In fact, Boas (2002) found, the Nuxalk and Comox had the highest percentage of common features of all tribes surveyed.

49 This observation is based on the assumption that the Nuxalk diverged north from an original Proto-Salish homeland in the lower Fraser River region.

50 The Comox were followed by the Nuu-chah-nulth, Nahwitti, Nuxalk, and Heiltsuk.

51 An example is the Ts'msyen story "The Origin of the Name of Chief Mountain," told by Mathew Gurney and Robert Stewart of Kincolith (Gingolx̱) to William Beynon (cited in Cove and MacDonald 1987, 340). See also McDonald (2003).

52 Significantly, Boas suggests that this last group (the Chinookan) was the entry point to the coast not only for this story but for others as well. A further example of more

northerly groups borrowing elements from the Chinookan is in the Nuu-chah-nulth name Tsítsikle, for the benevolent stranger who provided herring to a boy and his starving family (Boas 2002, 269–70). Boas identifies this name as a reduplicated form of the Chinookan name for the Wanderer, Tsikla.

53 Recounted in Cove and MacDonald (1987, 252, 254). See also the Nlaka'pamux story "The Hunter and the Goats" in Teit (1912, 261).

54 In some versions, it was a robin carrying a sprig of salmonberry (*Rubus spectabilis*).

55 Told by Kenneth Harris (Chief Ha<u>x</u> Bagwoot<u>x</u>w) (cited in Johnson 1997, 140).

56 See Swanton (1905, 212–13; 1908, 689–700) and Cove and MacDonald (1987, 80–1, 82–3, 167, 311).

57 This story was recounted to William Beynon by Mrs Heber (Lucille) Clifton and Mrs Robinson, both of the Eagle (L<u>a</u>xsgiik) clan of Sen'arhait in Kitamaat, who had married into the Gitga'at community of Hartley Bay. See also Barbeau (n.d., cited in McDonald 2003, 127–34).

CHAPTER THIRTEEN

1 The term used by Boas (1930 and elsewhere) is "prayers," but Kwakwaka'wakw historian and language specialist Dr Daisy Sewid-Smith (Mayanilth) (pers. comm., 1994) said that "words of praise" is a more accurate translation from the original Kwak'wala.

2 Examples can be seen in the north among the Koyukon (R.K. Nelson 1983), Yupiaq (Kawagley 1995), Yukon Dene (Cruikshank 1998, 2005), Tlingit (Emmons 1991; Thornton 1999, 2008), and Nisga'a (Wilp Wil<u>x</u>o'oskwhl Nis<u>g</u>a'a 1995; Vickers 2008); on the west coast of Vancouver Island among the Nuu-chah-nulth (E.R. Atleo 2004); in the Interior Plateau among the Nlaka'pamux (Hanna and Henry 1995), Tsilhqot'in (L.R. Smith 2008b), Secwepemc (R.E. Ignace 2008), Spokan (J.A. Ross 2011), and Sahaptin (Hunn, Selam, and family 1990); on the Great Plains among the Plains Cree (Kovach 2006), Lakota Sioux (J.E. Brown 1953), and Pawnee (P.V. Beck and Walters 1977); in eastern North America among the Mohawk (Wall 1993) and Anishenaabe (A.M. Miller 2010); in the Southwest among the Hopi (Loftin 1994) and Navajo (Diné) (Cajete 2000); in California among the Miwok, Klamath, and other Native Americans (M.K. Anderson 2005); and in Mexico and beyond among the Rarámuri and other peoples (Wyndham 2004; Salmón 2000).

3 See, for example, E.N. Anderson (1996), M.K. Anderson (2005), E.R. Atleo (2004, 2011), Berkes (2008), P.V. Beck and Walters (1977), J.E. Brown (1953, 1986), Cajete (2000), Callicott (1989), Senos et al. (2006), L.R. Smith (2008b), Turner and Atleo (1998), and Wall (1993).

4 Teit's notation under "flowers in general" recounted, "Flowers are the valuables of the earth or mtns and if they are plucked ruthlessly the earth sorrows or cries." Then, under "siekEm [*s-yíqm*] grass in general," he wrote, "Flowers, plants & grass especially the latter are the covering or blanket of the earth[.] If too much plucked or ruthlessly destroyed [the] earth [is] sorry and weeps[.] It rains or is angry and makes rain, fog & bad weather" (cited in Turner 2005, 20; and Turner, Thompson, et al. 1990, 54).

5 The Rarámuri, or Tarahumara, of the Sierra Madres in Mexico, for example, use a term, *iwígara*, that derives from the root word *íwi* ("soul, breath of life"). It pertains to everything that breathes or respires: plants, animals, humans, stones, the land – all of these share the same breath, and all have "souls." *Iwígara* embraces the total interconnectedness and the integration of all life, physical and spiritual, within the

Rarámuri universe, and it ties people to their songs and ceremonies, to their foods, and to the land that nourishes them (Salmón 2000).

6 This practice of offering a small gift or tobacco is a widespread tradition, for example, among Plains Cree medicine gatherers when they remove medicinal roots (Kovach 2006).

7 For example, the Tsilhqot'in are taught from an early age that one should not point at people because it is disrespectful. Xeni Gwet'in chief Roger William (2003) explained that the same situation holds for mountains: "In our ways, you know, you don't point at people; it's disrespect when you're pointing at people and they can't hear what you're saying. So when you point at *Ts̲'il?os̲* and *?Eniyud* – *?Eniyud*'s a lot meaner than *Ts̲'il?os̲* – but it would rain on you or snow, or you would have bad luck. And it usually happens when you're not ready. So ... at the least time you expected, it could snow on you in the middle of the summer, depending what the situation was." On Vancouver Island the Saanich Straits Salish also stress that one should never point at the mountains known as the Malahat because doing so will cause thunder; these mountains are the home of the Thunderbird (Elsie Claxton, pers. comm., 1998).

8 There was a similar taboo against harvesting giant California mussels (*Mytilus californianus*) while harvesting seaweed.

9 Other plants are called *maaya g̲alipliip* ("thunder berries") in Sm'algya̲x, the Ts'msyen language, and are apparently also associated with thunder: wild lily-of-the-valley (*Maianthemum dilatatum*), false Solomon's-seal (*Maianthemum racemosum*), and twistedstalk (*Streptopus* spp.) (Turner and Thompson 2006).

10 In mythical times Sword Fern (*Polystichum munitum*) was said to be a supernatural person who controlled the weather. He was said to have a hairy face, earrings of dentalia, and cheeks smeared with red ochre, and he featured in rituals to call the northwest wind (Boas 1921).

11 Pond-lily (*Nuphar lutea* ssp. *polysepala*) leaves tend to align themselves on the surface of a lake. When a southeast wind is blowing, they lie flat on the water; when the west wind blows, they lift off the surface, and "the whole lake would be moving." Long ago, whenever there was a bad storm, when the southeast or southwest winds had been blowing for several days, the Hesquiaht would take a pond-lily leaf and slam it down on the water, calling out "*Hachłiitł*! *Hachłiitł*!" ("West wind! West wind!"). This was said to calm the storm quickly and bring in the gentler west wind the next day (Turner and Efrat 1982, 70).

12 The letters of ethnographers Aurel Krause and Arthur Krause (1993) from their time in Tlingit country in 1882 describe some other instances of more recent transgressions: "The [Tlingit] Indians seem to protect a duck, the oldsquaw. One of the many causes given for the poor weather in February was that somebody killed an oldsquaw. The native name *ja-a-uná* is a good rendition of the melodic call of this duck" (156). "We learn that the Indians now accuse us of causing the bad weather. First we dragged a dead porcupine through the snow instead of carrying it, then we washed the mountain sheep in salt water. Today they want to burn us in effigy" (158). "The fish baskets laid out last night [at Deshu, Portage Bay] also yielded only an insignificant catch, so everybody was upset. Several times the word *ch'lakass* was heard. The presence of the *gutzkakon*, the stranger from far away, was definitely at fault for the low catch. Without doubt, his shooting of the raven, oldsquaws, and other sacred animals had caused the severe winter and heavy snowfall" (197). See also Emmons (1991).

13 Even in the Secwepemc area, where people apparently did not eat bracken fern, it was known in the eastern dialect as *pəp'əʔséstsʼyə* ("snake grass") (Sarah Deneault, pers. comm. to Marianne Ignace, 1995).

14 Told by Jimmy Williams of Kispiox to Marius Barbeau in 1920 (cited in Cove and MacDonald 1987, 278–80; see also 79–83, 167).

15 This story has similarities to a number of others, including the story of the Copper Salmon, told by Walter of the Rear-Town-People of Yan (cited in Swanton 1908, 689–700).

16 Cf. the Haida names for devil's-club, *ts'iihlinjaaw* (Skidegate dialect), and for gambling sticks, *ts'iihl*.

17 Devil's-club berries are called "highbush cranberries of the grizzly" in the Nuxalk language (Turner 1973).

18 For example, the Saanich (Straits Salish) name is *tuʔtuʔə́ləqəp* ("echo"), a word that has been used more recently for "telephone."

19 Potlatch ceremonies have many different components and vary considerably; one of the clearest accounts of the Kwakwaka'wakw potlatch complex is in the film *Smoke from His Fire* (Recalma-Clutesi 2007).

20 For *masru*, the Kotzebue Inuit of Alaska would leave a piece of dried fish: "We carefully opened these [mouse caches filled with *masru*] and emptied the *masru* onto our bag to sort; big ones into the sack and small ones and any peculiar ones back into the cache. Then we put a piece of dried fish into the cache and covered it back up. That was in trade and a thank you. It is always done that way and it is most important at freeze-up because there is less time for the mice to gather more food" (A. Jones 1983, 117).

21 Four is a sacred, "magic" number, which, some people say, corresponds to the four directions.

22 As noted previously, Boas's (1930) translation of the original Kwak'wala word as "prayer" is inaccurate; according to Dr Daisy Sewid-Smith (pers. comm., 1994), the term implies "words of praise."

23 The regalia, masks, rattles, and whistles used in the Hamatsa and other sacred ceremonies are themselves spiritual entities, to be handled only by those who are qualified and trained to do so.

24 See Kennedy and Bouchard (1983) for Sliammon/Tla'amin Coast Salish, E.R. Atleo (2004), and Turner and Berkes (2006). For example, de Laguna (1972, 102) reported for the Tlingit, "Animals were always treated with great respect. The hunter or fisherman appealed to them before capture; and after their death, followed certain observances that were supposed to propitiate their spirits. The folktales tell of punishment of those who even spoke disrespectfully or slightingly of the animals ... it was reasoned that all humans and animals have the same attributes, and differ only in form."

25 For descriptions of some of these ceremonies, see Hart (1976), Hunn and French (1981), Hunn, Selam, and family (1990), Hunn, Turner, and French (1998), Kuhnlein and Turner (1991), J.A. Ross (2011), and Turner, Bouchard, and Kennedy (1980).

26 Hill-Tout referred to this as "wild raspberry" but identified it with the name "*Rubus Nutkanus*," a synonym for *Rubus parviflorus* (thimbleberry), whose succulent shoots are most often eaten, along with those of *Rubus spectabilis* (salmonberry) (Turner 1995).

27 See Compton (1993b), Johnson (1997), Thornton (1999), Turner (1995, 1997a), and Turner, Ignace, and Ignace (2000).

28 George (2003) recalled that he and the other children would ask for some of the sprouts, but most were eaten by the women and were considered a special women's food.

29 In one example for fisheries, many First Salmon ceremonies lasted for four days, and over this period, no more salmon were allowed to be caught, thus ensuring a substantial run passed through before people started to catch and process their winter supply (E. Claxton and Elliott 1994; Turner and Berkes 2006).

30 According to Dr Earl Claxton Sr (pers. comm., 2005), the Saanich name for this plant, *q'əx̱mín*, is related to the term *sq'əx̱* ("to put a curse on someone") because of its role in protecting people from evil curses.

31 Following her seclusion, the girl was ceremonially greeted as a woman and undertook ritual washing and cleansing. For a Nlaka'pamux girl, among others, during her period of seclusion she was not allowed to eat fresh meat or salmon, and she was never allowed to go near a stream in which salmon swam, or they would disappear from that stream (Laforet, Turner, and York 1993; Maud 1978c, 55).

32 Even the next child born after twins was a "mystery" and underwent the same types of restrictions as his or her twin siblings (Teit 1909).

33 John Ross (2011) notes that before burning over the remaining tules, the women would gather the edible seeds from the yellow pond-lily.

34 John Ross (2011) reports a similar belief system for the Spokan, noting, "A recent widow was forbidden to pick any berries ... breaking this taboo could ruin an entire crop." Teit (1930, 175) explained, "The belief was that the ghost might follow the widow to the berry patches and harm other pickers, and blast the crop." Food was always shared with elder widows and widowers, so not being able to pick berries did not result in a food shortage for them.

35 California basket maker Jacquelyn Ross, of the Coastal Miwok, explained how Miwok weavers are taught to harvest their materials with respect and only from places that are "clean" and untouched by activities such as drinking or partying. Basket materials from such places would make baskets that would be unlucky for the people who used them. The materials for baby baskets, she said, were selected with especially great care, and the weaver was careful to think good thoughts and to think with love about the baby to be born so that the basket would protect the baby and surround it with a positive force (cited in Turner 1996). Similar sentiments have been expressed by basket weavers from northwestern North America, such as Mary Thomas (2001).

36 In the European tradition, the term "oak" in English derives from its name in Scandinavian languages, *ek* or *eg*, which, in turn, is related to the old Norse word for "reverence" or "respect," *igja* (Jonsson 1996).

37 One kind of tree in the forests of Thailand called *jan*, for example, can be cut down only if it is to be used to make a coffin for a king or other member of a royal family (Chinatat Nagashinha, Phranakhon Rajabhat University, pers. comm., 2006).

38 See E.R. Atleo (2004, 2011) and McIlwraith (1948); see also Colding and Folke (2001), L.R. Smith (2008b), and Turner and Berkes (2006).

CHAPTER FOURTEEN

1 See, for example, the writings of E.N. Anderson (1996), M.K. Anderson (2005), Berkes (2012), Callicott (1989), J. Ford and Martinez (2000), Gadgil, Berkes, and Folke (1993), and Minnis and Elisens (2000).

2 The ineffective methods of resource managers are described in numerous publications, including Berkes, Colding, and Folke (2003), Blaser (2004), T.J.S. Carlson and

Maffi (2004), W. Davis (2009), Mascia et al. (2003), Millennium Ecosystem Assessment (2003), and Stepp, Wyndham, and Zarger (2002).

3 As noted by Secwepemc language specialists Drs Ron Ignace and Marianne Ignace (pers. comm., 2011), the term "Mother Nature" is a Western term and does not occur in the Secwepemc language. Mary Thomas often used it metaphorically, however, to enable her to communicate effectively with those outside of her own culture about relationships with the land and environment.

4 As cited in tables 2-1 and 2-2, the following are examples of plants with evidently very long human associations in northwestern North America, based on their known more recent cultural importance, on their known presence in the region (or in northeastern Asia) in the Late Pleistocene and/or Early Holocene, and in some cases, on the pervasiveness of their names in the region's languages: spiny wood fern, bracken fern, kinnikinnick, saskatoon berry, seaside strawberry, chokecherry, gray currant, wild roses, thimbleberry, red elderberry, soapberry, blueberries, lingonberry, bog cranberry, highbush cranberry, fireweed, cow-parsnip, western dock, northern riceroot, sagebrush/wormwood (*Artemisia tilesii* and other spp.), alders, birches, spruces, lodgepole pine, poplars, willows, and cattail.

5 The obsidian is estimated to have been brought to the site about 9,800 years ago (E.J. Dixon 1999).

6 This is a critical lesson today as well; we cannot consider the production of coffee or apples or fish in isolation from the host of equipment we apply in the growing, harvesting, and processing of these resources, and when we think of energy expended, we need to consider all of these peripheral but integral parts of the system.

7 In part, as well, the immensity of these middens may be a factor of better preservation as more shell accumulates; other middens may also have existed earlier, to be inundated as relative sea levels rose (Moss 2011).

8 As Moss (2011) emphasizes, however, it cannot be claimed that these trends were universal or that they happened within a measured, orderly trajectory. There are many exceptions to such generalizations and many gaps in the archaeological record that blur and distort any picture.

9 A case in point is the term *s-ts'ik'* or *s-ts'ik*, identified by Kuipers (2002) as a Proto-Salish term for "nut," "acorn," and/or "fir or pine cone" (or seed). Derivatives of this term are found in some modern Coast Salish languages for "hazelnut" and in some modern Interior Salish languages for "whitebark pine seed." Forms of the term also occur in Ts'msyenic languages (Gitxsan, Nisg̱a'a) for "hazelnut" and in Athabaskan (Tsilhqot'in) for "whitebark pine seed." Others use it for "Garry oak acorns" and still others for conifer seeds generally.

10 These *Camassia* species include both common camas (*C. quamash*) and the larger, more robust Leichtlin's camas (*C. leichtlinii*).

11 There were, of course, positive outcomes from the new relationships, but as the stated goal of governments and the church was to assimilate Indigenous peoples into mainstream society, the negative consequences in terms of loss of languages, cultural practices, traditional foods, and overall health and well-being of Indigenous populations were immense (Duff 1997; Fisher 1992; Lutz 2008; Ommer et al. 2007; Suttles 1990b; Turner and Turner 2008).

12 *Camassia quamash* may also have existed during the Pleistocene in a few coastal refugia, such as the Brooks Peninsula on Vancouver Island, and possibly in western Washington (Tomimatsu, Kephart, and Vellend 2009).

13 According to Secwepemc elder Mary Thomas (pers. comm., 1994), ribs freshly extracted from a deer that has been pit-cooked can be hit in "just the right way" with

a maul or stone for them to shatter into multiple sharp slivers, which could then be used as awls, in basketry, for stitching deer hides, and for other purposes.

14 According to the locality of the cooking and cultural preferences (see chapter 5), the types of vegetation used to surround the food being cooked in earth ovens have also varied: salal and sword fern, kelp and other seaweed, skunk-cabbage leaves, Douglas-fir boughs, ponderosa pine needles, wild rose branches, or fireweed plants.

15 The numerous acorn-processing pits here were lined with hemlock boughs and marked at the corners with stakes, and they were evidently ingeniously designed to capture moving groundwater from a well-situated aquifer in order to leach the bitter tannins from the kernels of potentially millions of acorns (Croes, Fagan, and Zehendner 2009a, 2009b; Croes et al. 2007).

16 The research of Tomimatsu, Kephart, and Vellend (2009) is an example of DNA analyses that can help to shed light on the origins and affiliations of key ethnobotanical species, although this study did not reveal any evidence of human transport of camas bulbs.

17 The Nuu-chah-nulth and Ditidaht did, however, within the past century or so, harvest this seaweed for sale to Chinese and Japanese people in the Victoria area (see Turner, Thomas, et al. 1983).

18 Notable exceptions include hazelnuts, whitebark pine, and sometimes other conifer seeds, balsamroot seeds, and acorns, all of which were harvested and eaten within their ranges in the study area.

19 For example, wooden digging sticks with carved antler or wooden handles predominated in the Interior Plateau, whereas over most of the Northwest Coast, yew-wood root diggers with knobbed tops were more common.

20 Mary Thomas (pers. comm., 2001) recalled that the Secwepemc, at least, also made canoes of bundles of tule stalks (*Schoenoplectus acutus*), similar to those made by the Paiute and others living around the Great Basin wetlands, such as Stillwater Marsh in Nevada.

21 As well as detachable points joined to the main shaft with twined cordage, some of these spears comprised as many as three different kinds of wood, each with its own density, to yield a perfect balance for throwing (Turner, Thomas, et al. 1983).

22 Later, after Europeans arrived, the women adapted these same basketry techniques and materials to create items of interest to the newcomers: tea trays, cups and saucers, knitting baskets, and even coffee tables were created to cater to a new market for trade and sale.

23 The Coast Salish weavers also used dog hair and duck feathers in their weaving, and mountain goat wool was also spun and used for blankets of various types. Sometimes they added cattail and fireweed seed fluff to these materials in their spinning.

24 Basketry patterns reflect the weavers' close connections with the natural world, as many of them – even those that are seemingly simple geometric designs – are symbolic representations of patterns in nature: ferns, mountains, or animals of various kinds.

25 After the major food harvesting was finished, the basket materials were often dried and stored, to be soaked and worked on during the winter. Schlick (1994, 142) estimates, from the southern Interior Plateau, that it may have taken a woman two to three months to complete a woven bag of about 15 by 20 inches.

26 For example, Annie York suggested that a Nlaka'pamux household might include sixteen or more large storage baskets for dried berries, perhaps as many for dried roots, several others for harvesting and cooking, and one for whipping soapberries, as well as various bags (Turner, Thompson, et al. 1990).

27 Specific recipes for medicinal preparations are often considered a family's property, said to lose their effectiveness if too many people know about them.
28 As noted in chapter 7, there is a strong correlation among food plants, medicinal plants, and plants that have recognized poisonous properties, a phenomenon that is, again, not limited to northwestern North America but quite widespread. In fact, the majority of widely used medicines of classical Greek and Roman times were actually food species.
29 Two of the ingredients, saskatoon berry bark and trembling aspen, are the same; others, like chokecherry for the Dakelh and bitter cherry for the Saanich, are similar. Furthermore, each mixture contains one herbaceous species: sarsaparilla for the Dakelh medicine and licorice fern for the Saanich (Elsie Claxton, pers. comm., 1993; Poser 2008b; Turner and Hebda 1990).
30 Shamanistic healing may also involve the use of plants, either as administered by the shaman or as taken by the shaman as part of his spiritual preparation, along with other types of training.
31 Some medicines are known to be particularly powerful spiritually and have the strength to ward off illness and protect people from any kind of evil or bad luck; although the medicines themselves may differ from place to place, the concept of such plants having spiritual power is universal in the study area.
32 Saanich elder Violet Williams (cited in Turner and Hebda 2012, 30) recalled, "The old people used to say, 'Don't take a big chunk [of bark for medicine].' Just take a thin strip [about 3 to 5 by 30 centimetres], and then the tree will heal up, and the person who is being treated will heal up too, just like the tree." Another protocol that she and Elsie Claxton (ibid., 29–30) described was that a healer should always go out to gather her medicine in the morning, before eating anything. Gitga'at elder Helen Clifton (cited in Turner and Thompson 2006) added that when one is harvesting medicine, she should always focus on that task and try not to do anything else – just think about the person for whom the medicine is intended. Mary Thomas (pers. comm., 1994) said that for the best results, one should gather medicine only from remote places far from human activity so that the medicine will not be subject to profane influences.
33 For example, the practice of burning the skin with birch conk (*Inonotus obliquus*) and other substances to "draw out" underlying pain (Mary Thomas, pers. comm., 2001; L.M.J. Gottesfeld 1992b; Johnston 1997; Poser 2008b) is very similar to the moxibustion practised in traditional Chinese medicine and may, in fact, originate from common ancestral practices from the ancient past. The use of stinging nettle as a counter-irritant to treat arthritic pain is also likely quite old and is shared by Europeans and other Indigenous peoples across North America.
34 This need for dispersal is described in Ames and Maschner (1999), Deur (2002), Hayden (1997), and Moss et al. (2004).
35 Most perennials are capable of continued growth through the activity of meristems – areas of unspecialized tissues in roots, stems, bark, and buds that can differentiate into more specialized tissues. Meristems enable plants to grow new roots, underground stems, vascular tissues, bark, branches, shoots, or leaves on an ongoing basis and, as required, to repair and renew these organs if they are damaged. This capacity for regrowth is an ability evolved over eons, enabling these plants to withstand various kinds of disturbance, such as breakage from wind or water, damage from insects or fire, or browsing by herbivorous animals. Similarly, some ferns and flowering plants, as well as producing spores or seeds, are able to regenerate entire new individuals from vegetative propagules. For example, a single leaf

of stonecrop, or lavaberry, or a single tiny bulblet or sprout of northern riceroot or chocolate lily can grow roots and generate into an entire new plant. Human harvesting of such plants, if done with care, will actually trigger new growth and reproductive responses.

36 See, for example, M.K. Anderson (2005, 2009), Berkes (2012), Blackburn and Anderson (1993), Boyd (1999b), Deur and Turner (2005), J. Ford and Martinez (2000), Gill (2005a, 2005b), Hunn, Selam, and family (1990), and Minnis and Elisens (2000).

37 The Jilaa K̲uns (Creek Woman or Fog Woman) story of the Haida and Tlingit and the Ts'msyen story of the "perpetual winter," for example, not only embody the histories of volcanic eruptions and times of immense cold spells but also serve as lessons for people not to harm other creatures or cause needless suffering, as well as not to become arrogant and boastful toward powerful environmental entities.

38 Elinor Ostrom was awarded the 2009 Nobel Prize in Economic Sciences.

39 Today, as well, with the electronic means at our disposal, such meetings can be "virtual" since oral communications can be recorded and retransmitted at another place and time. Throughout most of history, this has not been the case.

40 Haisla women sometimes married into Ts'msyen communities, for example (Helen's mother-in-law, Lucille Clifton, was originally from Kitamaat, a Haisla community), and Cowichan (Hul'qumi'num) women might marry into the Saanich or Songhees (Straits Salish) community or vice versa (Kennedy 1995).

41 Such a situation occurred between the Nuxalk of the Bella Coola Valley and certain Dakelh Ulkatcho families whom they hosted over the winter, at least from the early 1800s, and from whom they learned skills like making birchbark canoes in exchange for allowing the Ulkatcho access to their salmon (Turner 1973).

42 The example of Interior Salish tree terms, cited in chapter 3, is a case in point. In general, trees that were widespread across the northern Interior Plateau (e.g., Rocky Mountain juniper and lodgepole pine) were found to have cognate terms between the three northern Interior Salish languages (Secwepemc, Nlaka'pamux, and Stl'atl'imx), whereas those with disjunct distributions (occurring on the coast and again in the Interior Wet Belt) tended to have Secwepemc names that were different from those of the other two languages and in some cases cognate with those of Okanagan, Salish (Flathead), and other Interior Salish groups of the southern Interior Plateau. This pattern suggests a loss of contact with original coastal trees like western hemlock, western redcedar, white pine, and Pacific yew as the Secwepemc moved across the landscape into the Interior Dry Belt and then a reconnection with these species after the original names had been forgotten (Turner, Ignace, and Compton 1998). This idea has not been confirmed, however, by other plants' names, attributes, and historical and current distributions (see chapter 3).

43 In some cases, the new tools took the form of the traditional ones, as was the case with metal halibut hooks, which replaced the original bentwood hooks but retained their overall shape and style.

44 Among these Indigenous scholars are Dr Louis Miranda (Squamish), Dr Margaret Siwallace (Nuxalk), Dr Mary Thomas and Dr Aimee August (both Secwepemc), Dr Earl Claxton Sr (YELḰÁTŦE) (Saanich Straits Salish), Dr John Elliott (STOLȻEŁ) and Dr Samuel Sam (both Saanich), Dr Arvid Charlie (Luschiim) (Hul'q̓umi'num), and Dr Daisy Sewid-Smith (Mayanilth) (Kwakwaka'wakw).

45 Among these Indigenous scholars are Dr E. Richard Atleo (Chief Umeek) (Nuu-chah-nulth), Chief Earl Maquinna George (Nuu-chah-nulth), Dr Pakki Chipps-Sawyer (Beecher Bay Straits Salish and Ditidaht), Dr Ronald Ignace (Secwepemc), Roger

John (St'at'imc), Dr Judith Thompson (Edōsdi) (Tahltan), Teresa Ryan (Ts'msyen), Linda Smith (Tsilhqot'in), Alestine Andre (Gwich'in), Nicholas Claxton (Saanich), and Dawn Smith (Nuu-chah-nulth).

46 Small Nuu-chah-nulth trinket baskets of wrapped twined construction with basket sedge (*Carex obnupta*) and "three-square" (*Schoenoplectus americanus*), which in the 1960s were being sold by the weavers from store to store in downtown Victoria for as little as fifty cents each, are valued at hundreds of dollars today, and a knobbed Maquinna hat of similar weaving style would have an economic worth of thousands of dollars.

47 The Haida have a cedar management plan that prescribes planting and protecting cedars all over Haida Gwaii where they have been logged so that, a thousand years from now, Haida carvers will be able to make their monumental canoes skim over the water, as they did a thousand years ago (Turner and Wilson 2008).

48 However, some of the Interior Plateau peoples ate pine mushrooms and some other types (A. Jones 1983; Kuhnlein and Turner 1991; Turner 1997a; Turner, Kuhnlein, and Egger 1985).

49 The Tsawout Seafood Festival has become an annual event, with pit-cooking and a re-enactment of the First Salmon ceremony observed on each occasion (up to and including 2014).

50 Kwaxsistalla was given the rights to perform the Atlágimma dance by the family of Evelyn Windsor of Bella Bella. The dance was performed for the Society of Ethnobiology in May 2010. The story underlying the figures was interpreted by Adam Dick's brother, Michael Dick, based on their mother's teachings.

51 In 1999 the Nisg̱a'a Lisms Government, the Province of British Columbia, and the Government of Canada signed a landmark "modern" treaty, giving the Nisg̱a'a rights to control and use a substantial portion of their original lands.

52 See Maa-nulth First Nations (2008) for the final agreement between the Huu-ay-aht, Ka:'yu:'k't'h'/Che:k'tles7et'h, Toquaht, Uchucklesaht, and Ucluelet First Nations and the provincial and federal governments. Sections of interest include chapter 6, "Cultural Activities," 70; chapter 20, "Heritage Sites," 228; chapter 21, "Harvest of Monumental Cedar and Cypress on Provincial Crown Land," 232; and chapters 22 to 24 on "Parks and Environment," more specifically chapter 24, "Gathering," 249.

53 Co-managed protected areas include Gwaii Haanas National Park Reserve, National Marine Conservation Area Reserve, and Haida Heritage Site on Haida Gwaii – perhaps the best example in Canada (see Parks Canada 2013) – as well as Gulf Islands National Park Reserve and Pacific Rim National Park Reserve. The Province of British Columbia in collaboration with First Nations has also developed a new "protected area" designation of "conservancy," in which traditional harvesting practices are condoned but not large-scale commercial harvesting or resource extraction (cited in Turner and Bitonti 2011).

54 This is a slow-growing herbaceous perennial, whose taproots in this case were massive, 3 to 4 centimetres across and up to 20 centimetres or more long, indicating that the plants were probably decades old.

55 Declaration on the Rights of Indigenous Peoples was formally adopted by the United Nations General Assembly during its 62nd Session in New York City on 13 September 2007.

56 See Nolan and Turner (2011). Chipps-Sawyer (2007) explores the use of multimedia in knowledge retention and transmission in the context of her own family history and of knowledge from various communities on the west coast of Vancouver Island.

57 See H.T. Lewis (1977, 1982, 1993), H.T. Lewis and Ferguson (1988), A.M. Miller (2010), A.M. Miller, Davidson-Hunt, and O'Flaherty (2008), Turner (1999), and Turner, Davidson-Hunt, and O'Flaherty (2003).

58 For example, Stuart Crawford (2007) proposes officially assigning the Secwepemctsín name *wíla* (or *wíle*) for *Bryoria fremontii*, which otherwise has only a descriptive common name, "tree hair" or "black tree lichen."

59 In Pohnpei, one of the Federated States of Micronesia, where people are facing similar health risks from erosion of their traditional food varieties, a "Go Local" campaign was initiated and has met with tremendous success in profiling and promoting foods that had almost been forgotten (Kuhnlein, Erasmus, and Spigelski 2009; Kuhnlein, Erasmus, Spigelski, et al. 2013).

60 See K.M.A. Chan et al. (2007), Colding and Folke (2001), Helin (2008), L. King (2004), and Wyllie-Echeverria and Cox (2000).

61 Many Indigenous people would argue, in any case, that *all* land is sacred.

62 An example of such initiatives is the recent "Idle No More!" movements across Canada and internationally (Anonymous 2013), intended to raise awareness of government, politicians, and the general public about the issues Canadian First Nations are facing, particularly in relation to treaty rights and government policy changes around environmental protection.

REFERENCES

Abeysirigunawardena, D.S., and I.J. Walker. 2008. "Sea Level Responses to Climatic Variability and Change in Northern British Columbia." *Atmosphere-Ocean* 46, no. 3: 277–96.

Acheson, Steven R. 1995. "In the Wake of the Iron People: A Case for Changing Settlement Strategies among the Kunghit Haida." *Journal of the Royal Anthropological Institute* 1, no. 2: 273–99.

– 1998. *In the Wake of the Ya'aats' Xaatgaay ['Iron People']: A Study of Changing Settlement Strategies among the Kunghit Haida*. Oxford, UK: British Archaeological Reports.

Acheson, Steven R., and Rebecca J. Wigen. 2002. "Evidence for a Prehistoric Whaling Tradition among the Haida." *Journal of Northwest Anthropology* 36, no. 2: 155–70.

Adams, Karen R., and Susan J. Smith. 2011. "Reconstructing Past Life-Ways with Plants I: Subsistence and Other Daily Needs." In *Ethnobiology*, ed. E.N. Anderson et al., 149–72. Hoboken, NJ: Wiley-Blackwell.

Adams, R.P. 2007. "*Juniperus maritima*, the Seaside Juniper, a New Species from Puget Sound, North America." *Phytologia* 89: 263–83.

Ager, Thomas A., and Lynn P. Ager. 1980. "Ethnobotany of the Eskimos of Nelson Island, Alaska." *Arctic Anthropology* 17, no. 1: 27–48.

Aiken, S.G., M.J. Dallwitz, L.L. Consaul, C.L. McJannet, L.J. Gillespie, R.L. Boles, G.W. Argus, J.M. Gillett, P.J. Scott, R. Elven, M.C. LeBlanc, A.K. Brysting, and H. Solstad. 2003. *Flora of the Canadian Arctic Archipelago: Descriptions, Illustrations, Identification, and Information Retrieval*. http://nature.ca/aaflora/data/index.htm (accessed 2 June 2013).

Alaska Native Language Center. 2011. *Indigenous Peoples and Languages of Alaska*. Map. Fairbanks, AK: Alaska Native Language Center, University of Alaska Fairbanks; Anchorage, AK: Institute of Social and Economic Research, University of Alaska Anchorage. http://www.uaf.edu/anlc (accessed 25 August 2013).

Albright, S.L. 1984. *Tahltan Ethnoarchaeology*. Burnaby, BC: Department of Archaeology, Simon Fraser University.

Alcorn, J.B., J. Bamba, S. Masiun, I. Natalia, and A. Royo. 2003. "Keeping Ecological Resilience Afloat in Cross-Scale Turbulence: An Indigenous Social Movement Navigates Change in Indonesia." In *Navigating the Dynamics of Social-Ecological Systems*, ed. Fikret

Berkes, Johan Colding, and Carl Folke, 299–327. Cambridge, UK: Cambridge University Press.

Alexander, Diana. 1992. "A Reconstruction of Prehistoric Land Use in the Mid-Fraser River Area Based on Ethnographic Data." In *A Complex Culture of the British Columbia Plateau: Traditional Stl'atl'imx Resource Use*, ed. Brian Hayden, 99–176. Vancouver, BC: UBC Press.

Ames, Kenneth M. 1991. "The Archaeology of the *Longue Durée*: Temporal and Spatial Scale in the Evolution of Social Complexity on the Southern Northwest Coast." *Antiquity* 85: 935–45.

– 1995. "Chiefly Power and Household Production on the Northwest Coast." In *Foundations of Inequality*, ed. T.D. Price and G.M. Feinman, 155–87. New York, NY: Plenum.

– 1996. "Life in the Big House: Household Labor and Dwelling Size on the Northwest Coast." In *People Who Lived in Big Houses: Archaeological Perspectives on Large Domestic Structures*, ed. G. Coupland and E.B. Banning, 131–50. Madison, WI: Prehistory.

– 2005. "Intensification of Food Production on the Northwest Coast and Elsewhere." In *"Keeping It Living": Traditions of Plant Use and Cultivation on the Northwest Coast of North America*, ed. Douglas Deur and Nancy J. Turner, 67–100. Seattle, WA: University of Washington Press; Vancouver, BC: UBC Press.

Ames, Kenneth M., and Alan G. Marshall. 1980. "Villages, Demography and Subsistence Intensification on the Southern Columbia Plateau." *North American Archaeologist* 2: 25–52.

Ames, Kenneth M., and Herbert D.G. Maschner. 1999. *Peoples of the Northwest Coast: Their Archaeology and Prehistory*. London, UK: Thames and Hudson.

Amoss, Pamela. 1978. *Coast Salish Spirit Dancing: The Survival of an Ancestral Religion*. Seattle, WA: University of Washington Press.

Anderson, E.N. 1996. *Ecologies of the Heart: Emotion, Belief and the Environment*. New York, NY: Oxford University Press.

Anderson, E.N., Deborah Pearsall, Eugene Hunn, and Nancy J. Turner, eds. 2011. *Ethnobiology*. Hoboken, NJ: Wiley-Blackwell.

Anderson, M. Kat. 1993a. "The Experimental Approach to Assessment of the Potential Ecological Effects of Horticultural Practices by Indigenous Peoples on California Wildlands." Phd diss., University of California, Berkeley.

– 1993b. "Native Californians as Ancient and Contemporary Cultivators." In *Before the Wilderness: Environmental Management by Native Californians*, ed. Thomas C. Blackburn and M. Kat Anderson, 151–74. Menlo Park, CA: Ballena.

– 1996. "Tending the Wilderness." *Restoration and Management Notes* 14, no. 2: 154–66.

– 2005. *Tending the Wild: Native American Knowledge and Management of California's Natural Resources*. Berkeley, CA: University of California Press.

– 2009. *The Ozette Prairies of Olympic National Park: Their Former Indigenous Uses and Management*. Port Angeles, WA: Olympic National Park.

Anderson, M. Kat, and Michael G. Barbour. 2003. "Simulated Indigenous Management: A New Model for Ecological Restoration in National Parks." *Ecological Restoration* 21, no. 4: 269–77.

Anderson, Robert B., Scott MacAulay, Bob Kayseas, and Kevin Hindle. 2008. "On Their Own Terms: Indigenous Communities, Development and the New Economy." In *Non-Market Entrepreneurship: Interdisciplinary Approaches*, ed. G.E. Shockley, P.M. Frank, and R.R. Stough, 89–126. Cheltenham, UK: Edward Elgar.

Andre, Alestine. 2006. "Nan T'aih Nakwits'inahtsih (The Land Gives Us Strength): The Medicine Plants Used by Gwich'in People of Canada's Western Arctic to Maintain Good Health and Well Being." MA thesis, University of Victoria.

Andre, Alestine, and Alan Fehr. 2000. *Gwich'in Ethnobotany: Plants Used by the Gwich'in for Food, Medicine, Shelter and Tools.* Inuvik, NT: Gwich'in Social and Cultural Institute, Tsiigehtchic, and Aurora Research Institute.

Andre, Alestine, Amanda Karst, and Nancy J. Turner. 2006. "Arctic and Subarctic Plants." In *Environment, Origins and Population*, ed. D.H. Ubelaker et al., 236–50. Washington, DC: Smithsonian Institution.

Andre, Alestine, Ruth Welsh, and Nancy J. Turner. 2003. "Looking After Our Elders: Healthcare and Well-being of the Elderly from the Perspective of Gwich'in and Other First Nations of Canada." In *Alternative Medicine and the Elderly*, ed. E. Paul Cherniack, 287–300. Bronx, NY: Springer-Verlag.

Anonymous. 2013. "Idle No More." http://en.wikipedia.org/wiki/Idle_No_More (accessed 1 April 2013).

Archambault, J. 2006. "Native Views of Origins." In *Handbook of North American Indians*, vol. 3, *Environment, Origins, and Population*, ed. D.H. Ubelaker, 4–15. Washington, DC: Smithsonian Institution.

Arima, E.Y., Denis St Claire, Louise Clamhouse, Joshua Edgar, Charles Jones, and John Thomas. 1991. *Between Ports Alberni and Renfrew: Notes on West Coast Peoples.* Ottawa, ON: Canadian Museum of Civilization.

Arlotto, Anthony. 1981. *Introduction to Historical Linguistics.* Lanham, MD: University Press of America.

Arnason, T., Richard J. Hebda, and Timothy Johns. 1981. "Use of Plants for Food and Medicine by Native Peoples of Eastern Canada." *Canadian Journal of Botany* 59: 2189–325.

Arnett, Chris. 1999. *Terror of the Coast: Land Alienation and Colonial War on Vancouver Island and the Gulf Islands, 1849–1863.* Vancouver, BC: Talonbooks.

Arnold, J.E. 1992. "Complex Hunter-Gatherer-Fishers of Prehistoric California: Chiefs, Specialists, and Maritime Adaptations of the Channel Islands." *American Antiquity* 57: 60–84.

Atleo, E. Richard (Chief Umeek). 2004. *Tsawalk: A Nuu-chah-nulth Worldview.* Vancouver, BC: UBC Press.

– 2011. *Principles of Tsawalk: An Indigenous Approach to Global Crisis.* Vancouver, BC: UBC Press.

Atleo, M.R. 2009. "Understanding Aboriginal Learning Ideology through Storywork with Elders." *Alberta Journal of Educational Research* 55, no. 4: 452–67.

Atwater, Brian F. 1992. "Geologic Evidence for Earthquakes during the Past 2000 Years along the Copalis River, Southern Coastal Washington." *Journal of Geophysical Research* 97, no. B2: 1901–19.

Automated Similarity Judgment Program. N.d. http://email.eva.mpg.de/~wichmann/ASJPHomePage.htm (accessed 21 August 2013).

Babcock, Marguerite. 1967. "The Gathering and Preparing of Camas in Southern Vancouver Island." Unpublished ms. Cited and quoted with permission of the author. Copy in possession of Nancy J. Turner.

Baker, Richard, Jocelyn Davies, and Elspeth Young. 2001. *Working on Country: Contemporary Indigenous Environmental Management in Australia.* Oxford, UK: Oxford University Press.

Balée, William L. 1994. *Footprints of the Forest: The Historical Ecology of Plant Utilization by an Amazonian People.* New York, NY: Columbia University Press.

Bandringa, R.W. 1999. "The Ethnobotany and Descriptive Ecology of Bitterroot, *Lewisia rediviva* Pursh (Portulacaceae), in the Lower Thompson River Valley, British Columbia: A Salient Root Food of the Nlaka'pamux First Nation." MSc thesis, University of British Columbia.

Banfield, William. 1860. Letter to Colonial Secretary, 23 February 1860, Colonial Correspondence, file 107. Victoria, BC: British Columbia Archives.

Bannister, Kelly. 2000. "Chemistry Rooted in Cultural Knowledge: Unearthing the Links between Antimicrobial Properties and Traditional Knowledge of Food and Medicinal Plant Resources of the Secwepemc (Shuswap) Aboriginal Nation." Phd diss., University of British Columbia.

– 2006. "Prophet River Ethnobotany: A Report on Traditional Plant Knowledge and Contemporary Concerns of the Prophet River First Nation." Prophet River, BC: Oil and Gas Commission and Prophet River First Nation.

Bannister, Kelly, and Sandra L. Peacock. 1998. "The Bite Is in the Bark: Pitcooked Balsamroot and the Food-Medicine Continuum." Paper presented at the 21st Annual Conference of the Society of Ethnobiology, Reno, Nevada, 17 April.

Bannister, Kelly, and Mary Thomas. Forthcoming. "The Importance of Cultural Knowledge in Understanding the Antimicrobial Chemistry of Balsamroot (*Balsamorhiza sagittata*)." In *Secwepemc People and Plants: Research Papers in Shuswap Ethnobotany*, ed. Marianne B. Ignace, Nancy J. Turner, and Sandra L. Peacock. Published by the Society of Ethnobiology in the monograph series Contributions in Ethnobiology.

Baptiste, G., and M. Wollstonecroft. 1997. "Plant Remains." In *Vancouver Island Highway Project, Victoria Approaches Archaeological Data Recovery DcRu–92*, vol. 1, 129–39. Report prepared for the British Columbia Ministry of Transportation and Highways and the British Columbia Archaeology Branch. Victoria, BC: Millennia Research.

Barbeau, Marius. N.d. Unpublished manuscripts by binder number. "The Eagle Clan," no. [56] 24; "The Haida Connections of Gitxon," nos 1–9; "Eagle Clans Adawx," nos 56–162. Ottawa: Canadian Museum of Civilization.

– 1961. *Tsimsyan Myths Illustrated.* Ottawa, ON: Department of Northern Affairs and National Resources.

Barber, Elizabeth W. 1995. *Women's Work: The First 20,000 Years: Women, Cloth, and Society in Early Times.* New York, NY: W.W. Norton.

Barnett, Homer G. 1955. *The Coast Salish of British Columbia.* Portland, OR: University of Oregon Press.

Barnett, James K. 2008. *Captain Cook in Alaska and the North Pacific.* Anchorage, AK: Todd Communications.

Barrie, J.V., K.W. Conway, H. Josenhans, J.J. Clague, R.W. Mathewes, and D.W. Fedje. 2005. "Late Quaternary Geology of Haida Gwaii and Surrounding Marine Areas." In *Haida Gwaii: Human History and Environment from the Time of Loon to the Time of the Iron People*, ed. D.W. Fedje and R.W. Mathewes, 7–20. Vancouver, BC: UBC Press.

Basso, K.H. 1996. *Wisdom Sits in Places: Landscape and Language among the Western Apache.* Albuquerque, NM: University of New Mexico Press.

Bates, Dawn, Thom Hess, and Vi Hilbert. 1994. *Lushootseed Dictionary.* Seattle: University of Washington Press.

Battiste, M., and J.Y. Henderson. 2000. *Protecting Indigenous Knowledge and Heritage: A Global Challenge.* Saskatoon, SK: Purich.

Beattie, O., B. Apland, E.W. Blake, J.A. Cosgrove, S. Gaunt, S. Greer, A.P. Mackie, K. Mackie, D. Straathof, V. Thorp, and P.M. Troffe. 2000. "The Kwädąy Dän Ts'ìnchį Discovery from a Glacier in British Columbia." *Canadian Journal of Archaeology* 24: 129–47.

Beck, Mary L. 1989. *Heroes and Heroines in Tlingit-Haida Legends.* Anchorage, AK, and Portland, OR: Alaska Northwest Books.

Beck, Peggy V., and Anna L. Walters, eds. 1977. *The Sacred: Ways of Knowledge Sources of Life.* Tsaile (Navajo Nation), AZ: Navajo Community College Press.

Beckwith, Brenda R. 1995. "An Evaluation of the Cultural Significance of Plants Used by the Coast Miwok of Marin and Southern Sonoma Counties with Emphasis on Olompoli State Historic Park." MSC thesis, California State University.
– 2004. "The Queen Root of This Clime: Ethnoecological Investigations of Blue Camas (*Camassia quamash, C. leichtlinii*; Liliaceae) Landscapes on Southern Vancouver Island, British Columbia." Phd diss., University of Victoria.
Bell, Jeff. 2008. "Ancient Settlement Found at Building Site." *Victoria Times Colonist*, 6 January.
Benyowski, Mauritius Augustus (Benyovszky, Móric August). 1790. *Memoirs and Travels of Mauritius Augustus Count de Benyowsky*. London, UK: Robinson. http://www.wdl.org/en/item/2547/zoom/#group=1&page=7&zoom=0.4021&centerX=0.5000&centerY=0.7958 (accessed 12 October 2013).
Beresford-Kroeger, Diana. 2010. *Arboretum Borealis: A Lifeline of the Planet*. New York, NY: Viking.
Berkes, Fikret. 1993. "Traditional Ecological Knowledge in Perspective." In *Traditional Ecological Knowledge: Concepts and Cases*, ed. J.T. Inglis, 1–9. Ottawa, ON: Canadian Museum of Nature and International Development Research Centre.
– 1999. *Sacred Ecology: Traditional Ecological Knowledge and Resource Management*. 1st ed. Philadelphia, PA: Taylor and Francis.
– 2012. *Sacred Ecology: Traditional Ecological Knowledge and Resource Management*. 3rd ed. Philadelphia, PA: Taylor and Francis.
– 2013. "Religious Traditions and Biodiversity." In *Encyclopedia of Biodiversity*, 2nd ed., vol. 6, ed. S.A. Levin, 380–8. Waltham, MA: Academic Press.
Berkes, Fikret, Johan Colding, and Carl Folke. 2000. "Rediscovery of Traditional Ecological Knowledge as Adaptive Management." *Ecological Applications* 10: 1251–62.
– eds. 2003. *Navigating Social-Ecological Systems: Building Resilience for Complexity and Change*. Cambridge, UK: Cambridge University Press.
Berkes, Fikret, and Carl Folke, eds. 1998. *Linking Social and Ecological Systems: Management Practices and Social Mechanisms for Building Resilience*. Cambridge, UK: Cambridge University Press.
Berlin, Brent. 1992. *Ethnobiological Classification: Principles of Categorization of Plant and Animals in Traditional Societies*. Princeton, NJ: Princeton University Press.
Berlin, Brent, Dennis E. Breedlove, and Peter H. Raven. 1973. "General Principles of Classification and Nomenclature in Folk Biology." *American Anthropologist* 75: 214–42.
Bernick, Kathryn, 1998a. *Basketry and Cordage from Hesquiat Harbour*. Victoria, BC: Royal BC Museum.
– ed. 1998b. *Hidden Dimensions: The Cultural Significance of Wetland Archaeology*. Vancouver, BC: UBC Press.
– 1998c. "Stylistic Characteristics of Basketry from Coast Salish Area Wet Sites." In *Hidden Dimensions: The Cultural Significance of Wetland Archaeology*, ed. Kathryn Bernick, 139–56. Vancouver, BC: UBC Press.
– 2003. "A Stitch in Time: Recovering the Antiquity of a Coast Salish Basket Type." In *Emerging from the Mist: Studies in Northwest Coast Culture History*, ed. R.G. Matson, G. Coupland, and Q. Mackie, 230–43. Vancouver, BC: UBC Press.
– 2007. "DhRp 52 Wet-Site Component Exploratory Investigations, July 2007." Unpublished report for Katzie Development Corporation–Archaeology, Pitt Meadows, British Columbia.
– 2013. "People-Wetland Interaction in Canada." In *The Oxford Handbook of Wetland Archaeology*, ed. F. Menotti and A. O'Sullivan, 71–81. Oxford, UK: Oxford University Press.

Bhattacharyya, J. 2012. "Knowing Naŝlhiny (Horse), Understanding the Land: Free-Roaming Horses in the Culture and Ecology of the Brittany Triangle and Nemiah Valley." Phd diss., University of Waterloo.
Bielawski, Ellen. 2007. *In Search of Ancient Alaska: Solving the Mysteries of the Past*. Anchorage, AK: Alaska Northwest Books.
Birchwater, Sage, and Ulkatcho and Nuxalk elders. 1993. *Ulkatcho Stories of the Grease Trail*. Anahim Lake, BC: Ulkatcho Culture Curriculum Committee.
Bisset, N.G. 1976. "Hunting Poisons of the North Pacific Region." *Lloydia* 39, nos 2–3: 87–124.
Black, Martha. 1989. "Looking for Bella Bella: The R.W. Large Collection and Heiltsuk Art History." *Canadian Journal of Native Studies* 9, no. 2: 273–91.
Black, M. Jean. 1994. "Plant Dispersal by Native North Americans in the Canadian Subarctic." In *The Nature and Status of Ethnobotany*, 2nd ed., ed. Richard I. Ford, 255–62. Ann Arbor, MI: Museum of Anthropology, University of Michigan.
Blackburn, Thomas C., and M. Kat Anderson, eds. 1993. *Before the Wilderness*. Menlo Park, CA: Ballena.
Blackman, Margaret B. 1982. *During My Time: Florence Edenshaw Davidson, a Haida Woman*. Seattle, WA, and London, UK: University of Washington Press.
– 1990. "Haida Traditional Culture." In *Handbook of North American Indians*, vol. 7, *Northwest Coast*, ed. Wayne P. Suttles, 240–60. Washington, DC: Smithsonian Institution.
Blackstock, M.D. 2001. *Faces in the Forest: First Nations Art Created on Living Trees*. Montreal, QC, and Kingston, ON: McGill-Queen's University Press.
Blanchette, Robert A., Brian D. Compton, Nancy J. Turner, and Robert L. Gilbertson. 1992. "Nineteenth Century Shaman Grave Guardians Are Carved *Fomitopsis officinalis* Sporophores." *Mycologia* 84, no. 1: 119–24.
Blaser, M., ed. 2004. *In the Way of Development: Indigenous Peoples, Life Projects and Globalization*. Ottawa, ON: IDRC Books.
Blumenthal, M., A. Goldberg, and J. Brinckmann, eds. 2000. *Herbal Medicine: Expanded Commission E Monographs*. Austin, TX: American Botanical Council; Newton, MA: Integrative Medicine Communications.
Boas, Franz. 1890. "First General Report on the Indians of British Columbia." *Report of the Fifty Nineth Meeting of the British Association for the Advancement of Science*: 801–55.
– 1898. *The Mythology of the Bella Coola Indians*. New York, NY: American Museum of Natural History.
– 1901. *Kathlamet Texts*. Washington, DC: Smithsonian Institution.
– 1902. *Tsimshian Texts*. Washington, DC: Government Printing Office.
– 1909. *The Kwakiutl of Vancouver Island*. New York, NY: American Museum of Natural History.
– 1916. *Tsimshian Mythology*. Washington, DC: Government Printing Office.
– 1918. *Kutenai Tales*. Washington, DC: Government Printing Office.
– 1921. *Ethnology of the Kwakiutl*. Based on data collected by George Hunt. Washington, DC: Smithsonian Institution.
– 1923. Bella Bella field notes. Boas Collection, 372, reel 1. Philadelphia, PA: American Philosophical Society.
– 1930. *The Religion of the Kwakiutl Indians*. New York, NY: Columbia University Press.
– 1934. *Geographical Names of the Kwakiutl Indians*. New York, NY: Columbia University Press.
– 1935. *Kwakiutl Culture as Reflected in Mythology*. New York, NY: G.E. Stechert.
– 1966. *Kwakiutl Ethnography*. Ed. H. Codere. Chicago, IL: University of Chicago Press.

– 2002. *Indian Myths and Legends from the North Pacific Coast of America: A Translation of Franz Boas's 1895 Edition of Indianische Sagen von der Nord Pacifischen Küste Amerikas.* Trans. Dietrich Bertz. Ed. Randy Bouchard and Dorothy Kennedy. Vancouver, BC: Talonbooks.
Bocquet, A., J.L. Brochier, A. Emery-Barbier, K. Lundstrom-Baudais, C. Orcel, and F. Vin. 1987. "A Submerged Neolithic Village: Charavines 'Les Baigneurs' in Lake Paladru, France." In *Wetlands in European Prehistory*, ed. J.M. Coles and A.J. Lawson, 33–54. Oxford, UK: Clarendon.
Boelscher, Marianne. 1988. *The Curtain Within: Haida Social and Mythical Discourse.* Vancouver, BC: UBC Press.
Bouchard, Randy. 1978. "Additional Information Concerning Sechelt Ethnobotany, Provided by Johnny Joe and Madeline Joe, with Reference to Turner, Timmers, and Bouchard (1972)." Unpublished ms. BC Indian Language Project, Victoria, British Columbia.
Bouchard, Randy, and Dorothy I.D. Kennedy. 1969–72. "Okanagan Stories (October 1969)." Unpublished ms. BC Indian Language Project, Victoria, British Columbia.
– 1973–78. "Mainland Comox Plant Names, and supplemental Notes (December 1973, May 1975)." Unpublished ms. BC Indian Language Project, Victoria, British Columbia.
– 1977a. "Additional Information Pertaining to Sechelt Ethnobotany, Provided by Sarah Silvey, Nov. 19–23, 1977." Unpublished ms. Cited with permission of the authors. Copy in possession of Nancy J. Turner.
– 1977b. "Memories of the Early Days." Story no. 7 in "Lillooet Stories." Unpublished ms. BC Indian Language Project, Victoria, British Columbia.
– eds. 1979. *Shuswap Stories.* Vancouver, BC: CommCept.
– 1990. *Clayoquot Sound Indian Land Use.* Report prepared for MacMillan Bloedel Limited, Fletcher Challenge Canada, and British Columbia Ministry of Forests. Victoria, BC: BC Indian Language Project.
Bouchard, Randy, and Nancy J. Turner. 1976. "Ethnobotany of the Squamish Indian People of British Columbia." Unpublished ms. BC Indian Language Project, Victoria, British Columbia.
Boyd, Robert T. 1990. "Demographic History, 1774–1874." In *Handbook of North American Indians*, vol. 7, *Northwest Coast*, ed. Wayne P. Suttles, 135–48. Washington DC: Smithsonian Institution.
– 1992. "Population Decline from Two Epidemics on the Northwest Coast." In *Disease and Demography in the Americas*, ed. J.W. Verano and D.H. Ubelaker, 249–55. Washington DC: Smithsonian Institution.
– 1996. *People of the Dalles: The Indians of Wascopam Mission.* Lincoln, NE: University of Nebraska Press.
– 1999a. *The Coming of the Spirit of Pestilence: Introduced Infectious Diseases and Population Decline among Northwest Coast Indians, 1774–1874.* Vancouver, BC: UBC Press.
– ed. 1999b. *Indians, Fire and the Land in the Pacific Northwest.* Corvallis, OR: Oregon State University Press.
Bracken, C. 1997. *The Potlatch Papers: A Colonial Case History.* Chicago, IL: University of Chicago Press.
Bringhurst, Robert, and William Reid. 1996. *The Raven Steals the Light.* Seattle, WA: University of Washington Press.
British Columbia. 1875. *Papers Connected with the Indian Land Question, 1850–1875.* Victoria, BC: Government Printer. (Reprinted 1987 as *Indian Land Question, 1850–1875, 1877.*)
– 1885. *Sessional Papers.* Victoria, BC: Government of British Columbia.
– 1887. *Sessional Papers.* Victoria, BC: Government of British Columbia.

– 1996. *Heritage Conservation Act [RSBC 1996] Chapter 187*. Victoria, BC: Queen's Printer. http://www.bclaws.ca/EPLibraries/bclaws_new/document/ID/freeside/00_96187_01 (accessed 6 October 2013).
British Columbia Ministry of Environment. 2007. *Environmental Trends in British Columbia: 2007*. Victoria, BC: Government of British Columbia. http://www.env.gov.bc.ca/soe (accessed 3 June 2013).
British Columbia Ministry of Forests. 1999a. *Biogeoclimatic Zones of British Columbia*. Map. Victoria, BC: Government of British Columbia.
British Columbia Ministry of Forests, Lands and Natural Resource Operations. 1994. "Western Redcedar." In *Tree Book: Learning to Recognize Trees of British Columbia*. 2nd ed. Victoria, BC: Government of British Columbia. http://www.for.gov.bc.ca/hfd/library/documents/treebook/westernredcedar.htm (accessed 2 October 2013).
– 2012. *A History of the Battle against the Mountain Pine Beetle, 2000–2012*. Victoria, BC: Government of British Columbia. http://www.for.gov.bc.ca/hfp/mountain_pine_beetle/Pine%20Beetle%20Response%20Brief%20History%20May%2023%202012.pdf (accessed 3 June 2013).
Brown, Annora. 1970. *Old Man's Garden*. Vancouver, BC: Evergreen.
Brown, Cecil H. 1984. *Language and Living Things: Uniformities in Folk Classification and Naming*. New Brunswick, NJ: Rutgers University Press.
– 1999. *Lexical Acculturation in Native American Languages*. Oxford, UK: Oxford University Press.
– 2010. "Development of Agriculture in Prehistoric Mesoamerica: The Linguistic Evidence." In *Pre-Columbian Foodways: Interdisciplinary Approaches to Food, Culture, and Markets in Ancient Mesoamerica*, ed. John E. Staller and Michael D. Carrasco, 71–108. New York, NY: Springer-Verlag.
Brown, Frank, and Kathy Brown, with Barbara Wilson, Pauline Waterfall, and Gloria Cranmer Webster. 2009. *Staying the Course, Staying Alive: Coastal First Nations Fundamental Truths*. Victoria, BC: Biodiversity BC.
Brown, Joseph Epes, ed. 1953. *The Sacred Pipe*. Norman, OK: University of Oklahoma Press.
– 1986. *The Spiritual Legacy of the American Indian*. New York: NY: Crossroad.
Brown, K.J., and Richard J. Hebda. 2002a. "Ancient Fires on Southern Vancouver Island, British Columbia, Canada: A Change in Causal Mechanisms at about 2,000 ybp." *Environmental Archeology* 7: 1–12.
– 2002b. "Origin, Development, and Dynamics of Coastal Temperate Conifer Rainforests of Southern Vancouver Island, Canada." *Canadian Journal of Forest Research* 32: 353–72.
– 2003. "Coastal Rainforest Connections Disclosed through a Late Quaternary Vegetation, Climate, and Fire History Investigation from the Mountain Hemlock Zone on Southern Vancouver Island, British Columbia, Canada." *Review of Palaeobotany and Palynology* 123: 247–69.
Brown, K.J., A.B. Nielsen, R.J. Fitton, and R.J. Hebda. 2008. "Postglacial Evolution and Spatial Differentiation of Seasonal Temperate Rainforest in Western Canada." *The Holocene* 18, no. 5: 715–27.
Brown, Robert. 1873. "The Indian Story of 'Jack and the Bean Stalk.'" In *The Races of Mankind, Being a Popular Description of the Characteristics, Manners and Customs of the Principal Varieties of the Human Family*, vol. 1, 131–2. New York, NY: Cassel, Petter and Galpin.
Budhwa, Rick. 2007. *The Importance, Traditional Use and Locations of Various Berry Species within the Gitxsan Traditional Territory*. Smithers, BC: Bulkley Valley Centre for Natural Resources Research and Management, British Columbia Ministry of Forests and Range – Skeena/Stikine Region, and Crossroads Cultural Resource Management.

Burley, D.V., S. Hamilton, and K. Fladmark. 1996. *Prophecy of the Swan: Fur Trade History of the Upper Peace River Valley, 1794–1823*. Vancouver, BC: UBC Press.
Burroughs, William James. 2005. *Climate Change in Prehistory: The End of the Reign of Chaos*. Cambridge, UK: Cambridge University Press.
Burton, Carla M.A. 2012. "Wilaat Hooxhl Nisga'ahl [Galdo'o] [Ýans]: Gik'uuhl-gi, Guuń-sa ganhl Angoogam̓ / Using Plants the Nisga'a Way: Past, Present and Future Use." Phd diss., University of Victoria.
Bye, Robert. 1986. "Medicinal Plants of the Sierra Madre: Comparative Study of Tarahumara and Mexican Market Plants." *Economic Botany* 40: 103–24.
Bye, Robert, and Edelmira Linares. 1986. "Ethnobotanical Notes from the Valley of San Luis, Colorado." *Journal of Ethnobiology* 6, no. 2: 289–306.
Byram, Scott, and David G. Lewis. 2001. "Ourigan: Wealth of the Northwest Coast." *Oregon Historical Quarterly* 102, no. 2: 126–57.
Cajete, G.A. 1994. *Look to the Mountain: An Ecology of Indigenous Education*. Skyland, NC: Kivaki.
– 2000. *Native Science: Natural Laws of Interdependence*. Sante Fe, NM: Clear Light.
Calder, James A., and Roy L. Taylor. 1968. *Flora of the Queen Charlotte Islands*. Part 1. Ottawa, ON: Research Branch, Department of Agriculture.
Caldwell, Megan E., Dana Lepofsky, Georgia Combes, Michelle Washington, John R. Welch, and John R. Harper. 2012. "A Bird's Eye View of Northern Coast Salish Intertidal Resource Management Features, Southern British Columbia, Canada." *Journal of Island and Coastal Archaeology* 7, no. 2: 219–33.
Callicott, J. Baird. 1989. "Traditional American Indian and Western European Attitudes toward Nature: An Overview." In *Defense of the Land Ethic: Essays in Environmental Philosophy*, 194–5. Albany, NY: State University of New York Press.
Campbell, Kenneth. 2005. *Persistence and Change: A History of the Ts'msyen Nation*. Prince Rupert, BC: First Nations Education Services.
Campbell, Sarah K., and Virginia L. Butler. 2010. "Archaeological Evidence for Resilience of Pacific Northwest Salmon Populations and the Socioecological System over the Last ~7,500 Years." *Ecology and Society* 15, no. 1: 17. http://www.ecologyandsociety.org/vol15/iss1/art17 (accessed 3 June 2013).
Canadian Pacific Railway. 1936. *B.C. Coast Steamship Service Brochure*. Victoria, BC: Canadian Pacific Railway.
Canoe and Kayak. 2008. "Tribal Journeys Paddle Concludes on Vancouver Island." 31 July. http://www.canoekayak.com/canoe/tribal-journeys-paddle-concludes-on-vancouver-island (accessed 3 June 2013).
Capes, Katherine H. 1964. *Contributions to the Prehistory of Vancouver Island*. Pocatello, ID: Idaho State University.
Carlson, Barry F., and Nancy Flett. 1989. *Spokane Dictionary*. Missoula, MT: University of Montana Press.
Carlson, Keith T. 1997. "Stólo Exchange Dynamics." *Native Studies Review* 11: 30–55.
– ed. 2001. *A Stó:lō–Coast Salish Historical Atlas*. Chilliwack, BC: Sto:lo Heritage Trust; Vancouver, BC: Douglas and McIntyre.
Carlson, Roy L. 1976. "Prehistory of the Northwest Coast." In *Indian Art Traditions of the Northwest Coast*, ed. Roy L. Carlson, 13–32. Burnaby, BC: Archaeology Press, Simon Fraser University.
– 1994. "Trade and Exchange in Prehistoric British Columbia." In *Prehistoric Exchange Systems in North America*, ed. T. Baugh and J. Ericson, 307–61. New York, NY: Plenum.

– 1996. "Introduction to Early Human Occupation in British Columbia." In *Early Human Occupation in British Columbia*, ed. Roy L. Carlson and Luke Dalla Bona, 3–10. Vancouver, BC: UBC Press.
Carlson, Thomas J.S., and Luisa Maffi, eds. 2004. *Ethnobotany and Conservation of Biocultural Diversity: Advances in Economic Botany.* Vol. 1. Bronx, NY: New York Botanical Garden Press.
Carrier Linguistic Committee. 2000. *Hanúyeh Ghun 'Útni-i: Plants of Carrier Country.* 1973. Reprint, Fort St James, BC: Central Carrier Language.
Celletti, Mark, dir. 2009. *Canoe Way: The Sacred Journey*. Documentary film. http://canoeway.org/Home_Page.html (accessed 8 July 2013).
Černjagina, O.A. 1999. "Fußnote 58." In *Georg Vil'gel'm Steller – Opisanie zemli Kamčatki*, 259. Petropavlovsk-Kamčatskij, Russia: Kamčatskij pečatnyj dvor.
Chamberlain, A.F. 1892. "Report on the Kootenay Indians of Southeastern British Columbia." In *Eighth Report on the Northwestern Tribes of Canada*, 545–615. London, UK: British Association for the Advancement of Science.
Chambers, K.J. 2001. "Autecology, Ethnobotany and Agronomy of *Balsamorhiza sagittata*: Northwestern Plateau, British Columbia." MSC thesis, University of Victoria.
Chambers, K.J., E. Small, P.M. Catling, N.J. Turner, P. Bowen, and C.P. Keller. 2002. "Poorly Known Economic Plants of Canada – 35: Arrow-Leaved Balsamroot *Balsamorhiza sagittata* (Pursh.) Nutt." *Canadian Botanical Association Bulletin* 35, no. 4: 40–5.
Chan, Laurie. 2009. "Is Local Food Poisoning Natives?" *Vancouver Province*, 1 March.
Chan, H.M., E. Khoury, M. Sedgemore, S. Sedgemore, and H.V. Kuhnlein. 1996. "Organochlorine Pesticides and Polychlorinated Biphenyl Congeners in Ooligan Grease: A Traditional Food Fat of British Columbia First Nations." *Journal of Food Composition Analysis* 9: 32–42.
Chan, Kai M.A., Robert M. Pringle, Jai Ranganathan, Carol L. Boggs, Yvonne L. Chan, Paul R. Ehrlich, Peter K. Haff, Nicole E. Heller, Karim Al-Khafaji, And Dena P. Macmynowski. 2007. "When Agendas Collide: Human Welfare and Biological Conservation." *Conservation Biology* 21, no. 1: 59–68.
Chance, David H., and Jennifer V. Chance. 1982. *Kettle Falls: 1971/1974.* Moscow, ID: University of Idaho.
Chandler, R.F., S.N. Hooper, and M.J. Harvey. 1982. "Ethnobotany and Phytochemistry of Yarrow, *Achillea millefolium*, Compositae." *Economic Botany* 36: 203–23.
Chandran, M.D. Subash, and Madhav Gadgil. 1993. "'Kans' – Safety Forests of Uttara Kannada." In *Geschichte der Kleinprivatwaldwirtschaft – Geschichte des Bauernwaldes*, ed. Helmut Brandl et al., 49–57. Freiburg, Germany: Forstliche Versuchs and Forschungsanstalt Baden-Württemberg.
Cheatham, Richard. 1988. *Late Archaeic Settlement Pattern in the Long Tom Sub-Basin, Upper Willamette Valley, Oregon.* Eugene, OR: University of Oregon Press.
Chiefs of the 'Namgis (Nimpkish) Nation. 1919. Letter from Alert Bay to Deputy Superintendent General of Indian Affairs, Ottawa, 6 April. Typescript viewed and transcribed by Nancy J. Turner in Kingcome Village, 2005.
Chipps-Sawyer, Allis Pakki. 2007. "Standing on the Edge of Yesterday: A Dilemma of Oral Knowledge in a West Coast Family." Phd diss., University of Victoria.
CIBA (California Indian Basketweavers Association). 2009. http://www.ciba.org (accessed 22 August 2013).
Claxton, Earl, Sr (YELḰÁTŦE), and John Elliott Sr (STOLȻEȽ). 1993. *The Saanich Year.* Saanichton, BC: School District No. 63, Saanich School Board.
– 1994. *Reef Net Technology of the Saltwater People.* Brentwood Bay, BC: Saanich Indian School Board.

Claxton, Nicholas. 2003. "The Douglas Treaty and WSÁNEC Traditional Fisheries: A Model for Saanich Peoples Governance." MA thesis, University of Victoria. http://web.uvic.ca/igov/research/pdfs/N%20Claxton,%20Final.pdf (accessed 25 August 2013).

Claxton v. Saanichton Marina Ltd. 1989. 3 C.N.L.R. 46 British Columbia Court of Appeal. http://www.mandellpinder.com/pdf/pdf-our-firm/saanichton-marina.pdf (accessed 22 August 2013).

Colding, Johan, Thomas Elmqvist, and Per Olsson. 2003. "Living with Disturbance: Building Resilience in Social-Ecological Systems." In *Navigating Social-Ecological Systems: Building Resilience for Complexity and Change*, ed. Fikret Berkes, Johan Colding, and Carl Folke, 163–86. Cambridge, UK: Cambridge University Press.

Colding, Johan, and Carl Folke. 2001. "Social Taboos: 'Invisible' Systems of Local Resource Management and Biological Conservation." *Ecological Applications* 11, no. 2: 584–600.

Cole, Douglas, and Bradley Lockner, eds. 1989. *The Journals of George M. Dawson: British Columbia, 1875–1878*. Vols 1–2. Vancouver, BC: UBC Press.

Collins, June McCormick. 1974. *Valley of the Spirits: The Upper Skagit Indians of Western Washington*. Seattle, WA, and London, UK: University of Washington Press.

Colville, Frederick V. 1902. "Wokas, a Primitive Food of the Klamath Indians." In *Report of the United States National Museum, Smithsonian Institution*, 725–39. Washington, DC: Smithsonian Institution.

Commission for Environmental Cooperation. 2009. *North American Environmental Atlas*. Montreal, QC: Commission for Environmental Cooperation. http://www.cec.org/Page.asp?PageID=924&SiteNodeID=495 (accessed 5 August 2013).

Compton, Brian D. 1993a. "The North Wakashan 'Wild Carrots': Clarificiation of Some Ethnobotanical Ambiguity in Pacific Northwest Apiaceae." *Economic Botany* 47, no. 3: 297–303.

– 1993b. "Upper North Wakashan and Southern Tsimshian Ethnobotany: The Knowledge and Usage of Plants and Fungi among the Oweekeno, Hanaksiala (Kitlope and Kemano), Haisla (Kitamaat) and Kitasoo Peoples of the Central and North Coast of British Columbia." Phd diss., University of British Columbia.

Connell, J.F., and J. Allen. 2004. "Dating the Colonization of Sahul (Pleistocene Australia-New Guinea): A Review of Recent Research." *Journal of Archaeological Science* 31: 835–53.

Connolly, Thomas J., and Pat Barker. 2004. "Basketry Chronology of the Early Holocene in the Northern Great Basin." In *Early and Middle Holocene Archaeology of the Northern Great Basin*, ed. D.L. Jenkins, Thomas J. Connolly, and C.M. Aikens, 241–50. Eugene, OR: University of Oregon Press.

Connolly, Thomas J., and William J. Cannon. 1999. "Comments on 'America's Oldest Basketry.'" *Radiocarbon* 41, no. 3: 309–13.

Corsiglia, J., and G. Snively. 1995. "Global Lessons from the Traditional Science of Long Resident Peoples." In *Thinking Globally about Mathematics and Science Education*, ed. G. Snively and A. McKinnon, 25–51. Vancouver, BC: Center for the Study of Curriculum and Instruction, University of British Columbia.

Council of the Haida Nation. 1990. *Yakoun, River of Life*. Massett, Haida Gwaii, BC: Council of the Haida Nation.

Coupland, G. 1989. "Warfare and Social Complexity on the Northwest Coast." In *Cultures in Conflict: Current Archaeological Perspectives*, ed. D. Tkaczuk and B. Vivian, 205–15. Calgary, AB: Archaeological Association of the University of Calgary.

Cove, John J., and George F. MacDonald, eds. 1987. *Tsimshian Narratives I: Tricksters, Shamans and Heroes*. Ottawa, ON: Canadian Museum of Civilization.

Cox, Ross. 1957. *The Columbia River, or Scenes and Adventures during a Residence of Six Years on the Western Side of the Rocky Mountains among Various Tribes of Indians hitherto Unknown; together with "A Journey across the American Continent."* Ed. Edgar I. Stewart and Jane R. Stewart. Original 1832. Norman: University of Oklahoma Press.

Craig, Juliet, and Robin Smith. 1997. *"A Rich Forest": Traditional Knowledge, Inventory and Restoration of Culturally Important Plants and Habitats in the Atleo River Watershed.* Final report of the Ahousaht Ethnobotany Project, 1996. Ahousaht, BC: Ahousaht Band Council; Tofino, BC: Long Beach Model Forest.

Crockford, S.J. 1997. *Osteometry of Makah and Coast Salish Dogs.* Burnaby, BC: Archaeology Press, Simon Fraser University.

Craig, Juliet. 1998. "'Nature Was the Provider': Traditional Ecological Knowledge and Inventory of Culturally Significant Plants and Habitats in the Atleo River Watershed, Ahousaht Territory, Clayoquot Sound." MSC thesis, University of Victoria.

Crawford, Gary W. 2008. "Macroremains Analysis: 1593–1598." In *Encyclopedia of Archaeology*, vol. 2, ed. Deborah M. Pearsall, 1593–8. New York, NY: Academic Press.

Crawford, Stuart. 2007. "Ethnolichenology of *Bryoria fremontii*: Wisdom of Elders, Population Ecology, and Nutritional Chemistry." MSC thesis, University of Victoria.

– Forthcoming. "The Ethnolichenology of *Wila* (*Bryoria fremontii*): An Important Edible Lichen of Secwepemc Country and Neighbouring Territories." In *Secwepemc People and Plants: Research Papers in Shuswap Ethnobotany*, ed. Marianne B. Ignace, Nancy J. Turner, and Sandra L. Peacock. Published by the Society of Ethnobiology in the monograph series Contributions in Ethnobiology.

Croes, Dale R. 1989. "Lachane Basketry and Cordage: A Technical, Functional and Comparative Study." *Canadian Journal of Archaeolology* 13: 165–205.

– 1992. "An Evolving Revolution in Wet Site Research on the Northwest Coast of North America." In *The Wetland Revolution in Prehistory*, ed. B.J. Coles, 99–111. Exeter, UK: Department of History and Archaeology, University of Exeter.

– 1995. *The Hoko River Archaeological Site Complex: The Wet/Dry Site (45CA213) 3,000–1,700 B.P.* Pullman, WA: Washington State University Press.

– 1997. "The North-Central Cultural Dichotomy on the Northwest Coast of North America: Its Evolution as Suggested by Wet-Site Basketry and Wooden Fish-Hooks." *Antiquity* 71: 594–615.

– 2003. "Northwest Coast Wet-Site Artifacts: A Key to Understanding Resource Procurement, Storage, Management, and Exchange." In *Emerging from the Mist: Studies in Northwest Coast Culture History*, ed. R.G. Matson, G. Coupland, and Q. Mackie, 51–75. Vancouver, BC: UBC Press.

– 2012. "Ancient Basketry of the Olympic Peninsula." In *From the Hands of a Weaver: Olympic Peninsula Basketry through Time*, ed. Jacilee Wray and the Olympic Peninsula Intertribal Cultural Advisory Committee, 142–55. Norman, OK: University of Oklahoma Press.

Croes, Dale R., John L. Fagan, and Maureen M. Zehendner. 2009a. "Assessing the Results of the 2006 and 2007 Investigations at Sunken Village Wet Site." *Journal of Wetland Archaeology* 9, no. 1: 185–201.

Croes, Dale R., John L. Fagan, and Maureen M. Zehendner, eds. 2009b. "Sunken Village, Sauvie Island, Oregon, USA: A Report on the 2006–2007 Investigations of National Historic Landmark Site 35MU4." *Journal of Wetland Archaeology* 9, no. 1: 1–216.

Croes, Dale R., Michael Martin, John L. Fagan, Maureen Zehendner, Michele Punkw, Kathleen Hawes, and Olivia Ness, with Confederated Tribes of Grand Ronde Community of Oregon, Confederated Tribes of Siletz Indians, and Confederated Tribes of Warm

Springs. 2007. "Sunken Village Wet Site (35MU4) Summary 2007 Field Report." *The Midden* 39: 1–10.
Croft, Shannon, and R.W. Mathewes. Forthcoming. "Barking up the Right Tree: Understanding Birch Bark Artifacts of Canadian Plateau Peoples in British Columbia." *BC Studies* 180 (Winter 2013–14).
Crosby, Alfred W. 1986. *Ecological Imperialism: The Biological Expansion of Europe, 900–1900*. Cambridge, UK: Cambridge University Press.
Crowell, Aron L. 1988. "Prehistory of Alaska's Pacific Coast." In *Crossroads of Continents: Cultures of Siberia and Alaska*, ed. W.E. Fitzhugh and A. Crowell, 130–341. Washington, DC: Smithsonian Institution.
Crowell, Aron L., Amy F. Steffian, and Gordon L. Pullar, eds. 2001. *Looking Both Ways: Heritage and Identity of the Alutiiq People*. Fairbanks, AK: University of Alaska Press.
Cruikshank, J. 1998. *The Social Life of Stories: Narrative and Knowledge in the Yukon Territory*. Lincoln, NE: University of Nebraska Press.
– 2005. *Do Glaciers Listen? Local Knowledge, Colonial Encounters, and Social Imagination*. Vancouver, BC: UBC Press; Seattle, WA: University of Washington Press.
CUJ (*Confederated Umatilla Journal*). 2008. "Putting First Foods First." Editorial. *Confederated Umatilla Journal* 11, no. 3: 1–3.
Cullis-Suzuki, Severn. 2007. "An Ethnoecological Study of the Kwakwaka'wakw Traditional Harvesting of Eelgrass, *Zostera marina* L.; Zosteraceae." MSc thesis, University of Victoria.
Curtis, Edward S. 1907. *The North American Indian*. Vol. 3, *The Haisla*. Evanston, IL: Northwestern University Library.
– 1911. *The North American Indian*. Vol. 7, *The Yakima. The Klickitat. Salishan Tribes of the Interior. The Kutenai*. Evanston, IL: Northwestern University Library. http://curtis.library.northwestern.edu/curtis/toc.cgi (accessed 25 August 2013).
– 1915. *The North American Indian*. Vol. 10, *The Kwakiutl*. Evanston, IL: Northwestern University Library. http://curtis.library.northwestern.edu/curtis/toc.cgi (accessed 25 August 2013).
– 1916. *The North American Indian*. Vol. 11, *Haida/Tlingit*. Evanston, IL: Northwestern University Library. http://curtis.library.northwestern.edu/curtis/toc.cgi (accessed 25 August 2013).
Cutright, Paul Russell. 2003. *Lewis and Clark: Pioneering Naturalists*. 2nd ed. Lincoln, NE: University of Nebraska Press.
Cybulski, J.S. 2006. "Skeletal Biology: Northwest Coast and Plateau." In *Handbook of North American Indians*, vol. 3, *Environment, Origins, and Population*, ed. D.H. Ubelaker, 532–47. Washington, DC: Smithsonian Institution.
Cybulski, J.S., D.E. Howes, J.C. Haggarty, and M. Eldridge. 1981. "An Early Human Skeleton from South-Central British Columbia: Dating and Bioarchaeological Inference." *Canadian Journal of Archaeology* 5: 49–59.
Daly, Richard. 2005. *Our Box Was Full: An Ethnography for the Delgamuukw Plaintiffs*. Vancouver, BC: UBC Press.
Darby, Melissa Cole. 1996. "Wapato for the People: An Ecological Approach to Understanding the Native American Use of *Sagittaria latifolia* on the Lower Columbia River." MA thesis, Portland State University.
– 2005. "The Intensification of Wapato (*Sagittaria latifolia*) by the Chinookan People of the Lower Columbia River." In *"Keeping It Living": Traditions of Plant Use and Cultivation on the Northwest Coast of North America*, ed. Douglas Deur and Nancy J. Turner, 194–217. Seattle, WA: University of Washington Press; Vancouver, BC: UBC Press.

Daugherty, Dick, and Janet Friedman. 1976. "An Introduction to Ozette Art." In *Indian Art Traditions of the Northwest Coast*, ed. Roy L. Carlson, 183–95. Burnaby, BC: Archaeology Press, Simon Fraser University.

David, Nicholas, and Carol Kramer. 2001. *Ethnoarchaeology in Action*. Cambridge, UK: Cambridge University Press.

Davidson, John. 1919. "Douglas Fir Sugar." *Canadian Field Naturalist* 33: 6–9.

– 1922. "The Prospects for Commercial Drug-Plant Cultivation in B.C." Lecture presented to the Second Annual Convention of the BC Branch of the Canadian Society of Technical Agriculturalists, Vancouver, BC. http://botanyjohn.org/gallery/v/procomdr/2007_565_0674.jpg.html (accessed 23 August 2013).

– 1942. *The Cascara Tree in British Columbia*. Victoria, BC: British Columbia Ministry of Agriculture.

Davidson-Hunt, Iain J. 2003. "Indigenous Lands Management, Cultural Landscapes and Anishinaabe People of Shoal Lake, Northwestern Ontario." *Environments* 31, no. 1: 21–41.

Davis, Alison, with Beatrice Wilson and Brian D. Compton. 1995. *Salmonberry Blossoms in the New Year: Some Culturally Significant Plants of the Haisla Known to Occur within the Greater Kitlope Ecosystem*. Kitamaat, BC: Kitimaat Village Council, Nanakila Press, and Ecotrust Canada.

Davis, Alison, and Tsay Tay Forestry Limited. 2008. *Sekani Ethnobotany: Traditional Role of Plants amongst the Sekani People*. Prince George, BC: Sekani Nation, Tsay Tay Forestry Limited, Axion Business Corporation, and Trafford Publishing.

Davis, Wade. 2001. *Light at the Edge of the World: A Journey through the Realm of Vanishing Cultures*. Vancouver, BC: Douglas and McIntyre; Washington, DC: National Geographic Society.

– 2009. *The Wayfinders: Why Ancient Wisdom Matters in the Modern World*. CBC Massey Lectures. Toronto, ON: House of Anansi.

Davis, Wade, and Thom Henley. 1990. *Penan: Voice for the Borneo Rainforest*. Vancouver, BC: Western Canada Wilderness Committee.

Dawson, George M. 1880. *On the Haida Indians of the Queen Charlotte Islands*. Montreal, QC: Geological Survey of Canada.

– 1892. "Notes on the Shuswap People of British Columbia." In *Proceedings and Transactions of the Royal Society of Canada for the Year 1891*, vol. 9, sec. 2, part 1: 3–44.

de Laguna, Frederica. 1972. *Under Mt. St. Elias: The History and Culture of The Yakutat Tlingit*. Washington, DC: Smithsonian Institution.

de Laguna, Frederica, Francis A. Riddell, Donald F. McGeein, Kenneth S. Lane, J. Arthur Freed, and Carolyn Osborne. 1964. *Archaeology of the Yakutat Bay Area, Alaska*. Washington, DC: Smithsonian Institution.

De Vries, Todd. 2012. "Haida Calendar." In Íihlx̱*aadas húus* x̱íinaangslaang/*We Haida's Are Coming Alive Again*. http://haidalanguage.blogspot.ca/2011/11/haida-calendar.html (accessed 11 October 2013).

Decosse, S.S. 1980. "Athapaskan-Tlingit Trade Relations: The Case for Balanced Reciprocity." MA thesis, University of Victoria.

Delgamuukw v. British Columbia. 1997. Docket 23799, Judgments of the Supreme Court of Canada, 11 December.

Demes, K.W., M.H. Graham, and T.S. Suskiewicz. 2009. "Phenotypic Plasticity Reconciles Incongruous Molecular and Morphological Taxonomies: The Giant Kelp, *Macrocystis* (Laminariales, Phaeophyceae), Is a Monospecific Genus." *Journal of Phycology* 45: 1266–9.

Densmore, F. 1939. *Nootka and Quileute Music*. Washington, DC: Smithsonian Institution.

Department of Indian Affairs. 1900. "Notes from Peter O'Reilly's Visit to Laxkibu Territory, Lowe Inlet, 22 July 1889." In *Annual Report*. Part 2. Ottawa, ON: Government of Canada.

Deur, Douglas. 1998. "Wetland Cultivation on the Northwest Coast of North America." MA thesis, Louisiana State University.

– 2000. "A Domesticated Landscape: Native American Plant Cultivation on the Northwest Coast of North America." Phd diss., Louisiana State University.

– 2002a. *Huckleberry Mountain Traditional-Use Study: Final Report*. Seattle, WA: National Park Service, United States Department of the Interior.

– 2002b. "Rethinking Precolonial Plant Cultivation on the Northwest Coast of North America." *Professional Geographer* 54: 140–57.

– 2005. "Tending the Garden, Making the Soil: Northwest Coast Estuarine Gardens as Engineered Environments." In *"Keeping It Living": Traditions of Plant Use and Cultivation on the Northwest Coast of North America*, ed. Douglas Deur and Nancy J. Turner, 296–330. Seattle, WA: University of Washington Press; Vancouver, BC: UBC Press.

– 2009. "'A Caretaker Responsibility': Revisiting Klamath and Modoc Traditions of Plant Community Management." In *Indigenous Resource Management: Past, Present and Future*, ed. Dana Lepofsky. Special issue, *Journal of Ethnobiology* 29, no. 2: 296–322.

Deur, Douglas, and Nancy J. Turner, eds. 2005. *"Keeping It Living": Traditions of Plant Use and Cultivation on the Northwest Coast of North America*. Seattle, WA: University of Washington Press; Vancouver, BC: UBC Press.

Deur, Douglas, Nancy J. Turner, Clan Chief Adam Dick (Kwaxsistalla), Daisy Sewid-Smith (Mayanilth), and Kim Recalma-Clutesi (Oqwilowgwa). 2013. "Subsistence and Resistance on the British Columbia Coast: Kingcome Village's Estuarine Gardens as Contested Space." In *Ethnobotany in British Columbia: Plants and People in a Changing World*, ed. Nancy J. Turner and Dana Lepofsky. Special issue, *BC Studies* 179: 13–37.

Deveau, Amy. 2010. "Kwakwaka'wakw Use of the Edible Seaweed Lheq'estén (*Porphyra abbottiae*) and Heavy Metal Bioaccumulation at Traditional Harvesting Sites in Queen Charlotte Strait." MSc thesis, University of Victoria.

Devereaux, F., and K. Kittredge. 2008. *Feasting for Change: Reconnecting to Food, Land and Culture*. Victoria, BC: Office of the Community Nutritionist for Aboriginal Health.

Dewhirst, John. 1978. "Nootka Sound: A 4,000 Year Perspective." *Provincial Archives of British Columbia: Sound Heritage* 7, no. 2: 1–29.

Diamond, Jared. 2005. *Collapse: How Societies Choose to Fail or Succeed*. New York, NY: Viking.

Dickson, J.H. 2000. "Bryology and the Iceman: Chorology, Ecology and Ethnobotany of the Mosses *Neckera complanata* Hedw. and *Neckera crispa* Hedw." In *The Iceman and His Natural Environment*, ed. Sigmar Bortenschlager and Klaus Oeggl, 77–81. Vienna, Austria: Springer Verlag.

Dickson, J.H., and P.J. Mudie. 2008. "The Life and Death of Kwäday Dän Ts'ìnchí, an Ancient Frozen Body from British Columbia: Clues from Remains of Plants and Animals." *Northern Review* 28: 27–50.

Dickson, J.H., K. Oeggl, and L.L. Handley. 2003. "The Iceman Reconsidered." *Scientific American* 288: 60–9.

Dickson, J.H., M.P. Richards, R.J. Hebda, P.J. Mudie, O. Beattie, S. Ramsay, N.J. Turner, B.J. Leighton, J.M. Webster, N.R. Hobischak, G.S. Anderson, P.M. Troffe, and R.J. Wigen. 2004. "Kwäday Dän Ts'ìnchí, the First Ancient Body of a Man from a North American Glacier: Reconstructing His Last Days by Intestinal and Biomolecular Analyses." *The Holocene* 14, no. 4: 481–6.

Dilbone, Megan. 2011. "The Nutritious Springtime Candy of People and Animals in British Columbia: Lodgepole Pine Cambium (*Pinus contorta* Douglas ex Louden var. *latifolia* Engelm. ex S. Watson)." MSC thesis, University of Victoria.

Dilbone, Megan, Nancy J. Turner, and Patrick von Aderkas. 2013. "Lodgepole Pine Cambium (*Pinus contorta* Dougl. ex Loud. var. *latifolia* Engelm. ex S. Wats.): A Springtime First Peoples' Food in British Columbia." *Ecology of Food and Nutrition* 52, no. 2: 130–47. http://dx.doi.org/10.1080/03670244.2012.706013 (accessed 8 June 2013).

Dillehay, T.D. 2000. *The Settlement of the Americas: A New Prehistory*. New York, NY: Basic Books.

Dillehay, T.D., C. Ramírez, M. Pino, M.B. Collins, J. Rossen, and J.D. Pino-Navarro. 2008. "Monte Verde: Seaweed, Food, Medicine, and the Peopling of South America." *Science* 320: 784–6.

Dixon, E. James. 1999. *Bones, Boats and Bison: Archeology and the First Colonization of Western North America*. Albuquerque, NM: University of New Mexico Press.

Dixon, May, Mary Palmantier, and Aert H. Kuipers. 1982. *A Western Shuswap Reader*. Leiden, Netherlands: University of Leiden.

Dolšak, Nives, and Elinor Ostrom. 2003. *The Commons in the New Millennium: Challenges and Adaptation*. Cambridge, MA: MIT Press.

Donald, Leland. 1997. *Aboriginal Slavery on the Northwest Coast of North America*. Berkeley, CA: University of California Press.

– 2003. "The Northwest Coast as a Study Area: Natural, Prehistory, and Ethnographic Issues." In *Emerging from the Mist: Studies in Northwest Coast Culture History*, ed. R.G. Matson, G. Coupland, and Q. Mackie, 289–327. Vancouver, BC: UBC Press.

Douglas, David. 1914. *Journal Kept by David Douglas during His Travels in North America, 1823–1827*. London, UK: Royal Horticultural Society.

Douglas, George W., G.B. Straley, D.V. Meidinger, and J. Pojar, eds. 1998–2002. *Illustrated Flora of British Columbia*. 8 vols. Seattle, WA: University of Washington Press; Victoria, BC: British Columbia Conservation Data Centre.

Dove, Carla J., P. Gregory Hare, and Marcy Heacker. 2005. "Identification of Ancient Feather Fragments Found in Melting Alpine Ice Patches in Southern Yukon." *Arctic* 58, no. 1: 38–43.

Downs, Ken. 2006. "The Tsimshian Homeland: An Ancient Cultural Landscape." MA thesis, Athabaska University.

Drabek, A.S., and K.R. Adams. 2004. *The Red Cedar of Afognak: A Driftwood Journey*. Afognak, AK: Native Village of Afognak.

Driscoll, Scott. 2003. "Camas-Bulb Restoration Honors Tribal Traditions." *Horizon Air Magazine*, November, 38–49.

Driver, Harold C. 1969. *Indians of North America*. Chicago, IL: University of Chicago Press.

Drucker, Philip. 1951. *The Northern and Central Nootkan Tribes*. Washington, DC: Smithsonian Institution.

Druehl, L. 2000. *Pacific Seaweeds: A Guide to Common Seaweeds of the West Coast*. Madeira Park, BC: Harbour.

Dubin, Lois S. 1999. *North American Indian Jewelry and Adornment from Prehistory to the Present*. New York, NY: Harry N. Abrams.

Duff, Wilson. 1956. *Prehistoric Stone Sculpture of the Fraser River and Gulf of Georgia*. Victoria, BC: British Columbia Provincial Museum.

– 1997. *The Indian History of British Columbia: The Impact of the White Man*. New ed. Original 1965. Victoria, BC: Royal BC Museum.

Duke, James A. 1985. *CRC Handbook of Medicinal Herbs*. Boca Raton, FL: CRC Press.

– 2009. "Achillea millefolium." In *Dr. Duke's Phytochemical and Ethnobotanical Databases.* http://sun.ars-grin.gov:8080/npgspub/xsql/duke/plantdisp.xsql?taxon=18 (accessed 23 August 2013).
Dundes, Alan. 1965. *The Study of Folklore.* Englewood Cliffs, NJ: Prentice-Hall.
Dunn, John. 1844. *History of the Oregon Territory and British North American Fur Trade, with an Account of the Habits and Customs of the Principal Native Traits on the Northern Continent.* London, UK: Edwards and Hughes.
Dunn, John Asher. 1978. *A Practical Dictionary of the Coast Tsimshian Language.* Ottawa, ON: National Museum of Man.
Easton, N. Alexander. 1990. "The Archaeology of Straits Salish Reef Netting: Past and Future Research Strategies." *Journal of Northwest Anthropology* 24, no. 2: 161–77.
Edwards, G.T. 1979. "Indian Spaghetti." *The Beaver* (Autumn): 4–11.
– 1980. "Bella Coola: Indian and European Medicines." *The Beaver* (Winter): 5–11.
Eidlitz, K. 1969. *Food and Emergency Food in the Circumpolar Area.* Uppsala, Sweden: Almqvist and Wiksells Boktryckeri AB.
Eldridge, M. 1997. *The Significance and Management of Culturally Modified Trees.* Final report prepared for Vancouver Forest Region and CMT Standards Steering Committee. http://www.for.gov.bc.ca/ftp/archaeology/external/!publish/web/culturally_modified_trees_significance_management.pdf (accessed 9 June 2013).
Elliott, Dave, Sr, and Janet Poth. 1990. *Saltwater People.* Saanich, BC: School District No. 63, Saanich School Board.
Ellis, D.W., and L. Swan. 1981. *Teachings of the Tides: Uses of Marine Invertebrates by the Manhousaht People.* Nanaimo, BC: Theytus Books.
Ellis, D.W., and S. Wilson. 1981. *The Knowledge and Usage of Marine Invertebrates by the Skidegate Haida People of the Queen Charlotte Islands.* Skidegate, BC: Queen Charlotte Islands Museum Society.
Elmendorf, William W. 1935–36. *Field Notes of Lakes Indian Culture and Ethnography Recorded from Lakes Informant Nancy Wynecoop.* Accession no. 934888. Victoria, BC: British Columbia Archives.
– 1960. *The Structure of Twana Culture.* Pullman, WA: Washington State University Research Studies.
– 1965. "Linguistic and Geographic Relations in the Northern Plateau Area." *Southwestern Journal of Anthropology* 21: 63–73.
Elms, Lindsay. 1996. "Woss Lake: And the Oolichan Trail." http://www.beyondnootka.com/articles/woss.html (accessed 13 October 2013).
Emmons, George T. 1911. *The Tahltan Indians.* Philadelphia, PA: University Museum, University of Pennsylvania.
– 1991. *The Tlingit Indians.* Ed. Fredrica de Laguna. Seattle, WA: University of Washington Press; New York, NY: American Museum of Natural History.
Enrico, John. 1991. *The Lexical Phonology of Masset Haida.* Fairbanks, AK: Alaska Native Language Center, University of Alaska Fairbanks.
– ed. 1995. *Skidegate Haida Myths and Histories.* Trans. John Enrico. Translation of John R. Swanton's *Haida Texts and Myths, Skidegate Dialect.* Skidegate, BC: Queen Charlotte Islands Museum Press.
Enrico, John, and Wendy B. Stuart. 1996. *Northern Haida Songs.* Lincoln, NE: University of Nebraska Press.
Erlandson, Jon M. 2002. "Anatomically Modern Humans, Maritime Migrations, and the Peopling of the New World." In *The First Americans: The Pleistocene Colonization of the New World,* ed. N. Jablonski, 59–92. San Francisco, CA: California Academy of Sciences.

Erlandson, Jon M., and Scott M. Fitzpatrick. 2006. "Oceans, Islands, and Coasts: Current Perspectives on the Role of the Sea in Human Prehistory." *Journal of Island and Coastal Archaeology* 1, no. 1: 5–32.

Erlandson, Jon M., Madonna L. Moss, and Mathew Des Lauriers. 2008. "Life on the Edge: Early Maritime Cultures of the Pacific Coast of North America." *Quaternary Science Reviews* 27: 2232–45.

Erlandson, Jon M., Michael H. Graham, Bruce J. Bourque, Debra Corbett, James A. Estes, and Robert S. Steneck. 2007. "The Kelp Highway Hypothesis: Marine Ecology, the Coastal Migration Theory, and the Peopling of the Americas." *Journal of Island and Coastal Archaeology* 2, no. 2: 161–74.

Erman, Adolph. 1848. *Reise um die Erde durch Nord-Asien und die beiden Oceane in den Jahren 1828, 1829 und 1830*. Vol. 1, no. 3, *Die Ochozker Küste, das Ochozker Meer und die Reisen auf Kamtschatka im Jahre 1829*. Berlin, Germany: Reimer.

Eshbaugh, W.H. 1993. "History and Exploitation of a Serendipitous New Crop Discovery." In *New Crops*, ed. J. Janick and J.E. Simon, 132–9. New York, NY: Wiley.

Etkin, Nina L. 2006. *Edible Medicines: An Ethnopharmacology of Food*. Tucson, AZ: University of Arizona Press.

Evans, Nicholas. 2010. *Dying Words: Endangered Languages and What They Have to Tell Us*. Hoboken, NJ: Wiley-Blackwell.

Fedje, Daryl W., and Tina Christensen. 1999. "Modelling Paleoshorelines and Locating Early Holocene Coastal Sites in Haida Gwaii." *American Antiquity* 64, no. 4: 635–52.

Fedje, Daryl W., Tina Christensen, H. Josenhans, J.B. McSporran, and J. Strang. 2005. "Millennial Tides and Shifting Shores: Archaeology on a Dynamic Landscape." In *Haida Gwaii: Human History and Environment from the Time of Loon to the Time of the Iron People*, ed. Daryl W. Fedje and R.W. Mathewes, 164–86. Vancouver, BC: UBC Press.

Fedje, Daryl W., H. Josenhans, J.J. Clague, J.V. Barrie, D.J. Archer, and J.R. Sout. 2005. "Hecate Strait Paleoshorelines." In *Haida Gwaii: Human History and Environment from the Time of Loon to the Time of the Iron People*, ed. Daryl W. Fedje and R.W. Mathewes, 21–37. Vancouver, BC: UBC Press.

Fedje, Daryl W., A.P. Mackie, R.J. Wigen, Q. Mackie, and C.R. Lake. 2005. "Kilgii Gwaay: An Early Maritime Site in the South of Haida Gwaii." In *Haida Gwaii: Human History and Environment from the Time of Loon to the Time of the Iron People*, ed. Daryl W. Fedje and R.W. Mathewes, 187–203. Vancouver, BC: UBC Press.

Fedje, Daryl W., and Q. Mackie. 2005. "Overview of Cultural History [of Haida Gwaii]." In *Haida Gwaii: Human History and Environment from the Time of Loon to the Time of the Iron People*, ed. Daryl W. Fedje and R.W. Mathewes, 154–86. Vancouver, BC: UBC Press.

Fedje, Daryl W., Martin P.R. Magne, and Tina Christensen. 2005. "Text Excavations at Raised Beach Sites in Southern Haida Gwaii and Their Significance to Northwest Coast Archaeology." In *Haida Gwaii: Human History and Environment from the Time of Loon to the Time of the Iron People*, ed. Daryl W. Fedje and R.W. Mathewes, 204–44. Vancouver, BC: UBC Press.

Fedje, Daryl W., and R.W. Mathewes, eds. 2005. *Haida Gwaii: Human History and Environment from the Time of Loon to the Time of the Iron People*. Vancouver, BC: UBC Press.

Feld, Steven, and Keith H. Basso, eds. 1996. *Senses of Place*. Santa Fe, NM: School of American Research Press.

Felger, Richard S., and Mary B. Moser. 1973. "Eelgrass (*Zostera marina* L.) in the Gulf of California: Discovery of Its Nutritional Value by the Seri Indians." *Science* 181: 355–6.

Fiedel, Stuart J. 2007. "Quacks in the Ice: Waterfowl, Paleoindians, and the Discovery of America." In *Foragers of the Terminal Pleistocene in North America*, ed. Renee B. Walker and Boyce N. Driskell, 1–14. Lincoln, NE: University of Nebraska Press.

Findlay, Andrew. 2009. "Pipeline Would Bring Tankers into B.C. Inlets." Strait.com. http://www.straight.com/news/pipeline-would-bring-tankers-bc-inlets (accessed 9 June 2013).
First People. 2009a. *Native American Indian Agreements and Treaties.* http://www.firstpeople.us/FP-Html-Treaties/Treaties.html (accessed 16 September 2013).
– 2009b. "Treaty with the Flatheads, etc, July 16, 1855." In *Native American Indian Agreements and Treaties.* http://www.firstpeople.us/FP-Html-Treaties/TreatyWithTheFlatheadsetc1855.html (accessed 16 September 2013).
– 2009c. "Treaty with the Makah, January 31, 1885." In *Native American Indian Agreements and Treaties.* http://www.firstpeople.us/FP-Html-Treaties/TreatyWithTheMakah1855.html (accessed 16 September 2013).
– 2009d. "Treaty with the Quínaíelt, July 1, 1855." In *Native American Indian Agreements and Treaties.* http://www.firstpeople.us/FP-Html-Treaties/TreatyWithTheQuinaielt1855.html (accessed 16 September 2013).
– 2009e. "Treaty with the Wallawalla, Cayuse, June 9, 1855." In *Native American Indian Agreements and Treaties.* http://www.firstpeople.us/FP-Html-Treaties/TreatyWithTheWallawallaCayuse1855.html (accessed 16 September 2013).
First Voices. 2000–11. "Language Legacies Celebrating Indigenous Cultures." http://www.firstvoices.com (accessed 9 June 2013).
Fisher, Robin. 1992. *Contact and Conflict: Indian-European Relations in British Columbia, 1774–1890.* 2nd ed. Original 1977. Vancouver, BC: UBC Press.
Fladmark, Knut. 1979. "Routes: Alternate Migration Corridors for Early Man in North America." *American Antiquity* 44: 55–69.
– 1986. "The First Americans: Getting One's Berings." *Natural History* 95, no. 11: 8–19.
Fleisher, Mark S. 1980. "The Ethnobotany of the Clallam Indians of Western Washington." *Northwest Anthropological Research Notes* 14, no. 2: 192–210.
Ford, Jesse, and Dennis Martinez, eds. 2000. *Traditional Ecological Knowledge, Ecosystem Science and Environmental Management.* Special issue, *Ecological Applications* 10, no. 5.
Ford, Richard I. 1979. "Paleoethnobotany in American Archaeology." In *Advances in Archaeological Method and Theory*, vol. 2, ed. Michael B. Schiffer, 285–336. New York, NY: Academic Press.
– ed. 1994. *The Nature and Status of Ethnobotany.* Ann Arbor, MI: Museum of Anthropology, University of Michigan.
– ed. 2001. *Ethnobiology at the Millennium.* Ann Arbor, MI: Museum of Anthropology, University of Michigan.
Fort Victoria Journal. 1846–50. Edited and transcribed by Graham Brazier et al. Hudson's Bay Company Archives, B.226/a/1. http://www.fortvictoriajournal.ca (accessed 17 September 2013).
Fortescue, Michael. 2010. "Yeniseian: Siberian Intruder or Remnant?" In *The Dene-Yeniseian Connection*, ed. James Kari and Ben A. Potter. Special issue, *Anthropological Papers of the University of Alaska* 5, ns: 310–15.
Foster, Rhonda, and Dale R. Croes. 2004. "Joint Tribal/College Wet Site Investigations: A Critical Need for Native American Expertise." *Journal of Wetland Archaeology* 4: 125–37.
Fowler, Catherine S. 1992. *In the Shadow of Fox Peak: An Ethnography of the Cattail-Eater Northern Paiute People of Stillwater Marsh.* Fallon, NV: Fish and Wildlife Service, Region 1, Stillwater National Wildlife Reserve, United States Department of the Interior.
– 2000. "'We Live by Them': Native Knowledge of Biodiversity in the Great Basin of Western North America." In *Biodiversity and Native America*, ed. Paul E. Minnis and Wayne J. Elisens, 99–123. Norman: OK: University of Oklahoma Press.

Fowler, Catherine S., and Dana Lepofsky. 2011. "Traditional Resource and Environmental Management." In *Ethnobiology*, ed. E.N. Anderson et al., 285–304. Hoboken, NJ: Wiley-Blackwell.

Freeman, Milton M.R., and Ludwig N. Carbyn, eds. 1988. *Traditional Knowledge and Renewable Resource Management*. Edmonton, AB: Boreal Institute for Northern Studies, University of Alberta.

French, David H. 1965. "Ethnobotany of the Pacific Northwest Indians." *Economic Botany* 19: 378–82.

– 1981. "Neglected Aspects of North American Ethnobotany." *Canadian Journal of Botany* 59, no. 11: 2326–30.

Friedman, Janet P. 1978. *Wood Identification by Microscopic Examination: A Guide for the Archaeologist on the Northwest Coast of North America*. Victoria, BC: British Columbia Provincial Museum.

– 2005. "Prehistoric Uses of Wood at the Ozette Archaeological Site." In *Ozette Archaeological Project Research Reports*, vol. 3, *Ethnobotany and Wood Technology*, ed. David Welchel, 99–212. Seattle, WA: Northwest Regional Office, National Park Service, United States Department of the Interior; Pullman, WA: Department of Anthropology, Washington State University.

Gadgil, Madhav, and Fikret Berkes. 1991. "Traditional Resource Management Systems." *Resource Management and Optimization* 8, 127–41.

Gadgil, Madhav, Fikret Berkes, and Carl Folke. 1993. "Indigenous Knowledge for Biodiversity Conservation." *Ambio* 22, nos 2–3: 151–6.

Galloway, Brent D. 1982. *Upriver Halq'eméylem Ethnobotany*. Sardis, BC: Stó:lö Sitel Curriculum Committee, Coqualeetza Education Training Centre.

– 2009. *Dictionary of Upriver Halkomelem*. Vol. 1. Berkeley and Los Angeles, CA: University of California Press. http://www.escholarship.org/uc/item/65r158r4 (accessed 9 June 2013).

Garfield, Viola E. 1966. "The Tsimshian and Their Neighbours." In *The Tsimshian Indians and Their Arts*, ed. Viola E. Garfield and Paul S. Wingert, 3–70. Vancouver, BC: Douglas and McIntyre.

Garibaldi, Ann. 1999. *Medicinal Flora of the Alaska Natives: A Compilation of Knowledge from Literary Sources of Aleut, Alutiiq, Athabascan, Eyak, Haida, Inupiat, Tlingit, Tsimshian and Yupik Traditional Healing Methods Using Plants*. Anchorage, AK: Alaska Natural Heritage Program, Environment and Natural Resources Institute, University of Alaska Anchorage. http://aknhp.uaa.alaska.edu/botany/medicinal-flora-of-the-alaska-natives (accessed 9 June 2013).

– 2003. "Bridging Ethnobotany, Autecology and Restoration: The Study of Wapato (*Sagittaria latifolia*; Alismataceae) in Interior British Columbia." MSC thesis, University of Victoria.

Garibaldi, Ann, and Nancy J. Turner. 2004. "Cultural Keystone Species: Implications for Ecological Conservation and Restoration." *Ecology and Society* 9, no. 3. http://www.ecologyandsociety.org/vol9/iss3/art1 (accessed 9 June 2013).

Garrick, David. 1998. *Shaped Cedars and Cedar Shaping: A Guidebook to Identifying, Documenting, Appreciating and Learning from Culturally Modified Trees*. Vancouver, BC: Western Canada Wilderness Committee.

Gavin, Daniel G., Linda B. Brubaker, and Kenneth P. Lertzman. 2003. "Holocene Fire History of a Coastal Temperate Rain Forest Based on Soil Charcoal Radiocarbon Dates." *Ecology* 84: 186–201.

Gedalof, Ze'ev, Marlow Pellatt, and Dan J. Smith. 2006. "From Prairie to Forest: Three Centuries of Environmental Change at Rocky Point, Vancouver Island, British Columbia." *Northwest Science* 80, no. 1: 34–46.
George, Earl Maquinna. 1997. "Living on the Edge: Nuu-chah-nulth History from an Ahousaht Chief's Perspective." MA thesis, University of Victoria.
– 2003. *Living on the Edge: Nuu-chah-nulth History from an Ahousaht Chief's Perspective.* Winlaw, BC: Sono Nis.
Gerdts, Donna B., Brian D. Compton, Leonard Edwards, and Charles Ulrich. 1997. *Hul'q'umin'um' Words: An English-to-Hul'q'umin'um' and Hul'q'umin'um'-to-English Dictionary.* Nanaimo, BC: Chemainus, Nanaimo, and Nanoose First Nations and Nanaimo School District No. 68.
Gibson, James R. 1992. *Otter Skins, Boston Ships, and China Goods: The Maritime Fur Trade of The Northwest Coast, 1785–1841.* Montreal, QC, and Kingston, ON: McGill-Queen's University Press.
Gilbert, M. Thomas P., Dennis L. Jenkins, Anders Götherstrom, Nuria Naveran, Juan J. Sanchez, Michael Hofreiter, Philip Francis Thomsen, Jonas Binladen, Thomas F.G. Higham, Robert M. Yohe II, Robert Parr, Linda Scott Cummings, Eske Willerslev, M. Reidla, E. Tamm, E. Axelsson, A. Gotherstrom, and P.F. Campos. 2008. "DNA from Pre-Clovis Human Coprolites in Oregon, North America." *Science* 320, no. 5877: 786–9.
Gill, Steven J. 1983. "Ethnobotany of the Makah and Ozette People, Olympic Peninsula, Washington (USA)." Phd diss., Washington State University.
– 2005a. "Botanical Remains from the Ozette Village Site (45CA24)." In *Ozette Archaeological Project Research Reports*, vol. 3, *Ethnobotany and Wood Technology*, ed. David Welchel, 55–94. Seattle, WA: Northwest Regional Office, National Park Service, United States Department of the Interior; Pullman, WA: Department of Anthropology, Washington State University.
– 2005b. "Plant Utilization by the Makah and Ozette People, Ozette Peninsula, Washington." In *Ozette Archaeological Project Research Reports*, vol. 3, *Ethnobotany and Wood Technology*, ed. David Welchel, 339–434. Seattle, WA: Northwest Regional Office, National Park Service, United States Department of the Interior; Pullman, WA: Department of Anthropology, Washington State University.
Gilman, C., and J. Ronda. 2003. *Lewis and Clark: Across the Divide.* Washington, DC: Smithsonian Institution.
Gilmore, Melvin R. 1931. "Dispersal by Indians a Factor in the Extension of Discontinuous Distribution of Certain Species of Native Plants." *Papers of the Michigan Academy of Science, Arts and Letters* 13: 89–94.
Gisday Wa and Delgam Uukw. 1989. *The Spirit in the Land: The Opening Statement of the Gitksan and Wet'suwet'en Hereditary Chiefs in the Supreme Court of British Columbia, May 11, 1987.* Gabriola Island, BC: Reflections.
Glavin, Terry, and the People of the Nemiah Valley. 1992. *Nemiah: The Unconquered Country.* Vancouver, BC: New Star Books.
Global Diversity Foundation. 2013. "Biocultural Diversity." http://www.global-diversity.org/resources/biocultural-diversity (accessed 13 October 2013).
Goddard, Pliny E. 1917. "The Beaver Indians." *Anthropological Papers of the American Museum of Natural History* 10, part 4: 201–94.
Gomes, Thiago. 2012. "Restoring Tl'chés: An Ethnoecological Restoration Study in Chatham Islands, British Columbia, Canada." MA thesis, University of Victoria.

Goring-Morris, N. 1995. "Complex Hunter-Gatherers at the End of the Palaeolithic (20,000–10,000 B.P.)." In *The Archaeology of Society in the Holy Land*, ed. T.E. Levy, 141–68. Leicester, UK: Leicester University Press.

Gorman, Martin W. [1905–15]. *List of Plants in the Vicinity of Portland, Oregon.* Undated ms. In Special Collections and University Archives, Knight Memorial Library, University of Oregon. Courtesy of John A. Christy, Oregon Natural Heritage Information Center, Oregon State University, and Angie Kimpo, City of Portland Watershed Revegetation Program, Bureau of Environmental Services.

Gott, B. 1983. "Murnong – *Microseris scapigera*: A Study of a Staple Food of Victorian Aborigines." *Australian Aboriginal Studies*: 2–17.

– 1999. "Cumbungi, *Typha* species: A Staple Aboriginal Food in Southern Australia." *Australian Aboriginal Studies*: 33–5.

Gottesfeld, Allen S., R.W. Mathewes, and Leslie Main Johnson Gottesfeld. 1991. "Holocene Debris Flows and Environmental History, Hazelton Area, British Columbia." *Canadian Journal of Earth Sciences* 28, no. 10: 1583–93.

Gottesfeld, Leslie Main Johnson (see also under Johnson and Johnson Gottesfeld). 1992a. "The Importance of Bark Products in the Aboriginal Economies of Northwestern British Columbia, Canada." *Economic Botany* 46: 148–57.

– 1992b. "Use of Cinder Conk (*Inonotus obliquus*) by the Gitksan of Northwestern British Columbia, Canada." *Journal of Ethnobiology* 12, no. 1: 153–6.

Gottesfeld, Leslie Main, and Beverly Anderson. 1988. "Gitksan Traditional Medicine: Herbs and Healing." *Journal of Ethnobiology* 8, no. 1: 13–33.

Greer, Sheila. 1995. *Skookum Stories on the Chilkoot/Dyea Trail.* Whitehorse, YT: Carcross–Tagish First Nation.

Gremillion, Kristen J., ed. 1997. *People, Plants, and Landscapes: Studies in Paleoethnobotany.* Tuscaloosa, AL: University of Alabama Press.

Griffin, Dennis, 2001. "Contributions to the Ethnobotany of the Cup'it Eskimo, Nunivak Island, Alaska." *Journal of Ethnobiology* 21, no. 2: 91–127.

Gülden, Werner Friedrich, ed. 1992. *Forschungsreise nach Kamtschatka: Reise und Erlebnisse des Johann Karl Ehrenfried Kegel von 1841–1847.* Köln [Weimar, Wien]: Böhlau.

Gunderson, L.H., and C.S. Holling, eds. 2002. *Panarchy: Understanding Transformations in Human and Natural Systems.* Washington, DC: Island.

Gunther, Erna. 1973. *Ethnobotany of Western Washington: The Knowledge and Use of Indigenous Plants by Native Americans.* 1945. Rev. ed., Seattle, WA: University of Washington Press.

Gustafson, Carl E., Delbert Gilbow, and Richard D. Daugherty. 1979. "The Manis Mastodon Site: Early Man on the Olympic Peninsula." *Canadian Journal of Archaeology* 3: 157–64.

Haeberlin, H.K., and Erna Gunther. 1930. *The Indians of Puget Sound.* Seattle, WA: University of Washington Publications in Anthropology.

Haggan, Nigel, Cameron Ainsworth, Tony Pitcher, U. Rashid Sumaila, and Johanna Heymans. 2007. "Life in the Fast Food Chain: Où Sont les Poissons d'Antan?" In *Resetting the Kitchen Table: Food Security, Culture, Health and Resilience in Coastal Communities*, ed. C.C. Parrish, Nancy J. Turner, and S. Solberg, 51–79. New York, NY: Nova Science.

Halber Suarez, Tania. 2012. "Tribal Journeys: An Integrated Voice Approach towards Transformative Learning." Phd diss., University of Victoria.

Hallett, Douglas, Dana Lepofsky, R.W. Mathewes, and Kenneth P. Lertzman. 2003. "11,000 Years of Fire History and Climate in the Mountain Hemlock Rain Forests of Southwestern British Columbia Based on Sedimentary Charcoal." *Canadian Journal of Forest Research* 33: 292–312.

Hamersley-Chambers, Fiona, and Nancy J. Turner. 2011. "Plant Use by Northwest Coast and Plateau Indigenous Peoples." In *The Subsistence Economies of Indigenous North American Societies*, ed. Bruce D. Smith, 65–82. Washington, DC: Smithsonian Institution. (Updated from Turner and Hamersley-Chambers 2006.)

Hammett, Julia E. 1997. "Interregional Patterns of Land Use and Plant Management in Native North America." In *People, Plants, and Landscapes: Studies in Paleoethnobotany*, ed. Kristen J. Gremillion, 195–216. Tuscaloosa, AL: University of Alabama Press.

Hanna, D., and M. Henry. 1995. *Our Tellings: Interior Salish Stories of the Nlha'kapmx People.* Vancouver, BC: UBC Press.

Hare, P. Gregory, Sheila Greer, Ruth Gotthardt, Richard Farnell, Vandy Bowyer, Charles Schweger, and Diane Strand. 2004. "Ethnographic and Archaeological Investigations of Alpine Ice Patches in Southwest Yukon, Canada." *Arctic* 57, no. 3: 260–72.

Hargus, Sharon. 2007. *Witsuwit'en Grammar: Phonetics, Phonology, Morphology*. Vancouver, BC: UBC Press.

Harper, J. Russell. 1971. *Paul Kane's Frontier*. Austin, TX: University of Texas Press.

Harrington, J.P. 1941. "Unpublished Field Notes on Nicola Valley Athapaskan/Thompson River Salish." Microfilm ed., reel 14. Washington, DC: Smithsonian Institution.

Harris, Douglas C. 2001. *Fish, Law, and Colonialism: The Legal Capture of Salmon in British Columbia*. Toronto, ON: University of Toronto Press.

Harris, R. Cole. 1994. "Voices of Disaster: Smallpox around the Strait of Georgia in 1782." *Ethnohistory* 41: 591–626.

– 1997. *The Resettlement of British Columbia: Essays on Colonialism and Geographical Change.* Vancouver, BC: UBC Press.

Harrison, K. David. 2007. *When Languages Die: The Extinction of the World's Languages and the Erosion of Human Knowledge*. Oxford, UK: Oxford University Press.

Hart, Jeff. 1974. "Plant Taxonomy of the Salish and Kootenai Indians of Western Montana." MA thesis, University of Montana.

– 1976. *Montana: Native Plants and Early Peoples.* Helena, MT: Montana Historical Society and Montana Bicentennial Administration.

– 1979. "The Ethnobotany of the Flathead Indians of Western Montana." *Botanical Museum Leaflets, Harvard University* 27, no. 10: 261–307.

Hart, Jeff, Nancy J. Turner, and Lawrence R. Morgan. 1981. *Ethnobotany of the Kootenai Indians of Western North America*. Cranbrook, BC: Report to the Kootenay Indian Area Council (later Ktunaxa/Kinbasket Tribal Council).

Hartley Bay School. 1997. *Hartley Bay School Presents to You Molks [Crabapples]*. Prince Rupert, BC: Hartley Bay School and School District No. 52.

Harvey, Chris. 2002–11. Languagegeek website. With links to FirstVoices and Indigenous Language Institute, language maps, phonetic terminology guide, fonts, and keyboard layouts. http://www.languagegeek.com (accessed 26 January 2013).

Hawkes, M.W. 1985. "*Palmaria hecatensis* sp. nov. (Rhodophyta, Palmariales) from British Columbia and Alaska with a Survey of Other *Palmaria* Species." *Canadian Journal of Botany* 63, no. 3: 474–82.

Hayden, Brian, ed. 1992. *A Complex Culture of the British Columbia Plateau: Traditional Stl'átl'imx Resource Use.* Vancouver, BC: UBC Press.

– 1996. "Pathways to Power: Principles for Creating Socioeconomic Inequalities." In *Foundations of Social Inequality*, ed. T.D. Price and G.M. Feinman, 15–85. New York, NY: Plenum.

– 1997. *The Pithouses of Keatley Creek: Complex Hunter-Gatherers of the Northwest Plateau.* Toronto, ON: Harcourt Brace.

Hayden, Brian, and June Ryder. 1991. "Prehistoric Cultural Collapse in the Lillooet Area." *American Antiquity* 56: 50–65.
Hayden, Brian, and Rick J. Schulting. 1997. "The Plateau Interaction Sphere and Late Prehistoric Cultural Complexity." *American Antiquity* 62, no. 10: 51–85.
Health Canada. 2006. "National Native Alcohol and Drug Abuse Program." http://www.hc-sc.gc.ca/fniah-spnia/substan/ads/nnadap-pnlaada-eng.php (accessed 30 June 2013).
– 2009. "First Nations and Inuit Health." http://www.hc-sc.gc.ca/fniah-spnia/diseases-maladies/index-eng.php (accessed 30 June 2013).
Heaton, Timothy H. N.d. *Ice Age Paleontology of Southeast Alaska*. http://orgs.usd.edu/esci/alaska (accessed 30 June 2013).
Hebda, Richard J. 1995. "British Columbia Vegetation and Climate History with Focus on 6 KA BP." *Geographie Physique et Quaternaire* 49: 55–79.
– 1998. "Environmental History [Montane Cordillera Ecozone]." In *Assessment of Species Diversity in the Montane Cordillera Ecozone*, ed. G.G.E. Scudder and I.M. Smith. Burlington, ON: Ecological Monitoring and Assessment Network. http://www.naturewatch.ca/eman/reports/publications/99_montane/environm/enviro3.html (accessed 13 October 2013).
Hebda, Richard J., and S.G. Frederick. 1990. "History of Marine Resources of the Northeast Pacific since the Last Glaciation." *Transactions of the Royal Society of Canada*, ser. 1, 1: 319–42.
Hebda, Richard J., Sheila Greer, and Alexander Mackie. 2011. *Teachings from Long Ago Person Found: Highlights from the Kwädąy Dän Ts'ìnchį Project*. Joint project of the Champagne and Aishihik First Nations. Victoria BC: British Columbia Archaeology Branch and Royal BC Museum. http://issuu.com/royalbcmuseum/docs/kdt_highlights (accessed 13 October 2013).
Hebda, Richard J., and R.W. Mathewes. 1984. "Holocene History of Cedar and Native Indian Cultures of the North American Pacific Coast." *Science* 225: 711–12.
Hebda, Richard J., M.G. Pellatt, R.W. Mathewes, D.W. Fedje, and S. Acheson. 2005. "Vegetation History of Anthony Island, Haida Gwaii, and Its Relationship to Climate Change and Human Settlement." In *Haida Gwaii: Human History and Environment from the Time of Loon to the Time of the Iron People*, ed. D.W. Fedje and R.W. Mathewes, 59–76. Vancouver, BC: UBC Press.
Hebda, Richard J., Nancy J. Turner, Sage Birchwater, Michèle Kay, and the Elders of Ulkatcho. 1996. *Ulkatcho Food and Medicine Plants*. Anahim Lake, BC: Ulkatcho Band.
Heffernan, M.C. 1995. "Diabetes and Aboriginal Peoples: The Haida Gwaii Diabetes Project in a Global Perspective." In *A Persistent Spirit: Towards Understanding Aboriginal Health in British Columbia*, ed. Peter H. Stephenson et al., 261–92. Victoria, BC: Department of Geography, University of Victoria.
Heiltsuk College and Staff. 1997. *Traditional Heiltsuk Food*. Y. Kathy Brown, project coordinator and editor/writer. Victoria, BC: British Columbia Ministry of Education, Skills and Training.
Heizer, Robert F. 1938. "Aconite Arrow Poison in the Old and New World." *Journal of the Washington Academy of Sciences* 28, no. 8: 358–64.
Helin, Calvin. 2008. *Dances with Dependency: Out of Poverty through Self-Reliance*. Woodland Hills, CA: Ravencrest.
Heller, Christine A. 1993 *Wild Edible and Poisonous Plants of Alaska*. 1976. Reprint, Fairbanks, AK: University of Alaska Fairbanks.

Helm, June, and William C. Sturtevant, eds. 1982. *Handbook of North American Indians*. Vol. 6, *Subarctic*. Washington, DC: Smithsonian Institution.
Heron, Katrina. 2008. "A Modest Proposal for Sustainable Eating." *San Francisco Chronicle*, 17 August 2008. http://www.sfgate.com/cgi-bin/article.cgi?f=/c/a/2008/08/17/INCU12BAV3.DTL (accessed 30 June 2013).
Hess, Thom. N.d. "Borrowed Words and British Columbia Prehistory." Unpublished ms. Department of Linguistics, University of Victoria.
– 1987. "Principles to Use in Identifying Loanwords and Possible Sources." Unpublished ms and course materials. Department of Linguistics, University of Victoria.
Heusser, C.J. 1960. *Late-Pleistocene Environments of North Pacific North America: An Elaboration of Late-Glacial and Postglacial Climatic, Physiographic, and Biotic Changes*. Special issue, *American Geographical Society* 35.
Hill, Christopher L. 2006. "Geologic Framework and Glaciation of the Western Area." In *Handbook of North American Indians*, vol. 3, *Environment, Origins, and Population*, ed. D.H. Ubelaker, 47–60. Washington, DC: Smithsonian Institution.
Hill-Tout, Charles. 1905. "Report on the Ethnology of the Stlatlumh [Lillooet] of British Columbia." *Journal of the Anthropological Institute of Great Britain and Ireland* 35: 126–218.
– [1910]. *Report on the Ethnology of the Síciatl [Sechelt] of British Columbia, a Coast Division of the Salish Stock*. Victoria, BC: British Columbia Archives.
Hoover, Alan L., and Peter L. Macnair. 2002. *The Magic Leaves: A History of Haida Argillite Carving*. Victoria, BC: Royal BC Museum.
Hopkinson, Jennifer, Peter H. Stephenson, and Nancy J. Turner. 1995. "Changing Traditional Diet and Nutrition of Aboriginal Peoples of Coastal British Columbia." In *A Persistent Spirit: Towards Understanding Aboriginal Health in British Columbia*, ed. Peter H. Stephenson et al., 129–66. Victoria, BC: Department of Geography, University of Victoria.
Howard, Patricia, ed. 2003. *Women and Plants: Case Studies on Gender Relations in Local Plant Genetic Resource Management*. London, UK: Zed Books.
Howay, F.W. 1914. *British Columbia: From the Earliest Times to the Present*. Vancouver, BC: S.J. Clarke.
Hudson's Bay Company. N.d. *Papers Relating to Fort Simpson, 1832–70*. Winnipeg, MA: Public Archives of Manitoba.
Huelsbeck, D.R. 1988. "The Surplus Economy of the Central Northwest Coast." In *Prehistoric Economies of the Northwest Coast*, ed. B.L. Isaac, 149–77. Greenwich, CT: Research in Economic Anthropology.
Hul'qumi'num Treaty Group. 2009. *International Human Rights Commission Admits Hul'qumi'num Treaty Group Case*. Press release. 1 December. http://www.hulquminum.bc.ca/news/media_releases_job_postings (accessed 30 June 2013).
Hultén, Eric. 1968. *Flora of Alaska and Neighboring Territories: A Manual of the Vascular Plants*. Stanford, CA: Stanford University Press.
Hunn, Eugene S. 1981. "On the Relative Contribution of Men and Women to Subsistence among Hunter-Gatherers of the Columbia Plateau: A Comparison with *Ethnographic Atlas* Summaries." *Journal of Ethnobiology* 1, no. 1: 124–34.
– 1996. "Columbia Plateau Place Names: What Can They Teach Us?" *Journal of Linguistic Anthropology* 6: 3–26.
– 1999. "The Value of Subsistence for the Future of the World." In *Ethnoecology: Situated Knowledge/Located Lives*, ed. Virginia D. Nazarea, 23–36. Tuscon, AZ: University of Arizona Press.

Hunn, Eugene S., and David H. French. 1981. "*Lomatium*: A Key Resource for Columbia Plateau Native Subsistence." *Northwest Science* 55, no. 2: 87–94.

Hunn, Eugene S., and Helen H. Norton. 1984. "Impact of Mt. St. Helens Ashfall on Fruit Yield of Mountain Huckleberry, *Vaccinium membranaceum*, Important Native American Food." *Economic Botany* 38, no. 1: 121–7.

Hunn, Eugene S., with James Selam and family. 1990. *Nch'i-Wana, "The Big River": Mid-Columbia Indians and Their Land*. Seattle, WA: University of Washington Press.

Hunn, Eugene S., Nancy J. Turner, and David H. French. 1998. "Ethnobiology and Subsistence." In *Handbook of North American Indians*, vol. 12, *Plateau*, ed. Deward E. Walker Jr, 525–45. Washington, DC: Smithsonian Institution.

Hunn, Eugene S., D.R. Johnson, P.N. Russell, and T.F. Thornton. 2003. "Huna Tlingit Traditional Environmental Knowledge, Conservation, and the Management of a 'Wilderness' Park." *Current Anthropology* 44: S79-S103.

Hunt, George. 1922. *Notes on Plant Names [Kwakwaka'wakw]*. Victoria, BC: British Columbia Archives.

Hutchinson, Ian, and Alan D. McMillan. 1997. "Archaeological Evidence for Village Abandonment Associated with Late Holocene Earthquakes at the Northern Cascadia Subduction Zone." *Quaternary Research* 48: 79–87.

Ignace, Marianne B., Ronald E. Ignace, and Celia Nord. 2009. "Digging Sticks, Woodworms and Lizards: The Practical and Symbolic Division of Labour in Secwepemc Society." Paper presented at the 32nd Annual Conference of the Society of Ethnobiology, New Orleans, Louisiana.

Ignace, Marianne B., Nancy J. Turner, and Ronald E. Ignace. Forthcoming. "Coyote, Grouse and Trees: Secwepemc Lessons about Ethnobiological Knowledge." In *Secwepemc People and Plants: Research Papers in Shuswap Ethnobotany*, ed. Marianne B. Ignace, Nancy J. Turner, and Sandra L. Peacock. Published by the Society of Ethnobiology in the monograph series Contributions in Ethnobiology.

Ignace, Marianne B., Nancy J. Turner, and Sandra Peacock, eds. Forthcoming. *Secwepemc People and Plants: Research Papers in Shuswap Ethnobotany*. Published by the Society of Ethnobiology in the monograph series Contributions in Ethnobiology.

Ignace, Ronald E. (Stsmél'ecqen). 2008. "Our Oral Histories Are Our Iron Posts: Secwepemc Stories and Historical Consciousness." Phd diss., Simon Fraser University.

Ignace, Ronald E., and Marianne B. Ignace. Forthcoming. "*Re tsuwet.s-kucw ne Secwepemcúl'ecw*: Secwepemc Resource Use and Sense of Place." In *Secwepemc People and Plants: Research Papers in Shuswap Ethnobotany*, ed. Marianne B. Ignace, Nancy J. Turner, and Sandra L. Peacock. Published by the Society of Ethnobiology in the monograph series Contributions in Ethnobiology.

Inglis, J.T., ed. 1993. *Traditional Ecological Knowledge: Concepts and Cases*. Ottawa, ON: International Program on Traditional Ecological Knowledge, International Development Research Centre.

Inglis, Richard. 1976. "Wet Site Distribution – The Northern Case GbTo–33 – The Lachane Site." In *The Excavation of Water-Saturated Archaeological Sites (Wet Sites) on the Northwest Coast of North America*, ed. Dale R. Croes, 158–85. Ottawa, ON: National Museums of Canada.

IPCC (Intergovernmental Panel on Climate Change). 2007. *Climate Change 2007: The Physical Science Basis*. http://www.ipcc.ch/publications_and_data/publications_ipcc_fourth_assessment_report_wg1_report_the_physical_science_basis.htm (accessed 30 June 2013).

Iskut Elders. 2005. *Traditional Medicines and Remedies of the Iskut People*. Iskut, BC: Iskut First Nation.

Ives, John W. 1990. *A Theory of Northern Athapaskan Prehistory*. Boulder, CO: Westview.
– 2010. "Dene-Yeniseian, Migration, and Prehistory." In *The Dene-Yeniseian Connection*, ed. James Kari and Ben A. Potter. Special issue, *Anthropological Papers of the University of Alaska* 5, ns: 324–34.
Jacks, Terry, dir. 2000. *"The Warmth of Love": The Four Seasons of Sophie Thomas*. Documentary film. http://sophiethomas.org/spring.htm (accessed 30 June 2013).
Jacobs, M., Jr, and M. Jacobs Sr. 1982. "Southeast Alaska Native Foods." In *Raven's Bones*, ed. A. Hope III, 112–30. Sitka, AK: Sitka Community Association.
Jaganova, V.P. 2004. "Rastenija, primenjaemye v bytu i v lekarstvennych celjach." In *Sochranenie i vozroždenie tradicionnych obrjadovych prazdnikov u beregovych korjakov (nymylanov)*, ed. E. Kasten, 21–2. Krasnodar, Russia: Kamšat.
Jakubov (also Yakubov), V.V., and O.A. Černjagina. 2004. *Katalog flory Kamčatki: Sosudistye rastenija*. Petropavlovsk-Kamčatskij, Russia: Kamčatpress.
Jenness, Diamond. N.d. "Coast Salish Field Notes." Ottawa, ON: Canadian Ethnology Service Archives, Canadian Museum of Civilization.
Jilek, Wolfgang. 1982. *Indian Healing: Shamanic Ceremonialism in the Pacific Northwest Today*. Surrey, BC: Hancock House.
Jochelson, Waldemar. 1975. *The Koryak*. 1908. Reprint, New York: AMS Press.
Johns, Timothy. 1996. *The Origins of Human Diet and Medicine*. Tuscon, AZ: University of Arizona Press.
Johns, Timothy, and I. Kubo. 1988. "A Survey of Traditional Methods Employed for the Detoxification of Plant Foods." *Journal of Ethnobiology* 8, no. 1: 81–129.
Johnson, Leslie Main (see also under Gottesfeld and Johnson Gottesfeld). 1994. "Wet'suwet'en Ethnobotany: Traditional Plant Uses." *Journal of Ethnobiology* 14, no. 2: 185–210.
– 1997. "Health, Wholeness, and the Land: Gitksan Traditional Plant Use and Healing." Phd diss., University of Alberta.
– 1999. "Aboriginal Burning for Vegetation Management in Northwest British Columbia." 1994. Reprinted in *Indians, Fire and the Land in the Pacific Northwest*, ed. Robert T. Boyd, 238–54. Corvallis, OR: Oregon State University Press.
– 2006. "Gitksan Medicinal Plants: Cultural Choice and Efficacy." *Journal of Ethnobiology and Ethnomedicine* 2: 29. http://www.ethnobiomed.com/content/2/1/29 (accessed 24 August 2013).
– 2010. *Trail of Story, Traveller's Path: Reflections on Ethnoecology and Landscape*. Edmonton, AB: Athabasca University Press.
Johnson, Leslie Main, and Eugene S. Hunn. 2010. "Landscape Ethnoecology: Reflections." In *Landscape Ethnoecology: Concepts of Biotic and Physical Space*, ed. Leslie Main Johnson and Eugene S. Hunn, 279–97. New York, NY, and Oxford, UK: Berghahn Books.
Johnson, Martha, ed. 1992. *Lore: Capturing Traditional Environmental Knowledge*. Ottawa, ON: International Development Research Centre.
Johnson Gottesfeld, Leslie Main. 1993. "Plants, Land and People: A Study of Wet'suwet'en Ethnobotany." MSC thesis, University of Alberta.
– 1994. "Conservation, Territory and Traditional Beliefs: An Analysis of Gitksan and Wet'suwet'en Subsistence, Northwest British Columbia, Canada." *Human Ecology* 22, no. 4: 443–65.
Johnson Gottesfeld, Leslie Main, and Dale Vitt. 1996. "The selection of *Sphagnum* for diapers by Indigenous North Americans." *Evansia* 13, no. 3: 103–7.
Jones, Anore. 1983. *Nauriat Nigiñaqtuat: Plants That We Eat*. Kotzebue, AK: Maniilaq Association.
Jones, Charles, and Steven Bosustow. 1981. *Queesto: Pacheenaht Chief by Birthright*. Penticton, BC: Theytus Books.

Jones, James T. 2002. "'We looked after all the Salmon Streams': A Preliminary Assessment of Traditional Heiltsuk Cultural Stewardship of Salmon and Salmon Streams." MA thesis, University of Victoria.
Jonsson L. 1996. "Oaks in Sweden." *International Oaks* 7: 32–9.
Joseph, Leigh. 2012. "Finding Our Roots: Ethnoecological Restoration of *Lhásem* (*Fritillaria camschatcensis* (L.) Ker-Gawl), an Iconic Plant Food in the Squamish River Estuary, British Columbia." MSC thesis, University of Victoria.
Juntunen, J.R., M.D. Dasch, and A.B. Rogers. 2005. *The World of the Kalapuya*. Philomath, OR: Benton County Historical Society and Museum.
Justice, Clive L. 2000. *Mr Menzies' Garden Legacy: Plant Collecting on the Northwest Coast*. Vancouver, BC: Cavendish Books.
Kake, Organized Village of. 1989. *Keex̱' Ḵwaan: In Our Own Words, Interviews of Kake Elders*. Kake, AK: Organized Village of Kake.
Kari, James. 2007. *Dena'ina Topical Dictionary*. Fairbanks, AK: Alaska Native Language Center.
– 2010. "The Concept of Geolinguistic Conservatism in Na-Dene Prehistory." In *The Dene-Yeniseian Connection*, ed. James Kari and Ben A. Potter. Special issue, *Anthropological Papers of the University of Alaska* 5, ns: 194–222.
Kari, James, and Ben A. Potter. 2010a. "The Dene-Yeniseian Connection: Bridging Asia and North America." In *The Dene-Yeniseian Connection*, ed. James Kari and Ben A. Potter. Special issue, *Anthropological Papers of the University of Alaska* 5, ns: 1–24.
– eds. 2010b. *The Dene-Yeniseian Connection*. Special issue, *Anthropological Papers of the University of Alaska* 5, ns.
Karst, Amanda. 2005. "The Ethnoecology and Reproductive Ecology of Bakeapple (*Rubus chamaemorus* L., Rosaceae) in Southern Labrador." MSC thesis, University of Victoria.
Karst, Amanda, and Nancy J. Turner. 2011. "Local Ecological Knowledge and Importance of Bakeapple (*Rubus chamaemorus* L.) in a Southeast Labrador Métis Community." *Ethnobiology Letters* 2: 6–18. http://ethnobiology.org/publications/ethnobiology-letters/2 (accessed 1 July 2013).
Kaska Tribal Council. 1997. *Guzagi K'úgé: Our Language Book: Nouns: Kaska, Mountain Slavey and Sekani*. Whitehorse, YT: Kaska Tribal Council.
Kawagley, A.O. 1995. *A Yupiaq Worldview: A Pathway to Ecology and Spirit*. Prospect Heights, IL: Waveland.
– 2003. "Nurturing Native Languages." In *Nurturing Native Languages*, ed. Jon Reyhner et al., vii–x. Flagstaff, AZ: Centre for Excellence in Education, Northern Arizona University.
Kay, Michèle S. 1993. "Environmental, Cultural and Linguistic Factors Affecting Ulkatcho (Carrier) Botanical Knowledge." MSC thesis, University of Victoria.
Keddie, G. 1988. "The Kootenay Lichen Pounder." *The Midden* 20, no. 1: 6–9.
– 1997. "Aboriginal Defensive Sites. Part 4: Local Sites Are Dated. Final Conclusions." *Discovery: Royal BC Museum News and Events* 25, no. 2: 5.
– 2003. *Songhees Pictorial: A History of the Songhees People as Seen by Outsiders, 1790–1912*. Victoria, BC: Royal BC Museum. See also Keddie's supplement to this book at http://www.sfu.ca/brc/virtual_village/coast_salish/Songhees.html (accessed 13 October 2013).
Keely, Patrick B. 1980. "Nutrient Composition of Selected Important Plant Foods of the Pre-contact Diet of the Northwest Native American People." MA thesis, University of Washington.

Keely, Patrick B., Charlene S. Martinsen, Eugene S. Hunn, and Helen H. Norton. 1982. "Composition of Native American Fruits in the Pacific Northwest." *Journal of the American Dietetic Association* 81: 568–72.
Kelly, Isabel T. 1932. "Ethnography of the Surprise Valley Paiute." *University of California Publications in American Archaeology and Ethnology* 31, no. 3: 67–210.
Kenady, S.M., M.C. Wilson, R.F. Schalk, and R.R. Mierendorf. 2011. "Late Pleistocene Butchered *Bison antiquus* from Ayer Pond, Orcas Island, Pacific Northwest: Age Confirmation and Taphonomy." *Quaternary International* 233, no. 2: 130–41.
Kennedy, Dorothy I.D. 1984. "A Quest for a Cure: A Case Study in the Use of Health Care Alternatives." *Culture* 6: 21–31.
– 1995. "Looking for Tribes in All the Wrong Places: An Examination of the Central Coast Salish Social Network." MA thesis, University of Victoria.
Kennedy, Dorothy I.D., and Randy Bouchard. 1983. *Sliammon Life, Sliammon Lands*. Vancouver, BC: Talonbooks.
– 1986. "Appendix 2: Indian History and Knowledge of the Aspen Grove to Peachland Corridor of the Coquihalla Highway." In *Coquihalla Highway Project: Merritt to Peachland, B.C.: Detailed Heritage Resource Inventory and Impact Assessment*, by Arcas Associates, 121–35. Victoria, BC: British Columbia Heritage Conservation Branch, British Columbia Ministry of Tourism and Design, and Survey Branch of British Columbia Ministry of Transportation and Highways.
– 1988. "Appendix 2: Notes on Sahhaltkum Ethnography and History." In *Archaeological Excavations at the Sahhaltkum Bridge Site (EeQw 30), Chase, BC*, vol. 2, by Arcas Associates, 241. Victoria, BC: British Columbia Ministry of Tourism, Recreation and Culture and British Columbia Ministry of Transportation and Highways.
– 1990. "Bella Coola." In *Handbook of North American Indians*, vol. 7, *Northwest Coast*, ed. Wayne P. Suttles, 323–39. Washington, DC: Smithsonian Institution.
– 1992. "Stl'átl'imx (Fraser River Lillooet) Fishing." In *A Complex Culture of the British Columbia Plateau: Traditional Stl'atl'imx Resource Use*, ed. Brian Hayden, 266–354. Vancouver, BC: UBC Press.
– 1994. *Ditidaht Stories*. Based on stories recorded by Ditidaht elders around 1973 and by Ditidaht elders interviewed by Randy Bouchard and Dorothy Kennedy between 1989 and 1994. Nitinat, BC: Ditidaht Band Council and Ditidaht Education Department.
– 2010. *The Lil'wat World of Charlie Mack*. Vancouver, BC: Talonbooks.
Kirch, Patrick V., and Roger C. Green. 2001. *Hawaiki, Ancestral Polynesia: An Essay in Historical Anthropology*. Cambridge, UK: Cambridge University Press.
K̲ii'iljuus (Barbara J. Wilson) and Heather Harris. 2005. "Tllsda X̲aaydas K̲'aaygang.nga: Long, Long Ago Haida Ancient Stories." In *Haida Gwaii: Human History and Environment from the Time of Loon to the Time of the Iron People*, ed. D.W. Fedje and R.W. Mathewes, 121–39. Vancouver, BC: UBC Press.
Kimmerer, R.W., and F.K. Lake. 2001. "Maintaining the Mosaic: The Role of Indigenous Burning in Land Management." *Journal of Forestry* 99: 36–41.
King, James. 1785. *A Voyage to the Pacific Ocean*. Vol. 3. London, UK: Nicol and Cadell.
King, Leslie. 2004. "Competing Knowledge Systems in the Management of Fish and Forests in the Pacific Northwest." *International Environmental Agreements: Politics, Law and Economics* 4: 161–77.
Kinkade, M. Dale. 1986. *Blackcaps and Musqueam*. Paper presented to the 21st International Conference on Salish and Neighboring Languages, Seattle, Washington, 14–16 August.
– 1989. "Prehistory of Salishan Languages." Paper presented at the 88th Annual Meeting of the American Anthropological Association, Washington, DC.

– 1991. *Upper Chehalis Dictionary.* Missoula, MT: University of Montana Press.
– 2004. *[Lower] Cowlitz Dictionary and Grammatical Sketch.* Missoula, MT: University of Montana Press.
Kinkade, M. Dale, and Anthony Mattina. 1996. "Discourse." In *The Handbook of North American Indians*, vol. 17, *Languages*, ed. Ives Goddard, 244–74. Washington, DC: Smithsonian Institution.
Kirk, Ruth, and Richard D. Daugherty. 2007. *Archaeology in Washington.* Seattle, WA: University of Washington Press.
Klinkenberg, Brian, ed. 2013. *E-Flora BC: Electronic Atlas of the Plants of British Columbia.* Vancouver, BC: Lab for Advanced Spatial Analysis, Department of Geography, University of British Columbia. http://www.geog.ubc.ca/biodiversity/eflora (accessed 1 July 2013).
Kitamaat Indians. 1897. "Petition of the Kitamaat Indians, British Columbia, November 10th, 1897." Unpublished ms. Papers related to Kitimat and Indians, George Henry Raley Collection, British Columbia Archives, HD R13 R13.11.
Konlande, J.E., and J.R.K. Robson. 1972. "The Nutritive Value of Cooked Camas as Consumed by Flathead Indians." *Ecology of Food and Nutrition* 2: 193–5.
Koppel, Tom. 2001. "Discovery in Cave Shows Ice Didn't Bury All of Alaska." *Vancouver Sun*, 13 December.
– 2003. *Lost World: Rewriting Prehistory – How New Science Is Tracing America's Ice Age Mariners.* New York, NY: Atria Books.
Kovach, M. 2006. "Searching for Arrowheads: An Inquiry into Approaches to Indigenous Research Using a Tribal Methodology with a Nêhiýaw Kiskêýihtamowin Worldview." Phd diss., University of Victoria.
Kramer, Stephenie. 2000. "Camas Bulbs, the Kalapuya, and Gender: Exploring Evidence of Plant Food Intensification in the Willamette Valley of Oregon." MSC thesis, University of Oregon.
Krasutsky, Pavel. 2002. "Non-Timber Forest Products and Implications for Forest Managers: Birch Bark Extractives." Natural Resources Research Institute, University of Minnesota-Duluth. http://www.ntfpinfo.us/docs/other/Krasutsky2002-BirchBarkExtractives.pdf (accessed 13 August 2013).
Krause, Aurel. 1956. *The Tlingit Indians.* Trans. Erna Gunther. German original 1885. Seattle, WA: University of Washington Press.
Krause, Aurel, and Arthur Krause. 1993. *To the Chukchi Peninsula and to the Tlingit Indians, 1881/1882: Journals and Letters by Aurel and Arthur Krause.* Trans. M. Krause McCaffrey. German original 1885. Fairbanks, AK: University of Alaska Press.
Krauss, Michael E. 2005. "Athabaskan Tone." In *Athabaskan Prosody*, ed. Sharon Hargus and Keren Rice, 51–136. Amsterdam and Philadelphia: John Benjamins.
Krauss, Michael E., and Victor Golla. 1982. "Northern Athapaskan Languages." In *Handbook of North American Indians*, vol. 6, *Subarctic*, ed. June Helm and William C. Sturtevant, 67–85. Washington, DC: Smithsonian Institution.
'Ksan, People of. 1980. *Gathering What the Great Nature Provided: Food Traditions of the Gitksan.* Vancouver, BC: Douglas and McIntyre; Seattle, WA: University of Washington Press.
Ktunaxa/Kinbasket Tribal Council, Michael Keefer, Peter McCoy, and Troy Hunter. 1999. *ʔakxam̓is q̓api qapsin/All Living Things: A Ktunaxa Ethnobotany Handbook.* Part 1. Cranbrook, BC: Ktunaxa/Kinbasket Tribal Council.
Kubiak-Martens, Lucyna. 1996. "Evidence for Possible Use of Plant Foods in Palaeolithic and Mesolithic Diet from the Site of Całowanie in the Central Part of the Polish Plain." *Vegetation History and Archaeobotany* 5, nos 1–2: 33–8.
Kuhnlein, Harriet V. 1989a. "Factors Influencing Use of Traditional Foods among the Nuxalk People." *Journal of the Canadian Dietetic Association* 50, no. 2: 102–8.

– 1989b. "Nutrient Values in Indigenous Wild Berries Used by the Nuxalk People of Bella Coola, British Columbia." *Journal of Food Composition and Analysis* 2: 28–36.
– 1990. "Nutrient Values in Indigenous Wild Green Plants and Roots Used by the Nuxalk People of Bella Coola, British Columbia." *Journal of Food Composition and Analysis* 3: 38–46.
– 1992. "Change in the Use of Traditional Food by the Nuxalk Native People of British Columbia." In *Perspectives on Dietary Change: Studies in Nutrition and Society*, ed. G.H. Pelto and L.A. Vargas. Special issue, *Ecology of Food and Nutrition* 27, nos 3–4: 259–82.
Kuhnlein, Harriet V., Bill Erasmus, Hilary Creed-Kanashiro, Lois Englberger, Chinwe Okeke, Nancy J. Turner, Lindsay Allen, and Lalita Bhattacharjee. 2006. "Indigenous Peoples' Food Systems for Health: Finding Interventions that Work." *Public Health Nutrition* 9, no. 8: 1013–19.
Kuhnlein, Harriet V., Bill Erasmus, and Dina Spigelski, eds. 2009. *Indigenous Peoples' Food Systems: The Many Dimensions of Culture, Diversity, and Environment for Nutrition and Health.* Montreal, QC: Centre for Indigenous Peoples' Nutrition and Environment, McGill University; Rome, Italy: Food and Agriculture Organization of the United Nations.
Kuhnlein, Harriet V., Bill Erasmus, Dina Spigelski, and Barbara Burlingame, eds. 2013. *Indigenous Peoples' Food Systems for Health: Interventions for Health Promotion and Policy.* Montreal, QC: Centre for Indigenous Peoples' Nutrition and Environment, McGill University; Rome, Italy: Food and Agriculture Organization of the United Nations.
Kuhnlein, Harriet V., and O. Receveur. 1996. "Dietary Change and Traditional Food Systems of Indigenous Peoples." *Annual Review of Nutrition* 16: 417–42.
Kuhnlein, Harriet V., and Nancy J. Turner. 1987. "Cow-parsnip (*Heracleum lanatum*): An Indigenous Vegetable of Native People of Northwestern North America." *Journal of Ethnobiology* 6, no. 2: 309–24.
– 1991. *Traditional Plant Foods of Canadian Indigenous Peoples: Nutrition, Botany and Use.* Philadelphia, PA: Gordon and Breach. (Reissued 2009 by the Food and Agriculture Organization of the United Nations, Rome, Italy, http://www.fao.org/wairdocs/other/ai215e/ai215e00.HTM [accessed 2 July 2013].)
Kuhnlein, Harriet V., Nancy J. Turner, and P.D. Kluckner. 1982. "Nutritional Significance of Two Important 'Root' Foods (Springbank Clover and Pacific Silverweed)." *Ecology of Food and Nutrition* 12: 89–95.
Kuipers, Aert H. 1967. *The Squamish Language*. Vol. 1. The Hague, Netherlands: Mouton.
– 1969. *The Squamish Language*. Vol. 2. The Hague, Netherlands: Mouton.
– 1974. *The Shuswap Language*. The Hague, Netherlands: Mouton.
– 1983. *Shuswap-English Dictionary*. Leuven, Belgium: Peeters.
– 1989. *A Report on Shuswap with a Squamish Lexical Appendix*. Paris, France: Peeters–Selaf.
– 2002. *Salish Etymological Dictionary*. Missoula, MT: University of Montana Press.
Lacourse, T., and R.W. Mathewes. 2005. "Terrestrial Paleoecology of Haida Gwaii and the Continental Shelf: Vegetation, Climate, and Plant Resources of the Coastal Migration Route." In *Haida Gwaii: Human History and Environment from the Time of Loon to the Time of the Iron People*, ed. Daryl W. Fedje and R.W. Mathewes, 38–58. Vancouver, BC: UBC Press.
Lacourse, T., R.W. Mathewes, and Richard J. Hebda. 2007. "Paleoecological Analyses of Lake Sediments Reveal Prehistoric Human Impact on Forests at Anthony Island UNESCO World Heritage Site, Queen Charlotte Island (Haida Gwaii), Canada." *Quaternary Research* 68: 177–83.
Laforet, Andrea. 1990. "Regional and Personal Style in Northwest Coast Basketry." In *The Art of Native American Basketry: A Living Legacy*, ed. F.W. Porter, 281–98. New York, NY: Greenwood.

– 1992. "Windows on Diversity: Northwest Coast Baskets in the Pitt Rivers Collection." In *Basketmakers: Meaning and Form in Native American Baskets*, ed. L. Mowat, H. Murphy, and P. Dransart, 37–50. Oxford, UK: Pitt Rivers Museum, University of Oxford.

Laforet, Andrea, Nancy J. Turner, and Annie York. 1993. "Traditional Foods of the Fraser Canyon Nle7képmx." In *American Indian Linguistics and Ethnography in Honor of Laurence C. Thompson*, ed. T. Montler and A. Mattina, 191–213. Missoula, MT: University of Montana Press.

Laforet, Andrea, and Annie York. 1998. *Spuzzum: Fraser Canyon Histories, 1808–1939*. Ottawa, ON: Canadian Museum of Civilization; Vancouver, BC: UBC Press.

Lake, F.K. 2004. "The Integration of Traditional Ecological Knowledge and Fire Ecology to Restore and Conserve Riparian Zone Biodiversity in the Klamath-Siskiyou Ecoregion." Phd diss., Oregon State University.

Lamb, W. Kaye, ed. 1960. *The Letters and Journals of Simon Fraser, 1806–1808*. Toronto, ON: Macmillan.

Lamont, S.M. 1977. "The Fisherman Lake Slave and Their Environment: A Story of Flora and Faunal Resources." MSC thesis, University of Saskatchewan.

Langsdorff, Georg Heinrich. 1812. *Bemerkungen auf einer Reise um die Welt in den Jahren 1803 bis 1807*. Frankfurt am Main, Germany: Wilman.

Lanner, Ronald. 1996. *Made for Each Other: A Symbiosis of Birds and Pines*. Oxford, UK: Oxford University Press.

Lantz, Trevor. 2001. "The Population Ecology and Ethnobotany of Devil's Club (*Oplopanax horridus* (Sm.) Torr & A. Gray ex. Mig: Araliaceae)." MSC thesis, University of Victoria.

Lantz, Trevor, and Nancy J. Turner. 2003. "Traditional Phenological Knowledge (TPK) of Aboriginal Peoples in British Columbia." *Journal of Ethnobiology* 23, no. 2: 263–86.

Lantz, Trevor, Kristina Swerhun, and Nancy J. Turner. 2004. "Devil's Club (*Oplopanax horridus*): An Ethnobotanical Review." *Herbalgram* 62: 33–48.

Leonti, Marco, Otto Sticher, and Michael Heinrich. 2003. "Antiquity of Medicinal Plant Usage in Two Macro-Mayan Ethnic Groups (México)." *Journal of Ethnopharmacology* 88: 119–24.

Leopold, Aldo. 1990. *A Sand County Almanac, with Essays on Conservation from Round River*. 1949. Reprint, New York, NY: Ballantine.

Lepofsky, Dana. 2002. "Plants and Pithouses: Archaeobotany and Site Formation Processes at the Keatley Creek Village Site." In *Hunter-Gatherer Archaeobotany: Perspectives from the Northern Temperate Zone*, ed. Sarah L.R. Mason and Jon G. Hather, 62–73. London, UK: Institute of Archaeology, University College of London.

– 2004. "Paleoethnobotany in The Northwest." In *People and Plants in Ancient Western North America*, ed. Paul E. Minnis, 367–464. Washington, DC: Smithsonian Books.

– 2007. "Open Letter, Heritage Conservation Branch of British Columbia, September 2, 2007." Copy in possession of the author.

– ed. 2009a. *Indigenous Resource Management: Past, Present and Future*. Special issue, *Journal of Ethnobiology* 29, no. 2.

– 2009b. "Traditional Resource Management: Past, Present and Future." In *Indigenous Resource Management: Past, Present and Future*, ed. Dana Lepofsky. Special issue, *Journal of Ethnobiology* 29, no. 2: 184–212.

Lepofsky, Dana, Michael Blake, Douglas Brown, Sandra Morrison, Nicole Oakes, and Natasha Lyons. 2000. "The Archaeology of the Scowlitz Site, Southwestern British Columbia." *Journal of Field Archaeology* 27: 391–416.

Lepofsky, Dana, D. Hallett, K. Washbrook, A. McHalsie, K.P. Lertzman, and R.W. Mathewes. 2005. "Documenting Precontact Plant Management on the Northwest Coast: An

Example of Prescribed Burning in the Central and Upper Fraser Valley, British Columbia." In *Keeping It Living: Traditions of Plant Use and Cultivation on the Northwest Coast*, ed. Douglas Deur and Nancy J. Turner, 218–39. Seattle, WA: University of Washington Press; Vancouver, BC: UBC Press.

Lepofsky, Dana, E.K. Heyerdahl, K.P. Lertzman, D. Schaepe, and B. Mierendorf. 2003. "Historical Meadow Dynamics in Southwest British Columbia: A Multidisciplinary Analysis." *Conservation Ecology* [now *Ecology and Society*] 7, no. 3. http://www.ecologyandsociety.org/vol7/iss3/art5 (accessed 2 July 2013).

Lepofsky, Dana, and Michael Lenert. N.d. "Choppers in Context: The Archaeology of the Maccallum Site, SW BC." Unpublished ms. In possession of the authors.

– 2005. *Report on the 2004 Excavations of the Maccallum Site (Dh Rk 2), Aggasiz, BC (Stó:lo Permit no. 2004–15)*. Victoria, BC: British Columbia Archaeology Branch and British Columbia Ministry of Small Business, Tourism and Culture.

Lepofsky, Dana, K.P. Lertzman, D. Hallett, and R.W. Mathewes. 2005. "Climate Change and Culture Change on the Southern Coast of British Columbia, 2400–1200 B.P.: An Hypothesis." *American Antiquity* 70: 267–93.

Lepofsky, Dana, and Kenneth P. Lertzman. 2008. "Documenting Ancient Plant Management in the Northwest of North America." *Botany* 86: 129–45.

Lepofsky, Dana, and Natasha Lyons. 2013. "'The Secret Past Life of Plants': Paleoethnobotany in British Columbia." In *Ethnobotany in British Columbia: Plants and People in a Changing World*, ed. Nancy J. Turner and Dana Lepofsky. Special issue, *BC Studies* 179: 39–83.

Lepofsky, Dana, Natasha Lyons, and Madonna L. Moss. 2003. "The Use of Driftwood on the North Pacific Coast: An Example from Southeast Alaska." *Journal of Ethnobiology* 23, no. 1: 143–54.

Lepofsky, Dana, Madonna L. Moss, and Natasha Lyons. 2001. "The Unrealized Potential of Paleoethnobotany in the Archaeology of Northwestern North America: Perspectives from Cape Addington, Alaska." *Arctic Anthropology* 38, no. 1: 48–59.

Lepofsky, Dana, and Sandra L. Peacock. 2004. "A Question of Intensity: Exploring the Role of Plant Foods in Northern Plateau Prehistory." In *Complex Hunter-Gatherers: Evolution and Organization of Prehistoric Communities on the Plateau of Northwestern North America*, ed. B. Prentiss and I. Kuijt, 115–39. Salt Lake City, UT: University of Utah Press.

Lepofsky, Dana, and B. Pegg. 1996. "Archaeological and Ethnographic Research in the Kowesas Watershed." In *The Kowesas Watershed Assessment*, 38–44. Portland, OR: Ecotrust.

Lepofsky, Dana, D.M. Schaepe, A.P. Graesch, M. Lenert, P. Ormerod, K.T. Carlson, J.E. Arnold, M. Blake, P. Moore, and J.J. Clague. 2009. "Exploring Stó:lō–Coast Salish Interaction and Identity in Ancient Houses and Settlements in the Fraser Valley, British Columbia." *American Antiquity* 74, no. 4: 595–626.

Lepofsky, Dana, Nancy J. Turner, and Harriet V. Kuhnlein. 1985. "Determining the Availability of Traditional Wild Plant Foods: An Example of Nuxalk Foods, Bella Coola, BC." *Ecology of Food and Nutrition* 16: 223–41.

Lertzman, Kenneth P., D. Gavin, D. Hallett, L. Brubaker, D. Lepofsky, and R.W. Mathewes. 2002. "Long-Term Fire Regime Estimated from Soil Charcoal in Coastal Temperate Rainforests." *Conservation Ecology* [now *Ecology and Society*] 6, no. 2. http://www.ecologyandsociety.org/vol6/iss2/art5 (accessed 2 July 2013).

Lévi-Strauss, Claude. 1975. *The Raw and the Cooked*. New York, NY: Harper and Row.

Lewis, Henry T. 1977. "Maskuta: The Ecology of Indian Fires in Northern Alberta." *Western Canadian Journal of Anthropology* 7, no. 1: 15–52.

– 1982. *A Time for Burning*. Edmonton, AB: Boreal Institute for Northern Studies, University of Alberta.

– 1993. "Patterns of Indian Burning in California: Ecology and Ethnohistory." Original 1973. In *Before the Wilderness: Environmental Management by Native Californians*, ed. Thomas C. Blackburn and M. Kat Anderson, 55–116. Menlo Park, CA: Ballena.

Lewis, Henry T., and T.A. Ferguson. 1988. "Yards, Corridors and Mosaics: How to Burn Boreal Forest." *Human Ecology* 16, no. 1: 57–77. (Reprinted in Boyd 1999b.)

Lewis, Meriwether, and William Clark. 1805. "October 24th Thursday 1805." In *The Journals of Lewis and Clark*. http://www.infoplease.com/t/hist/lewis-clark-journal/day529.html (accessed 2 October 2013).

– 1806. "Monday May 5 1806." In *The Journals of the Lewis and Clark Expedition*. http://lewisandclarkjournals.unl.edu/read/?_xmlsrc=1806-05-05.xml&_xslsrc=LCstyles.xsl (accessed 29 September 2013).

Lillard, Charles, ed. 1987. *The Nootka: Scenes and Studies of Savage Life*, by Gilbert Malcolm Sproat. Original 1868. Winlaw, BC: Sono Nis.

Lloyd, T. Abe. 2011. "Cultivating the *Taki'lakw*, the Ethnoecology of *Tleksem*, Pacific Silverweed ...: Lessons from Clan Chief Kwaxsistalla of the Dzawada7enuxw Kwakwaka'wakw of Kingcome Inlet." MSC thesis, University of Victoria.

Loewen, Dawn C. 1998. "Ecological, Ethnobotanical, and Nutritional Aspects of Yellow Glacier Lily, *Erythronium grandiflorum Pursh* (Liliaceae) in Western Canada." MSC thesis, University of Victoria.

Loftin, John D. 1994. *Religion and Hopi Life in the Twentieth Century: A Comprehensive Study of How the Sacred and Secular Interact in Modern Hopi Culture*. New York, NY: Longitude Books.

Loh, Jonathan, and David Harmon. 2005. "A Global Index of Biocultural Diversity." *Ecological Indicators* 5: 231–41.

Lopatin, Ivan A. 1945. *Social Life and Religion of the Indians in Kitimat, British Columbia*. Los Angeles, CA: University of Southern California Press.

Losey, Robert J. N.d. *Archaeology of the Par-Tee Site, Northern Oregon Coast*. http://www.ualberta.ca/~rlosey/partee/index.htm (accessed 2 July 2013).

Losey, Robert J., Nancy Stenholm, Patty Whereat-Phillips, and Helen Vallianatos. 2003. "Exploring the Use of Red Elderberry (*Sambucus racemosa*) Fruit on the Southern Northwest Coast of North America." *Journal of Archaeological Science* 30: 695–707.

Ludwin, R.S., R. Dennis, D. Carver, A.D. McMillan, R. Losey, J. Clague, C. Jonientz-Trisler, J. Bowechop, J. Wray, and K. James. 2005. "Dating the 1700 Cascadia Earthquake: Great Coastal Earthquakes in Native Stories." *Seismological Research Letters* 76, no. 2: 140–7.

Ludwin, R.S., C.P. Thrush, K. James, D. Buerge, C. Jonientz-Trisler, J. Rasmussen, K. Troost, and A. de los Angeles. 2005. "Serpent Spirit-Power Stories along the Seattle Fault." *Seismological Research Letters* 76, no. 4: 426–31.

Lundy, Doris. 1976. "Styles of Coastal Rock Art." In *Indian Art Traditions of the Northwest Coast*, ed. Roy L. Carlson, 89–97. Burnaby, BC: Archaeology Press, Simon Fraser University.

Lutz, John S. 1995. "Preparing Eden: Aboriginal Land Use and European Settlement." Paper presented to the 1995 Meeting of the Canadian Historical Association, Université de Québec à Montréal, Montreal, Quebec.

– 2008. *Makuk: A New History of Aboriginal-White Relations*. Vancouver, BC: UBC Press.

Lutz, John S., and B. Neis, eds. 2008. *Making and Moving Knowledge: Interdisciplinary and Community-Based Research in a World on the Edge*. Montreal, QC, and Kingston, ON: McGill-Queen's University Press.

Lyons, Natasha. 2000. "Investigating Ancient Socioeconomy in Stó:lō Territory: A Palaeoethnobotanical Analysis of the Scowlitz Site, Southwestern B.C." MA thesis, Simon Fraser University.

Maa-nulth First Nations. 2008. *Maa-Nulth First Nations Final Agreement.* http://www.maanulth.ca/downloads/treaty/2010_maa-nulth_final_agreement_english.pdf (accessed 2 July 2013).
MacDermot, J.H. 1949. "Food and Medicinal Plants Used by the Indians of British Columbia." *Canadian Medical Association Journal* 61: 177–83.
MacDonald, George. 1976. "Prehistoric Art of the Northern Northwest Coast." In *Indian Art Traditions of the Northwest Coast*, ed. Roy L. Carlson, 99–129. Burnaby, BC: Archaeology Press, Simon Fraser University.
– 1983. *Haida Monumental Art.* Vancouver, BC: UBC Press.
Mack, Cheryl A. 1992. "In Pursuit of the Wild *Vaccinium*: Huckleberry Processing Sites in the Southern Washington Cascades." *Archaeology in Washington* 4: 3–16.
Mack, Cheryl A., and Richard H. McClure. 1996. "The Fruits of Their Labors: *Vaccinium* processing in the Washington Cascades." Paper presented at the 49th Annual Northwest Anthropological Conference, 14–16 March, Moscow, Idaho.
– 2002. "*Vaccinium* Processing in the Washington Cascades." *Journal of Ethnobiology* 22, no. 1: 35–60.
Mackie, A., D.W. Fedje, Q. Mackie, and M. Florian. 2003. "Early Holocene Wood Artifacts from the Kilgii Gwaay Site, Southern Haida Gwaii." Paper presented at the 10th International Wetland Archaeology Research Project Conference, South Puget Sound Community College, Olympia, Washington, 1–5 April.
Mackie, Kjerstin. 2008. "Analysis, Documentation and Conservation of the Kwäday Dän Ts'ìnchi Robe." Kwäday Dän Ts'ìnchi Symposium, April, Royal BC Museum, Victoria, British Columbia.
Mackie, Q., and S. Acheson. 2005. "The Graham Tradition." In *Haida Gwaii: Human History and Environment from the Time of Loon to the Time of the Iron People*, ed. Daryl W. Fedje and R.W. Mathewes, 274–302. Vancouver, BC: UBC Press.
Mackie, Q., Daryl Fedje, Duncan McLaren, Nicole Smith and Iain McKechnie. 2011. "Early Environments and Archaeology of Coastal British Columbia." In *Trekking the Shore: Changing Coastlines and the Antiquity of Coastal Settlement*, ed. N.F. Bicho, J.A. Haws, and L.G. Davis, 51–103. New York, NY: Springer.
Mackie, R.S. 1984. "Colonial Land, Indian Labour and Company Capital: The Economy of Vancouver Island, 1849–1858." MA thesis, University of Victoria.
– 1997. *Trading beyond the Mountains: The British Fur Trade on the Pacific, 1793–1843.* Vancouver, BC: UBC Press.
Mackin, Nancy. 2004. "Nisga'a Architecture and Landscape Histories: Ecological Wisdom and Community-Led Design." Phd diss., University of British Columbia.
– 2009. "Ancient and Future Orchard Gardens in Nisga'a and Tsimshian Tradition." Paper presented at the 32nd Annual Conference of the Society of Ethnobiology, New Orleans, Louisiana.
MacKinnon, A., J. Pojar, and R. Coupé, eds. 1992. *Plants of Northern British Columbia.* Vancouver, BC, and Edmonton, AB: Lone Pine.
Maclachlan, Morag, ed. 1998. *The Fort Langley Journals, 1827–30.* Vancouver, BC: UBC Press.
Macoun, J. ("Rambler"). 1920. "Wild Flowers of District." *Sidney and Islands Review and Saanich Gazette*, 24 June 1920, 2.
Maffi, Luisa, and Ellen Woodley. 2010. *Biocultural Diversity Conservation: A Global Sourcebook.* London, UK: Earthscan and International Union for Conservation of Nature.
Magne, Martin P.R. 2007. "Appendix 1: Notes on Cambium-Stripped Lodgepole Pine." In *Athapaskan Migrations: The Archaeology of Eagle Lake, British Columbia*, ed. R.G. Matson and Martin P.R. Magne, 68–71. Tuscon, AZ: University of Arizona Press.

Malhi, Ripan S., Katherine E. Breece, Beth A. Schultz Shook, Frederika A. Kaestle, James C. Chatters, Steven Hackenberger, and David Glenn Smith. 2004. "Patterns of mtDNA Diversity in Northwestern North America." *Human Biology* 76, no. 1: 33–5.
Manley, W.F. 2009. "Postglacial Flooding of the Bering Land Bridge: A Geospatial Animation." http://instaar.colorado.edu/QGISL/bering_land_bridge (accessed 3 July 2013).
Mann, Charles C. 2005. *1491: New Revelations of the Americas before Columbus*. New York, NY: Knopf.
Marles, Robin J. 1984. "The Ethnobotany of the Chipewyan of Northern Saskatchewan." MSc thesis, University of Saskatchewan.
Marles, Robin J., C. Clavelle, L. Monteleone, N. Tays, and D. Burns. 2000. *Aboriginal Plant Use in Canada's Northwest Boreal Forest*. Vancouver, BC: UBC Press.
Marr, K.I., G.A. Allen, and Richard J. Hebda. 2008. "Refugia in the Cordilleran Ice Sheet of Western North America: Chloroplast DNA Diversity in the Arctic-Alpine Plant *Oxyria digyna*." *Journal of Biogeography* 35: 1323–34.
Marsden, Susan, ed. 1992. *Na Amwaaltga Ts'msiyeen: The Tsimshian, Trade, and the Northwest Coast Economy*. Prince Rupert, BC: Museum of Northern British Columbia.
Marshall, Alan G. 1977. "Nez Perce Social Groups: An Ecological Interpretation." Phd diss., Washington State University.
Marshall, Yvonne. 1993. "A Political History of the Nuu-chah-nulth People: A Case Study of the Mowachaht and Muchalaht Tribes." Phd diss., Simon Fraser University.
Martin, Gary J. 1995. *Ethnobotany: A Methods Manual*. London, UK: Chapman and Hall.
Martindale, A.R.C. 1999. "The River of Mist: Cultural Change in the Tsimshian Past." Phd diss., University of Toronto.
– 2003. "A Hunter-Gatherer Paramount Chiefdom: Tsimshian Developments through the Contact Period." In *Emerging from the Mist: Studies in Northwest Coast Culture History*, ed. R.G. Matson, G. Coupland, and Q. Mackie, 12–50. Vancouver, BC: UBC Press.
Martindale, A.R.C., and Susan Marsden. 2003. "Defining the Middle Period (3500 BP to 1500 BP) in Tsimshian History through a Comparison of Archaeological and Oral Records." *BC Studies* 138: 13–50.
Martinez, Dennis. 2008. "Indigenous Integrity." *Resurgence and Ecologist* 250: 1. http://www.resurgence.org/magazine/article2632-Indigenous-Integrity.html (accessed 3 July 2013).
Maschner, Herbert D.G., Matthew W. Betts, Joseph Cornell, Jennifer A. Dunne, Bruce Finney, Nancy Huntly, James W. Jordan, Aaron A. King, Nicole Misarti, Katherine L. Reedy-Maschner, Roland Russell, Amber Tews, Spencer A. Wood, and Buck Benson. 2009. "Introduction to the Biocomplexity of Sanak Island, Western Gulf of Alaska." *Pacific Science* 63, no. 4: 673–709.
Mascia, Michael B., Peter Brosius, Tracy Dobson, Bruce Forbes, Leah Horowitz, Margaret A. McKean, and Nancy J. Turner. 2003. "Conservation and the Social Sciences." Editorial. *Conservation Biology* 17, no. 3: 1–2.
Mastrogiuseppe, Ron, and Steve Mark. 1992. "A 'New' Date for Mount Mazama's Climactic Eruption." *Nature Notes from Crater Lake* 23: 1. http://www.craterlakeinstitute.com/online-library/nature-notes/vol23-mazama-eruption-date.htm (accessed 3 July 2013).
Mather, K. 2000. "Bunchgrass and Beef: Bunchgrass Ecosystems and the Early Cattle Industry in the Thompson Okanagan." http://142.36.5.21/thomp-ok/article-LL/index-beef.html (accessed 13 October 2013).
Mathews, B. 2010. "Balanophagy in the Pacific Northwest: The Acorn-Leaching Pits at the Sunken Village Wetsite." *Journal of Northwest Anthropology* 43: 125–40.

Mathews, D., and P. Dady. 2008. "Douglas Fir Culturally Modified Trees: Some Initial Considerations." Paper presented at the Northwest Anthropological Conference, Victoria, British Columbia, 24 April.

Matson, R.G. 2003. "The Coast Salish House: Lessons from Shingle Point, Valdes Island, British Columbia." In *Emerging from the Mist: Studies in Northwest Coast Culture History*, ed. R.G. Matson, G. Coupland, and Q. Mackie, 76–104. Vancouver, BC: UBC Press.

Matson, R.G., and Diana Alexander. 1990. "Potato Mountain: The Archaeology of Alpine Root Procurement." Paper presented at the Canadian Archaeological Association Meeting, Whitehorse, Yukon, May.

Matson, R.G., and G. Coupland. 1995. *The Prehistory of the Northwest Coast*. San Diego, CA: Academic Press.

Matson, R.G., G. Coupland, and Q. Mackie, eds. 2003. *Emerging from the Mist: Studies in Northwest Coast Culture History*. Vancouver, BC: UBC Press.

Matson, R.G., and Martin P.R. Magne. 2007. *Athapaskan Migrations: The Archaeology of Eagle Lake, British Columbia*. Tuscon, AZ: University of Arizona Press.

Maud, Ralph, ed. 1978a. *The Salish People: The Local Contribution of Charles Hill-Tout*. Vol. 1, *The Thompson and the Okanagan*. Vancouver, BC: Talonbooks.

– ed. 1978b. *The Salish people: The Local Contribution of Charles Hill-Tout*. Vol. 2, *The Squamish and the Lillooet*. Vancouver, BC: Talonbooks.

– ed. 1978c. *The Salish People: The Local Contribution of Charles Hill-Tout*. Vol. 3, *The Mainland Halkomelem*. Vancouver, BC: Talonbooks.

– ed. 1978d. *The Salish People: The Local Contribution of Charles Hill-Tout*. Vol. 4, *The Sechelt and the South-Eastern Tribes of Vancouver Island*. Vancouver, BC: Talonbooks.

Mauss, M. 1990. *The Gift: The Form and Reason for Exchange in Archaic Societies*. Trans. W.D. Halls. French original 1923–24. London, UK: Routledge.

Mauzé, M. 1998. "Northwest Coast Trees: From Metaphors in Culture to Symbols for Culture." In *The Social Life of Trees: Anthropological Perspectives on Tree Symbolism*, ed. L. Rival, 233–51. Oxford, UK: Berg.

Mayne, R.C. 1862. *Four Years in British Columbia and Vancouver Island: An Account of Their Forests, Rivers, Coasts, Gold Fields, and Resources for Colonization*. London, UK: John Murray. http://www.archive.org/stream/fouryearsinbritioomayn/fouryearsinbritioomayn_djvu.txt (accessed 3 July 2013).

McCune, Letitia M. 1999. "Antioxidants in Canadian Boreal Forest Indigenous Medicinal Plant Treatments in Relation to Non-insulin Dependent Diabetes Mellitus." Phd diss., McGill University.

McCune, Letitia M., and Timothy Johns. 2002. "Antioxidant Activity in Medicinal Plants Associated with the Symptoms of Diabetes Mellitus Used by the Indigenous Peoples of the North American Boreal Forest." *Journal of Ethnopharmacology* 82: 197–205.

McCutcheon, Allison R. 1996. "Ethnopharmacology of Western North American Plants with Special Focus on the Genus *Artemisia L.*" Phd diss., University of British Columbia.

McCutcheon, Allison R., S.M. Ellis, R.E.W. Hancock, and G.H.N. Towers. 1992. "Antibiotic Screening of Medicinal Plants of British Columbian Native Peoples." *Journal of Ethnopharmacology* 37: 213–23.

McDonald, James Andrew. 2003. *People of the Robin: The Tsimshian of Kitsumkalum: A Resource Book for the Kitsumkalum Education Committee and the Coast Mountain School District 82*. Terrace, BC: Coast Mountain School District and Alberta ACADRE Network.

– 2005. "Cultivating in the Northwest: Early Accounts of Tsimshian Horticulture." In *"Keeping It Living": Traditions of Plant Use and Cultivation on the Northwest Coast of North*

America, ed. Douglas Deur and Nancy J. Turner, 240–73. Seattle, WA: University of Washington Press; Vancouver, BC: UBC Press.

McGarry, Ronald C., and Pamela McGarry. 1999. "Please Pass the Strychnine: The Art of Victorian Pharmacy." *JAMC* 161, no. 12: 1556–8. http://www.cmaj.ca/cgi/reprint/161/12/1556 (accessed 3 July 2013).

McIlwraith, T.F. 1948. *The Bella Coola Indians.* 2 vols. Toronto, ON: University of Toronto Press.

McKechnie, Iain. 2007. "Investigating the Complexities of Sustainable Fishing at a Prehistoric Village on Western Vancouver Island, British Columbia, Canada." *Journal for Nature Conservation* 15: 208–22.

McKenna-McBride Commission. 1913–16. Unpublished evidence, exhibits, applications, and reports of the McKenna-McBride Commission. Victoria, BC: British Columbia Archives.

McKibben, Bill. 2007. "Introduction and After Growth." In *Deep Economy: The Wealth of Communities and the Durable Future*, 1–45. New York, NY: Times Books.

McMillan, Alan D. 1999. *Since the Time of the Transformers: The Ancient Heritage of the Nuu-chah-nulth, Ditidaht, and Makah.* Vancouver, BC: UBC Press.

– 2003. "Reviewing the Wakashan Migration Hypothesis." In *Emerging from the Mist: Studies in Northwest Coast Culture History*, ed. R.G. Matson, G. Coupland, and Q. Mackie, 244–60. Vancouver, BC: UBC Press.

McMillan, Alan D., and Ian Hutchinson. 2002. "When the Mountain Dwarfs Danced: Aboriginal Traditions of Paleoseismic Events along the Cascadia Subduction Zone of Western North America." *Ethnohistory* 49, no. 1: 41–6.

McNeary, Stephen. 1974. "The Traditional Economic and Social Life of the Niska of British Columbia." Unpublished final report. Ethnographic Field Research Contract Programme, National Museum of Man, Ottawa, Ontario.

– 1976. "'Where Fire Came Down': Social and Economic Life of the Niska." Phd diss., Bryn Mawr College.

Meidinger, Del, and Jim Pojar, eds. 1991. *Ecosystems of British Columbia.* Victoria, BC: British Columbia Ministry of Forests.

Meilleur, Brien A. 1979. "Speculations on the Diffusion of *Nicotiana quadrivalvis* Pursh to the Queen Charlotte Islands and Adjacent Alaskan Mainland." *Syesis* 12: 101–4.

Meilleur, Brien A., Eugene S. Hunn, and Rachel L. Cox. 1990. "*Lomatium dissectum* (Apiaceae): Multi-purpose Plant of the Pacific Northwest." *Journal of Ethnobiology* 10, no. 1: 1–20.

Mellott, Carla Rae. 2010. "Contemporary Perspectives on the Practical, Ethical, and Ritual Aspects of the *Tsìnlhqút'ín Súnt'ìny* (*Claytonia lanceolata*) Harvest on Tsinuzch'ed ('Potato Mountain'), British Columbia." MSC thesis, University of Victoria.

Meltzer, David J. 2009. *First Peoples in a New World: Colonizing Ice Age America.* Berkeley, CA: University of California Press.

Menzies, Charles R. 2006a. "Ecological Knowledge, Subsistence, and Livelihood Practices: The Case of the Pine Mushroom Harvest in Northwestern British Columbia." In *Traditional Ecological Knowledge and Natural Resource Management*, ed. Charles R. Menzies, 99–116. Lincoln, NE: University of Nebraska Press.

– ed. 2006b. *Traditional Ecological Knowledge and Natural Resource Management.* Lincoln, NE: University of Nebraska Press.

Merriwether, D. Andrew. 2006. "Mitochondrial DNA." In *Handbook of North American Indians*, vol. 3, *Environment, Origins, and Population*, ed. D.H. Ubelaker, 817–30. Washington, DC: Smithsonian Institution.

Mierendorf, Robert R. 1999. "Precontact Use of Tundra Zones of the Northern Cascade Range of Washington and British Columbia." *Archaeology in Washington* 7: 3–23.
Millennium Ecosystem Assessment. 2003. *Ecosystems and Human Well-Being: Current State and Trends.* Vol. 1. Washington, DC: Island.
Miller, Andrew M. 2010. "Living with Boreal Forest Fires: Anishinaabe Perspectives on Disturbance and Collaborative Forestry Planning, Pikangikum First Nation, Northwestern Ontario." Phd diss., University of Manitoba.
Miller, Andrew M., Iain J. Davidson-Hunt, and M. O'Flaherty. 2008. *Pikangikum First Nation and Forest Fire Management: Old and New Knowledge of Fire.* Edmonton, AB: Sustainable Forest Management Network, University of Alberta.
Miller, Jay. 1999. *Lushootseed Culture and the Shamanic Odyssey: An Anchored Radiance.* Lincoln, NE: University of Nebraska Press.
Minnis, Paul E., ed. 2004. *People and Plants in Ancient Western North America.* Washington, DC: Smithsonian Books.
Minnis, Paul E., and Wayne J. Elisens, eds. 2000. *Biodiversity and Native America.* Norman, OK: University of Oklahoma Press.
Minore, D. 1972. *The Wild Huckleberries of Oregon and Washington: A Dwindling Resource.* Portland, OR: Pacific Northwest Forest and Range Experiment Station, United States Department of Agriculture.
Mitchell, Donald H., and Leland Donald. 1988. "Archaeology and the Study of Northwest Coast Economies." In *Prehistoric Economies of the Northwest Coast*, ed. B.L. Isaac, 293–351. Greenwich, CT: JAI Press.
Moerman, Daniel E. 1996. "An Analysis of the Food Plants and Drug Plants of Native North America." *Journal of Ethnopharmacology* 52, 1: 1–22.
– 2003. *Native American Ethnobotany: A Database of Foods, Drugs, Dyes and Fibers of Native American Peoples Derived from Plants.* http://herb.umd.umich.edu (accessed 6 July 2013).
– 2009. *Native American Medicinal Plants: An Ethnobotanical Dictionary*. Portland, OR: Timber.
Montler, Timothy. 1991. *Saanich, North Straits Salish Classified Word List.* Ottawa, ON: Canadian Museum of Civilization.
Moran, Bridget. 1988. *Stoney Creek Woman, Sai'k'iz Ts'eke: The Story of Mary John.* Vancouver, BC: Tillicum Library.
Morice, Father Adrien-Gabriel. 1893. *Notes Archaeological, Industrial, and Sociological on the Western Denes, with an Ethnographic Sketch of the Same.* Toronto: Copp, Clark.
Morrison, Sandra L. 1997. "Household Archaeology at the Scowlitz Site, Fraser Valley, BC." MA thesis, University of British Columbia.
Moss, Madonna L. 2005. "Tlingit Horticulture: An Indigenous or Introduced Development?" In *"Keeping It Living": Traditions of Plant Use and Cultivation on the Northwest Coast of North America*, ed. Douglas Deur and Nancy J. Turner, 274–95. Seattle, WA: University of Washington Press; Vancouver, BC: UBC Press.
– 2008a. "Islands Coming out of Concealment: Travelling to Haida Gwaii on the Northwest Coast of North America." *Journal of Island and Coastal Archaeology* 3: 35–53.
– 2008b. "Outer Coast Maritime Adaptations in Southern Southeast Alaska: Tlingit or Haida?" *Arctic Anthropology* 45, no. 1: 41–60.
– 2011. *Northwest Coast: Archaeology as Deep History*. Washington, DC: Society for American Archaeology Press.
Moss, Madonna L., and J.M. Erlandson. 1992. "Forts, Refuge Rocks, and Defensive Sites: The Antiquity of Warfare along the North Pacific Coast of North America." *Arctic Anthropology* 29, no. 2: 73–90.

Moss, Madonna L., Dorothy M. Peteet, and Cathy Whitlock. 2007. "Mid-Holocene Culture and Climate on the Northwest Coast of North America." In *Climate Change and Cultural Dynamics: A Global Perspective on Mid-Holocene Transitions*, ed. David G. Anderson, Kirk A. Maasch, and Daniel H. Sandweiss, 491–529. San Diego, CA: Elsevier and Academic Press.

Moss, Madonna L., et al. 2004. *Archaeological Investigation of Cape Addington Rockshelter: Human Occupation of the Rugged Seacoast on the Outer Prince of Wales Archipelago, Alaska.* Eugene, OR: Museum of Natural History and Department of Anthropology, University of Oregon.

Mourning Dove (Christine Quintasket). 1933. "How Coyote Happened to Make the Black Moss Food." In *Coyote Stories*, 119–25. Caldwell, ID: Caxton.

Moziño, José Mariano. 1970. *Noticias de Nutka: An Account of Nootka Sound in 1792.* Ed. and trans. Iris H. Wilson Engstrand. Seattle, WA: University of Washington Press.

Mudie, Petra J., Sheila Greer, Judith Brakel, James H. Dickson, Clara Schinkel, Ruth Peterson-Welsh, Margaret Stevens, Nancy J. Turner, M. Shadow, and Rosalie Washington. 2005. "Forensic Palynology and Ethnobotany of *Salicornia* Species (Chenopodiaceae) in Northwest Canada and Alaska." *Canadian Journal of Botany* 83, no. 1: 111–23.

Museum of Anthropology, University of British Columbia. 1994. *First Peoples of British Columbia.* Map. Vancouver, BC: Museum of Anthropology, University of British Columbia.

Nabhan, Gary P. 1997. *Cultures of Habitat: On Nature, Culture and Story*. Washington, DC: Counterpoint.

– 2006. *Renewing Salmon Nation's Food Traditions.* Corvallis, OR: Oregon State University Press.

Nabhan, Gary P., and Ashley Rood. 2004. *Renewing America's Food Traditions: Bringing Cultural and Culinary Mainstays from the Past into the New Millennium.* Flagstaff, AZ: RAFT Project and Center for Sustainable Environments, Northern Arizona University.

Nang King.aay 'uwans (James Young). 2005. "Taadl, Nang Kilslass, and Haida." In *Haida Gwaii: Human History and Environment from the Time of Loon to the Time of the Iron People*, ed. Daryl W. Fedje and R.W. Mathewes, 140–4. Vancouver, BC: UBC Press.

Natcher, D.C. 2004. "Implications of Fire Policy on Native Land Use in the Yukon Flats, Alaska." *Human Ecology* 24, no. 4: 421–41.

Natcher, D.C., and C.G. Hickey. 2003. "Putting the Community Back into Community-Based Resource Management: A Criteria and Indicators Approach to Sustainability." *Human Organization* 61, no. 4: 350–63.

National Park Service, United States Department of the Interior. 2004. "Glacier Bay National Park and Preserve." http://www.nps.gov/akso/akarc/cr_glba.htm (accessed 13 July 2013).

Nazarea, Virginia D., ed. 1999. *Ethnoecology: Situated Knowledge/Located Lives.* Tuscon, AZ: University of Arizona Press.

Nelson, Jenny. 1990. "Memories." In *Yakoun: River of Life*, ed. Frank Collison, 9. Haida Gwaii, BC: Council of the Haida Nation, Massett.

Nelson, Richard K. 1983. *Make Prayers to the Raven: A Koyukon View of the Northern Forest.* Chicago, IL: University of Chicago Press.

Neufeld, David, and Frank Norris. 1996. "The Coastal Tlingit: Trading Trail to the Interior." In *Chilkoot Trail: Heritage Route to the Klondike*, 22–48. Madeira Park, BC: Harbour.

Newcombe, Charles F. 1897–1916. "Unpublished Field Notes on Northwest Coast Ethnography." Microfilm. Victoria, BC: Britsh Columbia Archives and Royal BC Museum.

– ed. 1923. *Menzies' Journal of Vancouver's Voyage April to October, 1792*. Victoria, BC: William H. Cullin.

Newton, R., and M. Moss. 1984. *The Subsistence Lifeway of the Tlingit People: Excerpts of Oral Interviews.* Juneau, AK: USDA Forest Service, Alaska Region.
Niblack, Albert P. 1890. *The Coast Indians of Southern Alaska and Northern British Columbia.* Washington, DC: Government Printing Office.
Nicholas, George P. 1998. "Wetlands and Hunter-Gatherer Land Use in North America." In *Hidden Dimensions: The Cultural Significance of Wetland Archaeology*, ed. Kathryn Bernick, 31–46. Vancouver, BC: UBC Press.
– ed. 2010. *Being and Becoming Indigenous Archaeologists.* Walnut Creek, CA: Left Coast. http://www.lcoastpress.com/book.php?id=277 (accessed 6 July 2013).
Nicholas, George P., Nancy Jules Bonneau, and Leisl Westfall. Forthcoming. "Long-Term Secwepemc Plant Use: Initial Results of Archaeological and Archaeobotanical Investigations in the Interior Plateau, British Columbia." In *Secwepemc People and Plants: Research Papers in Shuswap Ethnobotany*, ed. Marianne B. Ignace, Nancy J. Turner, and Sandra L. Peacock. Published by the Society of Ethnobiology in the monograph series Contributions in Ethnobiology.
Nicholas, George P., and A. Tryon. 1999. *Archaeological Investigations at EeRb 140 and 144: Interim Report.* Kamloops, BC: Simon Fraser University.
Nicholas, George P., Michèle Wollstonecroft, and Gladys Baptiste. 1997. *Long-Term Secwepemc Plant Use: Initial Results of Archaeobotanical Investigations in the Interior Plateau.* Report prepared for the Social Sciences and Humanities Research Council of Canada and the Kamloops Indian Band. Kamloops, BC: Simon Fraser University.
Nicols, Johanna. 2008. "Language Spread Rates and Prehistoric American Migration Routes." *Current Anthropology* 49, no. 6: 1109–17.
Nigh, Ronald, Steffan Igor Ayora Diaz, Mauricio Bellon, Brent Berlin, Elois Ann Berlin, Stephen B. Brush, Francisco Chapela, Les Field, Edward F. Fischer, Virginia Nazarea, and Claudia Weimann. 2002. "Maya Medicine in the Biological Gaze: Bioprospecting Research as Herbal Fetishism." *Current Anthropology* 43: 451–77.
Nisgha [Nisg̲a'a] School District. 1986. *Haniimag̲oonisgum Algax̲hl Nisg̲a'a: Nisgha Phrase Dictionary.* New Aiyansh, BC: School District No. 92 (Nisgha).
NNABA (Northwest Native American Basketweavers Association). 2009. http://www.nnaba.org (accessed 6 July 2013).
Nolan, Justin, and Nancy J. Turner. 2011. "Ethnobotany, the Study of People-Plant Relationships." In *Textbook for Ethnobiology*, ed. E.N. Anderson et al., 133–48. Hoboken, NJ: Wiley-Blackwell.
Nord, Celia. 2009. "Ceremony and Symbolism: Women's Imagery of the Canadian Plateau." Unpublished ms. Golder and Associates and Department of Archaeology, Simon Fraser University.
Norton, Helen H. 1979a. "The Association between Anthropogenic Prairies and Important Food Plants in Western Washington." *Northwest Anthropological Research Notes* 13, no. 2: 175–200.
– 1979b. "Evidence for Bracken Fern as a Food for Aboriginal Peoples of Western Washington." *Economic Botany* 33, no. 4: 384–96.
– 1981. "Plant Use in Kaigani Haida Culture: Correction of an Ethnohistorical Oversight." *Economic Botany* 35, no. 4: 434–49.
– 1985. "Women and Resources of the Northwest Coast: Documentation from the 18th and Early 19th Centuries." Phd diss., University of Washington.
Norton, Helen H., E.S. Hunn, C.S. Martinsen, and P.B. Keely. 1984. "Vegetable Food Products of the Foraging Economies of the Pacific Northwest." *Ecology of Food and Nutrition* 14: 219–28.

Nutritional Healing Safety Committee. 2009. *Traditional Foods of Vancouver Island First Nations: Final Report: Proceedings of Conference at the City of Nanaimo and Snuneymuxw Nation, April 17–18, 2009.* http://www.indigenousfoodsvi.ca/wp-content/uploads/2011/09/final-report–2009-tf-conference.pdf (accessed 6 July 2013).
Nuxalk Food and Nutrition Program. 1984. *Nuxalk Food and Nutrition Handbook.* Bella Coola, BC: Nuxalk Nation.
Oberg, K. 1973. *The Social Economy of the Tlingit Indians.* Seattle, WA: University of Washington Press.
O'Clair, Rita M., and Sandra C. Lindstrom. 2000. *North Pacific Seaweeds.* Auke Bay, AK: Plant Press.
O'Neill, Brian L., Thomas J. Connolly, and Dorothy E. Freidel. 2004. *A Holocene Geoarchaeological Record for the Upper Willamette Valley, Oregon: The Long Tom and Chalker Sites.* Eugene, OR: Museum of Natural History and Department of Anthropology, University of Oregon.
Oeggl, Klaus, Werner Kofler, Alexandra Schmidl, James H. Dickson, Eduard Egarter-Vigl, and Othmar Gaber. 2007. "The Reconstruction of the Last Itinerary of 'Ötzi,' the Neolithic Iceman, by Pollen Analyses from Sequentially Sampled Gut Extracts." *Quaternary Science Reviews* 26, nos 7–8: 853–61.
Olson, R.L. 1935. *Owikeno Field Notes.* Berkeley, CA: University Archives, Bancroft Library, University of California.
– 1936. *The Quinault Indians.* Seattle, WA: University of Washington Publications in Anthropology.
– 1954. "Social Life of the Owikeno Kwakiutl." *University of California Anthropological Records* 14, no. 3: 213–60.
Ommer, Rosemary E., and Nancy J. Turner. 2004. "Informal Rural Economies in History." *Labour/Le Travail: Journal of Canadian Labour Studies* 53: 127–57.
Ommer, Rosemary E., and Coasts Under Stress Research Team. 2007. *Coasts Under Stress: Restructuring and Social-Ecological Health.* Montreal, QC, and Kingston, ON: McGill-Queen's University Press.
Ormsby, Margaret, ed. 1976. *A Pioneer Gentlewoman in British Columbia: The Recollections of Susan Allison.* Vancouver, BC: UBC Press.
Ortiz, Alfonso. 1973. "Look to the Mountaintop." In *Essays on Reflection*, ed. E.G. Ward, 88–104. Boston, MA: Houghton-Mifflin.
Ostapkowicz, Joanna, Dana Lepofsky, Rich Schulting, and Albert S. McHalsie. 2002. "The Use of Cattail (*Typha latifolia* L.) Down as a Sacred Substance by the Interior and Coast Salish." *Journal of Ethnobiology* 21, no. 2: 77–90.
Östlund, L., L. Ahlberg, O. Zackrisson, I. Bergman, and S. Arno. 2009. "Bark-Peeling, Food Stress and Tree Spirits: The Use of Pine Inner Bark for Food in Scandinavia and North America." In *Indigenous Resource Management: Past, Present and Future*, ed. Dana Lepofsky. Special issue, *Journal of Ethnobiology* 29, no. 2: 94–112.
Ostrom, Elinor. 1990. *Governing the Commons: The Evolution of Institutions for Collective Action.* New York, NY: Cambridge University Press.
Overfield, Theresa, William W. Epstein, and Larry A. Gaudioso. 1980. "Eskimo Uses of *Artemisia tilesii* (Compositae)." *Economic Botany* 34, no. 2: 97–100.
Palmer, Andie Diane. 2005. *Maps of Experience: The Anchoring of Land to Story in Secwépemc Discourse.* Toronto, ON: University of Toronto Press.
Palmer, Gary. 1975a. "Cultural Ecology in the Canadian Plateau: Pre-contact to the Early Contact Period in the Territory of the Southern Shuswap Indians of British Columbia." *Northwest Anthropological Research Notes* 9, no. 2: 199–245.

– 1975b. "Shuswap Indian Ethnobotany." *Syesis* 8: 29–81.
Palmer, Gary, M. Dale Kinkade, and Nancy J. Turner. 2003. "The Grammar of Snchitsu'umshtsn (Coeur d'Alene) Plant Names, with Comparative Notes on Other Interior Salish Languages." *Journal of Ethnobiology* 23, no. 1: 65–100.
Pandey, Deep Narayan. 1998. *Ethnoforestry: Local Knowledge for Sustainable Forestry and Livelihood Security*. Udaipur and New Delhi, India: Himanshu.
Parish, Roberta, Ray Coupé, and Dennis Lloyd, eds. 1996. *Plants of the Southern Interior, British Columbia*. Vancouver, BC, and Edmonton, AB: Lone Pine.
Parks Canada. 2013. *Gwaii Haanas National Park Reserve, National Marine Protected Area Reserve, and Haida Heritage Site*. http://www.pc.gc.ca/pn-np/bc/gwaiihaanas/index.aspx (accessed 7 July 2013).
Parlee, Brenda, and Fikret Berkes. 2006. "Indigenous Knowledge of Ecological Variability and Commons Management: A Case Study on Berry Harvesting from Northern Canada." *Human Ecology* 34: 515–28.
Parlee, Brenda, Fikret Berkes, and Teetl'it Gwich'in Renewable Resources Council. 2005. "Health of the Land, Health of the People: A Case Study on Gwich'in Berry Harvesting in Northern Canada." *EcoHealth* 2: 127–37.
Parrish, C.C., Nancy J. Turner, and S. Solberg, eds. 2007. *Resetting the Kitchen Table: Food Security, Culture, Health and Resilience in Coastal Communities*. New York, NY: Nova Science.
Paul, Frances. 1991. *Spruce Root Basketry of the Alaska Tlingit*. 1944. Reprint, Sitka, AK: Sheldon Jackson Museum.
Peacock, Sandra L. 1998. "Putting Down Roots: The Emergence of Wild Plant Food Production on the Canadian Plateau." Phd diss., University of Victoria.
– 2008. "From Complex to Simple: Balsamroot, Inulin, and the Chemistry of Traditional Interior Salish Pit-Cooking Technology." *Botany* 86: 116–28.
Peacock, Sandra L., and Dana Lepofsky. 2009. "Paper on Hat Creek Earth Ovens from Ethnobiology Meetings." Paper presented at the 32nd Annual Conference of the Society of Ethnobiology, New Orleans, Louisiana.
Peacock, Sandra L., and Nancy J. Turner. 2000. "'Just Like a Garden': Traditional Plant Resource Management and Biodiversity Conservation on the British Columbia Plateau." In *Biodiversity and Native North America*, ed. Paul E. Minnis and Wayne J. Elisens, 133–79. Norman, OK: University of Oklahoma Press.
Pearsall, Deborah M., and Christine A. Hastorf. 2011. "Reconstructing Past Life-Ways with Plants II: Human-Environment and Human-Human Interactions." In *Ethnobiology*, ed. E.N. Anderson et al., 173–87. Hoboken, NJ: Wiley-Blackwell.
Pellatt, M.G., and R.W. Mathewes. 1997. "Holocene Tree Line and Climate Change on the Queen Charlotte Islands, Canada." *Quaternary Research* 48: 88–9.
Perego, Uga A., N. Angerhofer, M. Pala, A. Olivieri, H. Lancioni, B.H. Kashani, V. Carossa, J.E. Ekins, A. Gómez-Carballa, and G. Huber. 2010. *The Initial Peopling of the Americas: A Growing Number of Founding Mitochondrial Genomes from Beringia*. Cold Spring Harbor, NY: Cold Spring Harbor Laboratory Press and Genome Research CSH Press. http://genome.cshlp.org/content/20/9/1174 (accessed 7 July 2013).
Pielou, E.C. 1991. *After the Ice Age: The Return of Life to Glaciated North America*. Chicago, IL: University of Chicago Press.
Pieroni, Andrea. 1999. "Toxic Plants as Food Plants in the Traditional Uses of the Eastern Apuan Alps Region, North-West Tuscany, Italy." In *Il Cibo e Il Corpo: Dal Cibo alla Cultura, Dalla Cultura al Cibo* [Food and Body: From Food to Culture, From Culture to Food], ed. Antonio Guerci, 267–72. Genova, Italy: Erga edizioni.
– 2001. "Evaluation of the Cultural Significance of Wild Food Botanicals Traditionally Consumed in Northwestern Tuscany, Italy." *Journal of Ethnobiology* 21, no. 1: 89–104.

Pierotti, R. 2008. *The World Wolf Made: Indigenous Knowledge, Ecology, and Evolutionary Biology.* Lincoln, NE: University of Nebraska Press.

Pierotti, R., and C.A. Annett. 1985. *Breeding Biology and Nest Site Selection in Western Gulls on Alcatraz Island.* Report to the National Park Service, United States Department of the Interior.

Pojar, Jim, and Andy Mackinnon, eds. 1994. *Plants of Coastal British Columbia Including Washington, Oregon and Alaska.* Vancouver, BC, and Edmonton, AB: Lone Pine; Seattle, WA: University of Washington Press.

Pokotylo, David, and Patricia Froese. 1983. "Archaeological Evidence for Prehistoric Root Gathering on the Southern Interior Plateau of British Columbia: A Case Study from the Upper Hat Creek Valley." *Canadian Journal of Archaeology* 7: 127–57.

Porsild, A.E. 1953. "Edible Plants of the Arctic." *Arctic* 6, no. 1: 15–34.

Port Simpson Curriculum Committee. 1983. *Port Simpson Foods: A Curriculum Development Project.* Prince Rupert, BC: People of Port Simpson and School District No. 52.

Poser, W.J., comp. 2008a. *Nak'albun/Dzinghubun Whut'en Bughuni Stuart/Trembleur Lake Carrier Lexicon.* Vanderhoof, BC: Yinka Dene Language Institute.

– comp. 2008b. *Saik'uz Whut'en Hubughunek/Stoney Creek Carrier Lexicon.* Vanderhoof, BC: Saik'uz First Nation.

Poser, W.J., Carrier-Sekani Tribal Council, Lheidli T'enneh, and University of British Columbia. 1999. "Carrier Monosyllabic Noun Stems." http://citeseerx.ist.psu.edu/viewdoc/download?doi=10.1.1.20.3971&rep=rep1&type=ps (accessed 7 July 2013).

Posey, Darrell A. 1990. "The Science of the Mebêngôkre." *Orion Nature Quarterly* 9, no. 3: 16–23.

Potter, Ben A. 2010. "Archaeological Patterning in Northeast Asia and Northwest North America: An Examination of the Dene-Yeniseian Hypothesis." In *The Dene-Yeniseian Connection*, ed. James Kari and Ben A. Potter. Special issue, *Anthropological Papers of the University of Alaska* 5, ns: 138–67.

Powell, James V., and Fred Woodruff. 1973. "Appendix 1: Additions to the Quileute Entries." In *Ethnobotany of Western Washington*, by Erna Gunther, 51–2. Seattle, WA: University of Washington Press.

Powell, Jay. 1991. *Our World – Our Ways: T'aat'aaqsapa Cultural Dictionary.* Port Alberni, BC: Nuu-chah-nulth Tribal Council.

Powell, Joseph F., and Jerome C. Rose. 2004. "Report on the Osteological Assessment of the 'Kennewick Man' Skeleton (CENWW.97.Kennewick)." In *Kennewick Man*, ed. F.P. McManamon. Washington, DC: Archeology Program, National Park Service, United States Department of the Interior. http://www.cr.nps.gov/archeology/kennewick/powell_rose.htm (accessed 7 July 2013).

Prentiss, Anna Marie, Natasha Lyons, Lucille E. Harris, Melisse R.P. Burns, and Terrence M. Godin. 2007. "The Emergence of Status Inequality in Intermediate Scale Societies: A Demographic and Socio-Economic History of the Keatley Creek Site, British Columbia." *Journal of Anthropological Archaeology* 26: 299–327.

Prentiss, W.C., J.C. Chatters, M. Lenert, D.S. Clarke, and R.C. O'Boyle. 2005. "The Archaeology of the Plateau of Northwestern North America during the Late Prehistoric Period (3500–200 B.P.): Evolution of Hunting and Gathering Societies." *Journal of World Prehistory* 19: 47–118.

Prescott-Allen, C., and R. Prescott-Allen. 1986. *The First Resource: Wild Species in the North American Economy.* New Haven, CT: Yale University Press.

Pretty, Jules, Bill Adams, Fikret Berkes, Simone de Ferreira de Athayde, Nigel Dudley, Eugene Hunn, Luisa Maffi, Kay Milton, David Rapport, Paul Robbins, Eleanor Sterling,

Sue Stolton, Anna Tsing, Erin Vintinner, and Sarah Pilgrim. 2009. "The Intersections of Biological Diversity: Towards Integration." *Conservation and Society* 7, no. 2: 100–12.

Price, Margo L., Patricia M. Samford, and Vincas P. Steponaitis, eds. 2001. "Lesson 2.7: Experimental Archaeology: Making Cordage." In *Intrigue of the Past: North Carolina's First Peoples: A Teacher's Activity Guide for Fourth through Eighth Grades.* Chapel Hill, NC: Research Laboratories of Archaeology, University of North Carolina. http://rla.unc.edu/lessons/lesson/l207/l207.htm (accessed 7 July 2013).

Prieto, J.M., M.C. Recio, R.M. Giner, S. Máñez and J.L. Ríos. 2003. "Pharmacological Approach to the Pro- and Anti-inflammatory Effects of *Ranunculus sceleratus* L." *Journal of Ethnopharmacology* 89: 131–7.

Proctor, Katherine Y. 2013. "Renewing Central Coast Salish Camas (*Camassia leichtlinii* (Baker) Wats., *C. quamash* (Pursh) Greene; Liliaceaee) Traditions through Access to Protected Areas: An Ethnoecological Inquiry." MSC thesis, University of Victoria.

Pringle, H. 2002. "Out of the Ice: Who Was the Ancient Traveler Discovered in an Alpine Glacier?" *Canadian Geographic* (July–August): 57–64.

Pritchard, Allan. 1990. "Letters of a Victorian Naval Officer: Edmund Verney in British Columbia, 1862–65." *BC Studies* 86: 28–56.

Prober, Suzanne M., Michael H. O'Connor, and Fiona J. Walsh. 2011. "Australian Aboriginal Peoples' Seasonal Knowledge: A Potential Basis for Shared Understanding in Environmental Management." *Ecology and Society* 16, no. 2. http://www.ecologyandsociety.org/vol16/iss2/art12 (accessed 7 July 2013).

Prouty, G.L., Brian L. O'Neill, and Thomas J. Connolly. 1999. "Aboriginal Plant Use at the Chalker and Long Tom Sites." In *A Holocene Geoarchaeological Record for the Upper Willamette Valley, Oregon: The Long Tom and Chalker Sites*, ed. Brian L. O'Neill, Thomas J. Connolly, and Dorothy E. Friede, 249–60. Eugene, OR: Museum of Natural History and Department of Anthropology, University of Oregon.

Pukonen, Jen C. 2008. "The *l'aayaʕas* Project: Revitalizing Traditional Nuu-chah-nulth Root Gardens in Ahousaht, British Columbia." MSC thesis, University of Victoria.

Pyne, Stephen J. 1995. *World Fire: The Culture of Fire on Earth.* New York, NY: Holt.

– 2001. *Fire: A Brief History.* Seattle, WA: University of Washington Press.

Quilt, Charlie. 1982. *Lhin Nits'en Nanayidash.* Story recorded by Ivor Myers and transcribed by June Williams. Translated into English by Bella Alphonse in December 2003. Cited in L.R. Smith 2008b.

R. v. Bartleman. 1984. Between Her Majesty the Queen, Respondent, and Joseph D. Bartleman, Appellant, BC. J. # 1760. (Cited in Claxton 2003.)

R. v. White and Bob. 1964. 50 D.L.R. (2d) 613 (also reported: 52 W.W.R. 193). (Cited in Claxton 2003.)

Ramirez, Carlos R. 2007. "Ethnobotany and the Loss of Traditional Knowledge in the 21st Century." Editorial. *Ethnobotany Research and Applications* 5: 245–7.

Rath, John C. 1981. *A Practical Heiltsuk-English Dictionary with a Grammatical Introduction.* Vols 1–2. Ottawa, ON: National Museum of Man.

Ray, Vern F. 1932. *The Sanpoil and Nespelem: Salish Peoples of Northeastern Washington.* Seattle, WA: University of Washington Publications in Anthropology.

– 1938. *Lower Chinook Ethnographic Notes.* Seattle, WA: University of Washington Publications in Anthropology.

– 1939. *Cultural Relations in the Plateau of Northwestern America.* Los Angeles, CA: Frederick Hodge Anniversary Publication Fund, Southwest Museum.

Reagan, Albert B. 1917. "Archaeological Notes on Western Washington and Adjacent British Columbia." *Proceedings of the California Academy of Sciences* 7, 4th ser., no. 1: 1–32.

– 1934. "Plants Used by the Hoh and the Quileute Indians." *Transactions of the Kansas Academy of Science* 37: 55–70.
Recalma-Clutesi, Kim, dir. 2005. *Ancient Sea Gardens: Mystery of the Pacific Northwest.* Documentary film. Toronto, ON: Aquaculture Pictures.
– dir. 2007. *Smoke from His Fire.* Documentary film. Aboriginal Peoples Television Network and British Columbia's Knowledge Network.
Reed, A.W., and Ross Calman. 2008. *Earth, Ocean, Sky: Traditional Maori Stories.* New Zealand: Raupo and Penguin.
Reedy-Maschner, Katherine L., and Herbert D.G. Maschner, eds. 2012. *Sanak Island, Alaska: A Cultural and Natural History.* Presented to Sanak Corporation, the Pauloff Harbor Tribe, and the Descendants of the People of Sanak Island. Pocatello, ID: Idaho Museum of Natural History, Idaho State University.
Reich, David, Nick Patterson, Desmond Campbell, et al. 2012. "Reconstructing Native American Population History." *Nature* 488, no. 7411: 370–4.
Reid, Martine J., and Daisy Sewid-Smith. 2004. *Paddling to Where I Stand: Agnes Alfred, Quiqwasutinuxw Noblewoman.* Vancouver, BC: UBC Press.
Reinhart, Karl J., Donny L. Hamilton, and Richard H. Hevly. 1991. "Use of Pollen Concentration in Paleopharmacology: Coprolite Evidence of Medicinal Plants." *Journal of Ethnobiology* 11, no. 1: 117–32.
Renfrew, Jane M. 1973. *Palaeoethnobotany: The Prehistoric Food Plants of the Near East and Europe.* New York, NY: Columbia University Press.
Reyhner, Jon, ed. 1997. *Teaching Indigenous Languages.* Proceedings of the Fourth Annual Stabilizing Indigenous Languages Symposium. Flagstaff, AZ: Northern Arizona University. http://jan.ucc.nau.edu/~jar/TILpi.pdf (accessed 7 July 2013).
Richards, Michael P., Sheila Greer, Lorna T. Corr, Owen Beattie, Alexander Mackie, Richard P. Evershed, Al von Finster, and John Southon. 2007. "Radiocarbon Dating and Dietary Stable Isotope Analysis of Kwäday Dän Ts'ìnchí." *American Antiquity* 72, no. 4: 719–34.
Richards, Rebecca Templin. 1997. "What the Natives Know: Wild Mushrooms and Forest Health." *Journal of Forestry* (September): 5–10.
Richards, Rebecca Templin, and Susan J. Alexander. 2006. *A Social History of Wild Huckleberry Harvesting in the Pacific Northwest.* Portland, OR: United States Department of Agriculture, Forest Service, Pacific Northwest Research Station. http://www.treesearch.fs.fed.us/pubs/22152 (accessed 19 October 2013).
Richards, Thomas, and Michael Rousseau. 1987. *Late Prehistoric Cultural Horizons on the Canadian Plateau.* Burnaby, BC: Department of Archaeology, Simon Fraser University.
Richardson, Allan. 1982. "The Control of Productive Resources on the Northwest Coast of North America." In *Resource Managers: North American and Australian Hunter-Gatherers,* ed. N.M. Williams and Eugene S. Hunn, 93–112. Washington, DC: American Association for the Advancement of Science Selected Symposia Series.
Rindos, David. 1984. *The Origins of Agriculture: An Evolutionary Perspective.* New York, NY: Academic Press.
Ritch-Krc, Elizabeth M. "A Selection of Traditional Medicinal Remedies Important to Contemporary Carrier People in Their Treatment of Disease." MSC thesis, University of British Columbia.
Ritch-Krc, Elizabeth M., S. Thomas, Nancy J. Turner, and G.N. Towers. 1996. "Carrier Herbal Medicine: Traditional and Contemporary Plant Use." *Journal of Ethnopharmacology* 52, no. 2: 85–94.
Ritland, K., L. D. Meagher, D. G. W. Edwards, and Y. A. El-Kassaby. 2005. "Isozyme Variation and the Conservation Genetics of Garry Oak." *Canadian Journal of Botany* 83: 1478–87.

Robinson, Harry, and Wendy Wickwire, eds. 1994. *Nature Power: In the Spirit of an Okanagan Storyteller*. Vancouver, BC: Douglas and McIntyre.
Robinson, J.P., S.A. Harris, and B.E. Juniper. 2001. "Taxonomy of the Genus *Malus* Mill. (Rosaceae) with Emphasis on the Cultivated Apple, *Malus domestica* Borkh." *Plant Systematics and Evolution* 226, nos 1–2: 35–58.
Routson, Kanin J. 2012. "*Malus* Diversity in Wild and Agricultural Systems." Phd diss., University of Arizona.
Routson, Kanin J., G.M. Volk, C.M. Richards, S.E. Smith, G.P. Nabhan, and V. Wyllie de Echeverria. 2012. "Genetic Variation and Distribution of Pacific Crabapple." *Journal of the American Society of Horticultural Science* 137, no. 5: 325–32.
Ross, John Alan. 2011. *The Spokan Indians*. Spokane, WA: Michael J. Ross.
Ross, P.S. 2000. "Marine Mammals as Sentinels in Ecological Risk Assessment." *Human and Ecological Risk Assessment* 6: 29–46.
Rousseau, Michael, R. Muir, D. Alexander, J. Breffitt, S. Woods, K. Berry, and T. Van Gaalen. 1991. *Results of the 1989 Archaeological Investigations Conducted in the Oregon Jack Creek Locality, Thompson River Region, South Central British Columbia*. Victoria, BC: British Columbia Archaeology Branch and British Columbia Ministry of Tourism, Recreation and Culture.
Royal Commission on Indian Affairs for the Province of British Columbia. 1914. *Meeting with the Tsah-Waw-Tineuch [Tsawataineuk] Band or Tribe of Indians at Alert Bay, on Thursday, the 4th Day of June, 1914*. Copy in possession of M. Willie, Kingcome Inlet.
Rubin, Daniel. 1999. "Sm'algyax Language Renewal: Prospects and Options." In *Revitalizing Indigenous Languages*, ed. Jon Reyhner et al., 17–32. Flagstaff, AZ: Northern Arizona University. http://jan.ucc.nau.edu/~jar/RIL_3.html (accessed 8 July 2013).
Russell, Priscilla. 1991a. *English Bay and Port Graham Alutiiq Plantlore: An Ethnobotany of the Peoples of English Bay and Port Graham, Kenai Peninsula, Alaska*. Homer, AK: Pratt Museum; Anchorage, AK: Chugach Heritage Foundation and Alaska Native Plant Society.
– 1991b. *Ninilchik Plantlore: An Ethnobotany of the Ninilchik Dena'ina, Aleut, and Russian Peoples*. Ninilchik, AK: Ninilchik Traditional Council.
Russell Kari, Priscilla. 1987. *Tanaina Plantlore: Dena'ina K'et'una: An Ethnobotany of the Dena'ina Indians of Southcentral Alaska*. 1977. Reprint, Anchorage, AK: Alaska Regional Office, National Park Service, United States Department of the Interior; Fairbanks, AK: Alaska Native Language Centre, University of Alaska Fairbanks.
Ryan, J., M.D. Rabesca, D. Robie, and M. Johnson. 1994. *Traditional Dene Medicine: Part I (Report) and Part II (Database)*. Lac La Martre, NT: Dene Medicine Project.
Ryan, Teresa L. 2000. "Defining Cultural Resources: Science, Law, and Resource Management for Sweetgrass *Schoenoplectus pungens* in Grays Harbor, Washington." MSC thesis, Central Washington University.
Salick, Jan, and Nanci Ross, eds. 2009. *Indigenous Peoples and Climate Change*. Special issue, *Global Environmental Change* 19.
Salmón, Enrique. 2000. "Kincentric Ecology: Indigenous Perceptions of the Human-Nature Relationship." *Ecological Applications* 10, no. 5: 1327–32.
– 2012. *Eating the Landscape: American Indian Stories of Food, Identity, and Resilience*. Tuscon, AZ: University of Arizona Press.
Sandgathe, D.M., and Brian Hayden. 2003. "Did Neanderthals Eat Inner Bark?" *Antiquity* 77, no. 298: 709–18.
Sapir, Edward. 1913–14. Unpublished notes, notebook 17, page 23a, December 1913 to January 1914, roll 23. Microfilm from American Philosophical Society. Victoria, BC: British Columbia Archives.

– 1916. *Time Perspective in Aboriginal American Culture: A Study in Method.* Ottawa, ON: Government Printing Bureau.
– 1923. "The Phonetics of Haida." *International Journal of American Linguistics* 2: 143–158.
Satterfield, Terre, Leslie Robertson, Nancy J. Turner, and Anton Pitts. 2012. *Being Gitka'a'ata: A Baseline Report on Gitka'a'ata Ways of Life, a Statement of Cultural Impacts Posed by the Northern Gateway Pipeline, and a Critique of the ENGP Assessment Regarding Cultural Impacts.* Submission to the Joint Review Panel for Review of the Enbridge Northern Gateway Project.
Sauer, Hans. 1992. "Towards a Linguistic Description and Classification of the OE Plant Names." In *Words, Texts and Manuscripts: Studies in Anglo-Saxon Culture: Presented to Helmut Gneuss on the Occasion of his Sixty-Fifth Birthday*, ed. Michael Korhammer, 381–408. Cambridge, UK: D.S. Brewer.
Schaeffer, Claude E. 1940. "The Subsistence Quest of the Kutenai: A Study of the Interaction of Culture and Environment." Phd diss., University of Pennsylvania.
Schaepe, David M. 2001. "Village Arrangements and Settlement Patterns." In *A Stó:lō–Coast Salish Historical Atlas*, ed. Keith T. Carlson, 36–9. Chilliwack, BC: Stó:lō Heritage Trust; Vancouver, BC: Douglas and McIntyre.
Schaepe, David M., Albert (Sonny) McHalsie, Keith Thor Carlson, and Patricia Ormerod. 2001. "Changing Households, Changing Houses." In *A Stó:lō–Coast Salish Historical Atlas*, ed. Keith T. Carlson, 40–9. Chilliwack, BC: Sto:lo Heritage Trust; Vancouver, BC: Douglas and McIntyre.
Schlick, M.D. 1994. *Columbia River Basketry: Gift of the Ancestors, Gift of the Earth.* Seattle, WA: University of Washington Press.
Schlumbaum, Angela, Marrie Tensen, and Viviane Jaenicke-Després. 2008. "Ancient Plant DNA in Archaeobotany." *Vegetation History Archaeobotany* 17: 233–44.
Scholefield, E.O.S. 1914. *British Columbia: From the Earliest Times to the Present.* Vols 1–2. Vancouver, BC: S.J. Clarke.
Schroeder, K.B., David G. Smith, Mattias Jacobsson, Michael H. Crawford, Theodore Schurr, Simina Boca, Donald F. Conrad, Jonathan Pritchard, Raul Tito, Ripan Malhi, Ludmilla Osipova, Larissa Tarskaia, Sergey Zhadanov, and Jeffrey D. Wall. 2009. "Haplotypic Background of a Private Allele at High Frequency in the Americas." *Molecular Biology and Evolution* 26, no. 5: 995–1016.
Schulting, Rick J. 1994. "An Investigation of Mortuary Variability and Socioeconomic Status Differentiation on the Northwest Plateau." MA thesis, Simon Fraser University.
Scientific Panel for Sustainable Forest Practices in Clayoquot Sound. 1995. *First Nations' Perspectives on Forest Practices in Clayoquot Sound.* Report 3 (with appendices). Victoria, BC: Cortex Consulting and Government of British Columbia.
Scott, G. Richard, and Dennis O'Rourke. 2010. "Genes across Beringia: A Physical Anthropological Perspective on the Dene-Yeniseian Hypothesis." In *The Dene-Yeniseian Connection*, ed. James Kari and Ben A. Potter. Special issue, *Anthropological Papers of the University of Alaska* 5, ns: 119–37.
Senos, René, Frank Lake, Nancy J. Turner, and Dennis Martinez. 2006. "Traditional Ecological Knowledge and Restoration Practice in the Pacific Northwest." In *Encyclopedia for Restoration of Pacific Northwest Ecosystems*, ed. Dean Apostol, 393–426. Washington, DC: Island.
Sewid-Smith, Daisy (Mayanilth), and Clan Chief Adam Dick (Kwaxsistalla), interviewed by Nancy J. Turner. 1998. "The Sacred Cedar Tree of the Kwakwaka'wakw People." In *Stars Above, Earth Below: Native Americans and Nature*, ed. M. Bol, 189–209. Pittsburg, PA: Carnegie Museum of Natural History.

Shebitz, Daniela J., Sarah H. Reichard, and Peter W. Dunwiddy. 2009. "Ecological and Cultural Significance of Burning Beargrass Habitat on the Olympic Peninsula, Washington." *Ecological Restoration* 27, no. 3: 306–19.
Sheldon Jackson College. 1976. *English Tlingit Dictionary: Nouns*. Sitka, AK: Bilingual Program, Sheldon Jackson College.
Shuswap Nation Tribal Council. 1989. *The Shuswap: "One People with One Mind, One Heart and One Spirit."* Kamloops, BC: Shuswap Nation Tribal Council.
Smelcer, John E., James Kari, and Millie Buck. 2011. *Ahtna Noun Dictionary and Pronunciation Guide*. 2nd ed. Binghamton, NY: J.E. Smelcer. http://www.johnsmelcer.com/resources/Ahtna+Dictionary.pdf (accessed 28 September 2013).
Smith, Bruce D. 1989. "Origins of Agriculture in Eastern North America." *Science* 246, no. 4937: 1566–71.
– 2005. "Low-Level Food Production and the Northwest Coast." In *"Keeping It Living": Traditions of Plant Use and Cultivation on the Northwest Coast of North America*, ed. Douglas Deur and Nancy J. Turner, 37–66. Seattle, WA: University of Washington Press; Vancouver, BC: UBC Press.
– 2009. "Core Conceptual Flaws in Human Behavioral Ecology." Addendum to "Initial Formation of an Indigenous Crop Complex in Eastern North America at 3800 B.P.," by Bruce D. Smith and R.I. Yarnell. *Proceedings of the National Academy of Sciences* 106: 6561–6.
Smith, C. Earle. 1968. "The New World Centers of Origin of Cultivated Plants and the Archaeological Evidence." *Economic Botany* 22, no. 3: 253–66.
Smith, E.A., and M. Wishnie. 2000. "Conservation and Subsistence in Small-Scale Societies." *Annual Review of Anthropology* 29: 493–524.
Smith, Harlan I. 1920–24. Unpublished papers, vol. 1, 63. Ottawa, ON: National Museum of Civilization.
– 1928. *Materia Medica of the Bella Coola and Neighbouring Tribes of British Columbia*. Ottawa, ON: National Museum of Canada.
– 1997. *Ethnobotany of the Gitksan Indians of British Columbia*. Ed. Brian D. Compton, B. Rigsby, and M-L. Tarpent. Hull, QC: Canadian Museum of Civilization.
Smith, Linda R. 2008a. "*Sès Ts'íqí Ghàghìndá* 'The Bear Who Married a Woman.'" In "*Súwh-t?'éghèdúdính*: The *Tsìnlhqút'ín Nímính* Spiritual Path," 33–54. MA thesis, University of Victoria.
– 2008b. "*Súwh-t?'éghèdúdính*: The *Tsìnlhqút'ín Nímính* Spiritual Path." MA thesis, University of Victoria.
Snyder, Lynn M., and Jennifer A. Leonard. 2011. "The Diversity and Origin of American Dogs." In *The Subsistence Economies of Indigenous North American Societies: A Handbook*, ed. Bruce D. Smith, 525–42. Washington, DC: Smithsonian Institution.
Snively, Gloria, and John Corsiglia. 2001. "Discovering Indigenous Science: Implications for Science Education." *Science Education* 85, no. 1: 6–34.
Solomon, Gilbert. 2006. As quoted in Nancy J. Turner, "Tsilhqot'in Expert Report, January 2004." Submitted in court testimony, Supreme Court of British Columbia, *Tsilhqot'in Nation v. British Columbia*, 2007 BCSC 1700.
Speller, Camilla F., Kjerstin Mackie, Alexander Mackie, and Dongya Y. Yang. Forthcoming. "Ancient DNA Analysis of the *Kwädąy Dän Ts'ìnchį* Robe and Sewn Bag." In *Kwädąy Dän Ts'ìnchį Long Ago Person Found*, ed. Alexander Mackie et al. Victoria, BC: Royal BC Museum.
Spier, Leslie, ed. 1938. *The Sinkaietk or Southern Okanagan of Washington*. Menasha, WI: George Banta.

Spier, Leslie, and Edward Sapir. 1930. *Wishram Ethnography*. Seattle, WA: University of Washington Publications in Anthropology.

Sproat, Gilbert Malcolm. 1987. *The Nootka: Scenes and Studies of Savage Life*. Ed. Charles Lillard. Original 1868. Winlaw, BC: Sono Nis.

Spurgeon, Terry. 2001. "Wapato *(Sagittaria latifolia)* in Katzie Traditional Territory, Pitt Meadows, British Columbia." MA thesis, Simon Fraser University.

Ståhlberg, Sabira, and Ingvar Svanberg. 2010. "Gathering Food from Rodent Nests in Siberia." *Journal of Ethnobiology* 30, no. 2: 184–202.

Steedman, E.V., ed. 1930. *The Ethnobotany of the Thompson Indians of British Columbia*. Based on field notes of James Teit. Washington, DC: Smithsonian Institution.

Steller, Georg Wilhelm. 1767. "Nachricht von dem ökonomischen Gebrauch des wilden Bärenklaus (*Sphondylium*) und einiger anderer Kräuter und Gewächse bey den Kamtschadalen." *Stralsundisches Magazin* 1, no. 1766: 411–34.

Stepp, J.R., F.S. Wyndham, and R.J. Zarger, eds. 2002. *Ethnobiology and Biocultural Diversity: Proceedings of the Seventh International Congress of Ethnobiology*. Athens, GA: University of Georgia Press.

Stevens, Michelle, and M. Kat Anderson. 2006. "Indian Hemp, *Apocynum cannabinum* L." In *USDA, NRCS Plant Guide*. http://plants.usda.gov/plantguide/pdf/cs_apca.pdf (accessed 10 July 2013).

Stewart, Hilary. 1977. *Indian Fishing: Early Methods on the Northwest Coast*. Seattle, WA: University of Washington Press.

– 1984. *Cedar: Tree of Life to the Northwest Coast Indians*. Seattle, WA: University of Washington Press; Vancouver, BC: Douglas and McIntyre.

Stone, Anne C. 2006. "Ancient DNA." In *Handbook of North American Indians*, vol. 3, *Environment, Origins, and Population*, ed. D.H. Ubelaker, 840–7. Washington, DC: Smithsonian Institution.

Strecker, Lisa. 2007. "Die Ethnobotanik der Kamtschatka-Halbinsel: Erfassung und Beschreibung des Datenmaterials in präsowjetischen, nicht-russischen und einigen rezenten Quellen." MA thesis, University of Hamburg.

Strong, Emory. 1959, *Stone Age on the Columbia River*. Portland, OR: Binford and Mort.

Stryd, Arnoud H. 1978. "A Review of the Recent Activities Undertaken by the Lillooet Archaeological Project." *The Midden* 12, no. 2: 5–20.

– 1997. *Culturally Modified Trees of British Columbia: A Handbook for the Identification and Recording of Culturally Modified Trees*. Victoria, BC: British Columbia Ministry of Forests.

– 2001. *Culturally Modified Trees of British Columbia: A Handbook to the Identification and Recording of Culturally Modified Trees – Version 2*. Rev. 2nd ed. Victoria, BC: British Columbia Ministry of Forests.

Stryd, Arnoud H., and M. Eldridge. 1993. "CMT Archaeology in British Columbia: The Meares Island Studies." *BC Studies* 99: 184–234.

Stryd, Arnoud H., and Michael Rousseau. 1996. "The Early Prehistory of the Mid-Fraser-Thompson River Area." In *Early Man Occupations in British Columbia*, ed. Roy L. Carlson and Luke Dalla Bona, 177–204. Vancouver, BC: UBC Press.

Summers, Ingrid. 1993–94. "Itelmen Plant Names and Plant Uses: Kamchatka Ethnographic-Ecological Expedition, 1993–94." Unpublished paper. http://www.facultysite.sinanewt.on-rev.com/html/PlantVystavka.htm (accessed 11 July 2013).

Supreme Court of British Columbia. 2007. "*Tsilhqot'in Nation v. British Columbia*, 2007 BCSC 1700: Reasons for Judgment." http://www.courts.gov.bc.ca/jdb-txt/sc/07/17/2007bcsc1700.pdf (accessed 11 July 2013).

Suryawanshi, D.G. 2006. "Like Paper: Birch Bark and Its Chemical Composition." *Restaurator* 27: 103–13.

Sutherland, J.E., S.C. Lindstrom, W.A. Nelson, J. Brodie, M.D. Lynch, M.S. Hwang, H.-G. Choi, M. Miyata, N. Kikuchi, M.C. Oliveira, T. Farr, C. Neefus, A. Mols-Mortensen, D. Milstein, and K.M. Müller. 2011. "A New Look at an Ancient Order: Generic Revision of the Bangiales (*Rhodophyta*)." *Journal of Phycology* 47, no. 5: 1131–51.

Suttles, Wayne P. 1951a. "The Early Diffusion of the Potato among the Coast Salish." *Southwestern Journal of Anthropology* 7, no. 3: 272–88.

– 1951b. "The Economic Life of the Coast Salish of Haro and Rosario Straits." Phd diss., University of Washington.

– 1955. *Katzie Ethnographic Notes.* Victoria, BC: Department of Education, British Columbia Provincial Museum.

– 1974. *Coast Salish and Western Washington Indians: The Economic Life of the Coast Salish of Haro and Rosario Straits.* New York, NY: Garland.

– 1976. "Productivity and Its Constraints: A Coast Salish Case." In *Indian Art Traditions of the Northwest Coast*, ed. Roy L. Carlson, 67–87. Burnaby, BC: Archaeology Press, Simon Fraser University.

– 1987a. "Affinal Ties, Subsistence, and Social Classes among the Coast Salish." In *Coast Salish Essays*, ed. Wayne P. Suttles, 15–25. Vancouver, BC: Talonbooks; Seattle, WA: University of Washington Press.

– ed. 1987b. *Coast Salish Essays.* Vancouver, BC: Talonbooks; Seattle, WA: University of Washington Press.

– 1987c. "Coping with Abundance: Subsistence on the Northwest Coast." In *Coast Salish Essays*, ed. Wayne P. Suttles, 45–63. Vancouver, BC: Talonbooks; Seattle, WA: University of Washington Press.

– 1990a. "Central Coast Salish." In *Handbook of North American Indians*, vol. 7, *Northwest Coast*, ed. Wayne P. Suttles, 453–75. Washington, DC: Smithsonian Institution.

– ed. 1990b. *Handbook of North American Indians.* Vol. 7, *Northwest Coast.* Washington, DC: Smithsonian Institution.

– 2005. "Coast Salish Resource Management: Incipient Agriculture?" In *"Keeping It Living": Traditions of Plant Use and Cultivation on the Northwest Coast of North America*, ed. Douglas Deur and Nancy J. Turner, 181–93. Seattle, WA: University of Washington Press; Vancouver, BC: UBC Press.

Suttles, Wayne P., and William Elmendorf. 1963. "Linguistic Evidence for Salish Prehistory." In *Proceedings of the 1962 Annual Spring Meeting of the American Ethnological Society, Seattle*, ed. Viola E. Garfield and W.L. Chafe, 41–52. Seattle WA: University of Washington Press.

Sutton, Mark Q., and E.N. Anderson. 2009. *An Introduction to Cultural Ecology.* 2nd ed. Lanham, MD: Rowman and Littlefield.

Sutton, Mark Q., Kristin D. Sobolik, and Jill K. Gardner. 2010. *Paleonutrition.* Tuscon, AZ: University of Arizona Press.

Swan, James. 1869. *The Indians of Cape Flattery, at the Entrance to the Strait of Fuca, Washington Territory.* Washington, DC: Smithsonian Institution.

– 1972. *The Northwest Coast, or Three Years in Washington Territory.* 1857. Reprint, Seattle, WA: University of Washington Press.

Swanton, John R. 1905. *Haida Texts and Myths – Skidegate Dialect.* Washington, DC: Smithsonian Institution.

– 1908. *Haida Texts – Masset Dialect.* New York, NY: American Museum of Natural History.

– 1909. *Tlingit Myths and Texts*. Washington, DC: Government Printing Office.
– 1913. "A Haida Food Plant [Beach Lupine, *Lupinus littoralis*]." *American Anthropologist* 15: 543–4.
Swetnam, T.W. 1984. "Peeled Ponderosa Pine Trees: A Record of Inner Bark Utilization by Native Americans." *Journal of Ethnobiology* 4, no. 2: 177–90.
Swoboda, Leo John. 1971. "Lillooet Phonology, Texts and Dictionary." MA thesis, University of British Columbia.
Teit, James A. 1896–1918. "Plant Names and Other Notes on Thompson [Nlaka'pamux] Indians." Unpublished field notes. Microfilm. American Philosophical Society, Boas Collection, S 1b.3, Philadelphia, Pennsylvania.
– 1898. *Traditions of the Thompson River Indians of British Columbia*. Boston, MA, and New York, NY: American Folklore Society.
– 1900. *The Thompson Indians of British Columbia*. New York, NY: American Museum of Natural History.
– 1906a. *The Lillooet Indians*. New York, NY: American Museum of Natural History.
– 1906b. "Notes on the Tahltan Indians of British Columbia." In *Boas Anniversary Volume: Anthropological Papers Written in Honor of Franz Boas*, by Berthold Laufer and H.A. Andrews, 337–49. New York, NY: American Museum of Natural History.
– 1909. *The Shuswap*. Includes "Appendix: Notes on the Chilcotin Indians," 759–89. New York, NY: American Museum of Natural History.
– 1912. *Mythology of the Thompson Indians*. Ed. Franz Boas. New York, NY: G.E. Stechert.
– 1917. "Okanagon Tales." In *Folk-Tales of Salishan and Sahaptin Tribes*, ed. Franz Boas, 65–93. Lancaster, PA: G.E. Stechert and Company.
– 1919. "Tahltan Tales." *Journal of American Folklore* 32, no. 124: 185–250.
– 1930. *The Salishan Tribes of the Western Plateaus*. Ed. Franz Boas. Washington, DC: Smithsonian Institution.
– 1956. "Field Notes on the Tahltan and Kaska Indians: 1912–1915." *Anthropologica* 3: 39–170.
Tennant, Paul. 1990. *Aboriginal People and Politics: The Indian Land Question in British Columbia, 1849–1989*. Vancouver, BC: UBC Press.
Tepper, Leslie H., ed. 1991. *The Bella Coola Valley: Harlan I. Smith's Fieldwork Photographs, 1920–1924*. Ottawa, ON: Canadian Museum of Civilization.
Thom, Brian. 2005. "Coast Salish Senses of Place: Dwelling, Meaning, Power, Property and Territory in the Coast Salish World." Phd diss., McGill University.
Thom, Brian, and Karen Fediuk. 2009. "Coast Salish Food Security in the 21st Century." Paper presented at the Sustaining Food Conference, Kwantlen Polytechnic University, Surrey, British Columbia, 20 March.
Thomas, D. 2000. "Face Painting from the Sapir Collection." In *HuupuKwanum.Tupaat: Nuu-chah-nulth Voices, Histories, Objects and Journeys*, ed. A.L. Hoover, 172–201. Victoria, BC: Royal BC Museum.
Thomas, Mary. 2001. *The Wisdom of Dr. Mary Thomas*. Ed. R.J. Hood, Nancy J. Turner, and J. Infanti. Victoria, BC: School of Environmental Studies, University of Victoria.
Thomas, Mary, Nancy J. Turner, and Ann Garibaldi. Forthcoming. "'Everything Is Deteriorating': Environmental and Cultural Loss in Secwepemc Territory." In *Secwepemc People and Plants: Research Papers in Shuswap Ethnobotany*, ed. Marianne B. Ignace, Nancy J. Turner, and Sandra L. Peacock. Published by the Society of Ethnobiology in the monograph series Contributions in Ethnobiology.
Thommasen H.V. 1995. "Medicinal Plants of the Nuxalk Natives of Bella Coola." *British Columbia Medical Journal* 37: 403–8.

Thompson, Judith C. (Edōsdi). 2004. "Gitga'at Plant Project: The Intergenerational Transmission of Traditional Ecological Knowledge Using School Science Curricula." MA thesis, University of Victoria.
– 2012. "*Hede Kehe' Hotzi' Kahidi* – 'Our Ancestors Are in Us': Strengthening Our Voices through Language Revitalization from a Tahltan Worldview." Phd diss., University of Victoria.
Thompson, Laurence C., and M. Dale Kinkade. 1990. "Languages." In *Handbook of North American Indians*, vol. 7, *Northwest Coast*, ed. Wayne P. Suttles, 30–51. Washington, DC: Smithsonian Institution.
Thompson, Laurence C., and M. Terry Thompson. [1975–80]. "Klallam Language Notes." Unpublished ms. In possession of the authors.
– 1996. *Thompson River Salish Dictionary*. Missoula, MT: University of Montana Press.
Thoms, Alston V. 1989. "The Northern Roots of Hunter-Gatherer Intensification: Camas and the Pacific Northwest." Phd diss., Washington State University.
– 2008a. "Ancient Savannah Roots of the Carbohydrate Revolution of South-Central North America." *Plains Anthropologist* 53, no. 205: 121–36.
– 2008b. "The Fire Stones Carry: Ethnographic Records and Archaeological Expectations for Hot-Rock Cookery in Western North America." *Journal of Anthropological Archaeology* 27: 443–60.
– 2009. "Rocks of Ages: Propagation of Hot-Rock Cookery in Western North America." *Journal of Archaeological Science* 36: 573–91.
Thoreau, Henry David. 1854. *Walden, or Life in the Woods*. Boston, MA: Ticknor and Fields.
Thornton, Thomas F. 1999. "*Tleikw aaní*, the 'Berried' Landscape: The Structure of Tlingit Edible Fruit Resources at Glacier Bay, Alaska." *Journal of Ethnobiology* 19, no. 1: 27–48.
– 2008. *Being and Place among the Tlingit*. Seattle, WA: University of Washington Press.
– 2011. "Workshop on 'Cultivation of Marinescapes on the Pacific Northwest Coast.'" 2nd International Marine Conservation Congress, Victoria, British Columbia, 13 May.
Thornton, Thomas F., with contributions by Deborah McBride, Saurabh Gupta, Carcross-Tagish First Nations, Chilkat Indian Village, Chilkoot Indian Association, and Skagway Traditional Council. 2004. *Klondike Gold Rush National Historical Park: Ethnographic Overview and Assessment*. Fairbanks, AK: Alaska Regional Office, National Park Service, United States Department of the Interior.
Thwaites, R.G. 1969. *Original Journals of the Lewis and Clark Expedition, 1804–1806*. 8 vols. 1904–05. Reprint, New York, NY: Arno.
Timbrook, Jan. 2007. *Chumash Ethnobotany: Plant Knowledge among the Chumash People of Southern California*. Berkeley, CA: Heyday Books; Santa Barbara, CA: Santa Barbara Museum of Natural History.
Timmers, Jan A. 1977. *A Classified English-Sechelt Word-List*. Lisse, Netherlands: Peter de Ridder.
Tirone, Susan, Blythe Sheppard, Nancy J. Turner, Lois Jackson, Anne Marshall, and Catherine Donovan. 2007. "Celebrating with Food: Food and Cultural Identity: Feasts, Ceremonies and Celebrations." In *Resetting the Kitchen Table: Food Security, Culture, Health and Resilience in Coastal Communities*, ed. C.C. Parrish, Nancy J. Turner, and S. Solberg, 145–60. New York, NY: Nova Science.
Tomimatsu, Hiroshi, Susan R. Kephart, and Mark Vellend. 2009. "Phylogeography of *Camassia quamash* in Western North America: Postglacial Colonization and Transport by Indigenous Peoples." *Molecular Ecology* 18: 3918–28.
Townsend, Lise. 2009. "Urban Watershed Health and Resilience, Evaluated through Land Use History and Eco-hydrology in Swan Lake Watershed (Saanich, B.C.)." MSC thesis, University of Victoria.

Trager, G. 1939. "'Cottonwood' = 'Tree': A Southwestern Linguistic Trait." *International Journal of American Linguistics* 9: 117–18.

Trigger, Bruce G. 2008. *A History of Archaeological Thought.* 2nd ed. Cambridge, UK: Cambridge University Press.

Troffe, Peter M., Camilla F. Speller, Al von Finster, and Dongya Y. Yang. Forthcoming. "Identification of Sockeye Salmon from the *Kwädąy Dän Ts'ìnchį* Site." In *Kwädąy Dän Ts'ìnchį Long Ago Person Found*, ed. Alexander Mackie et al. Victoria, BC: Royal BC Museum.

Trosper, Ronald L. 2002. "Northwest Coast Indigenous Institutions That Supported Resilience and Sustainability." *Ecological Economics* 41: 329–44.

– 2003. "Resilience in Pre-contact Pacific Northwest Social Ecological Systems." *Conservation Ecology* [now *Ecology and Society*] 7, no. 3. http://www.consecol.org/vol7/iss3/art6 (accessed 12 July 2013).

– 2009. *Resilience, Reciprocity and Ecological Economics: Northwest Coast Sustainability*. London, UK, and New York, NY: Routledge.

Trusler, Scott, and Leslie Main Johnson. 2008. "'Berry Patch' as a Kind of Place: The Ethnoecology of Black Huckleberry in Northwestern Canada." *Human Ecology* 36: 553–68.

Ts'msyen Sm'algya̲x Authority. 2001. *Sm'algyax Dictionary.* Copy in possession of Helen Clifton, Hartley Bay, British Columbia. Viewed January 2001.

Ts'msyen Sm'algya̲x Authority and Sm'algya̲x Committee of School District 52. 2011. *Sm'algyax Living Legacy Talking Dictionary*. http://web.unbc.ca/~smalgyax (accessed 12 July 2013).

Turner, Katherine L. 2010. "Establishing Community Owned Eco-cultural Tourism: A Case Study of Bioeconomic Development in Hartley Bay, British Columbia." MA thesis, Natural Resources Institute, University of Manitoba.

Turner, Katherine L., and Chris P. Bitonti. 2011. "Conservancies in British Columbia, Canada: Bringing Together Protected Areas and First Nations' Interests." *International Indigenous Policy Journal* 2, no. 2. http://ir.lib.uwo.ca/iipj/vol2/iss2/3 (accessed 12 July 2013).

Turner, Nancy J. 1973. "Ethnobotany of the Bella Coola Indians of British Columbia." *Syesis* 6: 193–220.

– 1974. *Plant Taxonomic Systems and Ethnobotany of Three Contemporary Indian Groups of the Pacific Northwest (Haida, Bella Coola, and Lillooet).* Victoria, BC: British Columbia Provincial Museum.

– 1977. "Economic Importance of Black Tree Lichen (*Bryoria fremontii*) to the Indians of Western North America." *Economic Botany* 31: 461–70.

– 1978. "Plants of the Nootka Sound Indians as Recorded by Captain Cook." *Sound Heritage* 7, no. 1: 78–87.

– 1981. "Indian Use of *Shepherdia canadensis*, Soapberry, in Western North America." *Davidsonia* 12, no. 1: 1–14.

– 1982. "Traditional Use of Devil's-Club (*Oplopanax horridus*; Araliaceae) by Native Peoples in Western North America." *Journal of Ethnobiology* 2, no. 1: 17–38.

– 1984. "Counter-irritant and Other Medicinal Uses of Plants in Ranunculaceae by Native Peoples in British Columbia and Neighbouring Areas." *Journal of Ethnopharmacology* 11: 181–201.

– 1987. "General Plant Categories in Thompson and Lillooet, Two Interior Salish Languages of British Columbia." *Journal of Ethnobiology* 7, no. 1: 55–82.

– 1988a. "Ethnobotany of Coniferous Trees in Thompson and Lillooet Interior Salish of British Columbia." *Economic Botany* 42, no. 2: 177–94.

– 1988b. "'The Importance of a Rose': Evaluating the Cultural Significance of Plants in Thompson and Lillooet Interior Salish." *American Anthropologist* 90, no. 2: 272–90.
– 1989. "'All Berries Have Relations': Midlevel Folk Plant Categories in Thompson and Lillooet Interior Salish." *Journal of Ethnobiology* 9, no. 1: 69–110.
– 1992a. "'Just When the Wild Roses Bloom': The Legacy of a Lillooet Basket Weaver." *TEK TALK: A Newsletter of Traditional Ecological Knowledge* 1, no. 2: 2–5.
– 1992b. "Plant Resources of the Stl'atl'imx (Fraser River Lillooet) People: A Window into the Past." In *A Complex Culture of the British Columbia Plateau: Traditional Stl'atl'imx Resource Use*, ed. Brian Hayden, 405–69. Vancouver, BC: UBC Press.
– 1994. "Ethnobotanical Harvest Experiments." In *Report on the 1993 Kitlope Science Fieldwork*, ed. Peter Schoonmaker and Erin L. Kellogg, 11–20. Portland, OR: Ecotrust Institute; Kitamaat, BC: Nanakila Institute.
– 1995. *Food Plants of Coastal First Peoples*. Vancouver, BC: UBC Press; Victoria, BC: Royal BC Museum.
– 1996. "'Dans une Hotte': L'importance de la vannerie das l'économie des peuples chasseurs-pêcheurs-cueilleurs du Nord-Ouest de l'Amérique du Nord" ['Into a Basket Carried on the Back': Importance of Basketry in Foraging-Hunting-Fishing Economies in Northwestern North America]. In *Contemporary Ecological Anthropology: Theories, Methods and Research Fields*, ed. C. Lévesque. Special issue, *Anthropologie et Sociétiés* 20, no. 3: 55–84.
– 1997a. *Food Plants of Interior First Peoples*. 1978. Rev. ed., Vancouver, BC: UBC Press; Victoria, BC: Royal BC Museum.
– 1997b. "'Le fruit de l'ours': Les rapports entre les plantes et les animaux das les langues et les cultures amérindiennes de la Côte-Ouest" ['The Bear's Own Berry': Ethnobotanical Knowledge as a Reflection of Plant-Animal Interrelationships in Northwestern North America]. In *Des Plantes et des Animaux: Visions et Pratiques Autochtones*, ed. Pierre Beaucage. Special issue, *Recherches amérindiennes au Québec* 27, nos 3–4: 31–48.
– 1998. *Plant Technology of British Columbia First Peoples*. Vancouver, BC: UBC Press; Victoria: BC: Royal BC Museum.
– 1999. "'Time to Burn': Traditional Use of Fire to Enhance Resource Production by Aboriginal Peoples in British Columbia." In *Indians, Fire, and the Land in the Pacific Northwest*, ed. Robert T. Boyd, 185–218. Corvallis: OR Oregon State University Press.
– 2001. "'Doing It Right': Issues and Practices of Sustainable Harvesting of Non-timber Forest Products Relating to First Peoples in British Columbia." *BC Journal of Ecosystems and Management* 1, no. 1. http://jem.forrex.org/index.php/jem/article/view/215 (accessed 12 July 2013).
– 2003a. "The Ethnobotany of 'Edible Seaweed' (*Porphyra abbottiae* and Related Species; Rhodophyta: Bangiales) and Its Use by First Nations on the Pacific Coast of Canada." *Canadian Journal of Botany* 81, no. 4: 283–93.
– 2003b. "'Passing on the News': Women's Work, Traditional Knowledge and Plant Resource Management in Indigenous Societies of Northwestern North America." In *Women and Plants: Case Studies on Gender Relations in Local Plant Genetic Resource Management*, ed. P.L. Howard, 133–49. London, UK: Zed Books.
– 2004a. *Plants of Haida Gwaii: X̲aadaa Gwaay guud gina k̲'aws (Skidegate), X̲aadaa Gwaayee guu giin k̲'aws (Massett)*. Winlaw, BC: Sono Nis.
– 2004b. "Tsilhqot'in Expert Report, January 2004." Submitted in court testimony, Supreme Court of British Columbia, *Tsilhqot'in Nation v. British Columbia*, 2007 BCSC 1700.
– 2005. *The Earth's Blanket: Traditional Teachings for Sustainable Living*. Vancouver, BC: Douglas and McIntyre; Seattle, WA: University of Washington Press.

– 2006a. "Lessons from the Grandmothers: Women's Roles in Traditional Botanical Knowledge and Wisdom in Northwestern North America." In *Conference Proceedings of the Fourth International Congress of Ethnobotany*, ed. Z. Fusün Ertug, 27–38. Istanbul, Turkey: Yeditepe University.
– 2006b. "'Those Women of Yesteryear': Women and Production of Edible Seaweed (*Porphyra abbottiae*) in Coastal British Columbia, Canada." In *Conference Proceedings of the Fourth International Congress of Ethnobotany*, ed. Z. Fusün Ertug, 499–505. Istanbul, Turkey: Yeditepe University.
– 2008. "Kinship Lessons of the Birch." *Resurgence* 250: 46–8.
– 2009. "Plantes aromatiques dans l'alimentation et la médecine des régions Nord-Ouest de l'Amérique du Nord" [Aromatic Plants Used in Alimentation and Medicine in the Northwestern Regions of North America]. Trans. Paul Goetz. *Phytothérapie* 7, no. 3: 135–46. http://www.springerlink.com/content/w78764g5nw411048 (accessed 12 July 2013).
– 2014. "The Food/Medicine/Poison Triangle: Implications for Traditional Ecological Knowledge Systems of Indigenous Peoples of British Columbia, Canada." In *Social-Ecological Diversity and Traditional Food Systems: Opportunities from the Biocultural World*, ed. Ranjay K. Singh, Nancy J. Turner, Victoria Reyes-Garcia, and Jules Pretty, 1–31. New Delhi, India: New India Publishing Agency.
Turner, Nancy J., Yilmaz Ari, Fikret Berkes, Iain Davidson-Hunt, Z. Fusun Ertug, and Andrew M. Miller. 2009. "Cultural Management of Living Trees: An International Perspective." In *Indigenous Resource Management: Past, Present and Future*, ed. Dana Lepofsky. Special issue, *Journal of Ethnobiology* 29, no. 2: 237–70.
Turner, Nancy J., and E.R. Atleo (Chief Umeek). 1998. "Pacific North American First Peoples and the Environment." In *Traditional and Modern Approaches to the Environment on the Pacific Rim: Tensions and Values*, ed. H. Coward, 105–24. Albany, NY: State University of New York Press.
Turner, Nancy J., and Marcus A.M. Bell. 1971. "The Ethnobotany of the Coast Salish Indians of Vancouver Island." *Economic Botany* 25, no. 1: 63–104; 25, no. 3: 335–9.
– 1973. "The Ethnobotany of the Southern Kwakiutl Indians of British Columbia." *Economic Botany* 27, no. 3: 257–310.
Turner, Nancy J., and Fikret Berkes. 2006. "Coming to Understanding: Developing Conservation through Incremental Learning." In *Developing Resource Management and Conservation*, ed. Fikret Berkes and Nancy J. Turner. Special issue, *Human Ecology* 34, no. 4: 495–513.
Turner, Nancy J., Randy Bouchard, and Dorothy I.D. Kennedy. 1980. *Ethnobotany of the Okanagan-Colville Indians of British Columbia and Washington*. Victoria, BC: British Columbia Provincial Museum.
Turner, Nancy J., Randy Bouchard, Dorothy I.D. Kennedy, and Jan Van Eijk. 1987. "Plant Knowledge of the Stl'atl'imx (Lillooet) People of British Columbia." Unpublished ms. In possession of Nancy J. Turner, School of Environmental Studies, University of Victoria.
Turner, Nancy J., and Cecil H. Brown. 2004. "Grass, Hay, and Weedy Growth: Utility and Semantics of Interior Salish Botanical Terms." In *Studies in Salish Linguistics in Honor of M. Dale Kinkade*, ed. Donna B. Gerdts and Lisa Matthewson, 410–28. Missoula, MT: University of Montana Press.
Turner, Nancy J., and Carla M.A. Burton. 2010. "Soapberry: Unique Northwestern Foaming Fruit." In *Festscrift for Thomas M. Hess on the Occasion of His Seventieth Birthday*, ed. David Beck, 278–305. Bellingham, WA: Whatcom Museum.
Turner, Nancy J., Carla M.A. Burton, and Jan van Eijk. 2013. "Plants in Language and Classification among BC First Peoples." In *Ethnobotany in British Columbia: Plants and People*

in a Changing World, ed. Nancy J. Turner and Dana Lepofsky. Special issue, *BC Studies* 179: 135–58.

Turner, Nancy J., and Helen Clifton. 2006. "'The Forest and the Seaweed': Gitga'at Seaweed, Traditional Ecological Knowledge and Community Survival." In *Traditional Ecological Knowledge and Natural Resource Management*, ed. Charles R. Menzies, 65–86. Lincoln, NE: University of Nebraska Press.

– 2009. "'It's So Different Today': Climate Change and Indigenous Lifeways in British Columbia, Canada." In *Indigenous Peoples and Climate Change*, ed. Jan Salick and Nanci Ross. Special issue, *Global Environmental Change* 19: 180–90.

Turner, Nancy J., and W. Cocksedge. 2001. "Aboriginal Use of Non-timber Forest Products in Northwestern North America: Applications and Issues." *Journal of Sustainable Forestry* 16, nos 3–4: 31–57.

Turner, Nancy J., Iain J. Davidson-Hunt, and M. O'Flaherty. 2003. "'Living on the Edge': Ecological and Cultural Edges as Sources of Diversity for Social-Ecological Resilience." *Human Ecology* 31, no. 3: 439–63.

Turner, Nancy J., and Alison Davis. 1993. "'When Everything Was Scarce': The Role of Plants as Famine Foods in Northwestern North America." *Journal of Ethnobiology* 13, no. 2: 1–28.

Turner, Nancy J., Douglas Deur, and Dana Lepofsky. 2013. "Plant Management Systems of British Columbia's First Peoples." In *Ethnobotany in British Columbia: Plants and People in a Changing World*, ed. Nancy J. Turner and Dana Lepofsky. Special issue, *BC Studies* 179: 107–33.

Turner, Nancy J., Douglas Deur, and Carla Rae Mellott. 2011. "'Up on the Mountain': Ethnobotanical Importance of Montane Ecosystems in Pacific Coastal North America." *Journal of Ethnobiology* 31, no. 1: 4–43.

Turner, Nancy J., and Barbara S. Efrat. 1982. *Ethnobotany of the Hesquiat Indians of Vancouver Island*. Victoria BC: British Columbia Provincial Museum.

Turner, Nancy J., Leslie Main Johnson Gottesfeld, Harriet V. Kuhnlein, and Adolf Ceska. 1992. "Edible Wood Fern Rootstocks of Western North America: Solving an Ethnobotanical Puzzle." *Journal of Ethnobiology* 12, no. 1: 1–34.

Turner, Nancy J., R. Gregory, C. Brooks, L. Failing, and T. Satterfield. 2008. "From Invisibility to Transparency: Identifying the Implications [of Invisible Losses to First Nations Communities]." *Ecology and Society* 13, no. 2. http://www.ecologyandsociety.org/vol13/iss2/art7 (accessed 12 July 2013).

Turner, Nancy J., and Fiona Hamersley-Chambers. 2006. "Northwest Coast and Plateau Plants." In *Handbook of North American Indians*, vol. 3, *Environment, Origins, and Population*, ed. D.H. Ubelaker et al., 251–62. Washington, DC: Smithsonian Institution.

Turner, Nancy J., and Richard J. Hebda. 1990. "Contemporary Use of Bark for Medicine by Two Salishan Native Elders of Southeast Vancouver Island." *Journal of Ethnopharmacology* 29, no. 1: 59–72.

– 2012. *Saanich Ethnobotany: Culturally Important Plants of the W̱SÁNEĆ (Saanich) People of Southern Vancouver Island*. Victoria: Royal BC Museum.

Turner, Nancy J., Marianne B. Ignace, and Brian D. Compton. 1998. "Secwepemc (Shuswap) Tree Names: Key to the Past?" In *Salish Languages and Linguistics: Theoretical and Descriptive Pergspectives*, ed. Ewa Czaykowska-Higgins and M. Dale Kinkade, 387–417. Berlin, Germany: Mouton de Gruyter; New York, NY: Hawthorn.

Turner, Nancy J., Marianne B. Ignace, and Ronald E. Ignace. 2000. "Traditional Ecological Knowledge and Wisdom of Aboriginal Peoples in British Columbia." *Ecological Applications* 10, no. 10: 181–93.

Turner, Nancy J., Marianne B. Ignace, and Dawn C. Loewen, eds. Forthcoming. *Plants of the Secwepemc People*. Kamloops, BC: Shuswap Nation Tribal Council; Winlaw, BC: Sono Nis.

Turner, Nancy J., and Harriet V. Kuhnlein. 1982. "Two Important 'Root' Foods of the Northwest Coast Indians: Springbank Clover (*Trifolium wormskioldii*) and Pacific Silverweed (*Potentilla anserina* ssp. *pacifica*)." *Economic Botany* 36, no. 4: 411–32.

– 1983. "Camas (*Camassia* spp.) and Riceroot (*Fritillaria* spp.): Two Liliaceous 'Root' Foods of the Northwest Coast Indians." *Ecology of Food and Nutrition* 13: 119–219.

Turner, Nancy J., Harriet V. Kuhnlein, and Keith N. Egger. 1985. "The Cottonwood Mushroom (*Tricholoma populinum* Lange): A Food Resource of the Interior Salish Indian Peoples of British Columbia." *Canadian Journal of Botany* 65: 921–7.

Turner, Nancy J., and Meagan Kus. 2002. *Carrying a Culture: First Nations Basketry*. Exhibit guide. White Rock, BC: White Rock Museum and Archives.

Turner, Nancy J., and Dana Lepofsky, eds. 2013. *Ethnobotany in British Columbia: Plants and People in a Changing World*. Special issue, *BC Studies* 179.

Turner, Nancy J., and Dawn C. Loewen. 1998. "The Original 'Free Trade': Exchange of Botanical Products and Associated Plant Knowledge in Northwestern North America." *Anthropologica* 40: 49–70.

Turner, Nancy J., Łukasz Łuczaj, Paola Migliorini, Andrea Pieroni, Angelo L. Drean, Linda E. Sacchetti, and Maurizio G. Paoletti. 2011. "Edible and Tended Wild Plants, Traditional Ecological Knowledge and Agroecology." In *Towards a More Sustainable Agriculture*. Special issue, *Critical Reviews in Plant Sciences* 30, nos 1–2: 198–225. http://www.tandfonline.com/doi/abs/10.1080/07352689.2011.554492#.UeBxEsRza00 (accessed 12 July 2013).

Turner, Nancy J., Anne Marshall, Judith C. Thompson (Edōsdi), Robin June Hood, Cameron Hill, and Eva-Ann Hill. 2008. "'Ebb and Flow': Transmitting Environmental Knowledge in a Contemporary Aboriginal Community." In *Making and Moving Knowledge: Interdisciplinary and Community-Based Research in a World on the Edge*, ed. John S. Lutz and B. Neis, 45–63. Montreal, QC, and Kingston, ON: McGill-Queen's University Press.

Turner, Nancy J., and Rosemary Ommer. 2004. "'Our Food Is Our Medicine': Traditional Plant Foods, Traditional Ecological Knowledge and Health in a Changing Environment." *First Nations Nutrition and Health Conference Proceedings*, ed. A. Wong, 22–39. Vancouver, BC: Arbokem.

Turner, Nancy J., C.C. Parrish, and S. Solberg. 2007. "Introduction: Food Security in Coastal Communities in the Context of Restructuring: Coasts under Stress." In *Resetting the Kitchen Table: Food Security, Culture, Health and Resilience in Coastal Communities*, ed. C.C. Parrish, Nancy J. Turner, and S. Solberg, 1–12. New York, NY: Nova Science.

Turner, Nancy J., and Sandra Peacock. 2005. "'Solving the Perennial Paradox': Ethnobotanical Evidence for Plant Resource Management on the Northwest Coast." In *"Keeping It Living": Traditions of Plant Use and Cultivation on the Northwest Coast of North America*, ed. Douglas Deur and Nancy J. Turner, 101–50. Seattle, WA: University of Washington Press; Vancouver, BC: UBC Press.

Turner, Nancy J., Mark Plotkin, and Harriet V. Kuhnlein. 2013. "Integrity of Indigenous Food Systems: Global Environmental Challenges." In *Indigenous Peoples' Food Systems for Health: Interventions for Health Promotion and Policy*, ed. Harriet V. Kuhnlein, Bill Erasmus, Dina Spigelski, and Barbara Burlingame, 23–38. Montreal, QC: Centre for Indigenous Peoples' Nutrition and Environment, McGill University; Rome, Italy: Food and Agriculture Organization of the United Nations.

Turner, Nancy J., Kim Recalma-Clutesi, and Douglas Deur. 2013. *Back to the Clam Gardens*. Ecotrust Blog. http://www.ecotrust.org/indigenousaffairs/Back-to-the-Clam-Gardens.pdf (accessed 12 July 2013).

Turner, Nancy J., Colleen Robinson, Gideon Robinson, and Belle Eaton. 2012. "'To Feed All the People': Lucille Clifton's Fall Feasts for the Gitga'at Community of Hartley Bay, British Columbia." In *Explorations in Ethnobiology: The Legacy of Amadeo Rea*, ed. Marsha Quinlan and Dana Lepofsky, 322–63. Denton, TX: Society of Ethnobiology, Department of Geography, University of North Texas.

Turner, Nancy J., Leslie Saxon, and Judith C. Thompson (Edōsdi). 1997. Plants collected on Tahltan trip, 22–28 May. Collections and identifications by Nancy J. Turner. Samples at North Coast Tribal Council Education Centre.

Turner, Nancy J., Robin Y. Smith, and James T. Jones. 2005. "'A Fine Line between Two Nations': Ownership Patterns for Plant Resources among Northwest Coast Indigenous Peoples – Implications for Plant Conservation and Management." In *"Keeping It Living": Traditions of Plant Use and Cultivation on the Northwest Coast of North America*, ed. Douglas Deur and Nancy J. Turner, 151–80. Seattle, WA: University of Washington Press; Vancouver, BC: UBC Press.

Turner, Nancy J., and Roy L. Taylor. 1972. "A Review of the Northwest Coast Tobacco Mystery." *Syesis* 5: 249–57.

Turner, Nancy J., John Thomas (Tl'iishal), Barry F. Carlson, and Robert T. Ogilvie. 1983. *Ethnobotany of the Nitinaht Indians of Vancouver Island.* Victoria, BC: British Columbia Provincial Museum.

Turner, Nancy J., and Judith C. Thompson (Edōsdi), eds. 2006. *Plants of the Gitga'at People: 'Nwana'a lax Yuup.* Hartley Bay, BC: Gitga'at Nation; Victoria, BC: Coasts Under Stress Research Project and Cortex Consulting.

Turner, Nancy J., Laurence C. Thompson, M. Terry Thompson, and Annie Z. York. 1990. *Thompson Ethnobotany: Knowledge and Usage of Plants by the Thompson Indians of British Columbia.* Victoria, BC: Royal BC Museum.

Turner, Nancy J., Jan Timmers, and Randy Bouchard. 1972. "Sechelt Plant Names." Unpublished ms. In possession of Nancy J. Turner, School of Environmental Studies, University of Victoria.

Turner, Nancy J., and Katherine L. Turner. 2008. "'Where Our Women Used to Get the Food': Cumulative Effects and Loss of Ethnobotanical Knowledge and Practice." *Botany* 86, no. 1: 103–15.

Turner, Nancy J., and Patrick von Aderkas. 2009. *The North American Guide to Common Poisonous Plants and Mushrooms: How to Identify More Than 300 Toxic Plants Found in Homes, Gardens, and Open Spaces.* Portland, OR: Timber.

– 2012. "Sustained by First Nations: European Newcomers' Use of Indigenous Plant Foods in Temperate North America." *Acta Societatis Botanicorum Poloniae* 81, no. 4: 295–315. https://pbsociety.org.pl/journals/index.php/asbp/issue/view/88/showToc (accessed 12 July 2013).

Turner, Nancy J., and Barbara Wilson (K̲ii'iljuus). 2006. "'To Provide Living Plants for Study': The Value of Ethnobotanical Gardens and Planning the Qay'llnagaay Garden of Haida Gwaii." *Davidsonia* 16, no. 4: 111–25.

– 2008. "The Culture of Forests: Haida Traditional Knowledge and Forestry in the Twenty-First Century." In *Wild Foresting: Practicing Nature's Wisdom*, ed. A. Drengson and D.M. Taylor, 130–7. Washington, DC: Island.

Tyhurst, Robert. 1984. "The Chilcotin: An Ethnographic History." Unpublished ms. University of British Columbia.

Tyler, Varro E. 1987. *The Honest Herbal: A Sensible Guide to Herbs and Related Remedies.* Rev. ed. Philadelphia, PA: George F. Stickley.

U'mista Cultural Society, Juanita Pasco, and Brian D. Compton. 1998. *The Living World: Plants and Animals of the Kwakwa̲ka̲'wakw.* Alert Bay, BC: U'mista Cultural Society.

United Nations Commission on Environment and Development. 1987. *Our Common Future*. New York, NY: Oxford University Press.
United Nations. 1992. *Convention on Biological Diversity*. http://www.cbd.int/convention/convention.shtml (accessed 13 July 2013).
– 2007. *Declaration on the Rights of Indigenous Peoples*. http://www.un.org/esa/socdev/unpfii/documents/DRIPS_en.pdf (accessed 21 August 2013).
United States Department of Agriculture. 1999. "Ethnobotany, Culture, and Use of Great Camas." *Technical Notes: Plant Materials Technical Note No. 23*: 1–10. http://www.plant-materials.nrcs.usda.gov/pubs/orpmctn230999.pdf (accessed 13 July 2013).
Uu-a-thluk. 2008. *Čamus (Chum-us): West Coast Cooking Nuu-chah-nulth Style*. First Nations seafood cookbook. Endorsed by Nuu-chah-nulth Tribal Council [of *ha'wiih*, hereditary chiefs]. http://uuathluk.ca/wordpress/cookbook (accessed 13 July 2013).
Vajda, Edward J. 2010. "A Siberian Link with Na-Dene Languages." In *The Dene-Yeniseian Connection*, ed. James Kari and Ben A. Potter. Special issue, *Anthropological Papers of the University of Alaska* 5, ns: 33–99.
Vancouver, George. 1798. *A Voyage of Discovery to the Pacific Ocean and Round the World in the Years 1790–95*. Vol. 1. London, UK: G.G. and J. Robinson, Paternoster-Rowe, and J. Edwards, Pall Mall.
Vickers, Patricia June. 2008. "Ayaawx (Ts'msyen Ancestral Law): The Power of Transformation." Phd diss., University of Victoria.
von Dittmar, C. 1856. "Ueber die Koräken un die ihnen sehr nahe verwandten Tschuktschen." *Mélanges Russes Tirés du Bulletin Historico-philologique* 3: 1–48. http://www.siberian-studies.org/publications/PDF/Dittmar1858.pdf (accessed 8 June 2013).
von Kittlitz, Friedrich Heinrich. 1858. *Denkwürdigkeiten einer Reise nach dem russischen Amerika, nach Mikronesien und durch Kamtschatka*. Vol. 2. Gotha, Germany: Perthes.
Wainman, Newton, and R.W. Mathewes. 1987. "Forest History of the Last 12,000 Years Based on Plant Macrofossil Analysis of Sediment from Marion Lake, Southwestern British Columbia." *Canadian Journal of Botany* 65: 2179–87.
Wale, Vera. N.d. "Hlox̱s/Months – Gigeenix [Upriver Dialect]." In *Gitksan Dictionary*, tab 18, 3. Hazelton, BC: Gitksan-Wet'suwet'en Education Society.
Walker, Deward E., Jr, ed. 1998. *Handbook of North American Indians*. Vol. 12, *Plateau*. Washington DC: Smithsonian Institution.
Walker, I.J., and J.V. Barrie. 2004. "Geomorphology and Sea-Level Rise on one of Canada's Most 'Sensitive' Coasts: Northeast Graham Island, British Columbia." *Journal of Coastal Research* SI 39: 220–6.
Walker, I.J., and Marlow G. Pellatt. 2001. "Climate Change and Impacts in Southern British Columbia: A Palaeoenvironmental Perspective." Unpublished report. Climate Change Action Fund, Natural Resources Canada, Ottawa, Ontario.
Walker, Renee B., and Boyce N. Driskell, eds. 2007. *Foragers of the Terminal Pleistocene in North America*. Lincoln, NE: University of Nebraska Press.
Wall, Steve. 1993. *Wisdom's Daughters: Conversations with Women Elders of Native America*. New York, NY: Harper Collins.
Wallis, Wilson D. 1922. "Medicines Used by the Micmac Indians." *American Anthropologist* ns 24, no. 1: 24–30. http://www.jstor.org/stable/660063 (accessed 13 July 2013).
Wandsnider, Luann. 1997. "The Roasted and the Boiled: Food Composition and Heat Treatment with Special Emphasis on Pit-Hearth Cooking." *Journal of Anthropological Archaeology* 16: 1–48.

Watts, Annie, and Dolly Watts. 2007. *Where People Feast: An Indigenous People's Cookbook*. Vancouver, BC: Arsenal Pulp.

Weiser, A. 2006. "Exploring 10,000 Years of Human History on Ebey's Prairie, Whidbey Island, Washington." MA thesis, Simon Fraser University.

Weiser, A., and Dana Lepofsky. 2009. "Ancient Land Use and Management of Ebey's Prairie, Whidbey Island, Washington." In *Indigenous Resource Management: Past, Present and Future*, ed. Dana Lepofsky. Special issue, *Journal of Ethnobiology* 29, no. 2: 161–6.

Wennekens, A.J. 1985. "Traditional Plant Usage by Chugach Natives around Prince William Sound and on the Lower Kenai Peninsula, Alaska." MA thesis, University of Alaska Anchorage.

Whelchel, David, ed. 2005. *Ozette Archaeological Project Research Reports*. Vol. 3, *Ethnobotany and Wood Technology*. Seattle, WA: Northwest Regional Office, National Park Service, United States Department of the Interior; Pullman, WA: Department of Anthropology, Washington State University.

White, Richard. 1980. *Land Use, Environment and Social Change: The Shaping of Island County, Washington*. Seattle, WA: University of Washington Press.

– 1999. "Indian Land Use and Environmental Change, Island County, Washington." In *Indians, Fire, and the Land in the Pacific Northwest*, ed. Robert T. Boyd, 36–49. Corvallis, OR: Oregon State University Press.

Whitford, H.N., and R.D. Craig. 1918. *Forests of British Columbia*. Ottawa, ON: Commission of Conservation.

Wigen, R.J. 2005. "History of the Vertebrate Fauna in Haida Gwaii." In *Haida Gwaii: Human History and Environment from the Time of Loon to the Time of the Iron People*, ed. Daryl W. Fedje and R.W. Mathewes, 96–115. Vancouver, BC: UBC Press.

Wilkes, C. 1845. *Narrative of the United States Exploring Expedition, during the Years 1838, 1839, 1840, 1841, 1842*. Vol. 4. Philadelphia, PA: Lea and Blanchard.

William, Roger. 2003. "Affidavit." 9 September. Supreme Court of British Columbia, *Tsilhqot'in Nation v. British Columbia*, 2007 BCSC 1700.

Williams, Judith. 2006. *Clam Gardens: Aboriginal Mariculture on Canada's West Coast*. Vancouver, BC: New Star Books.

Williams, M.D. 1979. "The Harvesting of 'Sluckus' (*Porphyra perforata*) by the Straits Salish Indians of Vancouver Island." *Syesis* 12: 63–8.

Williams, Nancy M., and G. Baines, eds. 1993. *Traditional Ecological Knowledge: Wisdom for Sustainable Development*. Canberra, Australia: Centre for Resource and Environmental Studies, Australian National University.

Williams, Nancy M., and Eugene S. Hunn, eds. 1982. *Resource Managers: North American and Australian Hunter-Gatherers*. Boulder, CO: Westview.

Wilp Wilx̲o'oskwhl Nisg̲a'a. 1995. *Ayuukhl Nisg̲a'a*. Vol. 4, *The Land and Resources: Traditional Nisg̲a'a Systems of Land Use and Ownership*. New Aiyansh, BC: Nisg̲a'a Tribal Council.

Wilson, Robert L., and Catherine Carlson. 1980. *The Archaeology of Kamloops*. Burnaby, BC: Department of Archaeology, Simon Fraser University.

Witkowski, Stanley R., Cecil H. Brown, and Paul K. Chase. 1981. "Where Do Tree Terms Come From?" *Man* ns 16: 1–14.

Wollstonecroft, M. 2000. "The Fruit of Their Labour: A Palaeoethnobotanical Study of Site EeRb 140, a Multi-component Open-Air Archaeological Site on the B.C. Plateau." MA thesis, Simon Fraser University.

Wollstonecroft, M., and Gladys Baptiste. Forthcoming. "Linking the Ethnobotany and the Archaeology: Ancient Plant Remains from Site EeRb 140." In *Secwepemc People and Plants: Research Papers in Shuswap Ethnobotany*, ed. Marianne B. Ignace, Nancy J. Turner, and Sandra L. Peacock. Published by the Society of Ethnobiology in the monograph series Contributions in Ethnobiology.

Wollstonecroft, M., Peter R. Ellis, Gordon C. Hillman, and Dorian Q. Fuller. 2008. "Advances in Plant Food Processing in the Near Eastern Epipalaeolithic and Implications for Improved Edibility and Nutrient Bioaccessibility: An Experimental Assessment of *Bolboschoenus maritimus* (L.) Palla (Sea Club-Rush)." *Vegetation History and Archaeobotany* 17, supplement 1: S19–S27.

Work, John. 1945. *The Journal of John Work, January to October, 1835.* Ed. H.D. Dee. Victoria, BC: British Columbia Archives.

World Wildlife Fund. 2008. *Living Planet Report 2008.* Ed. Chris Hails. http://www.panda.org/about_our_earth/all_publications/living_planet_report (accessed 13 July 2013).

Wright, Robin, ed. 1991. *A Time of Gathering: Native Heritage in Washington State.* Seattle, WA: University of Washington Press.

Wyllie-Echeverria, S., and P.A. Cox. 2000. "Cultural Saliency as a Tool for Seagrass Conservation." *Biologia Marina Mediterranea* 7, no. 2: 421–4.

Wyllie de Echeverria, Victoria. 2013. "*Moolks* (Pacific crabapple, *Malus fusca*) on the North Coast of British Columbia: Knowledge and Meaning in Gitga'at Culture." MSc thesis, University of Victoria.

Wyndham, Felice S. 2004. "Learning Ecology: Ethnobotany in the Sierra Tarahumara, Mexico." Phd diss., University of Georgia.

Yamin-Pasternak, Sveta. 2011. "Ethnomycology: Fungi and Mushrooms in Cultural Entanglements." In *Ethnobiology*, ed. E.N. Anderson et al., 213–30. Hoboken, NJ: Wiley-Blackwell.

Yang, Andrew K. 2005. "Plant Names in Old and Middle English: Problems and Trends in Taxonomy." http://www.chass.utoronto.ca/~cpercy/courses/6361yang.htm (accessed 13 July 2013).

Yen, Douglas E. 1974. *The Sweet Potato and Oceania: An Essay in Ethnobotany*. Honolulu, HI: Bishop Museum Press.

York, Annie, Richard Daly, and Chris Arnett. 1993. *They Write Their Dreams on the Rock Forever: Rock Writings in the Stein River Valley of British Columbia.* Vancouver, BC: Talonbooks.

Yukon Executive Council. 2011. "Traditional Territories of Yukon First Nations and Settlement Areas of Inuvialuit and Tetlit Gwich'in." Map. http://www.eco.gov.yk.ca/landclaims/maps.html (accessed 13 July 2013).

Yukon Native Language Centre. 2011. "Yukon Native Languages." Map. http://www.ynlc.ca/languages (accessed 13 July 2013).

Zent, Stanford. 2009. "Methodology for Developing a Vitality Index of Traditional Environmental Knowledge (VITEK) for the Project 'Global Indicators of the Status and Trends of Linguistic Diversity and Traditional Knowledge.'" Terra Lingua. http://www.terralingua.org/vitek/wp-content/uploads/downloads/2012/03/VITEK_Report.pdf (accessed 13 July 2013).

Zhang, Linhai, Charles R. Brown, David Culley, Barbara Baker, Elizabeth Kunibe, Hazel Denny, Cassandra Smith, Neuee Ward, Tia Beavert, Julie Coburn, J.J. Pavek, Nora Dauenhauer, and Richard Dauenhauer. 2010. "Inferred Origin of Several Native American and Native Alaskan Potatoes from the Pacific Northwest and Southeast Alaska Using SSR Markers." *Euphytica* 174, no. 1: 15–29.

INDEX OF PLANT SPECIES

Italicized page numbers indicate figures and tables.

GENERAL INDEX

Italicized page numbers indicate figures and tables. Plants are also listed by taxonomical name in the Index of Plant Species.